T0073735

The Oxford Linear Algebra
for Scientists

The Oxford Linear Algebra for Scientists

Andre Lukas

University of Oxford

OXFORD
UNIVERSITY PRESS

Great Clarendon Street, Oxford, OX2 6DP,
United Kingdom

Oxford University Press is a department of the University of Oxford.
It furthers the University's objective of excellence in research, scholarship,
and education by publishing worldwide. Oxford is a registered trade mark of
Oxford University Press in the UK and in certain other countries

© Andre Lukas 2022

The moral rights of the author have been asserted

Impression: 2

All rights reserved. No part of this publication may be reproduced, stored in
a retrieval system, or transmitted, in any form or by any means, without the
prior permission in writing of Oxford University Press, or as expressly permitted
by law, by licence or under terms agreed with the appropriate reprographics
rights organization. Enquiries concerning reproduction outside the scope of the
above should be sent to the Rights Department, Oxford University Press, at the
address above

You must not circulate this work in any other form
and you must impose this same condition on any acquirer

Published in the United States of America by Oxford University Press
198 Madison Avenue, New York, NY 10016, United States of America

British Library Cataloguing in Publication Data

Data available

Library of Congress Control Number: 2022931100

ISBN 978–0–19–884491–4 (hbk)
ISBN 978–0–19–884492–1 (pbk)

DOI: 10.1093/oso/9780198844914.001.0001

Printed and bound by
CPI Group (UK) Ltd, Croydon, CR0 4YY

Links to third party websites are provided by Oxford in good faith and
for information only. Oxford disclaims any responsibility for the materials
contained in any third party website referenced in this work.

To my family

Preface

This textbook is based on a lecture course the author has given to first-year physics students at the University of Oxford and it profits from many years of tutoring physics students at Balliol College Oxford. It is an attempt to provide a modern introduction for students in the mathematical sciences into the classical subject of linear algebra.

Linear algebra is one of the basic disciplines of mathematics and it underlies many branches of more advanced mathematics, such as analysis in several variables and differential geometry. Put simply, linear structures are the building blocks for many more advanced constructions in mathematics. Frequently, linear algebra is also the first serious course in mathematics that undergraduate students have to face. It brings about the first rigorous proofs, which mark a definite departure from the standard high-school training of applying formulas and performing routine calculations. This transition can be difficult.

At the same time, linear algebra has many important applications in practically all areas of the mathematical sciences. The principle of linear superposition underlies a large variety of physical laws. Linearity is the leading approximation to systems near equilibrium so that even non-linear systems can often be described by linear counterparts. Methods of linear algebra also play an important role in computer and data sciences. For example, linear algebra is an important ingredient of artificial neural networks.

This ubiquity of linear algebra provides a unique opportunity for an introductory course in the subject. It can be used as an introductory course in mathematics, a case study in modern mathematics ideally suited to familiarize students with the axiomatic set-up and the systematic development of mathematical theories, and, at the same time, as a gateway into many areas of applied mathematics and science. The main purpose of the present book is to do justice to both of these aspects of linear algebra.

We hope such a dual approach will benefit both mathematicians and scientists. Particularly in the physical sciences, the widespread practice of teaching mathematics as a series of 'recipes', earmarked for certain applications, is highly unsatisfactory. The increasing importance of quantitative methods and of advanced mathematics means that beginning science students should develop an understanding of the structure of mathematics as well as its applications, without confusing these two sides of the subject. What is needed is a clear and coherent exposition of the overarching concepts, an uncompromising attitude towards mathematical rigour, while avoiding over-formalization, and a prompt connection to interesting, self-contained examples.

It came as a surprise to the author, while preparing a linear algebra lecture course, that textbooks following such an approach are lacking. There are, of course, numerous high-quality linear algebra textbooks with mathematical orientation (for example Blyth and Robertson 1975; Curtis 1996; Fischer 2010; Halmos 2017; Janich 1994; Lang 1996; Manin and Kostrikin 1989; Strang 1988). At the other end of the spectrum, there are many 'how-to-do' books available, which teach some practical aspects and applications of the subject but fail to present a logical exposition of the material.

The present book attempts to fill the gap between those extremes. It is aimed at beginning (first year) students in mathematics or the mathematical sciences, in particular in physics, engineering, and computer science. Few pre-requisites are required, other than basic numeracy and familiarity with basic mathematics (for example at the level of Lang 1998), as covered in the final years of most secondary schools. It presents a logical, mathematically coherent exposition of the standard material, including all relevant definitions and proofs, but avoids an overly formal approach. On the other hand, many examples and techniques for calculation which are essential for practical work with linear algebra have been included.

There are a number of starred chapters and sections (indicated by a *) which cover more advanced material, such as the Jordan normal form, the singular value decomposition, duality and tensors. These topics are covered for their mathematical or scientific relevance but they can be omitted at first reading. Numerous applications of linear algebra to problems in science are presented alongside, but clearly separated from the main text. Their style of presentation is usually less formal, focused on 'getting on' with the task at hand and arriving at an interesting result quickly and efficiently. They can be read independently and, ideally, provide a short, self-contained window into a topic in science, as well as an illustration of how linear algebra is applied. Throughout the text, we have included problems and their solutions. The reader is invited to work alongside the text, cover up solutions, and try for themselves — or to have a peek if they get stuck. The exercises at the end of each chapter include routine problems, somewhat more challenging problems marked with † and difficult and wide-ranging problems marked with ††.

The book is organized into 27 chapters, each around 10—15 pages, and, depending on the chapter, suitable for a 1- to 2-hour lecture slot. A short lecture course based on this book, omitting starred chapters and including a minimal amount of examples and applications, will take about 20 to 24 lectures. Depending on how many of the more advanced topics, the examples and applications are included, this can be extended to up to 36 lectures or more.

We hope that a student will gain from this book a good working knowledge of 'vectors and matrices' and its applications in science, as well as an appreciation of the structure and beauty of the subject of linear algebra.

Andre Lukas
Oxford 2021

Acknowledgements

I would like to thank the Balliol College physics students who have taught me, over many years, about the joys and difficulties of learning mathematics as a beginning science student. Many thanks to Kira Boehm, Daniel Karandikar, and Doyeong Kim for help with the typesetting of the original lecture notes and to Kira for also proofreading a large part of the present manuscript. I am grateful to Andrei Constantin for collaboration on an earlier version of this book. Many thanks also to Oxford University Press (OUP) for supporting this project and, in particular, to Sonke Adlung at OUP for much encouragement and help.

Cover

The cover picture has been created by the author, based on a discrete group of 4×4 matrices, in analogy with Exercise 13.5.

Acknowledgements

Contents

The Greek alphabet

lowercase	uppercase	name	lowercase	uppercase	name
α	A	alpha	ν	N	nu
β	B	beta	ξ	Ξ	xi
γ	Γ	gamma	o	O	omicron
δ	Δ	delta	π	Π	pi
ϵ	E	epsilon	ρ	P	rho
ζ	Z	zeta	σ	Σ	sigma
η	H	eta	τ	T	tau
θ	Θ	theta	υ	Υ	ypsilon
ι	I	iota	ϕ (φ)	Φ	phi
κ	K	kappa	χ	X	chi
λ	Λ	lambda	ψ	Ψ	psi
μ	M	mu	ω	Ω	omega

Mathematical symbols

symbol	meaning	symbol	meaning
$\mathbb{N}$	natural numbers	$\exists$	there exists
$\mathbb{Z}$	integers	$\forall$	for all
$\mathbb{Q}$	rational numbers	$\sum$	sum
$\mathbb{R}$	real numbers	$\prod$	product
$\mathbb{C}$	complex numbers	$\Rightarrow$	follows
$\mathbb{F}$	a field	$\Leftrightarrow$	equivalent
$\{\ldots\}$	a set	$\Re$	real part
$\{\}, \emptyset$	the empty set	$\Im$	imaginary part
$\in$	is an element of	$\cong$	isomorphic
$\notin$	not an element of	$\ncong$	not isomorphic
$\subset$	is a subset of	$\perp$	perpendicular
$\cup$	set union	$\times$	times
$\cap$	set intersection	$\oplus$	direct sum
∞	infinity	$\otimes$	tensor product
$\sim$	is related to	$\dagger$	Hermitian conjugation
$\rightarrow$	maps between	$\int$	integral
$\mapsto$	is mapped to	$=$	equal
$>$	greater	$:=$	defined to be equal
$\geq$	greater equal	$\overset{!}{=}$	required to be equal
$<$	less		
$\leq$	less equal		
$\circ$	composition		
$\vee$	or		
$\wedge$	and		

Mathematical notation and definitions

notation	meaning	defined in	page
$\mathbb{N}$	set of natural numbers	Ex. 2.1	13
$a \in A$	a is an element of set A	Sec. 2.1.1	14
$A \subset B$	A is a subset of B	Sec. 2.1.1	14
$A \cup B, A \cap B$	set union and intersection	Eq. (2.1)	14
$A \setminus B, \bar{B}$	set complement	Eq. (2.1)	14
$A \times B$	Cartesian product of sets	Eq. (2.4)	15
$a \sim b$	a and b are related	Def. 2.1	16
$A/\sim$	quotient of a set by a relation	Def. 2.3	16
$\mathbb{Z}$	set of integers	Ex. 2.3	17
$f : X \to Y$	function f between sets X, Y	Eq. 2.6	18
$\mathrm{Im}(f)$	image of a function f	Eq. 2.7	19
$f^{-1}(y)$	pre-image of y	Eq. (2.8)	19
$\mathrm{Gr}(f)$	graph of a function f	Eq. 2.9	19
$f \circ g$	composition of two functions	Sec. 2.3.2	20
id_X	identity map on a set X	Sec. 2.3.1	19
f^{-1}	inverse of a function	Sec. 2.3.4	22
$P \vee Q, P \wedge Q$	logical 'and' and 'or'	Tab. 2.1	24
$\bar{P}$	complement of a predicate	Eq. (2.18)	25
$P \Rightarrow Q$	P implies Q	Eq. (2.21)	26
$P \Leftrightarrow Q$	P and Q are equivalent	Eq. (2.22)	26
$\forall x \in X$	for all x in set X	Eq. (2.22)	26
$\exists x \in X$	there exists an x in X	Eq. (2.22)	26
$(G, \cdot)$	group G with multiplication $\cdot$	Def. 3.1	30
S_n	permutation group of n elements	Ex. 3.2	32
$\mathbb{Z}_n$	cyclic group of order n	Ex. 3.4	32
$\mathrm{sgn}(\sigma)$	sign of a permutation $\sigma \in S_n$	Def. 3.5	37
$(\mathbb{F}, +, \cdot)$	field $\mathbb{F}$ with addition and multiplication	Def. 4.1	40
$\sum_i a_i, \prod_i a_i$	sum and product of the a_i	Eq. (4.1)	41
$a > b, a \geq b$	a greater b, a greater equal b	Def. 4.2	41
$\mathbb{Q}$	set of rational numbers	Ex. 4.1	42
$\mathbb{R}$	set of real numbers	Ex. 4.2	43
$\mathbb{F}_p$	finite field with characteristic p	Ex. 4.3	43
$\mathbb{C}$	set of complex numbers	Sec. 4.3.1	44
$\Re(z), \Im(z)$	real and imaginary parts of $z \in \mathbb{C}$	Sec. 4.3.2	45
$\bar{z}, |z|$	complex conjugate and length of $z \in \mathbb{C}$	Eq. (4.11)	45
$\mathcal{P}(\mathbb{F})$	set of all polynomials over $\mathbb{F}$	Sec. 4.4	48
$\mathbb{F}^n$	coordinate vectors with entries in $\mathbb{F}$	Sec. 5.1.1	53
$\mathbf{e}_i$	standard unit vectors in $\mathbb{F}^n$	Eq. (5.6)	56
$(V, \mathbb{F}, +, \cdot)$	vector space V over a field $\mathbb{F}$	Def. 6.1	60
$\mathbf{v}, \mathbf{w}, \ldots$	vectors, elements of a vector space	Def. 6.1	60
$\mathbf{0}$	zero vector	Def. 6.1	60
$a, b, \ldots, \alpha, \beta, \ldots$	scalars in $\mathbb{F}$	Def. 6.1	60

χ_f	characterisitic polynomial of f	Eq. (19.6)	247
μ_f	minimal polynomial of f	Prop. 19.4	253
$\mathrm{tr}(A)$	trace of a matrix or linear map	Eq. (19.13)	251
$\langle \cdot, \cdot \rangle$	scalar product	Def. 22.2	288
$W^\perp$	orthogonal of a vector subspace	Eq. (22.20)	298
		Eq. (26.14)	368
$f^\dagger$	adjoint of a linear map f	Eq. (23.1)	303
$\mathrm{U}(V)$	unitary linear maps $V \to V$	Eq. (23.11)	309
$\mathrm{SU}(V)$	special unitary linear maps $V \to V$	Eq. (23.12)	309
$\mathrm{O}(n)$	orthogonal $n \times n$ matrices	Eq. (23.14)	310
$\mathrm{SO}(n)$	special orthogonal $n \times n$ matrices	Eq. (23.15)	310
$\mathrm{U}(n)$	unitary $n \times n$ matrices	Eq. (23.32)	313
$\mathrm{SU}(n)$	special unitary $n \times n$ matrices	Eq. (23.33)	314
$(\cdot, \cdot)$	bi-linear or sesqui-linear form	Def. 25.1	343
$\mathrm{O}(n_+, n_-)$	generalized orthogonal matrices	Eq. (25.15)	351
$\mathrm{SO}(n_+, n_-)$	generalized special orthogonal matrices	Eq. (25.16)	351
$\mathrm{U}(n_+, n_-)$	generalized unitary matrices	Eq. (25.17)	351
$\mathrm{SU}(n_+, n_-)$	generalized special unitary matrices	Eq. (25.18)	351
V^*	dual of a vector space V	Def. 26.1	363
$V \otimes W$	tensor space of vector spaces V, W	Def. 27.1	382
$\mathbf{v} \otimes \mathbf{w}$	tensor product of two vectors	Eq. (27.5)	383
$S^2 V, \wedge^2 V$	(anti-) symmetric rank two tensors	Eq. (27.16)	387
$\mathbf{v} \otimes_S \mathbf{w}$	symmetric tensor product	Eq. (27.18)	388
$\Lambda^q V^*$	set of alternating q forms	Eq. (27.34)	392
$\mathbf{v} \wedge \mathbf{w}$	wedge product	Eq. (27.18)	388
$V^{\otimes p} \otimes (V^*)^{\otimes q}$	set of (p, q) tensors	Eq. (27.32)	392
ΛV^*	outer algebra of V^*	Eq. (27.41)	394

List of applications

1
Linearity — an informal introduction

Linearity is a ubiquitous structure in mathematics and mathematical descriptions of natural phenomena. We begin this chapter by introducing the idea of linearity informally and by explaining, on an elementary level, the main reasons for its omnipresence. In the second part, we will characterize linearity more formally by introducing linear maps, an approach that is at the heart of the subject. Along the way, we illustrate a number of other key features of linear algebra, including linear equations and the relationship between linear maps and matrices.

1.1   Why linearity?

Summary 1.1 *An elementary reason for the foundational nature of linear algebra within mathematics is that, under suitable conditions, functions can be locally approximated by linear functions. This fact, together with the tendency of natural systems to reside near the minimum of their potential energy, explains the prevalence of linearity in scientific phenomena.*

For now, we will say informally that a situation exhibits *linearity* if we can identify a numerical output and a numerical input such that the former is proportional to the latter. The following is a well-known example of linearity:

Example 1.1 *(Hooke's law)*
Hooke's law states that the force exerted by a spring is proportional to its extension. More precisely, if x (x_0) is the length of the extended (unextended) spring and $\epsilon = x - x_0$ is the extension then the force is given by

$$F = -k\epsilon \,, \tag{1.1}$$

where k is a positive constant, referred to as the *spring constant*. It is also quite useful to consider the potential energy

$$V = \frac{1}{2}k\epsilon^2 \quad \Rightarrow \quad F = -\frac{dV}{d\epsilon} = -k\epsilon \tag{1.2}$$

stored in the spring whose (negative) first derivative gives the force, as indicated above.

Eq. (1.1) is, of course, an idealization and ceases to be valid if ϵ becomes too large (the spring is over-stretched) or too small (the spring is too compressed). An important lesson is that linearity in natural phenomena is usually only valid for a limited range of input values but broken outside this range. The many applications of Hooke's law show that, despite this limitation, the idea of linearity in natural phenomena is important — but needs to be accompanied by an understanding of its validity.

We will see shortly that (approximate) linearity in natural phenomena for a suitable range of input values is rather more general than the simple example of Hooke's law suggests. □

Mathematically, we can formulate linearity between a real-valued input and output by a *linear function* or *linear map* $f : \mathbb{R} \to \mathbb{R}$ (where $\mathbb{R}$ denotes the set of real numbers) given by

$$f(x) = ax \,, \tag{1.3}$$

for some real constant a. This is a rather simple function, so why is linearity such an important idea which underlies many areas of mathematics? The *Taylor series* for a suitably well-behaved but otherwise general function $f : \mathbb{R} \to \mathbb{R}$ around x_0 reads

$$f(x) - f(x_0) = f'(x_0)\,\epsilon + \cdots \tag{1.4}$$

where the prime denotes the derivative, $\epsilon = x - x_0$ is the deviation away from x_0 and the dots stand for terms with higher powers in ϵ. Evidently, in a sufficiently small neighbourhood of x_0, the variation of the function away from its value $f(x_0)$ at x_0 is well-described by a linear function. In other words, (sufficiently well-behaved) functions are locally (approximately) linear. This simple observation is one of the main reasons for the importance of linear algebra for many other fields of mathematics.

What about linearity in the natural sciences? The above discussion of the Taylor series suggests that Hooke's law in Example 1.1 might, in fact, not be as special as it seemed. To make this more precise, consider a simple system described by a single real number x and a potential energy function $V(x)$. The Taylor series of this potential function around x_0, this time to second order in the deviation $\epsilon = x - x_0$, reads

$$V(x) = V(x_0) + V'(x_0)\epsilon + \frac{1}{2}V''(x_0)\epsilon^2 + \cdots \,. \tag{1.5}$$

Taking the derivative gives the associated force

$$F = -\frac{dV}{d\epsilon} = F_0 - k\epsilon + \cdots \quad \text{where} \quad F_0 = -V'(x_0) \,, \quad k = V''(x_0) \,. \tag{1.6}$$

This shows that the change $F - F_0$ in the force is (approximately) linear. At first, this observation appears to be of little practical significance, since the constant force F_0 will drive our system away from x_0, thereby invalidating the expansion (1.5). However, natural systems tend to evolve towards values x_0 with minimal potential energy V. At such a minimum we have $F_0 = -V'(x_0) = 0$ and $k = V''(X_0) > 0$ and, in this case, the force (1.6) turns into a version of Hooke's law

$$F = -k\epsilon + \cdots \,, \tag{1.7}$$

with $k = V''(x_0)$ playing the role of the spring constant. In short, near points of minimal potential energy forces are (and remain) approximately linear and this accounts for the importance of linearity in natural phenomena (see Exercise 1.1).

1.2   Linearity, more abstractly

Summary 1.2 *Linearity of functions is characterized in a more abstract way and generalized to functions between vectors with two components. For this two-components case, we illustrate the relationship between linear functions and matrices. We discuss simple examples of linear equations and their solution structure, as one of the main motivations for developing linear algebra.*

Now that we have established the importance of linearity as a basic mathematical concept and as a widespread phenomenon in science, let us consider the idea from a more abstract point of view.

1.2.1   Linear functions

In Eq. (1.3) we have already defined what we mean by a linear map $f : \mathbb{R} \to \mathbb{R}$. Evidently, such a function satisfies the basic rules

$$f(x + \tilde{x}) = f(x) + f(\tilde{x}) , \qquad f(\alpha x) = \alpha f(x) , \tag{1.8}$$

for any real numbers x, $\tilde{x}$ and α. To see this explicitly from Eq. (1.3), for example for the first of the above rules, is quite straightforward:

$$f(x + \tilde{x}) \overset{(1.3)}{=} a(x + \tilde{x}) = ax + a\tilde{x} \overset{(1.3)}{=} f(x) + f(\tilde{x}) .$$

The conditions (1.8) really express the fact that linear maps are consistent with addition and multiplication of real numbers. Adding (or multiplying) first and then applying the linear map gives the same answer as carrying this out in reverse order. We can say that the two operations commute.

Conversely, any function which satisfies the rules (1.8) must be a linear function, that is, a function of the form (1.3). To show this we start with a function f which satisfies Eqs. (1.8) and define the number a by

$$a = f(1) . \tag{1.9}$$

By simply using the second condition (1.8), it follows that

$$f(x) = f(x\,1) = xf(1) = ax ,$$

and, hence, that f is indeed of the form (1.3). We also see that the conditions (1.8) are not independent and that the first one for addition can be deduced from the one for multiplication (see Exercise 1.3). Nevertheless, consistency of functions with addition and multiplication, as expressed by both Eqs. (1.8) is an important idea which will be used to define linearity more generally.

1.2.2   Linear equations

A common and extremely important class of problems is to find all solutions x to an equation of the form $f(x) = b$, where f is a function and b is a given number. If f is a linear function, given by $f(x) = ax$, then such an equation is called a *linear equation* and it is of the form

$$ax = b \, . \tag{1.10}$$

Obviously, the solution to this equation is $x = b/a$, provided that $a \neq 0$. But what if $a = 0$? Answers to this question from students can vary from 'x must be infinite' to 'b must be zero', neither of which can possibly be correct (the former because infinity is not a number, the latter since b is a priori given and may be non-zero). The point is that, even for a simple equation such as the above, we have to distinguish cases.

(1) $a \neq 0$: There is a unique solution $x = b/a$.

(2a) $a = 0$ and $b = 0$: Any value of x solves the equation.

(2b) $a = 0$ and $b \neq 0$: There is no solution since Eq. (1.10) becomes $0 = b$, which is false.

The important lesson is that solutions to linear equations have a somewhat complicated structure which depends on the values of the parameters (a and b in Eq. (1.10)).

1.2.3   Vectors with two components

So far, we have considered the simplest case of functions whose inputs and outputs are real numbers. But natural phenomena are frequently characterized by more than one variable. To describe such systems, we need to deal with lists of real numbers, or *vectors* as they are commonly called, and functions between them. To keep things simple for now, let us consider *coordinate vectors* $\mathbf{v}$ or $\mathbf{w}$ with two components, by which we mean columns

$$\mathbf{v} = \begin{pmatrix} v_1 \\ v_2 \end{pmatrix} , \qquad \mathbf{w} = \begin{pmatrix} w_1 \\ w_2 \end{pmatrix} \tag{1.11}$$

containing two real numbers v_1, v_2 or w_1, w_2. (Vectors will be denoted by bold-face letters.) The set of all such coordinate vectors with two components is denoted $\mathbb{R}^2$. (Later, we will, of course, generalize to coordinate vectors with an arbitrary number of components and, indeed, more abstract vectors which are not made up from components.) The reader has probably already come across such vectors and how to add them and multiply them with real numbers. The natural way to define these operations is to apply the rules for adding and multiplying real numbers component-wise, that is,

$$\mathbf{v} + \mathbf{w} := \begin{pmatrix} v_1 + w_1 \\ v_2 + w_2 \end{pmatrix} , \qquad \alpha\mathbf{v} := \begin{pmatrix} \alpha v_1 \\ \alpha v_2 \end{pmatrix} . \tag{1.12}$$

These two operations are referred to as *vector addition* and *scalar multiplication* and they are the key features of a *vector space*, as we will see later. Their geometrical interpretation is illustrated in Fig. 1.1.

1.2.4   Linearity for maps between vectors

What does it mean for functions $f : \mathbb{R}^2 \to \mathbb{R}^2$ whose inputs and outputs are two-coordinate vectors to be linear? This is where our re-formulation (1.8) of linearity

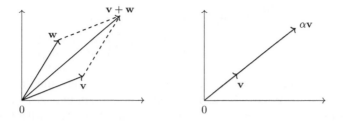

Fig. 1.1 Geometry of vector addition and scalar multiplication (for $\alpha > 1$) in $\mathbb{R}^2$.

helps. Motivated by these conditions, we declare that a function $f : \mathbb{R}^2 \to \mathbb{R}^2$ is called a *linear function* or *linear map* if it satisfies

$$f(\mathbf{v} + \mathbf{w}) = f(\mathbf{v}) + f(\mathbf{w}) , \qquad f(\alpha\mathbf{v}) = \alpha f(\mathbf{v}) , \qquad (1.13)$$

for all vectors $\mathbf{v}$ and $\mathbf{w}$ in $\mathbb{R}^2$ and all real numbers α. These conditions mean linear maps are compatible with vector addition and scalar multiplication (see Exercise 1.4). In general, functions which respect a certain algebraic structure are called *morphisms* and defining such functions is an important step in any mathematical build-up. Using this terminology, linear maps are the morphisms of vector spaces and, as we will see, they are central objects in linear algebra.

1.2.5   Linear maps and matrices

We have seen that linear maps between real numbers can be written as in Eq. (1.3) and that they are characterized by a single real number a. What do linear maps between coordinate vectors look like? Recall from Eq. (1.9) that we were able to recover a by applying the linear map to the number 1. For the two-component case, this suggests that we should apply the linear map to specific vectors. Simple vectors can be built by only using components 0 and 1 (the neutral elements of addition and multiplication) and this leads to the *standard unit vectors* in $\mathbb{R}^2$ defined as

$$\mathbf{e}_1 = \begin{pmatrix} 1 \\ 0 \end{pmatrix} , \qquad \mathbf{e}_2 = \begin{pmatrix} 0 \\ 1 \end{pmatrix} . \qquad (1.14)$$

Note that every vector $\mathbf{v}$ with components v_1, v_2 can be written in terms of $\mathbf{e}_1$ and $\mathbf{e}_2$ as

$$\mathbf{v} = v_1\mathbf{e}_1 + v_2\mathbf{e}_2 . \qquad (1.15)$$

This means the standard unit vectors are a particular example of a *basis*, an important idea in the theory of vector spaces which we will develop later. Inspired by Eq. (1.9), we consider the action $f(\mathbf{e}_1)$ and $f(\mathbf{e}_2)$ of the linear map on the standard unit vectors and write the result as

$$\begin{pmatrix} p \\ r \end{pmatrix} = f(\mathbf{e}_1) , \qquad \begin{pmatrix} q \\ s \end{pmatrix} = f(\mathbf{e}_2) , \qquad (1.16)$$

where p, q, r, and s are real numbers. Now we require only a small calculation to work out how the function f acts on an arbitrary vector $\mathbf{v}$:

$$f(\mathbf{v}) \overset{(1.15)}{=} f(v_1\mathbf{e}_1 + v_2\mathbf{e}_2) \overset{(1.13)}{=} v_1 f(\mathbf{e}_1) + v_2 f(\mathbf{e}_2)$$

$$\overset{(1.16)}{=} v_1 \begin{pmatrix} p \\ r \end{pmatrix} + v_2 \begin{pmatrix} q \\ s \end{pmatrix} \overset{(1.12)}{=} \begin{pmatrix} pv_1 + qv_2 \\ rv_1 + sv_2 \end{pmatrix} . \tag{1.17}$$

The final expression shows that the linear map is fully determined once we know the four numbers p, q, r, s. While a linear map between real numbers is specified by a single number, as in Eq. (1.3), a linear map between coordinate vectors with two components is specified by four real numbers. It is customary to arrange these four numbers into a 2×2 *matrix*

$$A = \begin{pmatrix} p & q \\ r & s \end{pmatrix} , \tag{1.18}$$

so that linear maps $\mathbb{R}^2 \to \mathbb{R}^2$ are in one-to-one correspondence with such 2×2 matrices with real entries. The action of the linear map f on a vector $\mathbf{v}$ is then symbolically written as

$$f(\mathbf{v}) = A\mathbf{v} := \begin{pmatrix} pv_1 + qv_2 \\ rv_1 + sv_2 \end{pmatrix} , \tag{1.19}$$

where the expression $A\mathbf{v}$ is referred to as the *multiplication of a matrix with a vector* and is defined by the right-hand side of Eq. (1.19). This definition means a matrix is multiplied with a vector row by row, with corresponding vector and row components multiplied and summed up.

The relationship between linear maps and matrices exemplified here is much more general. It is one of the central themes of linear algebra which will be developed systematically later. (For another example, see Exercise 1.5.)

1.2.6 Back to linear equations

With this understanding of coordinate vectors and linear maps between them, let us come back to linear equations, that is, equations of the form $f(\mathbf{x}) = \mathbf{b}$, where $\mathbf{b}$ is a given vector with components b_1 and b_2 and f is a linear map between two-coordinate vectors. We are interested in finding all vectors $\mathbf{x}$ with components x_1 and x_2 which satisfy this equation. If we describe f by a 2×2 matrix A, as above, the linear equation can also be written as $A\mathbf{x} = \mathbf{b}$.

Suppose, for concreteness, we choose the following matrix A and vector $\mathbf{b}$,

$$A = \begin{pmatrix} 3 & 1 \\ a & -1 \end{pmatrix} , \qquad \mathbf{b} = \begin{pmatrix} b \\ 1 \end{pmatrix} , \tag{1.20}$$

where a and b are real numbers. (For another example, see Exercise 1.6.) Then, using the definition of matrix-vector multiplication from Eq. (1.19), the linear equation becomes

$$A\mathbf{x} = \mathbf{b} \qquad \Leftrightarrow \qquad \left\{ \begin{array}{l} (E1): \ 3x_1 + x_2 = b \\ (E2): \ ax_1 - x_2 = 1 \end{array} \right\} , \tag{1.21}$$

that is, it turns into a system of two simultaneous linear equations in two variables x_1 and x_2. We can solve this system in the usual way by adding suitable multiples of the equations to eliminate one of the variables. In the present case, we can simply consider the sum of the two equations

$$(E1) + (E2): \quad (a+3)x_1 = b + 1 . \tag{1.22}$$

This result shows that, just as in the earlier case (1.10), we have to consider various cases:

(1) $a \neq -3$: There is a unique solution, $x_1 = (b+1)/(a+3)$ and $x_2 = (ab-3)/(a+3)$, obtained by dividing Eq. (1.22) by $a + 3$, inserting the result for x_1 into one of the Eqs. (1.21) and solving for x_2.

(2a) $a = -3$ and $b = -1$: In this case, Eq. (1.22) becomes trivial or, equivalently, the two equations in (1.21) become the same. This means the solution consists of all x_1 and x_2 which satisfy $x_2 = -3x_1 - 1$. This represents a line in the x_1-x_2 plane.

(2b) $a = -3$ and $b \neq -1$: There is no solution since Eq. (1.22) becomes $0 = b + 1$, which is false.

To summarize, the solution can be a single point, a line in the x_1-x_2 plane or there can be no solution at all, depending on the values of the parameters a and b. Even this simple example shows there is considerable structure in the solutions to linear equations. An important purpose of linear algebra is to understand this solution structure and also to provide efficient methods for solving linear equations. This will be covered in detail later on. We end this chapter with a real-world application of linear equations which illustrates some of the problems linear algebra should address.

Application 1.1 *Internet search algorithm*

Modern internet search engines order search results by assigning a page rank to each website. As we will see, finding the page rank can be formulated as a problem in linear algebra.

To explain the idea we start with a very simple network with four sites, labelled by $k = 1, 2, 3, 4$, and links indicated by arrows in the following diagram.

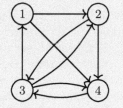

Each of the four sites has a certain number of links to the other sites (outgoing arrows). For example, site 1 links to sites 2 and 4 and, therefore, has $n_1 = 2$ links. Likewise, the number of links for the other sites are $n_2 = 2$, $n_3 = 3$ and $n_4 = 1$. Site 4 is linked to by all other sites and we express this mathematically by writing $L_4 = \{1, 2, 3\}$. Similarly, we have $L_1 = \{3\}$, $L_2 = \{1, 3\}$, and $L_3 = \{2, 4\}$.

We would like to assign real numbers x_1, x_2, x_3, x_4 to the sites which measure the popularity of the page. How should these numbers be obtained? A natural idea is that every incoming link should increase the popularity of a page by an amount proportional to the popularity of the remote page from which the link originates. For example, page 2 is linked to by pages 1 and 3, so x_2 should be increased by amounts proportional to x_1 and x_3. More specifically, we could say that $x_2 = x_1/2 + x_3/3$. Here, we have divided by 2 (3) since site 1 (3) has two (three) outgoing links. The underlying idea is that a link from a site with a small number of outgoing links is worth more than a link from a site with many outgoing links. Continuing along those lines with all four sites gives the equations

$$
\left.\begin{array}{l}
x_1 = \frac{x_3}{3} \\[4pt]
x_2 = \frac{x_1}{2} + \frac{x_3}{3} \\[4pt]
x_3 = \frac{x_2}{2} + \frac{x_4}{1} \\[4pt]
x_4 = \frac{x_1}{2} + \frac{x_2}{2} + \frac{x_3}{3}
\end{array}\right\}
\implies
\begin{pmatrix} x_1 \\ x_2 \\ x_3 \\ x_4 \end{pmatrix}
= \alpha
\begin{pmatrix} 4 \\ 6 \\ 12 \\ 9 \end{pmatrix}.
\tag{1.23}
$$

This is indeed a system of four linear equations for x_1, x_2, x_3, and x_4. The solution is easily obtained by adding suitable multiples of the four equations and it is shown in Eq. (1.23), on the right (α is a real number). The conclusion is that site 3 is the highest-ranked, followed by site 4.

The real internet has of course a very large number of sites, so we should formulate the problem for an arbitrary number, n, of sites. We label these sites by $k = 1, \ldots, n$ and denote their popularity by x_k. Site k has n_k links to other sites and it is linked to by certain sites whose labels we collect in a set L_k. With this notation the generalization of the linear system (1.23) can be written as

$$
x_k = \sum_{j \in L_k} \frac{x_j}{n_j},
\tag{1.24}
$$

where $k = 1, \ldots, n$. Note that the sum on the right-hand side runs over all pages j which link to page k. Eqs. (1.24) constitutes a system of n linear equations for the variables $x_1, \ldots, x_n$.

Solving potentially large linear systems, such as the above, requires more refined methods and a better understanding of their structure. Much of the course will be devoted to this task. The present application also raises more theoretical questions. It is evident that the system (1.24) always has the trivial solution where all $x_k = 0$, but of course this solution is not useful for the purpose of ranking sites. Is it an accident that the example (1.23) has non-trivial solutions or can this be guaranteed in general? We will return to this question when we have developed a deeper understanding of the structure of linear algebra.

1.3   Plan of the book

In the following Part I we start with the mathematical foundations: the basic mathematical language of sets and functions and the important algebraic structures of groups and fields which are both closely connected to vector spaces. A reader familiar with these basics, perhaps from their analysis course, can skip straight to Part II which introduces the algebraic structure of vector spaces, which is the arena and the main topic of linear algebra. Linear independence as well as basis and dimension of a vector space are the key concepts introduced in this part.

Some topics whose proper mathematical place is at a much later stage in the development of the subject are essential to a science student early in their course, for example in the context of mechanics or electromagnetism. These include the dot, cross, and triple product and simple geometric applications to lines and planes, as well as the ability to perform calculations efficiently. These and related topics are covered, at an elementary level, in Part III. As an added benefit, the material in this part also provides a source of examples for the remainder of the text.

The systematic development of the subject resumes in Part IV, where we introduce and analyse the morphisms of vector spaces — the linear maps. The formal development culminates in the rank theorem, a pivotal statement about the structure of linear maps. We will also explain the relation between linear maps and matrices, which we have already alluded to in Section 1.2.5, in general.

The main computational tools in linear algebra are algorithms to manipulate matrices. In Part V we develop some of these algorithms, related to row operations of matrices. Perhaps the most important application of linear algebra is to systems of linear equations. We will see how the results on the structure of linear maps allow us to understand the solution structure of systems of linear equations and how they can be solved in practice by using row operations of matrices. We also introduce determinants, another important tool for calculations as well as more abstract arguments. This concludes the basic development of the subject — the remainder of the book is devoted to more advanced topics in linear algebra.

Linear maps and matrices are complicated objects. Part VI deals with the question of how to cast linear maps and matrices into a simple, easy-to-handle form. Using the key ideas of eigenvalues and eigenvectors we will see that linear maps and matrices can often be diagonalized. Even when this is not possible a nearly diagonal form, called the Jordan normal form, can always be achieved.

In Part VII we resume — and generalize — the geometrical discussion from Part III by introducing scalar products on general vector spaces. Scalar products facilitate basic geometric notions such as the length of vectors, angles between vectors and orthogonality, and they generalize the dot product. Vector spaces with a scalar product, also called inner product vector spaces, have many applications — in fact, most scientific applications of linear algebra assume, explicitly or implicitly, the presence of a scalar product. A scalar product also singles out certain classes of linear maps — the self-adjoint and the unitary maps. They include the unitary and orthogonal matrices as well as rotation matrices which have many important applications. With new structure on vector spaces available it makes sense to re-visit the problem of diagonalizing linear maps. We also briefly cover a generalization of scalar products — symmetric bi-linear and Hermitian sesqui-linear forms — which make an appearance in certain scientific applications, for example, in the theory of relativity.

Our final topic, duality and tensors, in Part VIII is probably the most abstract one covered. However, dual and tensor vector spaces play an important role in more advanced mathematics as well as in many scientific applications and cannot be omitted.

Chapters and sections on more advanced topics which can be omitted at first reading are indicated by a *. Throughout the book, scientific applications of aspects of linear algebra have been included. They are clearly separated from the main development of the subject and presented on a grey background, in order to avoid confusion between mathematics and its scientific applications. They can be read separately from the main text and illustrate the wide range of linear algebra applications. Ideally, they also serve as a (small) window into an area of science.

Exercises

1.1 *Linearity near minima of potentials*
A physical system is described by a single real variable x and has potential energy $V(x) = \frac{1}{2}x^4 - x^2 + 1$. Find the minima x_0 of V. Show that for small deviations $\epsilon = x - x_0$ from each of the minima, the potential energy can be written as $V \simeq \frac{1}{2}k\epsilon^2 + \cdots$, with associated force $F = -k\epsilon$, and determine the constant k.

1.2 Is every function $\mathbb{R} \to \mathbb{R}$ whose graph is a straight line a linear function?

1.3 *Linear functions*
Let $f : \mathbb{R} \to \mathbb{R}$ be a function which satisfies $f(x\tilde{x}) = xf(\tilde{x})$ for all real x and $\tilde{x}$. Show that f also satisfies $f(x + \tilde{x}) = f(x) + f(\tilde{x})$ for all real x and $\tilde{x}$.

1.4 *Differential operators as linear maps*
Consider the 'differential' operator $D = \frac{d^2}{dx^2} + \frac{d}{dx}$ which maps (infinitely many times differentiable) functions f to $f'' + f'$. Show that D satisfies the linearity conditions $D(f + g) = D(f) = D(g)$ and $D(\alpha f) = \alpha D(f)$, where f and g are functions and α is a real number.

1.5 *Vectors with three components*
Generalize the discussion of Section 1.2.5 to linear maps $f : \mathbb{R}^3 \to \mathbb{R}^3$ between coordinate vectors with three components. In particular, use the three standard unit vectors

$$\mathbf{e}_1 = \begin{pmatrix} 1 \\ 0 \\ 0 \end{pmatrix}, \ \mathbf{e}_2 = \begin{pmatrix} 0 \\ 1 \\ 0 \end{pmatrix}, \ \mathbf{e}_3 = \begin{pmatrix} 0 \\ 0 \\ 1 \end{pmatrix}$$

to show that such a linear maps can be described by 3×3 matrices.

1.6 *Linear equations*
Consider a linear equation of the form $A\mathbf{x} = \mathbf{b}$ with

$$A = \begin{pmatrix} a & 1 \\ -1 & 2 \end{pmatrix}, \quad \mathbf{b} = \begin{pmatrix} 1 \\ b \end{pmatrix},$$

where $\mathbf{x}$ is a vector with components x_1 and x_2 and a and b are real numbers. Identify the different cases for the solution structure, depending on the values of a and b, and find the solution vectors $\mathbf{x}$ in each case.

1.7 *Internet search*
Following the notation of Application 1.1, consider a network with three sites, specified by the data $n_1 = 2$, $n_2 = 1$, $n_3 = 1$ and $L_1 = \{2\}$, $L_2 = \{1,3\}$, $L_3 = \{1\}$. Draw a graph for the network and write down the linear system (1.24) for this example. Show that its solution is

$$\begin{pmatrix} x_1 \\ x_2 \\ x_3 \end{pmatrix} = \alpha \begin{pmatrix} 2 \\ 2 \\ 1 \end{pmatrix},$$

where α is a real number.

Part I

Preliminaries

Before we can delve into the main subject, some preparation is required. We need to introduce the basic mathematical language of sets and functions, which is essential for the formulation not just of linear algebra but of any area of mathematics. The reader who is not yet familiar with these basics should go through this part carefully. However, in the interest of getting to the main subject quickly, we keep the exposition basic and concise.

In Chapter 2 we introduce sets, functions, and elements of logic — the basic language of mathematics. Groups are discussed in Chapter 3. They play an important role as one of the simplest algebraic structures in mathematics and they provide the mathematical framework for what scientists frequently refer to as symmetries. Group theory is a separate and vast topic in mathematics, but we will have to keep the discussion at a basic level and focus on those aspects which play a direct role in linear algebra. In particular, we will introduce the permutation groups and develop some of their properties. These results will be required for the definition of the determinant in Chapter 18.

Finally, in Chapter 4 we define fields, a pre-requisite for the definition of vector spaces, and derive a few conclusions from the field axioms. Fields are the standard arena for 'numerical' calculations, with the rational, real, and complex numbers being the most important examples. The field $\mathbb{Q}$ of rational numbers can be constructed from an equivalence relation on $\mathbb{Z}^2$, essentially a formal way of introducing fractions. We will indicate how this construction works and how it can be verified that $\mathbb{Q}$ does indeed satisfy the field axioms. The real numbers $\mathbb{R}$ are constructed as a 'completion' of $\mathbb{Q}$, obtained by augmenting the set with irrational numbers. This construction belongs into the realm of analysis and will only be described briefly. However, we will spend some time on setting up the complex numbers $\mathbb{C}$, as the reader might not yet be sufficiently familiar with them. This will become important whenever we work with vector spaces over the complex numbers.

2
Sets and functions

Sets and functions between sets provide the basic language of mathematics. More advanced topics, such as linear algebra, cannot be formulated properly without introducing this language first. This can be done in a strict, axiomatic manner, but here we adopt a less rigorous style to avoid creating a hurdle of formality early on.

2.1 Sets

Summary 2.1 *Sets are collections of objects called elements. There are three basic set operations, namely set union, set intersection, and set complements. The first two of these are associative, commutative and they satisfy distributive laws. The set complement converts between union and intersection. The Cartesian product of two sets is a set which contains all pairs of elements from the first and second set.*

2.1.1 (Non-) definition of sets

Intuitively, by a *set* we mean a collection of objects which are called *elements* or *members* of the set. A set can be specified by explicitly providing its elements and this is done using a notation with curly brackets, $\{\ldots\}$ so that, for example, $\{1, 2, 3\}$ is the set which contains 1, 2, and 3. For the purpose of this section, we use uppercase letters $A, B, \ldots$ to denote sets and lowercase letters $a, b, \ldots$ to denote their elements. For the *empty set* the symbol $\{\}$ or $\emptyset$ is used. By convention, all elements of a set are distinct so that repetitions of elements can be deleted, for example, $\{1, 2, 1\} = \{1, 2\}$. The order of elements in a set is irrelevant, for example $\{1, 3, 2\} = \{1, 2, 3\}$. Sets can also be elements of other sets; for example, we can consider the set $\{\{0, 2, 4, 6, 8\}, \{1, 3, 5, 9\}\}$ whose elements are the sets of even and odd numbers less than ten.

Example 2.1 *(Natural numbers)*
A foundational number set is given by the natural numbers $\mathbb{N} := \{0, 1, 2, 3, \ldots\}$. We will take this set for granted (although it can be defined by a set of axioms called Peano's axioms), and also take addition and multiplication of natural numbers as a given. □

As opposed to sets, *lists* provide ordered collections of objects, with repetitions allowed. We will need to use lists, for example, to describe vectors, and they are denoted by round brackets, $(\cdots)$. For example, $(1, 2, 1)$ is the list which consists of 1 and 2 with this particular order and multiplicity and we have $(1, 2, 1) \neq (1, 2)$ and $(1, 2, 1) \neq (1, 1, 2)$.

Lists can also be written as columns, rather than rows, of objects and this is the convention we will adopt for vectors.

Coming back to sets, set membership is indicated by the notation $a \in A$, read as 'a is an element of A', while non-membership is written as $a \notin A$, read as 'a is not an element of A'. We say that a set A is a *subset* of a set B, written as $A \subset B$, if every $a \in A$ is also an element of B. Two sets A and B are equal, written as $A = B$, if $A \subset B$ and $B \subset A$. This means equality of two sets can be established by showing that they are mutual subsets of each other, a technique frequently used in proofs. Subsets of a given set A can also be specified by a conditional notation of the form $\{a \in A \,|\, a \text{ satisfies a condition}\}$, read as 'the set of all a in A which satisfy the condition'. For example, the set $\{n \in \mathbb{N} \,|\, n = m^2 \text{ for a } m \in \mathbb{N}\}$ contains all square numbers.

2.1.2 Set operations

There are three main operations for sets, the *union*, the *intersection* and the *complement*, which are defined as follows (see also Fig. 2.1):

$$
\begin{aligned}
A \cup B &= \{x \in U \,|\, x \in A \text{ or } x \in B\} & \text{(union of } \dot{A} \text{ and } B) \\
A \cap B &= \{x \in U \,|\, x \in A \text{ and } x \in B\} & \text{(intersection of } A \text{ and } B) \\
A \setminus B &= \{a \in A \,|\, a \notin B\} & \text{(complement of } B \text{ within } A)
\end{aligned}
\tag{2.1}
$$

For the complement it is assumed that B is a subset of A. When it is understood from context what the set A is, the complement $A \setminus B$ is sometimes simply denoted by $\bar{B}$. Two sets A and B are called *disjoint* if their intersection is empty, that is, if $A \cap B = \emptyset$.

The union and intersection are associative and commutative operations, that is, they

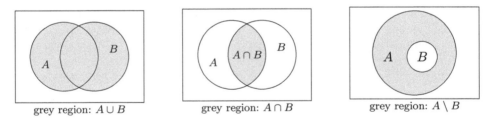

grey region: $A \cup B$ grey region: $A \cap B$ grey region: $A \setminus B$

Fig. 2.1 Union $A \cup B$ and intersection $A \cap B$ of two sets A and B and complement $A \setminus B$.

satisfy

$$
\begin{aligned}
A \cup (B \cup C) &= (A \cup B) \cup C & A \cap (B \cap C) &= (A \cap B) \cap C & \text{(associativity)} \\
A \cup B &= B \cup A & A \cap B &= B \cap A & \text{(commutativity)} .
\end{aligned}
$$

These rules follow directly from the definitions of union and intersection. Somewhat more involved are the following distributive laws which govern the relationship between union and intersection.

Proposition 2.1 *For any three sets A, B, and C, we have*

$$
A \cap (B \cup C) = (A \cap B) \cup (A \cap C), \qquad A \cup (B \cap C) = (A \cup B) \cap (A \cup C). \tag{2.2}
$$

Proof We prove the first of these equalities by showing that the left-hand side is a subset of the right-hand side and vice versa. We start with an arbitrary element $a \in A \cap (B \cup C)$ and we want to show it is contained in $(A \cap B) \cup (A \cap C)$. From $a \in A \cap (B \cup C)$, it follows that $a \in A$ and $a \in B \cup C$. The latter means that $a \in B$ or $a \in C$, so we have two cases. If $a \in A$ and $a \in B$ then $a \in A \cap B$. On the other hand, if $a \in A$ and $a \in C$ then $a \in A \cap C$. In either case, $a \in (A \cap B) \cup (A \cap C)$ and it follows that $A \cap (B \cup C) \subset (A \cap B) \cup (A \cap C)$.

Conversely, if $a \in (A \cap B) \cup (A \cap C)$ then $a \in A \cap B$ or $a \in A \cap C$. In the first case, $a \in A$ and $A \in B$ so that $a \in A \cap (B \cup C)$. In the second case, $a \in A$ and $a \in C$ and again it follows that $a \in A \cap (B \cup C)$. This shows that $(A \cap B) \cup (A \cap C) \subset A \cap (B \cup C)$ and completes the proof. We leave the proof of the second Eq. (2.2) as Exercise 2.1.
□

There are also rules for calculating with the complement.

Proposition 2.2 *For sets A, B, U with $A, B \subset U$, the complement in U satisfies*

$$\bar{\bar{A}} = A\,, \qquad \overline{A \cup B} = \bar{A} \cap \bar{B}\,, \qquad \overline{A \cap B} = \bar{A} \cup \bar{B}\,. \tag{2.3}$$

Proof We prove the second of these relations by mutual inclusion. An element $a \in \overline{A \cup B}$ is neither in A nor in B, so it must be in $\bar{A}$ and $\bar{B}$. It follows that $a \in \bar{A} \cap \bar{B}$ and, hence, that $a \in \overline{A \cup B} \subset \bar{A} \cap \bar{B}$. Conversely, an element $a \in \bar{A} \cap \bar{B}$ is contained in $\bar{A}$ and $\bar{B}$ and is, hence, neither in A nor in B. This means that $a \in \overline{A \cup B}$ so that $\bar{A} \cap \bar{B} \subset \overline{A \cup B}$. We leave the proof of the other equations as Exercise 2.2. □

By the size or *cardinality* of a set A we mean its number of elements, denoted by $|A|$. For example, the set $A = \{1, 3, 7, 9\}$ has cardinality $|A| = 4$. If the set A has an infinite number of elements, we write $|A| = \infty$.

2.1.3 New sets from old ones

There are a number of standard methods to construct new sets from given ones. The *power* set, denoted by 2^A, of a set A contains as its elements all subsets of A. If A is finite with cardinality $|A|$ then the power set has cardinality $2^{|A|}$, which motivates the notation. For example, for $A = \{1, 2\}$ we have

$$2^A = \{\{\}, \{1\}, \{2\}, \{1, 2\}\}\,.$$

The *Cartesian product* of two sets A and B, written as $A \times B$, consists of all pairs of elements, so

$$A \times B = \{(a, b) \,|\, a \in A \text{ and } b \in B\}\,. \tag{2.4}$$

For finite cardinalities $|A|$ and $|B|$ the cardinality of the Cartesian product is $|A \times B| = |A|\,|B|$. For example,

$$\{1, 2\} \times \{3, 4\} = \{(1, 3), (1, 4), (2, 3), (2, 4)\}\,.$$

The Cartesian product of a set A with itself is also denoted by $A^2 = A \times A$ and its elements are pairs of elements of A. More generally, we can take n Cartesian products of A with itself which is written as $A^n = A \times A \times \cdots \times A$. The elements of A^n consist of lists $(a_1, a_2, \ldots, a_n)$ of n elements of A. Such lists are also called *n-tuples*. For example, the set $\mathbb{N}^n$ consists of n-tuples of natural numbers.

2.2   Relations

Summary 2.2 *A relation between two sets is a subset of their Cartesian product which specifies which elements are considered to be related. An equivalence relation on a set is a specific relation which is reflexive, symmetric, and transitive. A set with an equivalence relation is partitioned into disjoint equivalence classes.*

2.2.1   Basic definitions

So far we have considered sets without any further structure. A simple but important way to introduce structure is by a *relation*.

Definition 2.1 *(Relation) A relation between two sets A and B is a subset R of $A \times B$. If $(a, b) \in R$ we write $a \sim b$ and we say that a and b are related. If $A = B$ we say that $R \subset A^2$ is a relation on A.*

This definition is a somewhat formal way of stating that a relation between sets A and B consists of pairs from the Cartesian product $A \times B$ which are 'declared' to be related. An important special class of relations are *equivalence relations*.

Definition 2.2 *(Equivalence relation) A relation on a set A is called an equivalence relation if it satisfies the following conditions for all $a, b, c \in A$.*

(i)	$a \sim a$	*(reflexivity)*
(ii)	$a \sim b$ *implies* $b \sim a$	*(symmetry)*
(iii)	$a \sim b$ *and* $b \sim c$ *implies* $a \sim c$	*(transitivity)*

For an equivalence relation every element is related to itself, relationship is symmetric and being related is 'passed on' (transitivity). The elements related to one another under an equivalence relation are collected in sets called *equivalence classes* which are defined as follows.

Definition 2.3 *(Equivalence class) Let A be a set with an equivalence relation and $a \in A$. The equivalence class $[a]$ is the subset of A which consists of all elements related to a, so $[a] = \{b \in A \mid b \sim a\}$. The set of all equivalence classes is called the quotient of A by $\sim$ and is denoted by $A/\sim$.*

The following example illustrates equivalence relations and classes.

Example 2.2 *(Equivalence relations and classes)*
Consider the set $\mathbb{N}$ of natural numbers. On this set, we define a relation by saying that two natural numbers n, m, are related, $n \sim m$, if $n + m$ is even. To see that this is an equivalence relation we have to check the three conditions in Def. 2.2. Since $n + n = 2n$ is always even we have $n \sim n$, meaning the relation is reflexive. Symmetry follows immediately because $n + m = m + n$. For transitivity, consider three numbers n, m, p with $n \sim m$ and $m \sim p$. This means $n + m$ and $m + p$ are both even, so their sum $(n + m) + (m + p) = n + p + 2m$ is even. But since $2m$ is even, this implies $n + p$ is even so that $n \sim p$.

What are the equivalence classes? Each two even numbers are related since their sum is even. The same is true for two odd numbers whose sum is also even. However, any even and any odd number are unrelated since their sum is odd. This shows that there are two equivalence classes, consisting of all the even and all the odd natural numbers, so that $\mathbb{N}/\sim = \{\{0, 2, 4, \ldots\}, \{1, 3, 5, \ldots\}\}$. □

2.2.2 Properties of equivalence relations

In Example 2.2, the set partitioned into disjoint equivalence classes. This feature is, in fact, general as shown in the following proposition:

Proposition 2.3 *Two equivalence classes are either equal or disjoint.*

Proof Consider a set A with an equivalence relation and two equivalence classes $[a]$ and $[b]$. If $[a] \cap [b] = \emptyset$ the statement is true so we can assume that there exists a $c \in [a] \cap [b]$. We want to show that $[a] \subset [b]$. To do this, start with an arbitrary $d \in [a]$, so that $d \sim a$. But since $c \in [a]$ we also have $a \sim c$ and transitivity implies that $d \sim c$. Further, from $c \in [b]$ we have $c \sim b$ and transitivity gives $d \sim b$, which shows that $d \in [b]$. Altogether, this means that $[a] \subset [b]$. The same argument, with the roles of $[a]$ and $[b]$ exchanged, can be repeated to show that $[b] \subset [a]$. It follows that $[a] = [b]$.
□

This statement means that an equivalence relation partitions a set A 'cleanly' into disjoint equivalence classes, as indicated in Fig. 2.2. Conversely, a partition of a set

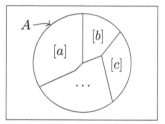

Fig. 2.2 An equivalence relation on a set A partitions the set into disjoint equivalence classes.

can be used to define an equivalence relation (see Exercise 2.4).The quotient $A/\sim$ is the set which consists of all these disjoint equivalence classes. Equivalence relations are a very useful tool for mathematical constructions, as the following example illustrates.

Example 2.3 *(Integers as equivalence classes)*
We have earlier introduced the natural numbers, $\mathbb{N}$, with addition and multiplication taken for granted. What about the integers? It turns out they can be constructed by introducing on the set $\mathbb{N}^2$ of integer pairs the equivalence relation

$$(n_1, n_2) \sim (m_1, m_2) \quad \text{if} \quad n_1 + m_2 = m_1 + n_2 . \tag{2.5}$$

By verifying the three properties in Def. 2.2, it can be shown that this is an equivalence relation (Exercise 2.3). What are the equivalence classes for this relation? For $n_1 \geq n_2$ we have $(n_1, n_2) \sim (n_1 - n_2, 0)$ and for $n_1 < n_2$ it follows that $(n_1, n_2) \sim (0, n_2 - n_1)$, so in each equivalence class we have an element of the form $(n, 0)$ or $(0, n)$. Two

such classes for different n are clearly inequivalent so that equivalence classes can be labelled by these elements. Intuitively, we identify the classes which contain $(n, 0)$ with the natural numbers n and the classes which contain $(0, n)$ with the negative natural numbers, also written as $-n$. In summary, we can define the integers, $\mathbb{Z}$, as $\mathbb{Z} = \mathbb{N}^2/\sim$.
□

2.3   Functions

Summary 2.3 *A function is a rule which assigns to each element of the domain exactly one element of the co-domain. Carrying functions out one after the other is referred to as function composition, which is an associative but not commutative operation. A function is called injective if the pre-image of each co-domain element contains at most one element. If the pre-image always contains at least one element it is called surjective. Functions which are both injective and surjective are called bijective. A function has an inverse function if and only if it is bijective.*

2.3.1   Definition of functions

A relation between two sets X and Y is generally not deterministic: an element of X may be related to more than one element of Y. In order to describe deterministic situations we require specific relations where every element of X is related to only one, unique element in Y. Such relations are described by *functions*.

Specifically, a *function f* from a set X to a set Y is a rule that assigns to every element of the source set X a *unique* element from the target set Y. The set X is also called the *domain* and the set Y the *co-domain* of the function. The input $x \in X$ of a function is also called its *argument* which is assigned by the function to its *value*, denoted by $f(x)$. In formal notation, the information about a function is summarized as follows:

$$f : X \to Y \qquad\qquad \text{or} \qquad\qquad X \xrightarrow{f} Y \qquad . \qquad\qquad (2.6)$$
$$x \mapsto f(x) \qquad\qquad\qquad\qquad\qquad x \xmapsto{f} f(x)$$

The top line reads 'f is a function with domain X and co-domain Y', while the bottom line means that the value of x under f is $f(x)$. It is sometimes useful to generalize the notation and allow a function $f : X \to Y$ to act on an entire subset $Z \subset X$ of the domain. This action is defined as $f(Z) := \{f(x) \mid x \in Z\}$ and it results in a subset of the co-domain which contains all the values f takes on Z. Depending on the context, a function may also be called a *map* and this term is normally used in linear algebra, where linear functions are referred to as *linear maps*. Note that it is important to distinguish the function f, which represents the rule of assignment, from its value $f(x)$ on a particular element $x \in X$. Two functions $f, g : X \to Y$ with the same domain and co-domain are called *equal*, written as $f = g$, if $f(x) = g(x)$ for all $x \in X$. The set of all values of a function $f : X \to Y$ is called the *image of the*

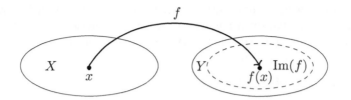

Fig. 2.3 Illustration of a function $f : X \to Y$.

function,

$$\mathrm{Im}(f) := f(X) = \{f(x) \,|\, x \in X\} \subset Y \,, \tag{2.7}$$

and this is a subset of the co-domain. For a given element $y \in Y$ of the co-domain the *pre-image* or *inverse image*

$$f^{-1}(y) := \{x \in X \,|\, f(x) = y\} \subset X \tag{2.8}$$

is the subset of all elements in the domain which are mapped to y. Some of these features are illustrated in Fig. 2.3.

On every set X there is a function $\mathrm{id}_X : X \to X$ defined by $\mathrm{id}_X(x) = x$, which maps every element to itself. This function is called the *identity map* or simply *identity* of X.

The connection between relations and functions can be understood in terms of the graph of a function $f : X \to Y$, defined by

$$\mathrm{Gr}(f) := \{(x, f(x)) \,|\, x \in X\} \subset X \times Y \,. \tag{2.9}$$

Intuitively, the graph consists of all pairs of arguments and values that should be drawn to visualize the function. It is a subset of the Cartesian product $X \times Y$ and, hence, defines a relation between X and Y (see Def. 2.1). This relation is deterministic in the sense discussed above, that is, every $x \in X$ is related to precisely one $y = f(x) \in Y$.

A common operation for functions is the restriction of the domain to a subset. Suppose, we start with a function $f : X \to Y$ and a subset $\tilde{X} \subset X$ of the domain. Then we can define a restricted function, denoted $f|_{\tilde{X}} : \tilde{X} \to Y$, by setting $f|_{\tilde{X}}(x) = f(x)$ for all $x \in \tilde{X}$, so that the values of f and its restriction are identical on the subset $\tilde{X}$.

Example 2.4 *(Functions)*
One way to specify a function is by an equation. For example, a linear function $f : \mathbb{R} \to \mathbb{R}$ is given by $f(x) = ax$ for a real number a. If $a \neq 0$, it follows that $\mathrm{Im}(f) = \mathbb{R}$ and otherwise, if $a = 0$, we have $\mathrm{Im}(f) = \{0\}$. The graph of this function consists of all points $\{(x, ax) \,|\, x \in \mathbb{R}\}$, which corresponds to a line through $(0, 0)$ with slope a.

The definition of a function can also involve case distinctions. For example,

$$g : \mathbb{R} \to \mathbb{R} \,, \qquad g(x) = \begin{cases} x & \text{for } x \geq 0 \\ 0 & \text{for } x < 0 \end{cases} \tag{2.10}$$

defines a piecewise linear function whose image is $\mathrm{Im}(g) = \mathbb{R}^{\geq 0} = \{x \in \mathbb{R} \,|\, x \geq 0\}$. The restriction $g|_{\mathbb{R}^{\geq 0}}$ of g to positive numbers is given by $g|_{\mathbb{R}^{\geq 0}}(x) = x$.

Of course, we can also define a function by simply providing its value for each element of the domain. For example, we can specify a function $h : \{1, 2, 3\} \to \{1, 2, 3\}$ by $h(1) = 1$, $h(2) = 3$ and $h(3) = 3$, so that its image is $\mathrm{Im}(h) = \{1, 3\}$. □

2.3.2   Composition of functions

Functions can be carried out one after the other, provided the co-domain of the first function is the same as the domain of the second function. This process is called *composition* of functions. More precisely, consider three sets X, Y, Z and two functions $f : X \to Y$ and $g : Y \to Z$. The composition $g \circ f : X \to Z$ is defined as $(g \circ f)(x) = g(f(x))$, that is, simply by evaluating the second function on the value of the first. The various mappings are summarized in the following diagram.

$$X \xrightarrow{\;f\;} Y \xrightarrow{\;g\;} Z$$
$$g \circ f$$

Note how the notation forces a reversal in the ordering of the two functions. While f acts first and g second, the fact that arguments are written to the right of the function symbol means the composition should be $g \circ f$.

Example 2.5 *(Composition of functions)*
Consider the functions defined by $f(x) = x^2$ and $g(x) = 2x + 3$, seen as functions $\mathbb{R} \to \mathbb{R}$. Then the composite function $(g \circ f)(x) = g(f(x)) = g(x^2) = 2x^2 + 3$ is simply obtained by 'inserting' one function into the other. In this case, we can also compose in the opposite order, $(f \circ g)(x) = f(g(x)) = f(2x + 3) = (2x + 3)^3$, which gives a different result. Function composition does not commute! □

Composition of maps is *associative* which means for three functions f, g, and h we have
$$f \circ (g \circ h) = (f \circ g) \circ h \,. \tag{2.11}$$
This property is easily verified using the definition of composition repeatedly.
$$(f \circ (g \circ h))(x) = f((g \circ h)(x)) = f(g(h(x))) = (f \circ g)(h(x)) = ((f \circ g) \circ h)(x)$$
The identity map acts as a *neutral element* of composition, in the sense that
$$f \circ \mathrm{id}_X = f \,, \qquad \mathrm{id}_Y \circ f = f \tag{2.12}$$
for every function $f : X \to Y$. This follows from $(f \circ \mathrm{id}_X)(x) = f(\mathrm{id}_X(x)) = f(x)$ and similarly for the other equation. The properties (2.11) and (2.12) of composition are quite important and, as we will see, are key ingredients in the definition of a group.

2.3.3   Properties of functions

There are a few important structural properties of functions which are summarized in the following definition.

Definition 2.4 *Let $f : X \to Y$ be a function with domain X and co-domain Y.*
(i) f is called injective if $f(x) = f(x')$ implies $x = x'$ for all $x, x' \in X$.
(ii) f is called surjective provided for all $y \in Y$ there exists an $x \in X$ with $y = f(x)$.
(iii) f is called bijective if it is injective and surjective.

An injective function (also called 'one-to-one' function) maps no two different elements of the domain to the same image or, in other words, the pre-image $f^{-1}(y)$ contains at most one element for all $y \in Y$. A surjective function (also called 'onto' function) 'reaches' every element of the domain, that is $\mathrm{Im}(f) = Y$, or equivalently, the pre-image $f^{-1}(y)$ consists of at least one element for all $y \in Y$. A bijective function combines these two properties, so that the pre-image $f^{-1}(y)$ consists of precisely one element for all $y \in Y$. These properties are illustrated in Fig. 2.4.

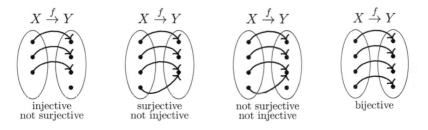

injective
not surjective

surjective
not injective

not surjective
not injective

bijective

Fig. 2.4 Illustration of injective, surjective, and bijective functions $f : X \to Y$.

Example 2.6 *(Basic function properties)*
The linear function $f : \mathbb{R} \to \mathbb{R}$ given by $f(x) = ax$ is bijective for $a \neq 0$. (It is surjective since $\mathrm{Im}(f) = \mathbb{R}$ and it is injective since $f(x) = f(x')$ implies $ax = ax'$ and, hence, $x = x'$.) For $a = 0$, every $x \in \mathbb{R}$ is mapped to 0 so in this case f it is neither injective nor surjective.

The piecewise linear function $g : \mathbb{R} \to \mathbb{R}$ in Eq. (2.10) is not injective (since all negative numbers are mapped to 0) and it is not surjective (since negative numbers are not images). However, if we modify g by restricting its co-domain, $g : \mathbb{R} \to \mathbb{R}^{\geq 0}$, (but with its values still defined by Eq. (2.10)) then it is surjective, although still not injective. $\qquad\square$

The next proposition asserts that the above properties of functions are preserved under composition.

Proposition 2.4 *For two injective (surjective) functions $f : X \to Y$ and $g : Y \to Z$ the composition $g \circ f : X \to Z$ is also injective (surjective). If f and g are bijective, then so is $g \circ f$.*

Proof First suppose that f and g are injective. We would like to show that $g \circ f$ is injective as well. Suppose that $(g \circ f)(x) = (g \circ f)(x')$ for $x, x' \in X$. From the definition of function composition, this means that $g(f(x)) = g(f(x'))$. Since g is injective, it follows that $f(x) = f(x')$ and injectivity of f then implies that $x = x'$. Hence $g \circ f$ is injective.

Now suppose that f and g are surjective. We want to show that $g \circ f$ is surjective as well. To do so, we start with a $z \in Z$ and try to construct an element in the pre-image $(g \circ f)^{-1}(z)$. Since g is surjective, there exists a $y \in Y$ such that $z = g(y)$. But f is also surjective, so we have an $x \in X$ with $y = f(x)$. Combining these statements gives $z = g(y) = g(f(x)) = (g \circ f)(x)$ and, hence, $g \circ f$ is surjective.

If f and g are bijective, then they are both injective and surjective. Then, from the previous statements, $g \circ f$ is injective and surjective and, hence, bijective. $\square$

Thanks to this proposition, we can say that the properties 'injective', 'surjective', and 'bijective' are preserved under function composition. As we will see, linearity is another such property which is preserved under composition (see Exercises 2.6).

We also note that the restriction $f|_{\bar{X}}$ of an injective function $f : X \to Y$ to a subset $\bar{X} \subset X$ of the domain is still injective. Indeed, if $f(x) = f(x')$ implies $x = x'$ for all $x, x' \in X$, then this is also implied for all $x, x' \in \bar{X}$. The analogous statement for surjective functions is, of course, false. Restricting the domain can lead to a smaller image, so that surjectivity can be lost.

Bijective maps allow us to be more precise about the notion of set cardinality. We say that a set X has cardinality n, written as $|X| = n$, if there exists a bijective map $X \to \{1, 2, \dots, n\}$. This means the elements of X can be indexed by integers so that the set can be written as $X = \{x_1, x_2, \dots, x_n\}$. Further, we say that the set has *countably infinite cardinality* if there exists a bijective map $X \to \mathbb{N}$, so the set can be written as $X = \{x_0, x_1, \dots\}$. If neither is the case, we say the cardinality is *non-countably infinite*.

2.3.4 The inverse function

Can the effect of a function be undone by an *inverse function*? We should first define what exactly we mean by this.

Definition 2.5 *For a function $f : X \to Y$ an inverse function is a function $g : Y \to X$, which satisfies $g \circ f = \mathrm{id}_X$ and $f \circ g = \mathrm{id}_Y$.*

A function $f : X \to Y$ need not have an inverse. For example, if f is not injective then there is a pre-image $f^{-1}(y)$ for some $y \in Y$ which contains at least two elements. In this case, it is not clear which of these two elements to choose for the value of the prospective inverse function with argument y. Likewise, if f is not surjective there exists an empty pre-image $f^{-1}(y)$ so in this case there is no candidate for the value of a prospective inverse function at y. Fortunately, being non-injective or non-surjective are the only obstructions to the existence of an inverse.

Proposition 2.5 *A function $f : X \to Y$ has an inverse if and only if f is bijective. In this case, the inverse, denoted by $f^{-1} : Y \to X$, is unique and bijective.*

Proof First assume that $f : X \to Y$ has an inverse $g : Y \to X$, hence $g \circ f = \mathrm{id}_X$ and $f \circ g = \mathrm{id}_Y$. We want to prove that f is bijective. To show that f is injective, start with $f(x) = f(x')$ and apply g from the left, so that $g(f(x)) = g(f(x'))$. But $g \circ f = \mathrm{id}_X$, hence $g(f(x)) = x$ and $g(f(x')) = x'$, which implies $x = x'$. To show that f is surjective, observe that for every $y \in Y$ there exists an $x = g(y) \in X$ with

$f(x) = f(g(y)) = y$. Hence f is bijective.

For the second part of the statement, assume that f is bijective. Define a map $g : Y \to X$ by $g(y) = x$, where x is the unique element in the pre-image $f^{-1}(y)$. Then g satisfies $(f \circ g)(y) = f(x) = y$ and $(g \circ f)(x) = g(y) = x$ and, hence, it is an inverse map for f.

Finally, to show uniqueness of the inverse map, consider two inverse maps $g : Y \to X$ and $\tilde{g} : Y \to X$. They satisfy $g \circ f(x) = x = \tilde{g} \circ f(x)$ for all $x \in X$. Since f is surjective, any $y \in Y$ can be written as $y = f(x)$ for some $x \in X$ and therefore $g(y) = \tilde{g}(y)$ for all $y \in Y$. Hence, $g = \tilde{g}$ and the inverse is unique. We leave the proof that f^{-1} is bijective as Exercise 3.1. □

Note that, by a slight abuse of notation, we are using the same symbol, f^{-1}, to denote the inverse image and the inverse function. It should be clear from the context which one is referred to. Since f^{-1} is also bijective, it has an inverse as well and it is intuitively clear that this inverse of the inverse must be the original function,

$$(f^{-1})^{-1} = f \; . \tag{2.13}$$

Formally, this follows because both f and $(f^{-1})^{-1}$ satisfy the conditions in Def. 2.5 for an inverse function to f^{-1} and uniqueness of the inverse function hence implies Eq. (2.13).

Combining Prop. 2.4 and Prop. 2.5, it is clear that the composition of two invertible functions is invertible. Moreover, the inverse of the composition can be worked out by the rule

$$(g \circ f)^{-1} = f^{-1} \circ g^{-1} \; . \tag{2.14}$$

Note the change of ordering in this formula which is indeed correct! For the proof note that both $(g \circ f)^{-1}$ and $f^{-1} \circ g^{-1}$ are an inverse to $g \circ f$, in the sense of Def. 2.5. Uniqueness of the inverse then leads to Eq. (2.14).

Example 2.7 *(Inverse function)*
The linear map $f : \mathbb{R} \to \mathbb{R}$ defined by $f(x) = ax$ for $a \neq 0$ is bijective and, hence, has a unique inverse. Clearly, the inverse function $f^{-1} : \mathbb{R} \to \mathbb{R}$ is given by $f^{-1}(x) = x/a$. Frequently, a function can be made bijective by modifying its domain or co-domain. For example, the quadratic function $f(x) = x^2$, seen as a function $f : \mathbb{R} \to \mathbb{R}$, is neither injective (since $f(x) = f(-x)$) nor surjective (since $f(x) \geq 0$). However, seen as a function $f : \mathbb{R}^{\geq 0} \to \mathbb{R}^{\geq 0}$ it is bijective. Its unique inverse $f^{-1} : \mathbb{R}^{\geq 0} \to \mathbb{R}^{\geq 0}$ is the square root function $f(x) = \sqrt{x}$. □

The above example shows that linear maps on $\mathbb{R}$ may or may not be bijective and, hence, may or may not have an inverse. Deciding whether a (more general) linear function has an inverse and how to compute it is an important problem which we will address in detail later.

2.4   Rudiments of logic

Summary 2.4 *Predicates are functions with co-domain* $\{0, 1\}$. *New predicates can be formed from given ones by the operations 'and', 'or' and the complement. These operations are closely related to the union, intersection and the complement of sets. Implications and quantifiers can be used to formulate conclusions and statements in terms of predicates. The structure of basic methods of proofs, such as direct proof, proof by contradiction, and proof by induction can be formulated in this language.*

Logic is an important foundational area of mathematics, which we cannot possibly do justice to in a short introduction. However, we need to develop our main subject systematically, including proofs, so introducing some elements of logic is unavoidable. We will keep the discussion short, focus on key ideas, explain notation and finish by discussing the logical structure of basic types of proofs.

2.4.1   Predicates and Boolean operations

For a set X a *Boolean function* or *predicate* on X is a function $P : X \to \{0, 1\}$. The idea is that elements $x \in X$ can be statements which are true if $P(x) = 1$ and false if $P(x) = 0$. Given two predicates, P and Q, we can define new predicates

$$P \vee Q, \qquad P \wedge Q, \tag{2.15}$$

read as 'P or Q' and 'P and Q', respectively, whose values are defined in Table 2.1. Note that the assignments in the table correspond to the 'intuitive' meaning of 'and'

Table 2.1 Logical operations 'or' and 'and' for predicates P and Q on a set X.

$P(x)$	$Q(x)$	$(P \wedge Q)(x)$	$(P \vee Q)(x)$	$(\bar{P} \vee Q)(x)$	$(\bar{Q} \vee P)(x)$
0	0	0	0	1	1
0	1	0	1	0	1
1	0	0	1	1	0
1	1	1	1	1	1

and 'or'. Both operations are associative and commutative, that is:

$$
\begin{array}{lll}
P \wedge (Q \wedge R) = (P \wedge Q) \wedge R & P \vee (Q \vee R) = (P \vee Q) \vee R & \text{(associativity)} \\
P \wedge Q = Q \wedge P & P \vee Q = Q \vee P & \text{(commutativity)}
\end{array}
$$

Commutativity is evident from Table 2.1 since the results in columns three and four do not depend on the ordering of the first two columns. Associativity can also be verified by a truth table, similar to Table 2.1 (Exercise 2.9). In addition to associativity and distributivity, there are two distributive laws which connect the two operations.

Proposition 2.6 *For three predicates P, Q, and R, we have*

$$P \wedge (Q \vee R) = (P \wedge Q) \vee (P \wedge R), \qquad P \vee (Q \wedge R) = (P \vee Q) \wedge (P \vee R). \tag{2.16}$$

Proof The proof can be accomplished by a truth table, going through all eight combinations for the values of P, Q, and R. As an example, suppose that $P(x) = Q(x) = 1$ and $R(x) = 0$. Then, from Table 2.1, we have $P \wedge (Q \vee R)(x) = 1$ and $(P \wedge Q) \vee (P \wedge R)(x) = 1$ which proves the first Eq. (2.16) for this case. The proof is completed by going through the other seven possibilities (Exercise 2.9). $\square$

The reader has probably noticed the formal similarity between the above rules for how to calculate with 'or' and 'and' and the rules for the union and intersection of sets, as discussed in Sec. 2.1.2. These similarities are, of course, not an accident and have to do with the fact that 'or' and 'and' have been used, then somewhat naively, to define the union and intersection of sets. The relationship can be made more precise by noting that there is a bijective correspondence between predicates on X and subsets of X. Concretely, we can assign to a predicate P on X the subset $X_P = \{x \in X \mid P(x) = 1\}$ of those elements in X, for which the predicate is true (also, see Exercise 2.8). Based on this correspondence, we can now define the union and intersection of sets

$$X_P \cup X_Q = X_{P \vee Q}, \qquad X_P \cap X_Q = X_{P \wedge Q}, \tag{2.17}$$

in terms of 'or' and 'and'. With these definitions, it is easy to derive the rules for calculating with sets from the above rules for predicates. For example, obtaining the first distributive law (2.2) for sets works, as follows:

$$X_P \cap (X_Q \cup X_R) \overset{(2.17)}{=} X_{P \vee (Q \wedge R)} \overset{(2.16)}{=} X_{(P \wedge Q) \vee (P \wedge R)} \overset{(2.17)}{=} (X_P \cap X_Q) \cup (X_P \cap X_R).$$

The *complement* $\bar{P}$ of a predicate P on X is defined as:

$$\bar{P}(x) = \begin{cases} 0 \text{ if } P(x) = 1 \\ 1 \text{ if } P(x) = 0 \end{cases}. \tag{2.18}$$

The complement satisfies a number of important relations, also known as *de Morgan's laws*.

Proposition 2.7 *(de Morgan's laws) The complement of predicates on X satisfies*

$$\bar{\bar{P}} = P, \qquad \overline{P \vee Q} = \bar{P} \wedge \bar{Q}, \qquad \overline{P \wedge Q} = \bar{P} \vee \bar{Q}. \tag{2.19}$$

Proof The first law, $\bar{\bar{P}} = P$, is obvious from the Def. (2.18) of the complement. The other two rules are verified by the truth table below:

P	Q	$P \vee Q$	$\bar{P} \wedge \bar{Q}$	$P \wedge Q$	$\bar{P} \vee \bar{Q}$
0	0	1	1	1	1
0	1	0	0	1	1
1	0	0	0	1	1
1	1	0	0	0	0

$\square$

The complement of predicates is closely related to the set complement, in much the same way the 'or' and 'and' operations are related to the union and intersection of sets. More precisely, we can now define the set complement in terms of the complement of predicates as

$$\overline{X_P} = X_{\bar{P}} \ . \tag{2.20}$$

With this definition, the rules for set complements in Prop. 2.2 immediately follow from de Morgan's laws (Exercise 2.10).

Finally, we need to introduce two simple pieces of terminology. A *tautology* is a predicate on X which returns true for all $x \in X$ and Table 2.1 shows that, for example, $P \vee \bar{P}$ is a tautology. A predicate which returns false for all $x \in X$ is called a *contradiction* and an example is provided by $P \wedge \bar{P}$.

2.4.2   Implications

Given two predicates P and Q on X, the *implication* connective allows us to form a new predicate, denoted by $P \Rightarrow Q$, and read as 'P implies Q' or 'Q follows from P'. It is defined by

$$(P \Rightarrow Q) = \bar{P} \vee Q \ . \tag{2.21}$$

Table 2.1 shows that $P \Rightarrow Q$ is true in all cases, except when P is true and Q is false. This makes sense, since an implication should only be called false if a true premise P leads to a false conclusion Q. An implication $P \Rightarrow Q$ is called *valid* if it is a tautology, that is, if $(P \Rightarrow Q)(x) = 1$ for all $x \in X$.

The predicate which tests mutual implication is written as $P \Leftrightarrow Q$, read as 'P equivalent to Q' and formally this can be defined by

$$(P \Leftrightarrow Q) = (P \Rightarrow Q) \wedge (Q \Rightarrow P) \ . \tag{2.22}$$

Table 2.1 shows (by taking a logical 'and' between the last two columns) that $(P \Leftrightarrow Q)(x)$ is true iff $P(x) = Q(x)$ and false otherwise. If $P \Leftrightarrow Q$ is a tautology we also say that P holds if and only if Q holds. The phrase 'if and only if' in this context is often abbreviated as 'iff'.

A simple calculation based on the definition (2.21), commutativity, and the first de Morgan law shows that:

$$(P \Rightarrow Q) = \bar{P} \vee Q = \bar{\bar{Q}} \vee \bar{P} = (\bar{Q} \Rightarrow \bar{P}) \ . \tag{2.23}$$

This result underlies the method of indirect proof (discussed in more detail later) by which the validity of $P \Rightarrow Q$ follows from the validity of $\bar{Q} \Rightarrow \bar{P}$.

2.4.3   Quantifiers

Logical statements often involve quantifiers, such as the quantifier 'or all', written as $\forall$, the quantifier 'there exists', written as $\exists$ and the quantifier 'there exists a unique' written as $\exists!$. Using this notation, most theorems can then be cast into one of the following forms:

$$(\forall x \in X)(P(x)) , \qquad (\exists x \in X)(P(x)) , \qquad (\exists! x \in X)(P(x)) . \tag{2.24}$$

From left to right, these statements should be read '$P(x)$ is true for all $x \in X$', 'there exists an $x \in X$, such that $P(x)$ is true', and 'there exists a unique $x \in X$ such that $P(x)$ is true'.

As an aside, we remark that quantifiers are also often used in the definition of sets. For example, the set

$$n\mathbb{Z} := \{k \in \mathbb{Z} \mid \exists m \in \mathbb{Z} : k = nm\}$$

consists of all integers which are multiples of n. (The colon in this expression is read as 'such that'.)

Sometimes, it is easier to prove that the negative of a statement is false rather than proving the original statement directly. In such cases, it is important to understand how to negate statements involving quantifiers. The general rule is that, under negation, the universal quantifier, $\forall$, turns into the existential one, $\exists$, and vice versa. More concretely, we have

$$\overline{(\forall x \in X)(P(x))} - (\exists x \in X)(\bar{P}(x)) , \qquad \overline{(\exists x \in X)(P(x))} = (\forall x \in X)(\bar{P}(x)) .$$

2.4.4 Patterns of proofs

We finish this section by discussing the logical structure of some common patterns of proof. A *direct proof* has the logical structure,

$$(P \wedge (P \Rightarrow Q)) \Rightarrow Q . \tag{2.25}$$

Using a truth table and the results from Table 2.1, it can be checked that this is tautological (see Exercise 2.11). Note that this formal expression captures what one would intuitively state as the structure of a direct proof: 'If P is true and if Q follows from P then Q is true.'

The corresponding expression for an *indirect proof* is

$$(\bar{Q} \wedge (\bar{Q} \Rightarrow \bar{P})) \Rightarrow \bar{P} . \tag{2.26}$$

It follows immediately from Eq. (2.25) by the replacements $P \to \bar{Q}$ and $Q \to \bar{P}$.

The logical structure of a *proof by contradiction* is

$$((\bar{Q} \Rightarrow P) \wedge (\bar{Q} \Rightarrow \bar{P})) \Rightarrow Q . \tag{2.27}$$

and, again, this is tautological (Exercise 2.11). A proof by contraction starts by assuming that Q is false. If this can be shown to imply both P and $\bar{P}$, a contradiction has been encountered and Q follows.

Example 2.8 *(Proof by contradiction)*
As an example of a proof by contradiction, we want to show that there are infinitely many prime numbers. This is the statement Q in Eq. (2.27). Then, the statement $\bar{Q}$ is that there are only finitely many prime numbers $(p_1, p_2, \ldots, p_n)$. It follows that every prime number is contained in the list $(p_1, p_2, \ldots, p_n)$ (the statement P). Consider the product $p = p_1 p_2 \cdots p_n$ and $q = p + 1$. If q is prime, then it is an additional prime number not in the list and the statement $\bar{P}$ follows. If q is not prime, then it contains a prime factor r. If this prime factor is in the list $(p_1, p_2, \ldots, p_n)$, then it is also a prime factor of p but p and q cannot have a common prime factor since $q - p = 1$. Hence, r cannot be in the list and again the statement $\bar{P}$ follows. We have now shown that $\bar{Q} \Rightarrow P$ and $\bar{Q} \Rightarrow \bar{P}$ are both valid, so from Eq. (2.27) it follows that Q holds. $\quad\square$

Another common type of proof is *proof by induction*. This arises in the context of predicates P on countably infinite sets $X = \{x_0, x_1, \ldots\}$. If we write $P_n = P(x_n)$ its logical structure is

$$(P_0 \vee ((\forall n \in \mathbb{N})(P_n \Rightarrow P_{n+1})) \Rightarrow ((\forall n \in \mathbb{N})P_n) . \tag{2.28}$$

While this may seem complicated at first it captures a simple idea. If the statement P_0 is true and every statement P_n implies its successor P_{n+1} then all statements P_n must be true.

Example 2.9 *(Proof by induction)*
We would like to prove a formula for the sum $S_n = 0 + 1 + 2 + \cdots + n$ and the claim is that

$$S_n = n(n+1)/2 . \tag{2.29}$$

Clearly, this claim is true for $n = 0$ (the 'basis' of the induction). Let us assume that it is true for n (the 'induction assumption'), so that $S_n = n(n+1)/2$. To show that it follows for $n + 1$, we carry out the calculation

$$S_{n+1} = 0 + 1 + \cdots + n + (n+1) = S_n + (n+1) = \frac{1}{2}n(n+1) + (n+1) = \frac{1}{2}(n+1)(n+2) ,$$

where the induction assumption has been used in the third step. The left- and right-hand sides of this equation are precisely the claim (2.29) with n replaced by $n + 1$. Hence, this shows that the statement is true for $n + 1$ and completes the 'induction step'. It follows that the statement holds for all $n \in \mathbb{N}$. $\quad\square$

Exercises

2.1 Prove the second Eq. (2.2) by showing mutual inclusion of the left- and right-hand sides.

2.2 Prove the first and third Eq. (2.3) by showing mutual inclusion of the left- and right-hand sides.

2.3 Show that the relation (2.5) is an equivalence relation on $\mathbb{N}^2$.

2.4 *Equivalence relations from partitions*
A set S is given as a union $S = \bigcup_i S_i$ of mutually disjoint subsets S_i. Define a relation on S by declaring $s, \tilde{s} \in S$ as related if they are contained in the same subset S_i. Show that this is an equivalence relation, which partitions S into the subsets S_i.

2.5 Show that the inverse of a bijective function is bijective.

2.6 *Composition of linear functions.*
(a) Let $f, g : \mathbb{R} \to \mathbb{R}$ be two linear functions. Show that their composition is also a linear function. Is there a difference between $f \circ g$ and $g \circ f$?
(b) Consider the same problem in two dimensions. Let $f, g : \mathbb{R}^2 \to \mathbb{R}^2$ be two linear functions. Show that their compositions $f \circ g$ and $g \circ f$ are linear as well. Write these functions as $f(\mathbf{x}) = A\mathbf{x}$ and $g(\mathbf{x}) = \tilde{A}\mathbf{x}$, where $\mathbf{x} \in \mathbb{R}^2$ is a vector with entries x_1, x_2, and

$$A = \begin{pmatrix} a & b \\ c & d \end{pmatrix}, \quad \tilde{A} = \begin{pmatrix} \tilde{a} & \tilde{b} \\ \tilde{c} & \tilde{d} \end{pmatrix}$$

are 2×2 matrices. Work out the two matrices, which describe $f \circ g$ and $g \circ f$. Are these matrices in general equal?

2.7 *'Simplifying' functions.*
Let $f : X \to Y$ be a function.
a) Let $g, h : Z \to X$ be functions and assume that f is injective. Show that $f \circ g = f \circ h$ implies that $g = h$.
b) Let $g, h : Y \to Z$ be functions and assume that f is surjective. Show that $(g \circ f = h \circ f)$ implies that $g = h$.

2.8 *Boolean functions and the power set*
Show that the number of Boolean functions on a set X with n elements is equal to 2^n. This is the same as the cardinality of the power set 2^X, defined as the set of subsets of X. Explain why this is not a coincidence.

2.9 Verify that the logical operations 'or' and 'and' are associative, using a truth table. Do the same for the distributive laws (2.16).

2.10 With the set operations defined as in Eqs. (2.17) and (2.20), show that Prop. 2.2 follows from de Morgan's laws in Prop. 2.7.

2.11 Prove that the statements (2.25) for a direct proof is tautological by completing the truth table in Table 2.1. Do the same with Eq. (2.27).

2.12 *Induction*
By using induction, show that $1 + 2^2 + 3^2 + \cdots + n^2 = n(n+1)(2n+1)/6$.

3
Groups

So far, the only structures on sets we have considered are relations. One of the simplest algebraic structures on a set is a *group* structure. Groups underly the definition of fields and vector spaces and for this reason alone we need to introduce them. Group theory is a large and diverse area of mathematics and we have to keep our discussion short, focusing on basics and some more specific aspects, which will become relevant later on. (For a dedicated introduction to group theory see, for example, Armstrong 2013.) Groups also provide the mathematical framework for symmetries which are immensely important for many areas of science, particularly in physics (see, for example, Cornwell 1997; Wybourne 1974). In the next section, we begin by defining groups, sub-groups, and the maps consistent with the group structure, the *group homomorphisms*. A number of simple examples will be presented as we go along. Permutation groups are required for the determinant (see Chapter 18) and will be examined in Section 3.2.

3.1 Definition and basic properties

Summary 3.1 *A group is a simple algebraic structure that consist of a set with a multiplication that is associative, has a neutral element, and an inverse for each group element. If the multiplication commutes the group is called Abelian. Examples of groups are all the bijective maps on a set, permutations, the integers with respect to addition, and the cyclic groups. Cartesian products of groups can be given a direct product group structure by component-wise multiplication. Maps between groups which are consistent with the group structure are called group homomorphisms.*

3.1.1 Definition

In Section 2.3 we have seen that functions $f : X \to X$ have a number of interesting properties. Their composition is associative, there exists a neutral element, the identity map, under composition, and, for bijective functions, there always exists a unique inverse. It makes sense to formalize these properties, and this leads to the definition of a *group*.

Definition 3.1 *(Group) A group $(G, \cdot)$ is a non-empty set G with an operation $\cdot :$ $G {\times} G \to G$, $(g_1, g_2) \mapsto g_1 {\cdot} g_2$ (called 'group multiplication'), which satisfies the following properties.*

(G1) $g_1 \cdot (g_2 \cdot g_3) = (g_1 \cdot g_2) \cdot g_3$ $\forall g_1, g_2, g_3 \in G$ *(associativity)*
(G2) $\exists e \in G : e \cdot g = g$ $\forall g \in G$ *((left) neutral element)*
(G3) $\forall g \in G \; \exists \tilde{g} \in G : \tilde{g} \cdot g = e$ *((left) inverse)*

If, in addition, $g_1 \cdot g_2 = g_2 \cdot g_1$ for all $g_1, g_2 \in G$, then the group is called Abelian.

Note that the term 'multiplication' in this definition refers to any operation with the stated properties, not just the usual multiplication of numbers. Groups can have a finite or infinite number of elements, as we will see, and in the former case G is called *finite* with *order* given by the cardinality $|G|$ (also see Exercise 3.3).

The above definition has a curious asymmetry, in that the neutral element and the inverse are postulated only when multiplied from the left. However, this is not a problem, as the following proposition shows.

Proposition 3.1 *For a group G, the left inverse is unique and is also a right inverse, so $\tilde{g} \cdot g = e$ implies $g \cdot \tilde{g} = e$. The left neutral element is unique and is also right neutral, so that $e \cdot g = g$ for all $g \in G$ implies $g \cdot e = g$.*

Proof   We begin by proving that every left inverse is also a right-inverse. For $g \in G$ consider a left-inverse $\tilde{g}$ so that $\tilde{g} \cdot g = e$. The inverse $\tilde{g}$ has its own left-inverse which we call $\tilde{\tilde{g}}$, so that $\tilde{\tilde{g}} \cdot \tilde{g} = e$. It follows that

$$ g \cdot \tilde{g} = e \cdot g \cdot \tilde{g} = \tilde{\tilde{g}} \cdot \underbrace{\tilde{g} \cdot g}_{=e} \cdot \tilde{g} = \tilde{\tilde{g}} \cdot \tilde{g} = e \,, $$

which completes the proof. We leave the other statements as Exercise 3.1.   □

Since the inverse for a given $g \in G$ is unique it is usually denoted by g^{-1}. The above proposition leads to two basics rules for calculating with the inverse, which mirror the rules (2.13) and (2.14) for the inverse of functions:

$$ (g^{-1})^{-1} = g \,, \qquad (g_1 \cdot g_2)^{-1} = g_2^{-1} \cdot g_1^{-1} \,. \tag{3.1} $$

The first of these follows from the fact that both g and $(g^{-1})^{-1}$ provide an inverse to g^{-1}. Hence, from the uniqueness of the inverse, they must be equal. Likewise, both $(g_1 \cdot g_2)^{-1}$ and $g_2^{-1} \cdot g_1^{-1}$ are inverse to $g_1 \cdot g_2$ and must be equal.

How does the group structure relate to the Cartesian product of sets? More concretely, for two groups $(G, \cdot)$ and $(\tilde{G}, \cdot)$ with neutral elements e and $\tilde{e}$, can the Cartesian product $G \times \tilde{G}$ be made into a group? The answer is 'yes' providing the multiplication on $G \times \tilde{G}$ is defined component-wise as:

$$ (g_1, \tilde{g}_1) \cdot (g_2, \tilde{g}_2) = (g_1 \cdot g_2, \tilde{g}_1 \cdot \tilde{g}_2) \,. \tag{3.2} $$

This leads to the *direct product group* $(G \times \tilde{G}, \cdot)$ with neutral element $(e, \tilde{e})$ and inverse $(g, \tilde{g})^{-1} = (g^{-1}, \tilde{g}^{-1})$. Associativity is clearly satisfied for the direct product group since it holds for each of the constituent groups. Further, if both $(G, \cdot)$ and $(\tilde{G}, \cdot)$ are Abelian groups, then so is the direct product group $(G \times \tilde{G}, \cdot)$. This construction is quite important and will re-appear in the context of fields and vector spaces later on. Of course it can be generalized to multiple factors. For example, for a group $(G, \cdot)$ the Cartesian product G^n can be made into a group by component-wise multiplication.

3.1.2 Examples of groups

It is now time to discuss a few important examples of groups.

Example 3.1 (*Bijective maps on a set*)
From Section 2.3, we know that all bijective maps $\varphi : X \to X$ on a set X form a group, denoted by $\mathrm{Bij}(X)$. The group multiplication is the composition of maps (which is associative), id_X is the neutral element, and the group inverse is the inverse map φ^{-1}. □

Example 3.2 (*Permutation groups*)
This is a special case of the previous example, where we consider all bijective maps on the set $X = \{1, 2, \ldots, n\}$. The resulting group is called the *permutation group S_n*. The idea is that a map $\sigma \in S_n$ corresponds to a permutation of $\{1, \ldots, n\}$ by permuting every $k \in \{1, \ldots, n\}$ to its image $\sigma(k)$. Permutation groups will be discussed in more detail in Section 3.2. □

Example 3.3 (*Integers*)
The natural numbers do not form a group with respect to either addition or multiplication. For addition we are missing the negative numbers that would provide an inverse, for multiplication there is no inverse because of the absence of fractions. On the integers $\mathbb{Z}$, constructed as a quotient of $\mathbb{N}^2$ (see Example 2.3), we can also introduce an addition and a multiplication, based on the corresponding operations on $\mathbb{N}$, by:

$$[(n_1, n_2)] + [(m_1, m_2)] = [(n_1 + n_2, m_1 + m_2)]$$
$$[(n_1, n_2)]\,[(m_1, m_2)] = [(n_1 m_1 + n_2 m_2, n_1 m_2 + n_2 m_1)] \,.$$

To see that this really corresponds to the familiar arithmetic on $\mathbb{Z}$, let us consider some examples. The sum of a positive number $[(n, 0)]$ and a negative number $[(0, m)]$ is $[(n, m)]$ which is indeed interpreted as the difference $n - m$. A product of two negative numbers $[(0, n)], [(0, m)]$ gives $[(nm, 0)]$, which corresponds to a natural number.

The integers $(\mathbb{Z}, +)$ with the above addition do form an Abelian group, with neutral element $0 = [(0, 0)]$ and inverse $-n = [(0, n)]$ for any $n \in \mathbb{Z}$. However, the integers with multiplication do not form a group since the multiplicative inverse is still missing. □

Example 3.4 ($\mathbb{Z}_n$)
The set $\mathbb{Z}_n := \{0, 1, \ldots, n - 1\}$ can be made into a group $(\mathbb{Z}_n, \cdot)$ by defining the 'multiplication'

$$k \cdot k' = (k + k') \bmod n \,, \tag{3.3}$$

where $k \bmod n$ denotes the remainder of the division of k by n. This forms an Abelian group with neutral element 0 and inverse $n - k$ for $k \in \mathbb{Z}_n$ which is also referred to as *cyclic group* of order n. One way to specify the group multiplication for a group with a finite number of elements is by a multiplication table. For example, from Eq. (3.3), the multiplication table for $\mathbb{Z}_3 = \{0, 1, 2\}$ is:

·	0	1	2
0	0	1	2
1	1	2	0
2	2	0	1

(See also Exercise 3.6.) □

3.1.3 Sub-groups

A standard step in the build-up of any algebraic structure is the introduction of the sub-structure. In the case of groups, this leads to the notion of *sub-groups*, defined as follows:

Definition 3.2 *(Sub-group) A subset $H \subset G$ of a group G is called a sub-group if it forms a group by itself under the multiplication defined on G.*

To check that a subset $H \subset G$ is a sub-group it is enough to verify that H contains the neutral element e, that is, contains the inverse h^{-1} for all $h \in H$ and that it is closed under group multiplication, so $h_1, h_2 \in H$ implies $h_1 \cdot h_2 \in H$.

Every group G contains two trivial sub-groups, namely the group $\{c\}$ which consists of the neutral element only and the whole group G. All other sub-groups are called *proper sub-groups*.

Example 3.5 *(Sub-groups)*
Consider the group $(\mathbb{Z}_4, \cdot)$, the cyclic group of order four, as defined in Example 3.4. Then $H = \{0, 2\}$ forms a sub-group, since it is closed under the group multiplication (3.3), contains the neutral element 0 and an inverse for each of its elements (as 2 is its own inverse). □

3.1.4 Group homomorphisms

The final step of the general set-up is to define the maps which are consistent with the group structure, the *group homomorphisms*.

Definition 3.3 *(Group homomorphism) A map $f : G \to \tilde{G}$ between two groups G and $\tilde{G}$ is called a group homomorphism if:*

$$f(g_1 \cdot g_2) = f(g_1) \cdot f(g_2) \tag{3.4}$$

for all $g_1, g_2 \in G$.

Group homomorphisms are simply maps between groups which commute with the group multiplication. In other words, group elements and their images under f multiply in the same way. As we will see in Chapter 12, linear maps are group homomorphisms with respect to vector addition.

As every map, group homomorphisms $f : G \to \tilde{G}$ have an image, $\text{Im}(f) \subset \tilde{G}$ which is a subset of the co-domain group. But there is another interesting set, the kernel, which can be defined because of the existence of a special group element, the neutral element.

Definition 3.4 *(Kernel of a group homomorphism) The kernel of a group homomorphism* $f : G \to \tilde{G}$ *is defined as* $\mathrm{Ker}(f) = f^{-1}(\tilde{e}) = \{g \in G \,|\, f(g) = \tilde{e}\} \subset G$, *where* $\tilde{e}$ *is the neutral element of* $\tilde{G}$.

This means the kernel is a subset of the domain group which consists of all group elements mapped to the co-domain neutral element. Injectivity and surjectivity of a group homomorphism can be phrased in terms of its image and kernel as explained in the following proposition:

Proposition 3.2 *(Properties of group homomorphisms) A group homomorphism* $f : G \to \tilde{G}$ *has the following properties:*

(i) $f(e) = \tilde{e}$ *so that* $e \in \mathrm{Ker}(f)$
(ii) $f(g^{-1}) = f(g)^{-1}$ *for all* $g \in G$
(iii) f *surjective* $\Leftrightarrow \mathrm{Im}(f) = \tilde{G}$
(iv) f *injective* $\Leftrightarrow \mathrm{Ker}(f) = \{e\}$
(v) $\mathrm{Im}(f)$ *is a sub-group of* $\tilde{G}$ *and* $\mathrm{Ker}(f)$ *is a sub-group of* G.

Proof (i) $f(e) = f(e \cdot e) = f(e) \cdot f(e)$ and multiplying both sides with $f(e)^{-1}$ implies that $f(e) = \tilde{e}$.
(ii) Clearly, $f(g)^{-1}$ is an inverse for $f(g)$ but so is $f(g^{-1})$ since $f(g^{-1}) \cdot f(g) = f(g^{-1} \cdot g) = f(e) = \tilde{e}$. The claim then follows from the uniqueness of the inverse.
(iii) This is clear from the definitions of the image and surjectivity.
(iv) Let f be injective and $g \in \mathrm{Ker}(f)$, so that $f(g) = \tilde{e}$. From part (i) it follows that $f(g) = \tilde{e} = f(e)$ and injectivity implies that $g = e$. Hence, the kernel only contains the neutral element, that is, $\mathrm{Ker}(f) = \{e\}$. Conversely, assume that $\mathrm{Ker}(f) = \{e\}$. We want to show that f is injective, so we start with $g_1, g_2 \in G$ satisfying $f(g_1) = f(g_2)$. Then $\tilde{e} = f(g_1) \cdot f(g_2)^{-1} = f(g_1) \cdot f(g_2^{-1}) = f(g_1 \cdot g_2^{-1})$, so that $g_1 \cdot g_2^{-1}$ is in the kernel. Hence $g_1 \cdot g_2^{-1} = e$ or $g_1 = g_2$, which shows that f is injective.
(v) This is left as Exercise 3.2. $\qquad\square$

A bijective group homomorphism $f : G \to \tilde{G}$ is also called a *group isomorphism*. Isomorphisms are of great importance in mathematics and the isomorphisms of vector spaces — the bijective linear maps — are a central theme of linear algebra, which we will develop in more detail later on. Recall that we have earlier defined what it means for two sets to be equal. Isomorphisms provide us with a different, structural notion of set equality. The existence of a group isomorphism implies that domain and co-domain groups are equal with respect to their group structures. This means that, by virtue of Eq. (3.4), the elements of the domain group and their images in the co-domain multiply 'in the same way'. Two groups G and $\tilde{G}$ connected by a group isomorphism are called *isomorphic*, written as $G \cong \tilde{G}$. All these features, including image and kernel, as well as the statements of Prop. 3.2 have their counterparts for linear maps, as we will see in Chapter 12.

Example 3.6 *(Group homomorphisms)*
Consider the map $f : \mathbb{Z} \to \mathbb{Z}_n$ defined by $f(k) = k \bmod n$. This is a group homomorphism, since:

$$f(k + k') = (k + k') \bmod n = [(k \bmod n) + (k' \bmod n)] \bmod n = f(k) + f(k') \,.$$

Its image is the entire co-domain, $\mathrm{Im}(f) = \mathbb{Z}_n$, so the map is surjective. The kernel, $\mathrm{Ker}(f) = n\mathbb{Z}$, consists of all multiples of n. Hence, f is not injective. □

Example 3.7 *(Group isomorphism)*
The set $G = \{1, -1\}$ forms a group under regular multiplication of integers. The map $f : \mathbb{Z}_2 \to G$ defined by $f(k) = (-1)^k$ is a group isomorphism, which shows that $G \cong \mathbb{Z}_2$. □

3.2 Permutation groups

> **Summary 3.2** *The permutation group S_n consists of all bijective maps on the set $\{1, \ldots, n\}$, with multiplication given by map composition. Transpositions are permutation which swap two numbers while leaving all others unchanged. Every permutation can be written as a product of transpositions. If the number of transpositions required is even, then the transposition is called even and the sign of the permutation is $+1$. Otherwise, the permutation is called odd and its sign is -1. The sign map is a group homomorphism from S_n to the cyclic group $\{\pm 1\}$.*

Permutation groups S_n and some of their properties will be required later, in the context of determinants. Here, we prepare the ground for these applications and also illustrate some of the general ideas around groups.

3.2.1 Calculating with permutations

The permutation groups S_n have already been defined in Example 3.2. Permutations $\sigma \in S_n$ are sometimes written as

$$\sigma = \begin{bmatrix} 1 & 2 & \cdots & n \\ \sigma(1) & \sigma(2) & \cdots & \sigma(n) \end{bmatrix}$$

indicating a permutation which permutes the numbers in the top row to the corresponding numbers in the bottom row. Group multiplication is composition of maps, the group identity is the identity map (the trivial permutation which leaves everything unchanged) and the inverse of $\sigma \in S_n$ is the inverse permutation σ^{-1} which 'undoes' the effect of the original permutation.

The simplest non-trivial example is the group S_2 of permutations of the set $\{1, 2\}$ which consists of two elements (see also Exercise 3.5):

$$S_2 = \left\{ e = \begin{bmatrix} 1 & 2 \\ 1 & 2 \end{bmatrix}, \tau = \begin{bmatrix} 1 & 2 \\ 2 & 1 \end{bmatrix} \right\}. \tag{3.5}$$

Clearly, this group is Abelian. What about the higher permutation groups? Consider the two permutations

$$\tau_1 = \begin{bmatrix} 1\ 2\ 3\ \cdots \\ 1\ 3\ 2\ \cdots \end{bmatrix}, \qquad \tau_2 = \begin{bmatrix} 1\ 2\ 3\ \cdots \\ 3\ 2\ 1\ \cdots \end{bmatrix},$$

in S_n which swap $(1,2)$ and $(2,3)$, respectively, while leaving all other numbers unchanged. Such simple permutations which only swap two numbers are called *transpositions*. Note that transpositions τ are their own inverse since $\tau \circ \tau = e$.

A quick calculation,

$$\tau_1 \circ \tau_2 = \begin{bmatrix} 1\ 2\ 3\ \cdots \\ 2\ 3\ 1\ \cdots \end{bmatrix} \neq \tau_2 \circ \tau_1 = \begin{bmatrix} 1\ 2\ 3\ \cdots \\ 3\ 1\ 2\ \cdots \end{bmatrix}, \tag{3.6}$$

shows that S_n is not Abelian for $n > 2$. The calculation underlying Eq. (3.6) is perhaps somewhat unfamiliar. To see, for example, that $\tau_2 \circ \tau_1$ maps 3 to 2 first note that $\tau_1(3) = 2$ and $\tau_2(2) = 2$. Combining these two statements gives $(\tau_2 \circ \tau_1)(3) = \tau_2(\tau_1(3)) = \tau_2(2) = 2$.

3.2.2 Permutations in terms of transpositions

It is intuitively clear that S_n has $n!$ elements and that every permutation can be written as a composition of transpositions. These statements are proved in the following proposition.

Proposition 3.3 *The permutation group S_n has $n! := 1 \cdot 2 \cdots n$ elements. Every permutation in S_n can be written as a composition of transpositions.*

Proof We can prove these statements by induction in n. Both statements are clearly true for S_2, as Eq. (3.5) shows. Now assume that they are true for S_n. Next consider the permutations $A_k := \{\sigma \in S_{n+1} \,|\, \sigma(n+1) = k\} \subset S_{n+1}$, which map $n+1$ to k and also the transposition $\tau \in S_{n+1}$ which swaps $n+1$ with k. Then, the permutations $\tau \circ \sigma$, where $\sigma \in A_k$ leave $n+1$ unchanged and can, hence, be identified with permutations in S_n. Then by the induction assumption, $\tau \circ \sigma$ can then be written in terms of transpositions, $\tau \circ \sigma = \tau_1 \circ \cdots \circ \tau_k$, so that $\sigma = \tau \circ \tau_1 \circ \cdots \tau_k$. This completes the induction for the statement about transpositions.

Further, by the induction assumption we have $|A_k| = n!$ for all $k = 1, \ldots, n+1$ and since $S_{n+1} = A_1 \cup \cdots \cup A_{n+1}$ is the disjoint union it follows that $|S_{n+1}| = (n+1)n! = (n+1)!$. $\qquad\square$

The number of transpositions required to build up a certain permutation σ is not unique. For example, if $\sigma = \tau_1 \circ \cdots \circ \tau_k$ can be written in terms of the k transpositions τ_i it is equally well given by a composition of the $k+2$ transpositions $\sigma = \tau_1 \circ \cdots \circ \tau_k \circ \tau \circ \tau$, with τ any transposition. (To see this, recall that $\tau^2 = e$ for any transposition τ.) However, while the number of transpositions required to generate a certain permutation is not unique, it is always either even or odd, as we will now show.

3.2.3 The sign of permutations

It is important to distinguish between *even* and *odd permutations* and the formal way to do this is by introducing the *sign of a permutation*.

Definition 3.5 *The sign of a permutation $\sigma \in S_n$ is defined as*

$$\text{sgn}(\sigma) = \prod_{i>j} \frac{\sigma(i) - \sigma(j)}{i - j} \,, \tag{3.7}$$

where the product runs over all $i, j \in \{1, \ldots, n\}$ with $i > j$. A permutation σ is called even if $\text{sgn}(\sigma) = 1$ and it is called odd if $\text{sgn}(\sigma) = -1$.

The formula (3.7) might be somewhat confusing at first but it is, in fact, easy to understand intuitively. First note that the products of numerators and the products of denominators in Eq. (3.7) consist of the same factors, up to signs and, therefore the value of the sign functions is indeed ± 1. The number of -1 factors in the product (3.7) corresponds to the number of pairs (i, j) with $i > j$ for which $\sigma(i) < \sigma(j)$, so where the 'natural' order of a pair (i, j) is changed by the permutation. If this number is even the permutation is called even, and odd otherwise. The sign satisfies the following important property:

Theorem 3.1 $\text{sgn}(\sigma \circ \rho) = \text{sgn}(\sigma) \, \text{sgn}(\rho)$ *for all $\sigma, \rho \in S_n$.*

Proof

$$\text{sgn}(\sigma \circ \rho) = \prod_{i>j} \frac{\sigma(\rho(j)) - \sigma(\rho(i))}{j - i} = \prod_{i>j} \frac{\sigma(\rho(j)) - \sigma(\rho(i))}{\rho(j) - \rho(i)} \prod_{i>j} \frac{\rho(j) - \rho(i)}{j - i}$$

$$= \prod_{\rho(i)>\rho(j)} \frac{\sigma(\rho(j)) - \sigma(\rho(i))}{\rho(j) - \rho(i)} \text{sgn}(\rho) = \text{sgn}(\sigma) \, \text{sgn}(\rho) \,.$$

$\square$

In fact, this proposition says that the sign function $\text{sgn} : S_n \to \{\pm 1\} \cong \mathbb{Z}_2$ defines a group homomorphism. The kernel of this homomorphism, which consists of all even permutations, forms a sub-group of S_n (see Prop. 3.2), which is called the *alternating group* A_n. Also, since $1 = \text{sgn}(e) = \text{sgn}(\sigma \circ \sigma^{-1}) = \text{sgn}(\sigma) \, \text{sgn}(\sigma^{-1})$, we conclude that

$$\text{sgn}(\sigma^{-1}) = \text{sgn}(\sigma)^{-1} \,, \tag{3.8}$$

so a permutation and its inverse always have the same sign. Also note that the sign of a transposition is always negative, since Eq. (3.7) has precisely one negative factor in this case.

Suppose a permutation $\sigma \in S_n$ can be written in terms of k transpositions τ_i as $\sigma = \tau_1 \circ \cdots \circ \tau_k$. Then, from Theorem (3.1), the *sign of the permutation* is given by

$$\text{sgn}(\sigma) = (-1)^k \,. \tag{3.9}$$

This means the number of transpositions required to build up the permutation σ must always be even for an even permutation and odd otherwise, as advertised earlier.

Problem 3.1 *(Even and odd permutations in S_3)*

Write down the elements of the permutation group S_3, and determine the even and odd permutations and the alternating group A_3.

Solution: The permutation group S_3 of the set $\{1, 2, 3\}$ has $3! = 6$ elements, which consist of the identity e, three transpositions τ_i and two further permutations, σ and $\tilde{\sigma}$, given by

$$\tau_1 = \begin{bmatrix} 1\ 2\ 3 \\ 1\ 3\ 2 \end{bmatrix}, \quad \tau_2 = \begin{bmatrix} 1\ 2\ 3 \\ 3\ 2\ 1 \end{bmatrix}, \quad \tau_3 = \begin{bmatrix} 1\ 2\ 3 \\ 2\ 1\ 3 \end{bmatrix}, \quad \sigma = \begin{bmatrix} 1\ 2\ 3 \\ 2\ 3\ 1 \end{bmatrix}, \quad \tilde{\sigma} = \begin{bmatrix} 1\ 2\ 3 \\ 3\ 1\ 2 \end{bmatrix}. \quad (3.10)$$

The three transpositions are of course odd permutations. Comparison with Eq. (3.6) shows that $\sigma = \tau_1 \circ \tau_2$ and $\tilde{\sigma} = \tau_2 \circ \tau_1$. Hence, σ and $\tilde{\sigma}$ are both even permutations and the alternating group $A_3 \subset S_3$ of even permutations is $A_3 = \{e, \sigma, \tilde{\sigma}\}$. (See also Exercise 3.7.)

Exercises

($\dagger$=challenging, $\dagger\dagger$=difficult, wide-ranging)

3.1 Prove the statements in Prop. 3.1.

3.2 Prove the statement (v) in Prop. 3.2, by showing that $\mathrm{Im}(f)$ and $\mathrm{Ker}(f)$ contain the neutral element, the inverse, and are closed under group multiplication.

3.3 *Order of sub-groups*
Consider a finite group G with sub-group $H \subset G$. We can define a relation on G by saying that $g, \tilde{g}$ are related if $g^{-1}\tilde{g} \in H$. Show that this defines an equivalence relation whose equivalence classes can be written as $gH = \{gh \mid h \in H\}$. Why do all these equivalence classes have the same number of elements? Show that the order of H must divide the order of G.

3.4 *The group of linear functions.*
Consider the set of all linear functions $f : \mathbb{R} \to \mathbb{R}$ that have an inverse. Show that this set together with the group multiplication defined as function composition forms a group.

3.5 *Permutation matrices*
(a) Consider the group $S_2 = \{e, \tau\}$ of permutations of the set $\{1, 2\}$, as given in Eq. (3.5). Show that the map f defined by

$$f(e) = \begin{pmatrix} 1\ 0 \\ 0\ 1 \end{pmatrix}, \quad f(\tau) = \begin{pmatrix} 0\ 1 \\ 1\ 0 \end{pmatrix}$$

satisfies the group homomorphism property (3.4). Also show that $f(\tau)$ permutes the standard unit vectors, that is, $f(\tau)(\mathbf{e}_1) = \mathbf{e}_2$ and $f(\tau)(\mathbf{e}_2) = \mathbf{e}_1$.
(b)† Attempt an analogous construction by starting with the permutation group S_3 and using 3×3 matrices.

3.6 *Classification of finite groups.*
(a) How many different group structures modulo group isomorphisms are out there? This depends on the number of elements of the group. If the group G has only one element, then $G = \{e\}$, so the group consists of the unit only. If the group has two elements, say $G = \{e, a\}$, show that the multiplication law is uniquely determined by the following table:

$\cdot$	e	a
e	e	a
a	a	e

Show that this group is isomorphic to $\mathbb{Z}_2$ and to S_2.
(b)† Find all possible group structures for a group $G = \{e, a, b\}$ with three different elements.

3.7 *Sub-groups of S_3*[†]

Find all proper sub-groups of S_3. Show that, apart from the alternating group A_3, there are three sub-groups of order two which are all isomorphic to S_2.

3.8 *Normal sub-groups*[††]

A sub-group $H \subset G$ is called normal if $gH = Hg$ for all $g \in G$.

(a) Show that the quotient $G/H :=$ $G/\sim$ with the equivalence relation $\sim$ from Exercise 3.3 can be given a group structure if H is a normal sub-group.

(b) Show that the kernel of a group homomorphism $f : G \to \tilde{G}$ is a normal sub-group of G.

(c) For a group homomorphism $f : G \to \tilde{G}$ show that $G/\mathrm{Ker}(f)$ is isomorphic to $\mathrm{Im}(f)$.

4
Fields

Fields are a much more complicated algebraic structures than groups. They consist of a set with two Abelian group structures, one referred to as addition, the other as multiplication, plus a compatibility requirement, called distributive law, between them. The well-known rules for calculating with numbers are, in fact, the rules for calculating in a field. Here we are putting these rules on axiomatic ground and also prepare for the definition of vector spaces, which relies on the one for fields.

We begin by defining fields — since we have introduced groups already, this is rather easy — and derive some of the implications from this definition. Examples of fields, including the important cases of rational numbers $\mathbb{Q}$ and real numbers $\mathbb{R}$, are introduced next. We devote a bit more space to the field $\mathbb{C}$ of complex numbers, which is perhaps less familiar to the reader. Finally, we present a few basic facts about polynomials, which will be required for the discussion of eigenvalues and eigenvectors in Part VI.

4.1 Fields and their properties

Summary 4.1 *Fields are algebraic structures with two operations, referred to as addition and multiplication, which both form an Abelian group and are connected by a distributive law. An order on a field provides a notion of 'less' and 'greater' which is consistent with addition and multiplication.*

4.1.1 Definition

Definition 4.1 *(Field)* *A field $(\mathbb{F}, +, \cdot)$ is a non-empty set with two operations*

$$+ : \mathbb{F} \times \mathbb{F} \to \mathbb{F} \qquad \cdot : \mathbb{F} \times \mathbb{F} \to \mathbb{F}$$
$$(a, b) \mapsto a + b \qquad (a, b) \mapsto ab \,,$$

called 'addition' and 'multiplication', which satisfy the following for all $a, b, c \in \mathbb{F}$.

(F1) $(\mathbb{F}, +)$ is an Abelian group with neutral element 0 and inverse $-a$.
(F2) $(\mathbb{F} \setminus \{0\}, \cdot)$ is an Abelian group with neutral element 1 and inverse a^{-1}.
(F3) $1 \neq 0$.
(F4) The distributive law $a(b + c) = ab + ac$ holds.

In short, a field combines two Abelian groups, which are linked by a distributive law. All the 'standard' rules for calculating which the reader is probably familiar with can be derived from the above axioms.

4.1.2 Some conclusions from the field axioms

Let us consider a few examples of simple conclusions from the field axioms.

Claim $0\,a = 0$ for all $a \in \mathbb{F}$.

Proof $0\,a = (0+0)a \overset{(F4)}{=} 0a + 0a$ and adding $-(0a)$ to both sides leads to the claim.

This property implies that we cannot find any element in the field whose product with 0 gives 1. Hence, 0 has no multiplicative inverse and this explains why 0 has been removed in the definition of the multiplicative group. It is also the explanation for axiom (F3). If $1 = 0$, then it follows that $a = 1\,a = 0\,a = 0$, so that the field only consists of a single element, $0 = 1$. The purpose of axiom (F3) is to exclude this trivial possibility.

Claim If $a\,b = 0$ then $a = 0$ or $b = 0$.

Proof If $a = 0$ we are done. If $a \neq 0$ we can multiply $ab = 0$ with the inverse a^{-1} to obtain $b = a^{-1}0 = 0$, where the last step follows from the previous claim.

This statement provides the basis for saying that a vanishing product implies the vanishing of (at least) one of its factors. It also implies that two non-zero elements in a field can never multiply to zero.

Claim $(-a)b = -(ab)$ for all $a, b \in \mathbb{F}$.

Proof $0 = 0\,b = (a + (-a))b \overset{(F4)}{=} ab + (-a)b \Rightarrow (-a)b = -(ab)$

Also note that the sum $a + (-b)$ is often written as $a - b$, so subtraction is defined in terms of addition and the additive inverse. In the same spirit, division is defined by $a \div b := ab^{-1}$, for $b \neq 0$. Many more simple and well-known relations of this kind follow from the field axioms (see Exercises 4.1 and 4.2) and they will be taken for granted from now on.

It is often convenient to write multiple sums and products in terms of the more concise summation and product notation.

$$a_1 + a_2 + \cdots + a_n = \sum_{i=1}^{n} a_i \,, \qquad a_1\, a_2 \cdots a_n = \prod_{i=1}^{n} a_i \,. \tag{4.1}$$

Using the summation notation, the distributive law can be generalized to (see Exercise 4.3):

$$\left(\sum_i a_i \right) \left(\sum_j b_j \right) = \sum_{i,j} a_i b_j \,. \tag{4.2}$$

4.1.3 Order on fields

It is often necessary to have a notion of 'less' or 'greater' on a field and this is defined by an order.

Definition 4.2 *(Order on fields) A field* $(\mathbb{F}, +, \cdot)$ *is called ordered with respect to an order* $>$ *if*

(O1) For all $a \in \mathbb{F}$ precisely one of $a > 0$, $a = 0$ and $-a > 0$ is true.
(O2) $a > 0$ and $b > 0$ implies $a + b > 0$.
(O3) $a > 0$ and $b > 0$ implies $ab > 0$.

We say that $a > b$, read as 'a is greater than b', if $a - b > 0$. This is also written as $b < a$, which reads 'b is less than a'. Further $a \geq b$ ($b \leq a$) means that $a > b$ or $a = b$ and is read as 'a greater or equal b' ('b less or equal a').

Perhaps surprisingly, not all fields admit an order. For example, the complex numbers cannot be ordered, as we will see. For fields with an order (such as the rational and real numbers), all the 'usual' rules for working with inequalities can be derived from the above axioms (see Exercise 4.4) and we will take these rules for granted from now on. Here is a simple example of a conclusion from the order axioms.

Claim For $a \neq 0$ we have $a^2 > 0$.

Proof From Exercise 4.1 we know that $(-a)^2 = a^2$. Since $a \neq 0$, (O1) implies that either $a > 0$ or $-a > 0$ so that the statement follows from (O3).

An immediate consequence is that $1 = 1^2 > 0$

The ordering axiom (O1) in Def. 4.2 facilitates introduction of the *absolute value* or *modulus* $|a|$ of a number $a \in \mathbb{F}$ by

$$|a| := \begin{cases} a & \text{if} \quad a > 0 \text{ or } a = 0 \\ -a & \text{if} \quad -a > 0 \end{cases} . \qquad (4.3)$$

An order on a field $\mathbb{F}$ also allows us to define intervals, for example

$$[a, b] := \{x \in \mathbb{F} \,|\, a \leq x \leq b\} , \qquad [a, b) := \{x \in \mathbb{F} \,|\, a \leq x < b\} .$$

Note that a square bracket indicates that the boundary is part of the interval, while a round bracket indicates the boundary is excluded.

4.2 Examples of fields

Summary 4.2 *Important examples of ordered fields are the rational numbers $\mathbb{Q}$ and the real numbers $\mathbb{R}$. There are also fields with a finite number of elements but they cannot be ordered.*

Example 4.1 *(Rational numbers)*
We can introduce the rational numbers as a quotient $\mathbb{Q} = \mathbb{Z} \times \mathbb{Z}^{\neq 0} / \sim$, where the equivalence relation $\sim$ is defined by $(p_1, q_1) \sim (p_2, q_2)$ if $p_1 q_2 = p_2 q_1$. Intuitively, if $p_1 q_2 = p_2 q_1$ then $p_1/q_1 = p_2/q_2$ so this equivalence relation identifies two pairs (p_1, q_1) and (p_2, q_2) if they represent the same fraction. Correspondingly, an equivalence class $[(p, q)]$ under this relation is also written as a fraction p/q, where $(p, q) \in [(p, q)]$ is any pair of integers in the class. Addition and multiplication on $\mathbb{Q}$ are defined by

$$[(p_1, q_1)] + [(p_2, q_2)] = [(p_1 q_2 + p_2 q_1, q_1 q_2)], \quad [(p_1, q_1)][(p_2, q_2)] = [(p_1 p_2, q_1 q_2)] , \quad (4.4)$$

where the components are added and multiplied according to the rules in $\mathbb{Z}$. It is a straightforward, although somewhat tedious exercise (see Exercise 4.5) to show that these definitions satisfy all the field axioms in Def. 4.1. The neutral element of addition is $[(0,1)]$ and the additive inverse of $[(p,q)]$ is $[(-p,q)]$. For multiplication the neutral element is $[(1,1)]$ and the inverse of $[(p,q)]$ is $[(q,p)]$.

Although Eqs. (4.4) might seem unusual at first, they do formalize the well-known rules for how to add and multiply fractions. The first Eq. (4.4) instructs us to bring the two fractions to the same denominator $(q_1 q_2)$ and then add the numerators. The second Eq. (4.4) simply says that fractions are multiplied by multiplying numerators and denominators.

To define an order on $\mathbb{Q}$, we note that we would call a fraction positive if both numerator and denominator have the same sign. This motivates the definition

$$[(p,q)] > 0 \quad \Leftrightarrow \quad pq \in \mathbb{N}^{\neq 0} , \tag{4.5}$$

which can indeed be shown to satisfy the order axioms in Def. 4.2 (see Exercise 4.6). $\square$

Example 4.2 *(Real numbers)*

The real numbers $\mathbb{R}$ can be constructed as limits of sequences of rational numbers. This is really a topic in analysis and will not be discussed explicitly here (see Exercise 4.18). Intuitively, $\mathbb{R}$ is obtained from $\mathbb{Q}$ by 'filling in the gaps' with irrational numbers. This construction implies that all real numbers can be approximated by rational numbers to arbitrary accuracy. This allows extending the definitions (4.4) of addition and multiplication on $\mathbb{Q}$ to $\mathbb{R}$. It can then be shown that $(\mathbb{R}, +, \cdot)$ is a field. In a similar way, the order (4.5) can be extended to $\mathbb{R}$. In the following, we will take this field and its order structure for granted (see Exercise 4.18). $\square$

Example 4.3 *(Finite fields)*

There exist 'unusual' fields with a finite number of elements which satisfy all the requirements in Def. 4.1. Consider the sets $\mathbb{F}_p = \mathbb{Z}_p = \{0, 1, \ldots, p-1\}$ with (prospective) addition and multiplication defined by:

$$a + b := (a + b) \bmod p , \quad a \cdot b := (ab) \bmod p . \tag{4.6}$$

We already know from Example 3.4 that $(\mathbb{F}_p, +)$ is an Abelian group. What about multiplication? Consider $\mathbb{F}_4$ where $2 \cdot 2 = (2\,2) \bmod 4 = 0$. We know that in a field two non-zero elements can never multiply to zero, so $\mathbb{F}_4$ with the above addition and multiplication cannot be a field. We can avoid this problem by demanding that p be a prime number and it turns out that $(\mathbb{F}_p, +, \cdot)$ for p prime is indeed a field.

The fields $\mathbb{F}_p$ do not have an order. In an ordered field we always have $1 > 0$ and, hence, $p - 1 = 1 + 1 + \cdots 1 > 0$. But $(p - 1) + 1 = 0$ in contradiction to axiom (O2) in Def. 4.2.

The simplest example of such a finite field is $(\mathbb{F}_2 = \{0, 1\}, +, \cdot)$. Since every field must contain the neutral elements 0 and 1, this is indeed the smallest field. From the definitions (4.6) its addition and multiplication tables are:

+	0	1
0	0	1
1	1	0

·	0	1
0	0	0
1	0	1

Note that, taking into account the mod 2 operation, in this field we have $1+1 = 0$. Since the elements of $\mathbb{F}_2$ can be viewed as the two states of a bit, this field has important applications in computer science and in coding theory. (See also Exercise 4.7 and Application 14.1.) □

4.3 The complex numbers

Summary 4.3 *Complex numbers $\mathbb{C}$ are of the form $z = a + ib$, where $a, b \in \mathbb{R}$ and i is the imaginary unit. As a set, they can be identified with the two-dimensional coordinate vectors $\mathbb{R}^2$. The complex numbers form a field with component-wise addition and complex multiplication defined by $i^2 = -1$. Complex conjugation $z = a + ib \mapsto \bar{z} = a - ib$ is a field automorphism which is used to write down the multiplicative inverse and to define the length of a complex number. Component vectors $\mathbb{R}^n$ with $n > 2$ and component-wise addition cannot be given a field structure. This motivates the introduction of vector spaces.*

In Section 3.1 we have introduced the direct product of groups, whereby the Cartesian product of two groups can be given a group structure by component-wise multiplication. Is there a similar construction for fields? For concreteness, we will address this question for the Cartesian product $\mathbb{R}^2$.

4.3.1 Construction of complex numbers

We have seen in Eq. (1.15) that elements of $\mathbb{R}^2$ can be written in terms of the standard unit vectors $\mathbf{e}_1$ and $\mathbf{e}_2$, so that a vector with components a and b can be written as $a\mathbf{e}_1 + b\mathbf{e}_2$. Alternatively, we can write such a pair of real numbers as a formal sum $a + ib$, where, for the time being, i is merely a symbol, called the *imaginary unit*. These formal sums form the set of *complex numbers*

$$\mathbb{C} := \{a + ib \,|\, a, b \in \mathbb{R}\} \,, \tag{4.7}$$

which are identified with vectors in $\mathbb{R}^2$ via the bijective map $a + ib \mapsto a\mathbf{e}_1 + b\mathbf{e}_2$. For now, this is just a different way of writing $\mathbb{R}^2$. But can $\mathbb{C}$ be turned into a field? From the construction of the direct product group, we know that $\mathbb{C}$ can be turned into an Abelian group $(\mathbb{C}, +)$ by component-wise addition

$$(a + ib) + (c + id) := (a + c) + i(b + d) \,, \tag{4.8}$$

with neutral element 0 and the inverse of $a + ib$ given by $-a - ib$.

At first it seems we can follow the same idea for multiplication, but there is a problem. For the multiplicative group $(\mathbb{R} \setminus \{0\}, \cdot)$ of a field, we have to remove 0 (see Def. 4.1).

Hence, the direct product group construction leads to a group structure on $(\mathbb{R} \setminus \{0\})^2$, with component-wise multiplication. However, elements of the form $(a,0)$ and $(0,b)$ are not contained in $(\mathbb{R} \setminus \{0\})^2$, so proceeding in this way does not tell us how we should multiply such elements. Worse, extending component-wise multiplication to such elements gives $(a,0)(0,b) = (0,0)$, so two non-zero numbers multiply to zero, a feature which is excluded for a field. The direct product construction is not the right way forward for the multiplicative group.

The key to defining multiplication on $\mathbb{C}$ is to impose the relation $i^2 = -1$. Then, the distributive law enforces

$$(a + ib)(c + id) = ac - bd + i(ad + bc) \ . \tag{4.9}$$

The neutral element for this multiplication is 1 and it is easy to check that an inverse is given by

$$(a + ib)^{-1} = \frac{a - ib}{a^2 + b^2} \ . \tag{4.10}$$

The remaining axioms in Def. 4.1 can be checked as well (see Exercise 4.8). We conclude that $(\mathbb{C}, +, \cdot)$ with addition and multiplication defined as in Eqs. (4.8) and (4.9) is a field.

We recall that, in an ordered field, we have $a^2 > 0$ for all $a \neq 0$. Since, $i^2 = -1$, we see that $(\mathbb{C}, +, \cdot)$ cannot be ordered.

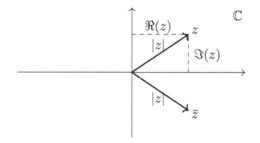

Fig. 4.1 The complex plane $\mathbb{C}$, a complex number z, its complex conjugate $\bar{z}$, its real- and imaginary parts $\Re(z)$, $\Im(z)$, and its length $|z|$.

4.3.2 Complex conjugation

For a complex number $z = a + ib \in \mathbb{C}$, a is called the *real part* of z, written as $a = \Re(z)$, while b is called the *imaginary part* and is written as $b = \Im(z)$. The *complex conjugate*, $\bar{z}$, and the *length* or *complex modulus*, $|z|$, are defined by (see Fig. 4.1)

$$\bar{z} = \Re(z) - i\,\Im(z) \ . \qquad |z| = \sqrt{z\bar{z}} = \sqrt{\Re(z)^2 + \Im(z)^2} \ . \tag{4.11}$$

One way to motivate these definitions is from the formula (4.10) for the inverse, where the complex conjugate and the length (square) appear in the numerator and denominator, respectively. In fact, the multiplicative inverse (4.10) can be written as

$$z^{-1} = \frac{\bar{z}}{|z|^2} \, . \tag{4.12}$$

Complex conjugation satisfies the following important properties.

Proposition 4.1 *For any two numbers* $z, w \in \mathbb{C}$ *we have*

$$\overline{z + w} = \bar{z} + \bar{w} \, , \qquad \overline{zw} = \bar{z}\bar{w} \, . \qquad \bar{\bar{z}} = z \, . \tag{4.13}$$

Proof   Write $z = a + ib$ and $w = c + id$.

$$\overline{z + w} = \overline{(a + c) + i(b + d)} = (a + c) - i(b + b) = a - ib + c - id = \bar{z} + \bar{w}$$
$$\overline{zw} = \overline{(ac - bd) + i(ad + bc)} = (ac - bd) - i(ad + bc) = (a - ib)(c - id) = \bar{z}\bar{w}$$
$$\bar{\bar{z}} = \overline{a - ib} = a + ib = z \, .$$

$\square$

The content of the first two equations (4.13) can be stated by saying that complex conjugation is a *field homomorphism* (see Exercise 4.11). In fact, complex conjugation is bijective, so it is a *field automorphism*. It turns out that the complex numbers only have two (continuous) field automorphisms, the identity and complex conjugation (see Exercise 4.11), and this fact can be viewed as a motivation for introducing complex conjugation.

Two immediate conclusions (Exercise 4.12) from the above proposition are

$$|zw| = |z|\,|w| \, , \qquad |z| = |\bar{z}| \, . \tag{4.14}$$

In other words, lengths of complex numbers multiply and a complex number and its conjugate have the same length.

Problem 4.1   *(Calculating with complex numbers)*

For the complex number $z = 2 + 4i$, work out $\Re(z)$, $\Im(z)$, $|z|^2$ and z^{-1}. With $w = 3 - 5i$, compute $z + w$ and zw and express each in the standard form $a + ib$. Finally, write $\zeta = (1 + i)/(2 - i)$ in standard form $a + ib$ and, hence, find its real and imaginary parts.

Solution: The real and imaginary parts, length, and inverse of $z = 2 + 4i$ are given by

$$\Re(z) = 2 \, , \qquad \Im(z) = 4 \, , \qquad |z|^2 = 2^2 + 4^2 = 20 \, , \qquad z^{-1} = \frac{\bar{z}}{|z|^2} = \frac{2 - 4i}{20} = \frac{1}{10} - \frac{i}{5} \, .$$

For its sum and product with $w = 3 - 5i$, we have

$$z + w = (2 + 4i) + (3 - 5i) = 5 - i \, , \qquad zw = (2 + 4i)(3 - 5i) = 26 + 2i \, .$$

Finally, to find the standard form of $\zeta = (1+i)/(2-i)$, all we have to do is multiply numerator and denominator by the complex conjugate, $2 + i$, of the denominator. We get

$$\zeta = \frac{1 + i}{2 - i} = \frac{(1 + i)(2 + i)}{(2 - i)(2 + i)} = \frac{1 + 3i}{5} = \frac{1}{5} + \frac{3}{5}i \, ,$$

so that $\Re(\zeta) = 1/5$ and $\Im(\zeta) = 3/5$. (See also Exercise 4.13.)

Problem 4.2 *(Polar decomposition for complex numbers)*

Show that every non-zero complex number $z \in \mathbb{C}$ can be uniquely written as $z = r\zeta$, where r is real positive and $\zeta \in \mathbb{C}$ has length one.

Solution: We can certainly write any complex number in this form by setting $r = |z|$ and $\zeta = z/|z|$. Conversely, if $z = r\zeta$ with $r > 0$ and $|\zeta| = 1$ it follows that $|z|^2 = r^2|\zeta|^2 = r^2$. Hence, $r = |z|$ is the unique solution for r and this uniquely determines $\zeta = z/r = z/|z|$.

If we take some properties of trigonometric functions and the exponential function for granted (proving these is really a task in analysis) we know that every complex number ζ with $|\zeta| = 1$ can be written as $\zeta = \cos(\theta) + i\sin(\theta) = e^{i\theta}$ for a unique $\theta \in [0, 2\pi)$. Hence, the polar decomposition of a complex number can also be cast into the form

$$z = re^{i\theta} , \tag{4.15}$$

where $r = |z| \geq 0$ is the length of z and $\theta = \arg(z) \in [0, 2\pi)$ is called the *argument* of z.

4.3.3   Beyond $\mathbb{R}^2$

Given that $\mathbb{R}^2$ can be turned into a field, it is natural to ask if the same can be accomplished for $\mathbb{R}^n$, when $n > 2$. We can certainly turn $\mathbb{R}^n$ into an Abelian group by component-wise addition. However, having made this choice, it is then impossible to define a generalization of the multiplication (4.9) which satisfies all the field axioms.

This break-down of the field structure for $\mathbb{R}^n$ with $n > 2$ is one of the main motivations for introducing vector spaces. They can be viewed as the 'next best thing' when the field structure is not available. Vector spaces will be defined in the next section.

4.4   Basics of polynomials

Summary 4.4 *Polynomial division is an important algorithm for calculating with polynomials. If a polynomial p has a zero at $x = a$, polynomial division can be used to show that $p(x) = (x - a)s(x)$. A zero a of p is said to have multiplicity m if $p(x) = (x - a)^m s(x)$ with $s(a) \neq 0$. A polynomial fully factorizes if the multiplicities of all zeros add up to the degree of the polynomial. Polynomials over the real numbers may or may not fully factorize, depending on the example. The fundamental theorem of algebra states that every non-constant polynomial over $\mathbb{C}$ has a zero and this implies that all polynomials over $\mathbb{C}$ fully factorize.*

4.4.1   Basics and polynomial division

Understanding the structure of polynomials is one of the tasks of algebra (see, for example, Lang 2000), and lies outside the realm of linear algebra. However, we will see in Part VI that the characteristic polynomial of a linear map is a central object in the theory of eigenvalues and eigenvectors. For this reason we collect a few basic facts on polynomials as required for our discussion later on.

By a *polynomial* over a field $\mathbb{F}$, we mean a function $p : \mathbb{F} \to \mathbb{F}$ of the form

$$p(x) = c_n x^n + c_{n-1} x^{n-1} + \cdots + c_1 x + c_0 , \qquad (4.16)$$

where $c_i \in \mathbb{F}$. The set of all such polynomials is also denoted by $\mathcal{P}(\mathbb{F})$. If $c_n \neq 0$, then we say that p has *degree* $\deg(p) = n$ and if $c_n = 1$ then p is called a *monic polynomial*. A *monomial* is a special polynomial of the form x^k. The polynomial p is called *reducible* if it can be written as a product $p = q\,r$ of two polynomials $p, r \in \mathcal{P}(\mathbb{F})$ with positive degree and is otherwise called *irreducible*. Whether a polynomial is reducible depends on the field $\mathbb{F}$. For example, $p(x) = x^2 + 1$ is irreducible as a polynomial over $\mathbb{R}$ but it is reducible over $\mathbb{C}$ since $p(x) = (x+i)(x-i)$. An important algorithm for calculation with polynomials is *polynomial division*.

Theorem 4.1 *(Polynomial division) Let p, q be polynomials over the field $\mathbb{F}$. Then, there exist unique polynomials r, s with $\deg(r) < \deg(q)$ such that $p = sq + r$.*

Proof This follows from the standard algorithm for polynomial division as, for example, explained in Lang 2000; Lang 1998. □

4.4.2 Zeros and multiplicity

We call $a \in \mathbb{F}$ a *zero* of the polynomial $p \in \mathcal{P}(\mathbb{F})$ if $p(a) = 0$. In this case, the polynomial division theorem with $q(x) = x - a$ implies that

$$p(x) = (x - a)s(x) + r(x) . \qquad (4.17)$$

Since $\deg(r) < \deg(q) = 1$, we conclude that r must be a degree 0 polynomial, that is, a constant. Inserting $x = 0$ into the equation (4.17) then immediately leads to $r = 0$, so that

$$p(x) = (x - a)s(x) . \qquad (4.18)$$

Here, s must be a polynomial of degree $n - 1$ (or else p would not be of degree n). If $s(a) \neq 0$ then a is called a *simple zero* of p, otherwise, if $s(a) = 0$, we can split a further factor $x - a$ off from s and repeat this process until the remaining polynomial is non-zero at $x = a$. The maximal number of factor $x - a$ obtained in this way is called the *multiplicity* of the zero a and it is formally defined as follows:

Definition 4.3 *(Multiplicity of a polynomial zero) A polynomial p is said to have a zero with multiplicity m at $x = a$ if $p(x) = (x - a)^m s(x)$, where s is a polynomial with $s(a) \neq 0$.*

4.4.3 Factorization

Suppose, for a degree n polynomial p, we have a complete list of pairwise different zeros $a_1, \ldots, a_k$ with multiplicities $m_1, \ldots, m_k$. By iterating the above arguments, it is then easy to see that p can be written as

$$p(x) = (x - a_1)^{m_1} \cdots (x - a_k)^{m_k} s(x) , \qquad (4.19)$$

where $\deg(s) = n - \sum_{i=1}^{k} m_i$. We say that p *fully factorizes* if s is a constant polynomial or, equivalently, if $\sum_{i=1}^{k} m_i = n$, that is, if the multiplicities of all pairwise different zeros add up to the degree. A fully factorizing degree n polynomials p can, hence, be written as

$$p(x) = c(x - a_1)^{m_1} \cdots (x - a_k)^{m_k} \quad \text{where} \quad \sum_{i=1}^{k} m_i = n , \tag{4.20}$$

and $c \in \mathbb{F}$. This is a very convenient form for p but whether it can always be achieved depends on the choice of field $\mathbb{F}$.

Over $\mathbb{R}$, it is easy to write down polynomials which have no (real) zeros at all, for example $p(x) = x^2 + 1$. There are also polynomials over $\mathbb{R}$ with zeros which still do not fully factorize. For example, $p(x) = x^3 - x^2 + x - 1$ has a zero at $x = 1$ and can be written as $p(x) = (x - 1)(x^2 + 1)$ which shows that there are no further zeros over $\mathbb{R}$. Hence, it does not fully factorize. On the other hand, the polynomial $p(x) = x^3 - 3x - 2$ has zeros at 2 and -1 with multiplicities 1 and 2, respectively and, hence, fully factorizes as $p(x) = (x - 2)(x + 1)^2$. The main message is that for polynomials over $\mathbb{R}$ the situation very much depends on the specific example.

Fortunately, things are much clearer for polynomials over $\mathbb{C}$. The main statement is the famous *fundamental theorem of algebra*, originally due to Gauß.

Theorem 4.2 *Every non-constant polynomial over $\mathbb{C}$ has a zero.*

Proof The most straightforward proofs can be found in the context of complex analysis (see, for example, Lang 2013). □

This means every polynomial over $\mathbb{C}$ can be written in the form (4.18) and if the polynomial s in this equation is non-constant we can apply the theorem again and split off another factor. Iterating this shows that all polynomials over $\mathbb{C}$ fully factorize.

An interesting special case is a polynomial p over $\mathbb{C}$, as in Eq. (4.16), with real coefficients c_i. If a is a zero of such a polynomial then complex conjugating and using Prop. 4.1 shows that $p(\bar{a}) = 0$. The conclusion is that for such polynomials with real coefficients zeros are either real or they come in complex conjugate pairs.

Exercises

(†=challenging, ††=difficult, wide-ranging)

4.1 *Some conclusions from field axioms*
Let $(\mathbb{F}, +, \cdot)$ be a field.
(a) Show that $-a = (-1)a$ for all $a \in \mathbb{F}$, where -1 is the additive inverse of 1.
(b) Show that $(-a)(-b) = ab$ for all $a, b \in \mathbb{F}$.

4.2 *Fractions*
For a field $(\mathbb{F}, +, \cdot)$, $a \in \mathbb{F}$ and $b \in \mathbb{F} \backslash \{0\}$ define a fraction by

$$\frac{a}{b} := a \div b = ab^{-1} .$$

Show that the usual rules for adding and multiplying fractions follow from the field axioms.

4.3 For a field $(\mathbb{F}, +, \cdot)$ show that $a(b_1 + b_2 + \cdots + b_n) = ab_1 + ab_2 + \cdots ab_n$, for example by induction in n. Use this result to prove Eq. (4.2).

4.4 For a field $(\mathbb{F}, +, \cdot)$ with an order, show that $a > 0$ implies $a^{-1} > 0$ for all

$a \in \mathbb{F}$. Also show that $a > b$ and $c > 0$ implies $ac > bc$.

4.5 Show that $(\mathbb{Q}, +, \cdot)$ with addition and multiplication defined as in Eq. (4.4) is a field.

4.6 Show that Eq. (4.5) defines an ordering on the field $(\mathbb{Q}, +, \cdot)$.

4.7 *The field $\mathbb{F}_p$*
Find the addition and multiplication table of the field $(\mathbb{F}_3, +, \cdot)$. For $\mathbb{F}_2$ show that $(a+b)^2 = a^2 + b^2$ and for $\mathbb{F}_3$ show that $(a+b)^3 = a^3 + b^3$. Generalize this and show that, for the field $\mathbb{F}_p$, we have $(a+b)^p = a^p + b^p$. (The equation you always, secretly, wanted to be true.)

4.8 Show that $\mathbb{C}$ with addition (4.8) and multiplication (4.9) forms a field.

4.9 In $\mathbb{C}$, show that the equation $z^2 = -1$ has precisely the solutions $z = \pm i$.

4.10 Find the solutions $z \in \mathbb{C}$ to the equation $z^n = 1$, where $n = 1, 2, \ldots$. (Hint: Use the polar decomposition of complex numbers.)

4.11 *Field automorphisms*[†]
A bijection $f : \mathbb{F} \to \mathbb{F}$ on a field $(\mathbb{F}, +, \cdot)$ is called a field automorphism if $f(a+b) = f(a) + f(b)$ and $f(ab) = f(a)f(b)$ for all $a, b \in \mathbb{F}$.
(a) Show that $f(0) = 0$, $f(1) = 1$, $f(-a) = -f(a)$ and $f(a^{-1}) = f(a)^{-1}$ for all $a \in \mathbb{F}$.
(b) Show that the only field automorphism on the rational numbers $\mathbb{Q}$ is the identity. (This means the identity is the only continuous field automorphism on $\mathbb{R}$.)
(c) Show that the only (continuous) field automorphisms on $\mathbb{C}$ are the identity and complex conjugation.

4.12 Derive the properties (4.14) of the length of a complex number from Prop. 4.1.

4.13 *Calculating with complex numbers*
Find the complex conjugate, the length and the inverse of the complex numbers $1+i$, $2-3i$, $-2-2i$ and $4-3i$. Convert the complex number $1/i$, $2/(1+i)$ and $(1-i)/(1+i)$ into standard form $a + ib$.

4.14 *Complex linear maps*
Show that the map $f : \mathbb{C} \to \mathbb{C}$ defined

by $f(z) = wz$ (for a fixed $w \in \mathbb{C}$) is linear, that is, it satisfies Eqs. (1.8) but with complex instead of real numbers. Write $z = x_1 + ix_2$ and find the 2×2 matrix which describes the action of f on the vector with components x_1 and x_2.

4.15 *The group $U(1)$*
Show that the set $U(1) = \{z \in \mathbb{C} \,|\, |z| = 1\}$ of complex numbers with length one forms a group, with complex multiplication as the group multiplication.

4.16 *Polynomial zeros*
Let $p \in \mathcal{P}(\mathbb{F})$ be a fully factorizing degree n polynomial

$$p(x) = c_n x^n + c_{n-1} x^{n-1} + \cdots + c_1 x + c_0 \,,$$

whose zeros a_i have multiplicities m_i, where $i = 1, \ldots, k$. Derive formulae for the sum $\sum_{i=1}^{k} m_i a_i$ and the product $\prod_{i=1}^{k} a_i^{m_i}$ of the zeros in terms of the coefficients c_i of p.

4.17 *Sign of polynomial zeros*
Let $p(x) = x^n + c_{n-1} x^{n-1} + \cdots + c_1 x + c_0$ be a fully factorizing polynomial over $\mathbb{R}$. Show that all zeros of p are negative iff all coefficients c_i are positive. (Hint: Find formulae for the coefficients c_i in terms of the zeros a_i.)

4.18 *Construction of the real numbers*[††]
If you do not know already, find out what a Cauchy series is and consider Cauchy series in $\mathbb{Q}$.
(a) Define a relation on Cauchy series in $\mathbb{Q}$ by which two series (a_i) and (b_i) are related if $(a_i - b_i)$ is a Cauchy series. Show that this is an equivalence relation.
(b) Use the equivalence relation from (a) to define $\mathbb{R}$ as the set of equivalence classes of Cauchy series.
(c) Show that $\mathbb{R}$ defined in this way forms a field.

4.19 *Arithmetics in $\mathbb{F}_p$*[††]
For the finite fields $\mathbb{F}_p$ from Example 4.3, write code in your favourite programming language which implements the addition and multiplication in Eq. (4.6). Keep p fixed but arbitrary.

Part II

Vector spaces

We are now ready to begin the systematic build-up of linear algebra, by introducing vector spaces, the arena within which linear algebra takes place. For the beginning student, the development of linear algebra often represents the first encounter with modern mathematics and the formality of the approach can come as a shock.

In order to ease into the subject, we begin with coordinate vectors in $\mathbb{F}^n$, that is, vectors which contain an arbitrary number, n, of elements of a field $\mathbb{F}$. They generalize the coordinate vectors with two real entries we have already briefly encountered in the introduction. We will introduce addition and scalar multiplication of such coordinate vectors, discuss their structural properties and some of the practical aspects of calculation.

Motivated by the properties of coordinate vectors, we present the general definition of *vector spaces* in Chapter 6, along with many examples to illustrate its scope. Of course these include coordinate vectors, but also more surprising examples, such as spaces made up from certain classes of functions. Many of these seemingly exotic vector spaces have scientific applications. Following the standard route, we also define the sub-structure for vector spaces, the *vector subspaces*, and the associated morphisms which are called *linear maps*.

A central notion in linear algebra is that of *linear independence*. It leads up to the ideas of *basis* and *dimension* of a vector space which we discuss in Chapter 7. A basis of a vector space is a list of linearly independent vectors which 'spans' the entire vector space. Given a basis, every vector can be represented in terms of *coordinates*. We will also see that, for a given vector space, the number of vectors in a basis is unique. This number is called the *dimension* of the vector space. In scientific applications, selecting a basis is sometimes referred to as a choice of 'coordinate system' and it is frequently one of the first steps towards setting up a mathematical model. The notion of basis is, therefore, crucial for the theoretical development of linear algebra and for many of its applications.

5

Coordinate vectors

In this chapter we introduce coordinate vectors with an arbitrary number of components from a general field $\mathbb{F}$, along with their addition and scalar multiplication. These are mathematically well-motivated generalizations of the two-coordinate vectors discussed in the introduction and they provide us with the intuition for how to define general vector spaces. In Chapter 7 we will see that such general vector spaces can always be described by coordinate vectors once a basis has been chosen.

But do we really need to care about vectors with an arbitrary number of components in an arbitrary field when it comes to scientific applications? After all, such applications are often tied to physical space whose description requires only three real components. The answer is an emphatic 'yes' — the scope of linear algebra applications is much wider than it might initially appear.

For example, describing the motion of n point masses in space requires vectors with $3n$ components. In relativity, space and time are combined and described by vectors with four components. In Application 1.1 we have seen that internet search requires vectors with a large number of real components. Also, the choice of field is not always confined to the real numbers. Quantum mechanics requires vectors with complex components (see Applications 26.1, 26.2, 26.4). Even vectors based on the seemingly exotic finite fields from Example 4.3 have applications, for example, in coding theory (see Application 14.1).

5.1 Basic definitions

Summary 5.1 *Coordinate vectors with n components are elements of $\mathbb{F}^n$, where $\mathbb{F}$ is a field. The two basic operations, vector addition and scalar multiplication, for such vectors are defined component-wise. We list the rules for calculating with these operations.*

5.1.1 Definition of coordinate vectors

In Section 1.2.3 we have already introduced coordinate vectors with two real components as well as their addition and scalar multiplication. These definitions can be readily generalized to vectors with n components, taken from a general field $\mathbb{F}$. The reader still uncomfortable with this level of abstraction may well replace $\mathbb{F}$ by the real numbers $\mathbb{R}$ (or the complex numbers $\mathbb{C}$) throughout.

In mathematical parlance, coordinate vectors are elements of the Cartesian product

$\mathbb{F}^n$, that is, they are n-tuples of numbers from the field $\mathbb{F}$. They are denoted by low-ercase bold-face letters and explicitly written as *column vectors*, for example:

$$\mathbf{v} = \begin{pmatrix} v_1 \\ \vdots \\ v_n \end{pmatrix} , \qquad \mathbf{w} = \begin{pmatrix} w_1 \\ \vdots \\ w_n \end{pmatrix} . \tag{5.1}$$

Here $v_1, \ldots, v_n \in \mathbb{F}$ and $w_1, \ldots, w_n \in \mathbb{F}$ are called the *components* of $\mathbf{v}$ and $\mathbf{w}$, respectively. We will often use *index notation* to refer to a vector and write the components of a vector $\mathbf{v}$ collectively as v_i, where the index i takes the values $i = 1, \ldots, n$. Elements of the underlying field $\mathbb{F}$ are also referred to as *scalars*.

We adhere to the common convention of writing coordinate vectors as columns but note that they could also be written as rows. In fact, it is useful to introduce the operation of *transposition* of a vector which converts a column vector into a row vector and vice versa. It is denoted by a superscript T attached to a vector so that, for a column vector $\mathbf{v}$, the vector $\mathbf{v}^T$ is a row vector with the same components. Using this notation, we will occasionally write column vectors $\mathbf{v}$ in Eq. (5.1) as $\mathbf{v} = (v_1, \ldots, v_n)^T$, in order to save space.

5.1.2 Addition and scalar multiplication

In Eq. (1.12) we have defined component-wise addition and scalar multiplication of vector in $\mathbb{R}^2$ and these definitions straightforwardly generalize to $\mathbb{F}^n$. Specifically, we define vector addition $+ : (\mathbb{F}^n, \mathbb{F}^n) \to \mathbb{F}^n$ and scalar multiplication $(\mathbb{F}, \mathbb{F}^n) \to \mathbb{F}^n$ for the vectors in Eq. (5.1) and a scalar $\alpha \in \mathbb{F}$ by

$$(\mathbf{v}, \mathbf{w}) \mapsto \mathbf{v} + \mathbf{w} := \begin{pmatrix} v_1 + w_1 \\ \vdots \\ v_n + w_n \end{pmatrix} , \qquad (\alpha, \mathbf{v}) \mapsto \alpha\mathbf{v} := \begin{pmatrix} \alpha v_1 \\ \vdots \\ \alpha v_n \end{pmatrix} . \tag{5.2}$$

Addition and multiplication of the components are of course those defined in the field $\mathbb{F}$. The intuitive interpretation of these operations has already been indicated in Fig. 1.1.

It is sometimes useful and efficient to express the above definitions in index notation, where they take the form

$$(\mathbf{v} + \mathbf{w})_i := v_i + w_i , \qquad (\alpha\mathbf{v})_i := \alpha v_i . \tag{5.3}$$

Here, the subscript i on the left-hand side indicates the ith component of the vector enclosed in brackets.

Problem 5.1 *Vector addition and scalar multiplication in $\mathbb{R}^3$*

Work out the vector sum of the $\mathbb{R}^3$ vectors $\mathbf{v} = (1, -2, 5)^T$ and $\mathbf{w} = (-4, 1, -3)^T$ and the scalar multiple of $\mathbf{v}$ by $\alpha = 3$.

Solution: The sum and the scalar multiple are given by

$$\mathbf{v} + \mathbf{w} = \begin{pmatrix} 1 \\ -2 \\ 5 \end{pmatrix} + \begin{pmatrix} -4 \\ 1 \\ -3 \end{pmatrix} = \begin{pmatrix} -3 \\ -1 \\ 2 \end{pmatrix}, \quad \alpha \mathbf{v} = 3 \begin{pmatrix} 1 \\ -2 \\ 5 \end{pmatrix} = \begin{pmatrix} 3 \\ -6 \\ 15 \end{pmatrix}. \tag{5.4}$$

The coordinate vector with all components equal to 0 (where 0 is the neutral element of addition in the field $\mathbb{F}$) is called the *zero vector* and it is denoted by the bold-face symbol

$$\mathbf{0} := \begin{pmatrix} 0 \\ \vdots \\ 0 \end{pmatrix}, \tag{5.5}$$

in order to distinguish it from the number 0.

5.1.3 Calculating with coordinate vectors

In Section 3.1 we have explained how the Cartesian product G^n of a group G can be made into a group by component-wise multiplication. If we apply this construction to the Abelian group $(\mathbb{F}, +)$ of addition in the field $\mathbb{F}$, it leads to a group structure on $\mathbb{F}^n$ whose 'multiplication' is evidently given by vector addition. Hence, we already know that $(\mathbb{F}^n, +)$ forms an Abelian group. Its neutral element is the zero vector $\mathbf{0}$, since $\mathbf{v} + \mathbf{0} = \mathbf{v}$ for all $\mathbf{v} \in \mathbb{F}^n$. Further, from $\mathbf{v} + (-\mathbf{v}) = \mathbf{0}$ it follows that $-\mathbf{v}$ is the inverse of a vector $\mathbf{v} \in \mathbb{F}^n$. There are a few further rules for how to calculate with coordinate vectors, related to scalar multiplication, which are listed in the following proposition.

Proposition 5.1 *For any coordinate vectors* $\mathbf{v}, \mathbf{w} \in \mathbb{F}^n$ *and any scalars* $\alpha, \beta \in \mathbb{F}$ *vector addition and scalar multiplication on* $\mathbb{F}^n$ *satisfy the following rules:*

(V0)	$(\mathbb{F}, +)$ *is an Abelian group with neutral element* $\mathbf{0}$ *and inverse* $-\mathbf{v}$.	*(Abelian group)*
(V1)	$\alpha(\mathbf{v} + \mathbf{w}) = \alpha\mathbf{v} + \alpha\mathbf{w}$	*(distributivity I)*
(V2)	$(\alpha + \beta)\mathbf{v} = \alpha\mathbf{v} + \beta\mathbf{v}$	*(distributivity II)*
(V3)	$(\alpha\beta)\mathbf{v} = \alpha(\beta\mathbf{v})$	*(multiplicative associativity)*
(V4)	$1\mathbf{v} = \mathbf{v}$	*(multiplicative neutral element)*

Proof We already know from the general construction of direct product groups in Section 3.1 that $(\mathbb{F}^n, +)$ is an Abelian group, so (V0) holds. The other rules follow directly by combining the axioms in Def. 4.1 for calculating in a field with the component-wise definition (5.2) of vector addition and scalar multiplication. For example, (V2) can be shown by

$$(\alpha + \beta)\mathbf{v} \overset{(5.2)}{=} \begin{pmatrix} (\alpha + \beta)v_1 \\ \vdots \\ (\alpha + \beta)v_n \end{pmatrix} = \begin{pmatrix} \alpha v_1 + \beta v_1 \\ \vdots \\ \alpha v_n + \beta v_n \end{pmatrix} \overset{(5.2)}{=} \begin{pmatrix} \alpha v_1 \\ \vdots \\ \alpha v_n \end{pmatrix} + \begin{pmatrix} \beta v_1 \\ \vdots \\ \beta v_n \end{pmatrix} \overset{(5.2)}{=} \alpha\mathbf{v} + \beta\mathbf{v}.$$

Using index notation and Eqs. (5.3) the same proof can also be written, somewhat more concisely, as

$$((\alpha + \beta)\mathbf{v})_i \overset{(5.3)}{=} (\alpha + \beta)v_i = \alpha v_i + \beta v_i \overset{(5.3)}{=} (\alpha \mathbf{v} + \beta \mathbf{v})_i .$$

As a further example, the proof of (V3), using index notation, reads

$$((\alpha\beta)\mathbf{v})_i \overset{(5.3)}{=} (\alpha\beta)v_i = \alpha(\beta v_i) \overset{(5.3)}{=} \alpha(\beta \mathbf{v})_i \overset{(5.3)}{=} (\alpha(\beta \mathbf{v}))_i .$$

The proofs of the remaining rules are analogous and are left as Exercise 5.1. $\qquad\square$

The above rules for calculating with coordinate vectors motivate the general definition of vector spaces which will be introduced in the next chapter.

5.2 Standard unit vectors

Summary 5.2 *The existence of two special elements in a field $\mathbb{F}$, the neutral elements of addition and multiplication, facilitates the definition of the standard unit vectors $\mathbf{e}_i$, where $i = 1, \ldots, n$, in $\mathbb{F}^n$. All vectors in $\mathbb{F}^n$ can be written in terms of standard unit vectors. Standard unit vectors can also be used to carry out vector addition and scalar multiplication.*

5.2.1 Definition of standard unit vectors

Recall that a field has two special elements, the neutral elements 0 and 1 of addition and multiplication, respectively. This fact allows us to define a special set of vectors in $\mathbb{F}^n$ by using only these two neutral elements as components. These vectors are called the *standard unit vectors*, denoted $\mathbf{e}_i \in \mathbb{F}^n$, where $i = 1, \ldots, n$, and they are defined as

$$\mathbf{e}_i := \begin{pmatrix} 0 \\ \vdots \\ 0 \\ 1 \\ 0 \\ \vdots \\ 0 \end{pmatrix} \leftarrow i^{\text{th}} \text{ position} . \tag{5.6}$$

Note that these are n vectors and the ith component of $\mathbf{e}_i$ is equal to 1 while all other components are equal to 0. Every vector $\mathbf{v} \in \mathbb{F}^n$ with components v_i can be written in terms of the standard unit vectors as

$$\mathbf{v} = v_1 \mathbf{e}_1 + \cdots + v_n \mathbf{e}_n = \sum_{i=1}^{n} v_i \mathbf{e}_i . \tag{5.7}$$

Problem 5.2 *Vectors in terms of standard unit vectors*

Write the vectors $(-3, 4)^T$, $(-7, 0, 3)^T$ and $(1, 2, -1, 3)^T$ in terms of standard unit vectors.

Solution:

$$\begin{pmatrix} -3 \\ 4 \end{pmatrix} = -3\mathbf{e}_1 + 4\mathbf{e}_2 \; , \qquad \begin{pmatrix} -7 \\ 0 \\ 3 \end{pmatrix} = -7\mathbf{e}_1 + 3\mathbf{e}_3 \; , \qquad \begin{pmatrix} 1 \\ 2 \\ -1 \\ 3 \end{pmatrix} = \mathbf{e}_1 + 2\mathbf{e}_2 - \mathbf{e}_3 + 3\mathbf{e}_4 \; .$$

5.2.2 Calculating with standard unit vectors

Vector additions and scalar multiplications can also be carried out in terms of standard unit vectors. With two vectors $\mathbf{v} = \sum_{i=1}^{n} v_i \mathbf{e}_i$ and $\mathbf{w} = \sum_{i=1}^{n} w_i \mathbf{e}_i$, we have

$$\mathbf{v} + \mathbf{w} = \sum_{i=1}^{n} v_i \mathbf{e}_i + \sum_{i=1}^{n} w_i \mathbf{e}_i = \sum_{i=1}^{n} (v_i + w_i) \mathbf{e}_i \; , \qquad \alpha \mathbf{v} = \alpha \sum_{i=1}^{n} v_i \mathbf{e}_i = \sum_{i=1}^{n} (\alpha v_i) \mathbf{e}_i \; .$$

To write these expressions in their final form, we have used the general rules from Prop. 5.1, notably associativity and commutativity from (V0), as well as (V2) and (V3).

In scientific applications, the case $n = 3$ is important for the description of physical space and in this context the three standard unit vectors are sometime denoted by $\mathbf{i}$, $\mathbf{j}$, and $\mathbf{k}$, so that

$$\mathbf{i} := \mathbf{e}_1 = \begin{pmatrix} 1 \\ 0 \\ 0 \end{pmatrix} \; , \quad \mathbf{j} := \mathbf{e}_2 = \begin{pmatrix} 0 \\ 1 \\ 0 \end{pmatrix} \; , \quad \mathbf{k} := \mathbf{e}_3 = \begin{pmatrix} 0 \\ 0 \\ 1 \end{pmatrix} \; . \tag{5.8}$$

Problem 5.3 *Calculating in terms of standard unit vectors*

Add the vectors $\mathbf{v} = (1, -2, 5)^T = \mathbf{i} - 2\mathbf{j} + 5\mathbf{k}$, $\mathbf{w} = (-4, 1, -3)^T = -4\mathbf{i} + \mathbf{j} - 3\mathbf{k}$ and work out the scalar multiple of $\mathbf{v}$ with $\alpha = 3$ in standard unit vector notation.

Solution:

$$\mathbf{v} + \mathbf{w} = (\mathbf{i} - 2\mathbf{j} + 5\mathbf{k}) + (-4\mathbf{i} + \mathbf{j} - 3\mathbf{k}) = -3\mathbf{i} - \mathbf{j} + 2\mathbf{k} \; .$$
$$\alpha \mathbf{v} = 3(\mathbf{i} - 2\mathbf{j} + 5\mathbf{k}) = 3\mathbf{i} - 6\mathbf{j} + 15\mathbf{k} \; .$$

See Exercise 5.3 for further examples.

Exercises

(†=challenging)

5.1 *Rules for coordinate vectors*
Proof (V1), (V3), and (V4) from Prop. 5.1 by using the definition (5.2) of vector addition and scalar multiplication, together with the axioms for a field. Also, carry out the same proofs using the definitions (5.3) and index notation.

5.2 *Standard unit vectors*
Write the $\mathbb{R}^4$ vectors $\mathbf{v} = (1, -2, 3, -5)^T$ and $\mathbf{w} = (0, -1, 1, -3)^T$ in terms of standard unit vectors.

5.3 *Calculating with standard unit vectors*
Carry out all mutual sums of the vectors $\mathbf{v} = 5\mathbf{i} + 2\mathbf{j} - 3\mathbf{k}$, $\mathbf{u} = \mathbf{i} - 4\mathbf{j} - \mathbf{k}$ and $\mathbf{w} = -2\mathbf{i} + 7\mathbf{j}$ in $\mathbb{R}^3$. Scalar multiply these vectors with $\alpha = -3$.

5.4 *Complex vectors*
Write the $\mathbb{C}^3$ vectors $\mathbf{v} = (1 + i, 5i, 3)^T$ and $\mathbf{w} = (2 - 3i, 1 - i, 2i)^T$ in terms if the standard unit vectors $\mathbf{i}$, $\mathbf{j}$, and $\mathbf{k}$. Work out the sum $\mathbf{v} + \mathbf{w}$ and the scalar multiplications $\alpha\mathbf{v}$ and $\alpha\mathbf{w}$, where $\alpha = 2 - i$.

5.5 *Vector space $\mathbb{F}_2^n$* †
Consider the vector space $\mathbb{F}_2^n$, based on the finite field $\mathbb{F}_2$ introduced in Example 4.3. How many elements does $\mathbb{F}_2^n$ have? Show that there is a bijective map between predicates on $\{1, \ldots, n\}$ and $\mathbb{F}_2^n$. Express vector addition in $\mathbb{F}_2^n$ in terms of 'and' and 'or' operations for the corresponding predicates.

6

Vector spaces

At this point it might seem that we have gone far enough in abstraction, having introduced coordinate vectors with an arbitrary number of components and from an arbitrary field. What else could be needed, particularly in scientific applications of linear algebra?

It turns out that there are sets of objects, quite unlike coordinate vectors, which nevertheless follow the same algebraic rules as coordinate vectors. Moreover, many of these have important scientific applications. For example, on the set of all real-valued functions $[a, b] \rightarrow \mathbb{R}$ on an interval $[a, b]$ addition and scalar multiplication can be defined in a way that follows the rules for coordinate vectors in Prop. 5.1. Function spaces of this kind are important in quantum mechanics. It makes sense to define a general algebraic structure that captures all examples similar to coordinate vectors: the *vector space*.

6.1 Basic definitions

Summary 6.1 *Abstract vector spaces* $(V, \mathbb{F}, +, \cdot)$ *are introduced, where* V *is a set of vectors and* $\mathbb{F}$ *is a field. There are two operations, vector addition,* $+$*, and scalar multiplication,* $\cdot$*, which are subject to a list of axioms. Vector subspaces are non-empty subsets of vector spaces which are closed under vector addition and scalar multiplication and they form vector spaces in their own right. Linear maps are the morphism of vector spaces, that is, they are the maps consistent with the vector space structure.*

6.1.1 Vector space axioms

In the previous chapter we have studied coordinate vectors in $\mathbb{F}^n$ and their properties. While these coordinate vectors play an important role in linear algebra, the modern approach to the subject is more general. Rather than defining vectors by 'what they are', they are defined by the properties they should satisfy. This means we are looking for an axiomatic definition of vector spaces, in analogy with the definitions of groups and fields in Chapters 3 and 4. The structure of this definition is very much inspired by what we have found for coordinate vectors. A vector space consists of a pair $(V, \mathbb{F})$, where V is a set whose elements are called *vectors* and $\mathbb{F}$ is a field, with elements called *scalars*. There are two operations, *vector addition* and *scalar multiplication*, which are required to satisfy a list of axioms. These axioms are, in fact, precisely the rules for calculations with column vectors listed in Prop. 5.1. Putting all this together, the formal definition of vector spaces is as follows:

Definition 6.1 *A vector space $(V, \mathbb{F}, +, \cdot)$ consists of a set V (with elements called vectors), a field $\mathbb{F}$ (with elements called scalars) and the two operations*

$$
\begin{array}{llll}
\text{vector addition:} & +: V \times V & \to \quad V, & (\mathbf{v}, \mathbf{w}) \quad \mapsto \quad \mathbf{v} + \mathbf{w} \\
\text{scalar multiplication:} & \cdot: (\mathbb{F}, V) & \to \quad V, & (\alpha, \mathbf{v}) \quad \mapsto \quad \alpha\mathbf{v}
\end{array}
$$

For all $\mathbf{v}, \mathbf{w} \in V$ and for all $\alpha, \beta \in \mathbb{F}$, these operations satisfy the following rules:

(V0)	$(V, +)$ *is an Abelian group with*	*(Abelian group)*
	neutral element $\mathbf{0}$ and inverse $-\mathbf{v}$	
(V1)	$\alpha(\mathbf{v} + \mathbf{w}) = \alpha\mathbf{v} + \alpha\mathbf{w}$	*(distributivity I)*
(V2)	$(\alpha + \beta)\mathbf{v} = \alpha\mathbf{v} + \beta\mathbf{v}$	*(distributivity II)*
(V3)	$(\alpha\beta)\mathbf{v} = \alpha(\beta\mathbf{v})$	*(multiplicative associativity)*
(V4)	$1\mathbf{v} = \mathbf{v}$	*(multiplicative neutral element)*

The neutral element $\mathbf{0}$ of vector addition is called the zero vector.

Note that this definition does not specify the nature of vectors. In particular, it is not assumed that they are made up from components. The expression $-\mathbf{v}$ does not imply any particular operation, such as multiplication by -1 — it is merely the symbol used for the additive inverse of a vector $\mathbf{v}$. In Prop. 6.1 we will see that the additive inverse $-\mathbf{v}$ is, in fact, obtained by multiplying $\mathbf{v}$ with -1 but this needs to be proved. The choice of field $\mathbb{F}$ is an important part of the definition of a vector space — it determines from which set the scalars are taken. Instead of using the somewhat cumbersome notation $(V, \mathbb{F}, +, \cdot)$ we will frequently just talk about a vector space V over (the field) $\mathbb{F}$.

6.1.2 Implications of vector space axioms

There are a few simple rules for calculating with vectors which are obvious for component vectors (there they follow immediately from the component-wise definitions of vector addition and scalar multiplication (5.2)) but in the present abstract case they have to be derived from the above axioms. A few such rules are covered in the following proposition.

Proposition 6.1 *For a vector space V over $\mathbb{F}$, the following rules hold for all $\mathbf{v} \in V$:*

(i)	$-(-\mathbf{v}) = \mathbf{v}$
(ii)	$0\mathbf{v} = \mathbf{0}$
(iii)	$\alpha\mathbf{0} = \mathbf{0}$ *for all $\alpha \in \mathbb{F}$*
(iv)	$(-1)\mathbf{v} = -\mathbf{v}$
(v)	$\alpha\mathbf{v} = \mathbf{0} \Rightarrow \alpha = 0$ *or $\mathbf{v} = \mathbf{0}$*

Proof (i) This follows from the fact that $(V, +)$ is a group and the first rule (3.1) for the group inverse.

(ii) Since $0\mathbf{v} = (0 + 0)\mathbf{v} \stackrel{(V2)}{=} 0\mathbf{v} + 0\mathbf{v}$ and $0\mathbf{v} = 0\mathbf{v} + \mathbf{0}$, it follows that $0\mathbf{v} = \mathbf{0}$.

(iii) $\alpha\mathbf{0} \stackrel{(ii)}{=} \alpha(0\,\mathbf{0}) \stackrel{(V3)}{=} (\alpha\,0)\mathbf{0} = 0\,\mathbf{0} \stackrel{(ii)}{=} \mathbf{0}$.

(iv) Since $\mathbf{0} \stackrel{(ii)}{=} 0\mathbf{v} = (1 + (-1))\mathbf{v} \stackrel{(V2),(V4)}{=} \mathbf{v} + (-1)\mathbf{v}$ and $\mathbf{0} = \mathbf{v} + (-\mathbf{v})$, it follows that $(-1)\mathbf{v} = -\mathbf{v}$.

(iv) If $\alpha = 0$ we are done. If $\alpha \neq 0$, then multiplying with α^{-1} gives $\mathbf{0} \overset{(iii)}{=} \alpha^{-1}\mathbf{0} = \alpha^{-1}(\alpha\mathbf{v}) \overset{(V3)}{=} (\alpha^{-1}\alpha)\mathbf{v} = 1\mathbf{v} \overset{(V4)}{=} \mathbf{v}$. $\qquad\square$

We can also generalize the distributive laws (V1) and (V2) to an arbitrary number of summands. Specifically, for vectors $\mathbf{v}, \mathbf{v}_1, \ldots, \mathbf{v}_n \in V$ and scalars $\alpha, \alpha_1, \ldots, \alpha_n \in \mathbb{F}$ we have (see Exercise 6.1):

$$\alpha \sum_{i=1}^{n} \mathbf{v}_i = \sum_{i=1}^{n} \alpha\mathbf{v}_i , \qquad \left(\sum_{i=1}^{n} \alpha_i\right) \mathbf{v} = \sum_{i=1}^{n} \alpha_i\mathbf{v} . \tag{6.1}$$

6.1.3 Vector subspaces

The first step after setting up a new algebraic structure is to introduce the correspond-ing 'sub-structure'. In the case of groups we have introduced the notion of sub-groups and for vector spaces we would like to define *vector subspaces*. These are subsets of vector spaces which form vector spaces in their own right. The formal definition is as follows:

Definition 6.2 *A non-empty subset $W \subset V$ of a vector space V is a vector subspace provided it satisfies the following conditions.*

(S1) For all $\mathbf{w}_1, \mathbf{w}_2 \in W$ we have $\mathbf{w}_1 + \mathbf{w}_2 \in W$.
(S2) For all $\mathbf{w} \in V$ and $\alpha \in \mathbb{F}$ we have $\alpha\mathbf{w} \in W$.

In other words, a vector subspace is a non-empty subset of a vector space which is closed under vector addition and scalar multiplication.

This definition does imply immediately that a vector subspace $W \subset V$ is a vector space (over the same field $\mathbb{F}$ that underlies V) in its own right, with vector addition and scalar multiplication defined by restriction from V to W.

To verify this, we first note that vector addition and scalar multiplication are closed operations on W from Def. 6.2. Further, all vector space axioms in Def. 6.1 which are merely rules for calculation are satisfied on W, simply because they are satisfied on V. We only have to be careful about the existence of the neutral element and the inverse. While these are certainly present in V it is not immediately clear they are contained in W. However, for any $\mathbf{w} \in W$ we know, combining Prop. 6.1 (ii) and Def. 6.2 (S2), that $\mathbf{0} = 0\,\mathbf{w} \in W$ and, hence, the zero vector is contained in W. Further, for $\mathbf{w} \in W$, we have from Prop. 6.1 (iii) and Def. 6.2 (S2), that $-\mathbf{w} = (-1)\mathbf{w} \in W$ so, from (S2), the inverse vector is contained in W.

Every vector space V has two trivial vector subspaces: the vector space $\{\mathbf{0}\}$, which consists of the zero vector, and the whole space V. All other vector subspaces are called *proper* and we will consider examples soon.

6.1.4 Linear Maps

The next step in the general build-up of the theory is to introduce the morphisms of vector spaces which are also called *linear maps*. Linear maps are to vector spaces what group homomorphisms are to groups. In the same way that group homomorphism are defined by being compatible with the group multiplication (see Eq. (3.4)), linear maps are those maps which are consistent with vector addition and scalar multiplication. In fact, the analogy is even closer since a vector space forms an Abelian group with respect to addition. Therefore, an obvious requirement for a map to be linear is that it is a group homomorphism relative to this additive group structure. A second requirement arises from consistency with scalar multiplication and this leads to the following definition.

Definition 6.3 *A map* $f : V \to W$ *between two vector spaces* V *and* W *over the same field* $\mathbb{F}$ *is called linear if*

(L1) $f(\mathbf{v}_1 + \mathbf{v}_2) = f(\mathbf{v}_1) + f(\mathbf{v}_2)$
(L2) $f(\alpha\mathbf{v}) = \alpha f(\mathbf{v})$

for all $\mathbf{v}, \mathbf{v}_1, \mathbf{v}_2 \in V$ *and for all* $\alpha \in \mathbb{F}$.

Note that the addition on the left-hand side of (L1) is carried out in V, while the one on the right-hand side is carried out in W. Likewise, the scalar multiplication with α in (L2) is in V on the left and in W on the right. For this to make sense V and W have to be vector spaces over the same field, as we have indeed required. It is sometimes useful to combine (L1) and (L2) into the single, equivalent linearity condition

$$f(\alpha_1\mathbf{v}_1 + \alpha_2\mathbf{v}_2) = \alpha_1 f(\mathbf{v}_1) + \alpha_2 f(\mathbf{v}_2)\,, \tag{6.2}$$

for all $\mathbf{v}_1, \mathbf{v}_2 \in V$ and all $\alpha_1, \alpha_2 \in \mathbb{F}$.

For now we are content having introduced the general idea of linearity of a map. Linear maps and their relation to matrices will be systematically discussed in Part IV.

6.1.5 Algebras

As we will see, some of the vector spaces we will come across carry an additional multiplication between vectors. Such vector spaces with multiplication are also called an *algebra*, a structure formally defined as follows:

Definition 6.4 *An algebra* $(V, \mathbb{F}, +, \cdot, *)$ *is a vector space* $(V, \mathbb{F}, +, \cdot)$ *with a multiplication* $* : V \times V \to V$ *which satisfies the following properties, for all* $\mathbf{v}_1, \mathbf{v}_2, \mathbf{w} \in V$ *and all* $\alpha_1, \alpha_2 \in \mathbb{F}$.

(i) $(\alpha_1\mathbf{v}_1 + \alpha_2\mathbf{v}_2) * \mathbf{w} = \alpha_1(\mathbf{v}_1 * \mathbf{w}) + \alpha_2(\mathbf{v}_2 * \mathbf{w})$ *(linear in first argument)*
(ii) $\mathbf{w} * (\alpha_1\mathbf{v}_1 + \alpha_2\mathbf{v}_2) = \alpha_1(\mathbf{w} * \mathbf{v}_1) + \alpha_2(\mathbf{w} * \mathbf{v}_2)$ *(linear in second argument)*

If there is a $\mathbf{e} \in V$ *with* $\mathbf{e} * \mathbf{v} = \mathbf{v} * \mathbf{e} = \mathbf{v}$ *for all* $\mathbf{v} \in V$ *the algebra is called an algebra with unit. If the product* $*$ *is associative, the algebra is called an associative algebra.*

In short, an algebra is a vector space with a multiplication which is bi-linear. We will not investigate algebras systematically but, occasionally, it will be useful to point to the above definition when we come across examples of algebras. One simple such example is the vector space $\mathbb{R}^3$ with the cross product as multiplication, which we discuss in Section 10.

6.2 Examples of vector spaces

Summary 6.2 *Standard examples of vector spaces are the coordinate vectors* $(\mathbb{F}^n, \mathbb{F}, +, \cdot)$. *There are also less conventional coordinate vectors, such as* $(\mathbb{C}^n, \mathbb{R}, +, \cdot)$, *where the vector components and the scalars are taken from different fields. The matrices* $\mathcal{M}_{n,m}(\mathbb{F})$ *of size* $n \times m$ *with entries in a field* $\mathbb{F}$ *form a vector space. The set of all functions* $X \to V$ *from a set* X *into a vector space* V *can be given a vector space structure. Many interesting function vector spaces arise as special cases or vector subspaces from this construction.*

6.2.1 Coordinate vector spaces

Coordinate vectors have motivated the Definition 6.1 so it should not come as a surprise that they provide examples of vector spaces.

Example 6.1 *($\mathbb{F}^n$ as a vector space over $\mathbb{F}$)*

Coordinate vectors with n components taken from a field $\mathbb{F}$ form a vector space $(\mathbb{F}^n, \mathbb{F}, +, \cdot)$, with vector addition and scalar multiplication defined component-wise, as in Eq. (5.2). This follows immediately by comparing the rules for calculating with coordinate vectors listed in Prop. 5.1 with the vector space axioms in Def. 6.1. The most commonly used fields are $\mathbb{F} = \mathbb{Q}, \mathbb{R}, \mathbb{C}$, but the finite fields $\mathbb{F}_p$, introduced in Example 4.3, can also be relevant. Also note that the field $\mathbb{F}$ forms a vector space (of vectors with one component) over itself. □

Example 6.2 *(Unusual coordinate vector spaces)*

While the coordinate vector spaces from Example 6.1 are the most commonly used ones, there are more exotic constructions where the vector components and the scalars are taken from different fields. For example, instead of the vector space $(\mathbb{C}^n, \mathbb{C}, +, \cdot)$ with complex vector components and complex scalars, we can also consider the space $(\mathbb{C}^n, \mathbb{R}, +, \cdot)$ with complex vector components but real scalars. Indeed, scalar multiplication of complex coordinate vectors with real numbers, defined component-wise, makes perfect sense and satisfies the required axioms (V1)–(V4) in Def. 6.1. The two vector spaces $(\mathbb{C}^n, \mathbb{C}, +, \cdot)$ and $(\mathbb{C}^n, \mathbb{R}, +, \cdot)$ are quite different despite the vectors being taken from the same set. In the following, unless otherwise stated, we will think of $\mathbb{F}^n$ as a vector space over the field $\mathbb{F}$. □

6.2.2 Matrices and matrix vector spaces

Coordinate vector spaces are based on defining vector addition and scalar multiplication component-wise but it should be clear that it is not essential for the components to be arranged in a column. They might be arranged in a row, in a rectangle, or even in a triangle for that matter. As long as we decide that addition works by adding components in the same position and scalar multiplication by multiplying every component with same the scalar, such objects can be given a vector space structure.

Rectangular arrangements of numbers from a field $\mathbb{F}$, with n rows and m columns, are called $n \times m$ *matrices* (with entries in $\mathbb{F}$) and they are written as

$$A = \begin{pmatrix} A_{11} & \cdots & A_{1m} \\ \vdots & & \vdots \\ A_{n1} & \cdots & A_{nm} \end{pmatrix} , \qquad B = \begin{pmatrix} B_{11} & \cdots & B_{1m} \\ \vdots & & \vdots \\ B_{n1} & \cdots & B_{nm} \end{pmatrix} . \tag{6.3}$$

The numbers $A_{ij}, B_{ij} \in \mathbb{F}$ are called the components or *entries* of the matrix. It is convenient (although slightly abusing notation) to denote the entire matrix and its entries by the same letter, as we have done above. Just like vectors, matrices can be written in index notation as the collection, A_{ij}, of their entries which are now labelled by two indices, $i = 1, \ldots, n$ and $j = 1, \ldots, m$. The set of all $n \times m$ matrices with entries in $\mathbb{F}$ is denoted by $\mathcal{M}_{n,m}(\mathbb{F})$. Note that $n \times 1$ matrices in $\mathcal{M}_{n,1}(\mathbb{F})$ are column vectors while $1 \times n$ matrices in $\mathcal{M}_{1,n}(\mathbb{F})$ are row vectors, each with n components in $\mathbb{F}$.

Example 6.3 *(Matrix vector spaces)*

The set $\mathcal{M}_{n,m}(\mathbb{F})$ can be made into a vector space over $\mathbb{F}$ by defining addition and scalar multiplication of matrices component-wise as

$$A + B := \begin{pmatrix} A_{11} + B_{11} & \cdots & A_{1m} + B_{1m} \\ \vdots & & \vdots \\ A_{n1} + B_{n1} & \cdots & A_{nm} + B_{nm} \end{pmatrix} , \qquad \alpha A := \begin{pmatrix} \alpha A_{11} & \cdots & \alpha A_{1m} \\ \vdots & & \vdots \\ \alpha A_{n1} & \cdots & \alpha A_{nm} \end{pmatrix} . \tag{6.4}$$

In index notation, the same definitions can be written as

$$(A + B)_{ij} := A_{ij} + B_{ij} , \quad (\alpha A)_{ij} := \alpha A_{ij} , \tag{6.5}$$

where the subscript on the left-hand side indicates that the entry (ij) is extracted from the matrix in the bracket. Of course, these definitions satisfy all the basic rules of vector addition and scalar multiplication listed in Def. 6.1, for exactly the same reasons coordinate vectors do. The zero 'vector' is the matrix with all entries equal to zero. □

We can pursue the analogy between coordinate vectors and matrices even further by introducing the analogue of the standard unit vectors. These are the *standard unit matrices*

$$E_{(ij)} = \begin{pmatrix} 0 & \cdots & 0 & 0 & 0 & \cdots & 0 \\ \vdots & & \vdots & \vdots & \vdots & & \vdots \\ 0 & \cdots & 0 & 0 & 0 & \cdots & 0 \\ 0 & \cdots & 0 & 1 & 0 & \cdots & 0 \\ 0 & \cdots & 0 & 0 & 0 & \cdots & 0 \\ \vdots & & \vdots & \vdots & \vdots & & \vdots \\ 0 & \cdots & 0 & 0 & 0 & \cdots & 0 \end{pmatrix} \leftarrow i^{\text{th}} \text{ row} , \tag{6.6}$$

$$\underset{\underset{j^{\text{th}} \text{ column}}{\uparrow}}{}$$

where $i = 1, \ldots, n$ and $j = 1, \ldots, m$ and the '1' appears in the ith row and j^{th} column with all other entries zero. (Note that here the indices i, j label the nm different matrices, rather than entries of a matrix. To emphasize this fact, they have been enclosed in brackets.) Every $n \times m$ matrix A can be written in terms of the standard unit matrices as

$$A = \sum_{i=1}^{n} \sum_{j=1}^{m} A_{ij} E_{(ij)} , \tag{6.7}$$

and addition and scalar multiplication can be expressed as

$$A + B = \sum_{i,j} (A_{ij} + B_{ij}) E_{(ij)} , \qquad \alpha A = \sum_{i,j} \alpha A_{ij} E_{(ij)} , \tag{6.8}$$

in complete analogy with Eqs. (5.8) for coordinate vectors.

Problem 6.1 *(Addition and scalar multiplication of matrices)*

Add the 2×2 matrices

$$A = \begin{pmatrix} 1 & -2 \\ 3 & -4 \end{pmatrix} , \qquad B = \begin{pmatrix} 0 & 5 \\ -1 & 8 \end{pmatrix} .$$

and work out the scalar multiple of A with $\alpha = 3$. Write A and B in terms of standard unit matrices and work out $A + B$ and αA using this notation.

Solution: Using matrix notation, the sum and the scalar multiple are given by

$$A + B = \begin{pmatrix} 1 & -2 \\ 3 & -4 \end{pmatrix} + \begin{pmatrix} 0 & 5 \\ -1 & 8 \end{pmatrix} = \begin{pmatrix} 1 & 3 \\ 2 & 4 \end{pmatrix} , \qquad \alpha A = 3 \begin{pmatrix} 1 & -2 \\ 3 & -4 \end{pmatrix} = \begin{pmatrix} 3 & -6 \\ 9 & -12 \end{pmatrix} .$$

In terms of the standard unit matrices, A and B can be written as

$$A = E_{(11)} - 2E_{(12)} + 3E_{(21)} - 4E_{(22)} , \qquad B = 5E_{(12)} - E_{(21)} + 8E_{(22)} .$$

In this language, their sum and the scalar multiple of A with $\alpha = 3$ are given by

$$A + B = E_{(11)} + 3E_{(12)} + 2E_{(21)} + 4E_{(22)} , \qquad \alpha A = 3E_{(11)} - 6E_{(12)} + 9E_{(21)} - 12E_{(22)} .$$

6.2.3 Vector spaces of functions

We begin by describing a very general construction of function vector spaces. Start with an arbitrary set X as well as a vector space V over the field $\mathbb{F}$ and define the set of all functions $\mathcal{F}(X, V) := \{g : X \to V\}$ from X to V. On this set, we can introduce an addition and a scalar multiplication by

$$(g + h)(x) := g(x) + h(x) , \qquad (\alpha g)(x) := \alpha g(x) , \tag{6.9}$$

where $g, h \in \mathcal{F}(X, V)$ are functions, $\alpha \in \mathbb{F}$ and $x \in X$. Note that the right-hand sides of these expressions are simply vector additions and scalar multiplications on the given vector space V, so they are well-defined. Moreover, Eqs. (6.9) satisfy all the vector space axioms in Def. 6.1 simply because these axioms are satisfied in V. The null 'vector' is the function whose value is the zero vector for all $x \in X$.

In conclusion, for a vector space V over $\mathbb{F}$, the function space $\mathcal{F}(X, V)$ with vector addition and scalar multiplication defined 'point-wise', as in Eq. (6.9), is a vector space over the same field $\mathbb{F}$. Many interesting function vector spaces can be obtained from this construction, by choosing specific sets X or specific vector spaces V or by restricting to certain vector subspaces. The following examples illustrate the range of possibilities.

Example 6.4 *(Coordinate vectors as functions)*
Coordinate vectors can be obtained from the above construction. To see this, choose the set $X = \{1, 2, \ldots, n\}$ and the vector space $V = \mathbb{F}$. Functions $g : \{1, 2, \ldots, n\} \to \mathbb{F}$ are specified by the n-tuples $(g(1), g(2), \ldots, g(n))$ of all their values and can, hence, be identified with vectors in $\mathbb{F}^n$. □

Example 6.5 *(Functions on a real interval)*
Choose the set $X = [a, b] \subset \mathbb{R}$ to be an interval on the real line and $V = \mathbb{F}$. The resulting space of functions, $\mathcal{F}([a, b], \mathbb{F})$, consists of all $\mathbb{F}$-valued functions, that is, typically real- or complex-valued functions, on the interval $[a, b]$. Vector addition and scalar multiplication, as defined in Eq. (6.9), really just amount to 'naive' addition and scalar multiplication of functions. Consider, for example, the two functions $g, h \in \mathcal{F}([a, b], \mathbb{F})$ defined by $g(x) = 2x^2 + 3x - 1$ and $h(x) = -2x + 4$. Their vector sum and the scalar multiple of g by $\alpha = 4$ are given by

$$(g + h)(x) = (2x^2 + 3x - 1) + (-2x + 4) = 2x^2 + x + 3$$
$$(\alpha g)(x) = 4(2x^2 + 3x - 1) = 8x^2 + 12x - 4 .$$

The vector space $\mathcal{F}([a, b], \mathbb{F})$ has many interesting vector subspaces, some of which we now discuss. □

Example 6.6 *(Continuous and differentiable functions)*
Let us focus on the space $\mathcal{F}([a, b], \mathbb{F})$ with $\mathbb{F} = \mathbb{R}$ or $\mathbb{F} = \mathbb{C}$, so on real- or complex-valued functions on the interval $[a, b]$. From Def. 6.2, any property of such real- or complex-valued functions which is preserved under the addition and scalar multiplication (6.9) can be used to obtain a vector subspace of $\mathcal{F}([a, b], \mathbb{F})$. For example, the sum of two continuous functions and the scalar multiple of a continuous function are both continuous, so the space $\mathcal{C}([a, b], \mathbb{F})$ of continuous $\mathbb{F}$-valued functions on the interval $[a, b]$ is a vector subspace of $\mathcal{F}([a, b], \mathbb{F})$. The same goes for the space $\mathcal{C}^k([a.b], \mathbb{F})$ of k times continuously differentiable functions. (See, for example, Lang 1997 for the relevant proofs.)

Example 6.7 *(Polynomials)*
The set $\mathcal{P}(\mathbb{F})$ of polynomials is a vector space over $\mathbb{F}$, a vector subspace of $\mathcal{F}(\mathbb{F}, \mathbb{F})$. Indeed, the sum of two polynomials and the scalar multiple of a polynomial are again

polynomials. We can restrict further to the set $\mathcal{P}_k(\mathbb{F})$ of all polynomials with degree less equal k. Since addition and scalar multiplication of such polynomials does not increase the degree beyond k, this set forms a vector subspace of $\mathcal{P}(\mathbb{F})$. □

Example 6.8 (*Solutions to differential equations*)

Many scientific problems involve solving differential equations of the form

$$p(x)\frac{d^2 g}{dx^2} + q(x)\frac{dg}{dx} + r(x)g = 0 , \qquad (6.10)$$

where $p, q, r \in \mathcal{C}([a, b], \mathbb{R})$ are fixed functions. The task is to find all functions $g \in \mathcal{C}^2([a, b], \mathbb{R})$ which satisfy this equation. Eq. (6.10) is referred to as a *second order, linear, homogeneous differential equation*. Here, the term 'second order' indicates that the highest derivative of g which appears is the second, 'linear' means there are no terms of quadratic or higher order in g and 'homogeneous' means there is no term independent of g. These properties immediately imply that for two solutions $g, h \in \mathcal{C}^2([a, b], \mathbb{R})$ to this equation, also the sum $g + h$ and scalar multiples αg, where $\alpha \in \mathbb{R}$, are solutions. This means the solutions to the differential equation (6.10) form a vector subspace of $\mathcal{C}^2([a, b], \mathbb{R})$.

A simple example is the differential equation

$$\frac{d^2 g}{dx^2} + g = 0 , \qquad (6.11)$$

which is obviously solved by $g(x) = \cos(x)$ and $g(x) = \sin(x)$. Since the solution space forms a vector space, we know that the functions $\alpha \cos(x) + \beta \sin(x)$ for arbitrary $\alpha, \beta \in \mathbb{R}$ also solve the equation. Of course, this can also be checked explicitly by inserting $g(x) = \alpha \cos(x) + \beta \sin(x)$ into Eq. (6.11). □

This list of examples hopefully illustrates the strength of the general approach. Much of what follows will only be based on the general Definition 6.1 of a vector space and, hence, will apply to all of the above examples and many more.

Exercises

6.1 *Vector space rules*
Show that Eqs. (6.1) follow from the vector space axioms, for example, by induction in n.

6.2 *Coordinate vector subspaces*
Which of the following sets constitute vector subspaces of the given vector space? Provide reasoning in each case.
(a) All vectors $(x, y, z)^T \in \mathbb{R}^3$ satisfy-

ing $x = y = 2z$.
(b) All vectors $(x, y)^T \in \mathbb{R}^2$ satisfying $x^2 + y^2 = 1$.

6.3 *Adding and multiplying 2×2 matrices*
What is the sum of the matrices

$$A = \begin{pmatrix} 1 & -2 \\ 3 & -1 \end{pmatrix} , \quad B = \begin{pmatrix} 0 & -5 \\ 2 & -8 \end{pmatrix}$$

and their scalar multiple with $\alpha = -3$?

6.4 *Adding and multiplying larger matrices*
Add the matrices

$$\begin{pmatrix} 1 & 0 & -2 & 4 \\ -1 & -3 & 0 & 8 \\ -4 & 4 & 2 & 2 \end{pmatrix}, \quad \begin{pmatrix} 0 & -5 & 3 & 0 \\ 1 & 7 & 0 & -5 \\ 1 & 2 & -3 & 0 \end{pmatrix}$$

and scalar multiply them with $\alpha = 2$.

6.5 *Matrices with complex entries*
Add the matrices

$$A = \begin{pmatrix} 2 & 1-i \\ i & 4i \end{pmatrix}, \quad B = \begin{pmatrix} 4-2i & -i \\ -2-i & -i \end{pmatrix}.$$

and scalar multiply them with $\alpha = 2 - 3i$.

6.6 *Matrix vector spaces*
Consider the vector space $\mathcal{M}_{n,n}(\mathbb{F})$ of $n \times n$ matrices A with entries in $\mathbb{F}$. Are the following subsets vector subspaces? Provide reasoning in each case.
(a) All A with $A_{11} = 0$.
(b) All A with $A_{11} = 1$.
(c) All A satisfying $\sum_{i,j=1}^{n} A_{ij} = 0$.

6.7 *(Anti-) symmetric 2×2 matrices*
A 2×2 matrix $A \in \mathcal{M}_{2,2}(\mathbb{F})$ given by

$$A = \begin{pmatrix} a & b \\ c & d \end{pmatrix},$$

is called symmetric if $c = b$. It is called anti-symmetric if $c = -b$ and $a = d = 0$. Show that the symmetric (anti-symmetric) 2×2 matrices form a vector subspace of $\mathcal{M}_{2,2}(\mathbb{F})$.

6.8 *Function vector subspaces*
Which of the following sets constitute vector subspaces of the vector space of real-valued functions $f : \mathbb{R} \to \mathbb{R}$? Provide reasoning in each case.

(a) Even functions, that is, functions satisfying $f(x) = f(-x)$ for all $x \in \mathbb{R}$.
(b) Odd functions, that is, functions satisfying $f(x) = -f(-x)$ for all $x \in \mathbb{R}$.
(c) Functions satisfying $f(0) = 0$.
(d) Functions satisfying $f(0) = 1$.

6.9 *Polynomial vector spaces*
Which of the following sets are vector subspaces of the vector space $\mathcal{P}_2(\mathbb{R})$ of at most quadratic polynomials in x? Provide reasoning in each case.
(a) All polynomials of the form $ax + b$.
(b) All polynomials of the form $(x+b)^2$.

6.10 *Linear maps*
Consider a vector space V over $\mathbb{F}$, a scalar $\alpha \in \mathbb{F}$, and a non-zero vector $\mathbf{u} \in V$.
(a) Show that the map $f : V \to V$ defined by $f(\mathbf{v}) = \alpha\mathbf{v}$ is linear.
(b) Show that the map $f : V \to V$ defined by $f(\mathbf{v}) = \mathbf{u} + \mathbf{v}$ is not linear.

6.11 *Linear maps for functions*
Consider the vector space $\mathcal{C}^\infty([a,b])$ of infinitely many times differentiable functions and $p \in \mathcal{C}^\infty([a,b])$.
(a) Show that the map $F : \mathcal{C}^\infty([a,b]) \to \mathcal{C}^\infty([a,b])$ defined by multiplication with p, so $F(g)(x) = p(x)g(x)$, is linear.
(b) Show that the map $D : \mathcal{C}^\infty([a,b]) \to \mathcal{C}^\infty([a,b])$ defined by differentiation, so $D(g)(x) = g'(x)$, is linear.

6.12 Let $f : V \to W$ be a linear map.
(a) Show that f maps the zero vector of V to the zero vector of W.
(b) Show that $f(-\mathbf{v})$ is the additive inverse of $f(\mathbf{v})$.

6.13 *More function vector spaces*
Use the construction of Section 6.2.3 to find at least three further vector spaces of functions.

7
Elementary vector space properties

Now that we have set up vector spaces in general, we can start to develop the subject systematically. For the remainder of Part II we will be working with a general vector space V over a field $\mathbb{F}$, except in some of the examples or if stated otherwise.

The first important concept we introduce is that of *linear independence* of a (finite) list of vectors. Roughly speaking, a list of vectors is called linearly independent if none of the vectors can be expressed in terms of the others. Linear independence allows us to introduce the notion of a *basis* of a vector space: a list of linearly independent vectors which 'spans" the vector space.

Bases are absolutely crucial for both the theory of vector spaces and their applications. As we will see, the number of vectors in a basis of a given vector space is fixed and this number is called the *dimension* of the vector space. It turns out that every vector space $(V, \mathbb{F}, +, \cdot)$ of dimension n is isomorphic to the coordinate vector space $(\mathbb{F}^n, \mathbb{F}, +, \cdot)$, so that every vector in V can be described by a unique vector in $\mathbb{F}^n$, whose components are the coordinates relative to a chosen basis of V. Apart from the theoretical insight, this also provides a practical way of computing with abstract vector spaces by using coordinates relative to a basis.

7.1 Linear independence

Summary 7.1 *The most general algebraic expressions in a vector space are referred to as linear combinations. The set of all linear combinations of a given list of vectors is called the span of these vectors and it forms a vector subspace. A finite number of vectors is called linearly independent if none of their linear combinations, except the trivial one, gives the zero vector. Otherwise, the vectors are called linearly dependent. A set of vectors is linearly dependent if and only if one of them can be written as a linear combination of the others.*

7.1.1 Linear combinations and span

We are working with a general vector space V over the field $\mathbb{F}$. Given that we have two operations, vector addition and scalar multiplication, at our disposal the most general algebraic expression is of the form

$$\alpha_1 \mathbf{v}_1 + \cdots + \alpha_k \mathbf{v}_k = \sum_{i=1}^{k} \alpha_i \mathbf{v}_i \,, \tag{7.1}$$

where $\mathbf{v}_1, \ldots, \mathbf{v}_k \in V$ and $\alpha_1, \ldots, \alpha_k \in \mathbb{F}$ are k vector and scalars, respectively. An expression of the form (7.1) is called a *linear combination* of the vectors $\mathbf{v}_1, \ldots, \mathbf{v}_k$. The set of all linear combinations of given vectors $\mathbf{v}_1, \ldots, \mathbf{v}_k$ is called the *span* of these vectors, and it is written as

$$\mathrm{Span}(\mathbf{v}_1, \ldots, \mathbf{v}_k) := \left\{ \sum_{i=1}^{k} \alpha_i \mathbf{v}_i \,\middle|\, \alpha_i \in \mathbb{F} \right\}. \tag{7.2}$$

Proposition 7.1 *For any vectors $\mathbf{v}_1, \ldots, \mathbf{v}_k \in V$, the span, $\mathrm{Span}(\mathbf{v}_1, \ldots, \mathbf{v}_k)$, is a vector subspace of V.*

Proof All we need to do is to verify that the span satisfies the conditions in Def. 6.2. Consider two vectors $\mathbf{u}, \mathbf{w} \in \mathrm{Span}(\mathbf{v}_1, \ldots, \mathbf{v}_k)$ in the span. By definition of the span, this means they can be written as linear combinations

$$\mathbf{u} = \sum_{i=1}^{k} \alpha_i \mathbf{v}_i, \quad \mathbf{w} = \sum_{i=1}^{k} \beta_i \mathbf{v}_i,$$

for suitable scalars $\alpha_i, \beta_i \in \mathbb{F}$. Their sum and the scalar multiple of $\mathbf{u}$ with $\alpha \in \mathbb{F}$ are then given by

$$\mathbf{u} + \mathbf{w} = \sum_{i=1}^{k} (\alpha_i + \beta_i) \mathbf{v}_i, \quad \alpha \mathbf{u} = \sum_{i=1}^{k} (\alpha \alpha_i) \mathbf{v}_i \tag{7.3}$$

and are, hence, both contained in the span. This shows that the conditions (S1) and (S2) of Def. 6.2 are indeed satisfied. $\qquad\square$

This result means that the span provides us with a way of generating vector subspaces. The span has a straightforward geometric interpretation, at least for coordinate vectors with real entries. The span of a single vector $\mathbf{v} \in \mathbb{R}^n$ consists of all scalar multiples of this vector and, hence, can be thought of as the line through $\mathbf{0}$ which contains $\mathbf{v}$. The span of two vectors $\mathbf{u}, \mathbf{v} \in \mathbb{R}^n$ (which are not multiples of each other) represents the plane through $\mathbf{0}$ which contains both vectors. More generally, spans of column vectors are lines, planes and their higher-dimensional analogues through the 'origin' $\mathbf{0}$. We will be more precise about this later but for now just present an example.

Example 7.1 *(The span of vectors)*

(a) Here is an example for the span of a single vector in $\mathbb{R}^2$ (see Fig. 7.1):

$$\mathbf{v} = \begin{pmatrix} 3 \\ 2 \end{pmatrix} \quad \Rightarrow \quad \mathrm{Span}(\mathbf{v}) = \{\alpha \mathbf{v} \,|\, \alpha \in \mathbb{R}\} = \left\{ \begin{pmatrix} 3\alpha \\ 2\alpha \end{pmatrix} \,\middle|\, \alpha \in \mathbb{R} \right\}.$$

(b) For a simple example in $\mathbb{R}^3$ consider the span of the first two standard unit vectors $\mathrm{Span}(\mathbf{e}_1, \mathbf{e}_2) = \{x\mathbf{e}_1 + y\mathbf{e}_2 \,|\, x, y \in \mathbb{R}\}$ which, of course, corresponds to the x–y plane.

(c) For a more complicated example in $\mathbb{R}^3$, define the two vectors $\mathbf{v} = (-1, 2, 1)^T$ and

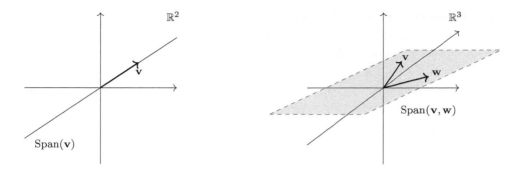

Fig. 7.1 The span of a single vector in $\mathbb{R}^2$ (left) and the span of two vectors in $\mathbb{R}^3$ (right).

$\mathbf{w} = (2, 1, 0)^T$. Their span is given by

$$\text{Span}(\mathbf{v}, \mathbf{w}) = \{\alpha\mathbf{v} + \beta\mathbf{w} \mid \alpha, \beta \in \mathbb{R}\} = \left\{ \begin{pmatrix} -\alpha + 2\beta \\ 2\alpha + \beta \\ \alpha \end{pmatrix} \mid \alpha, \beta \in \mathbb{R} \right\},$$

which describes a plane through $\mathbf{0}$ (see Fig. 7.1). $\qquad\square$

7.1.2 Linear independence

Two different sets of vectors can lead to the same span. Consider vectors $\mathbf{v}_1, \ldots, \mathbf{v}_k$ and a linear combination $\mathbf{w} = \sum_{i=1}^{k} \alpha_i \mathbf{v}_i$. It should be intuitively clear (and will be shown below) that

$$\text{Span}(\mathbf{v}_1, \ldots, \mathbf{v}_k, \mathbf{w}) = \text{Span}(\mathbf{v}_1, \ldots, \mathbf{v}_k) \,, \tag{7.4}$$

so removing $\mathbf{w}$ leaves the span unchanged. How can we decide whether a given set of vectors is minimal, in the sense that no vector can be removed without changing the span? This question leads to the concept of *linear independence* which is central to the subject. Formally, it is defined as follows:

Definition 7.1 *Let V be a vector space over a field $\mathbb{F}$ and $\alpha_i \in \mathbb{F}$ scalars. The vectors $\mathbf{v}_1, \ldots, \mathbf{v}_k \in V$ are called linearly independent if the equation*

$$\sum_{i=1}^{k} \alpha_i \mathbf{v}_i = \mathbf{0} \tag{7.5}$$

implies that all $\alpha_i = 0$. Otherwise, the vectors are called linearly dependent.

Recall from Prop. 6.1 (iii) that $\alpha\mathbf{0} = \mathbf{0}$ for all $\alpha \in \mathbb{F}$. This means any list of vectors which contains the zero vector allows for a non-trivial solution to Eq. (7.5) and is, hence, linearly dependent.

7.1.3 Properties of linearly independent vectors

It may not be immediately obvious how the definition of linear independence relates to our problem of finding a minimal set of vectors for a given span. The connection is made by the following statement.

Proposition 7.2 *The vectors $\mathbf{v}_1, \ldots, \mathbf{v}_k$ are linearly dependent iff one vector $\mathbf{v}_i$ can be written as a linear combination of the others.*

Proof The proof is rather simple but note that there are two directions to show.

'$\Rightarrow$': Assume that the vectors $\mathbf{v}_1, \ldots, \mathbf{v}_k$ are linearly dependent so that the equation $\sum_{i=1}^{k} \alpha_i \mathbf{v}_i = 0$ has a solution with at least one $\alpha_i \neq 0$. Say, $\alpha_1 \neq 0$, for simplicity. Then we can solve for $\mathbf{v}_1$ to get

$$\mathbf{v}_1 = -\frac{1}{\alpha_1} \sum_{i>1} \alpha_i \mathbf{v}_i \,, \tag{7.6}$$

and, hence, we have expressed $\mathbf{v}_1$ as a linear combination of the other vectors.

'$\Leftarrow$': Now assume one vector, say $\mathbf{v}_1$, can be written as a linear combination of the others so that $\mathbf{v}_1 = \sum_{i>1} \beta_i \mathbf{v}_i$. Then it follows that $\sum_{i=1}^{n} \alpha_i \mathbf{v}_i = 0$ with $\alpha_1 = 1 \neq 0$ and $\alpha_i = -\beta_i$ for $i > 1$. Hence, the vectors are linearly dependent. $\square$

So for a linearly dependent set of vectors we can write (at least) one vector as a linear combination of the others. Removing this vector from the list leaves the span unchanged, as in Eq. (7.4). A linearly independent set is one which cannot be further reduced in this way, so is 'minimal' in this sense. The following proposition states this more formally.

Proposition 7.3 *For vectors $\mathbf{v}_1, \ldots, \mathbf{v}_k \in V$ the following statements are equivalent.*
(i) $\mathbf{v}_1, \ldots, \mathbf{v}_k$ are linearly dependent.
(ii) One vector from $\mathbf{v}_1, \ldots, \mathbf{v}_k$ can be removed without changing the span.

Proof '(i) $\Rightarrow$ (ii)': If the vectors $\mathbf{v}_1, \ldots, \mathbf{v}_k$ are linearly dependent then one of the vectors, say $\mathbf{v}_k$, can be written as a linear combination $\mathbf{v}_k = \sum_{i=1}^{k-1} \beta_i \mathbf{v}_i$ of the others. A vector $\mathbf{v} \in \mathrm{Span}(\mathbf{v}_1, \ldots, \mathbf{v}_k)$ can then be written as

$$\mathbf{v} = \sum_{i=1}^{k} \alpha_i \mathbf{v}_i = \sum_{i=1}^{k-1} \alpha_i \mathbf{v}_i + \alpha_k \mathbf{v}_k = \sum_{i=1}^{k-1} (\alpha_i + \alpha_k \beta_i) \mathbf{v}_i \,.$$

This shows that $\mathbf{v} \in \mathrm{Span}(\mathbf{v}_1, \ldots, \mathbf{v}_{k-1})$, so $\mathrm{Span}(\mathbf{v}_1, \ldots, \mathbf{v}_k) \subset \mathrm{Span}(\mathbf{v}_1, \ldots, \mathbf{v}_{k-1})$. The reverse inclusion, $\mathrm{Span}(\mathbf{v}_1, \ldots, \mathbf{v}_{k-1}) \subset \mathrm{Span}(\mathbf{v}_1, \ldots, \mathbf{v}_k)$, holds trivially since it is always possible to set the scalar in front of $\mathbf{v}_k$ to zero. Hence, equality of the two sets and (ii) follows.

'(ii) $\Rightarrow$ (i)': Say that $\mathbf{v}_k$ is the vector which can be removed without changing the span, so that $\mathrm{Span}(\mathbf{v}_1, \ldots, \mathbf{v}_k) = \mathrm{Span}(\mathbf{v}_1, \ldots, \mathbf{v}_{k-1})$. Then $\mathbf{v}_k \in \mathrm{Span}(\mathbf{v}_1, \ldots, \mathbf{v}_{k-1})$ so that $\mathbf{v}_k = \sum_{i=1}^{k-1} \beta_i \mathbf{v}_i$ for some scalars β_i. This means $\mathbf{v}_k$ can be written as a linear combination of the other vectors and it follows from Prop. 7.2 that $\mathbf{v}_1, \ldots, \mathbf{v}_k$ are linearly dependent. $\square$

7.1.4 Examples for linear independence

Let us illustrate the idea of linear independence with a number of examples and exercises.

Example 7.2 *(Linear independence of standard unit vectors)*

Consider the standard unit vectors $\mathbf{e}_1, \ldots, \mathbf{e}_n \in \mathbb{F}^n$. Eq. (7.5) for this case reads

$$\sum_{i=1}^{n} \alpha_i \mathbf{e}_i = \begin{pmatrix} \alpha_1 \\ \vdots \\ \alpha_n \end{pmatrix} \overset{!}{=} \mathbf{0} .$$

This is only solved if all $\alpha_i = 0$ and this means the standard unit vectors are linearly independent. □

Problem 7.1 *(Linear independence in $\mathbb{R}^3$)*

Show that the $\mathbb{R}^3$ vectors $\mathbf{v}_1 = (0, 1, 1)^T$, $\mathbf{v}_2 = (0, 1, 2)^T$ and $\mathbf{v}_3 = (1, 1, -1)^T$ are linearly independent.

Solution: Again, using Eq. (7.5), we have

$$\alpha_1 \mathbf{v}_1 + \alpha_1 \mathbf{v}_2 + \alpha_3 \mathbf{v}_3 = \begin{pmatrix} \alpha_3 \\ \alpha_1 + \alpha_2 + \alpha_3 \\ \alpha_1 + 2\alpha_2 - \alpha_3 \end{pmatrix} \overset{!}{=} \mathbf{0} .$$

The first entry leads to $\alpha_3 = 0$ and combining the other two entries (setting $\alpha_3 = 0$) implies $\alpha_1 = \alpha_2 = 0$. Therefore the three vectors are linearly independent.

Problem 7.2 *(Linear dependence in $\mathbb{R}^3$)*

Show that the three $\mathbb{R}^3$ vectors $\mathbf{v}_1 = (-2, 0, 1)^T$, $\mathbf{v}_2 = (1, 1, 1)^T$ and $\mathbf{v}_3 = (0, 2, 3)^T$ are linearly dependent.

Solution: Forming a general linear combination gives

$$\alpha_1 \mathbf{v}_1 + \alpha_1 \mathbf{v}_2 + \alpha_3 \mathbf{v}_3 = \begin{pmatrix} -2\alpha_1 + \alpha_2 \\ \alpha_2 + 2\alpha_3 \\ \alpha_1 + \alpha_2 + 3\alpha_3 \end{pmatrix} \overset{!}{=} \mathbf{0} .$$

This set of equations clearly has non-trivial solutions, for example $\alpha_1 = 1$, $\alpha_2 = 2$, $\alpha_3 = -1$, so that the vectors are linearly dependent. Alternatively, this could have been inferred from Prop 7.2 by noting that $\mathbf{v}_3 = \mathbf{v}_1 + 2\mathbf{v}_2$.

Example 7.3 *(Linear independence for up to three vectors)*

Let us discuss linear dependence for systems of one, two, and three vectors. To understand linear independence for a single vector $\mathbf{v} \in V$, we have to consider solutions $\alpha \in \mathbb{F}$ to the equation $\alpha \mathbf{v} = \mathbf{0}$. We know from Prop. 6.1 (iii) that $\alpha \mathbf{0} = \mathbf{0}$ for all $\alpha \in \mathbb{F}$. This means that the zero vector is linearly dependent. On the other hand, for

$\mathbf{v} \neq \mathbf{0}$ the equation $\alpha \mathbf{v} = \mathbf{0}$ is only solved by $\alpha = 0$, as Prop. 6.1 (v) asserts. Any non-zero vector is, therefore, linearly independent.

Next, consider two (non-zero) linearly dependent vectors $\mathbf{u}, \mathbf{v}$. From Prop. 7.2 this means that one can be written as a linear combination of the other, for example $\mathbf{u} = \alpha \mathbf{v}$. Hence, two linearly dependent vectors are scalar multiples of each other, that is, they belong to the same line through $\mathbf{0}$.

Analogously, for three linearly dependent vectors $\mathbf{u}, \mathbf{v}, \mathbf{w}$, one can be expressed as a linear combination of the other two, for example, $\mathbf{u} = \alpha \mathbf{v} + \beta \mathbf{w}$. This means that $\mathbf{u} \in \mathrm{Span}(\mathbf{v}, \mathbf{w})$. $\qquad\square$

Problem 7.3 *(Linear independence for polynomials)*

Consider the space $\mathcal{P}(\mathbb{F})$ of polynomials (see Example 6.6) with $\mathbb{F} = \mathbb{R}$ or $\mathbb{F} = \mathbb{C}$. Show that the monomials $1, x, x^2, \ldots, x^k \in \mathcal{P}(\mathbb{F})$ are linearly independent.

Solution: For linear independence we have to show that the equation

$$\sum_{i=0}^{k} \alpha_i x^i = 0 . \tag{7.7}$$

only has the trivial solution $\alpha_i = 0$ for all $i = 1, \ldots, k$. To do this we should recall that the 'zero vector' is the function identical to zero so we are looking for the solutions α_i which solve Eq. (7.7) for all $x \in \mathbb{R}$. This means if Eq. (7.7) is satisfied for certain α_i, then so are derivatives of Eq. (7.7). Taking the ith derivative and then setting $x = 0$ immediately implies that $\alpha_i = 0$. Hence, Eq. (7.7) only has the trivial solution and we conclude that the monomials are linearly independent.

Problem 7.4 *(Solutions to differential equations)*

In Example 6.8 we have explained that the solutions to homogeneous, linear second order differential equations form a vector space. A simple example of such a differential equation is

$$\frac{d^2 g}{dx^2} = -g .$$

Show that the solutions $g(x) = \sin(x)$ and $g(x) = \cos(x)$ to this differential equation are linearly independent.

Solution: Using Eq. (7.5) we should start with $\alpha \sin(x) + \beta \cos(x) = 0$. We are looking for pairs (α, β) which solve this equation for all x. (Recall that the zero vector in a function vector space is the zero function.) Hence, we can constrain the allowed values for α and β by choosing particular x values. Setting $x = 0$ we learn that $\beta = 0$ and setting $x = \pi/2$ it follows that $\alpha = 0$. Hence, sin and cos are linearly independent.

7.2 Basis and dimension

Summary 7.2 *A basis of a vector space is a list of vectors which are linearly independent and which span the entire vector space. Every vector can be written as a unique linear combination of the vectors in a basis. The coefficients in such a linear combination are called the coordinates of the vector relative to the basis. The dimension of a vector space is the number of vectors in a basis. Every finitely spanned vector space has a basis and, hence, a well-defined dimension.*

7.2.1 Basis and coordinates

For a vector space V, it is useful to have a 'minimal' number of vectors which still span the entire space. Such vectors are called a *basis* and this important notion is formally defined as follows:

Definition 7.2 *For a vector space V, a list $(\mathbf{v}_1, \ldots, \mathbf{v}_n)$ of vectors $\mathbf{v}_i \in V$ is called a basis of V if*

(B1) $\mathbf{v}_1, \ldots, \mathbf{v}_n$ are linearly independent
(B2) $V = \mathrm{Span}(\mathbf{v}_1, \ldots, \mathbf{v}_n)$.

It is clear from condition (B2) that every vector in V can be written as a linear combination of the basis vectors but, what is more, for a given vector this linear combination is unique.

Proposition 7.4 *The vectors $(\mathbf{v}_1, \ldots, \mathbf{v}_n)$ form a basis of V if and only if every vector $\mathbf{v} \in V$ can be written as a unique linear combination*

$$\mathbf{v} = \sum_{i=1}^{n} \alpha_i \mathbf{v}_i \,. \tag{7.8}$$

Proof '$\Rightarrow$' Assume that $(\mathbf{v}_1, \ldots, \mathbf{v}_n)$ is a basis of V. From (B2) this means every vector $\mathbf{v}$ can be written as a linear combination of the $\mathbf{v}_i$. It remains to show uniqueness. To do this, we write $\mathbf{v}$ as two linear combinations

$$\mathbf{v} = \sum_{i=1}^{n} \alpha_i \mathbf{v}_i = \sum_{i=1}^{n} \beta_i \mathbf{v}_i \quad \Rightarrow \quad \sum_{i=1}^{n} (\alpha_i - \beta_i)\mathbf{v}_i = 0 \,,$$

with coefficients α_i and β_i and show that the coefficients must be equal. Indeed, taking the difference leads to the equations on the right-hand side and, from linear independence of the basis, it follows that all $\alpha_i - \beta_i = 0$, so that $\alpha_i = \beta_i$ for all $i = 1, \ldots, n$.

'$\Leftarrow$' Now assume every vector $\mathbf{v}$ can be written as a unique linear combination of $\mathbf{v}_1, \ldots, \mathbf{v}_n$. This means that $\mathbf{v}_1, \ldots, \mathbf{v}_n$ span V and (B2) follows. We still need to show condition (B1), that is linear independence. To do this, we note that, just as any other vector, the zero vector can be written as a linear combination $\mathbf{0} = \sum_i \alpha_i \mathbf{v}_i$. This equation is satisfied if all $\alpha_i = 0$ and from uniqueness this must be the only possibility. Hence, the vectors $\mathbf{v}_1, \ldots, \mathbf{v}_n$ are indeed linearly independent. $\square$

The scalars α_i in Eq. (7.8) are called the *coordinates* of $\mathbf{v}$ relative to the basis $(\mathbf{v}_1, \ldots, \mathbf{v}_n)$. They can be organized into a vector $\boldsymbol{\alpha} \in \mathbb{F}^n$ with entries α_i, called the *coordinate vector* for $\mathbf{v}$.

7.2.2 Examples of bases and coordinates

Prop. 7.4 is the mathematical basis for a process routinely used in science and often referred to as 'choosing a coordinate system'. What is really meant by this is choosing a basis of the vector space and representing vectors by their coordinates relative to this basis. Let us illustrate this with a few examples.

Example 7.4 *(Standard unit vectors as a basis)*
The standard unit vectors $\mathbf{e}_1, \ldots, \mathbf{e}_n \in \mathbb{F}^n$ form a basis of $\mathbb{F}^n$. Indeed, from Example 7.2 we already know they are linearly independent. Moreover, every vector $\mathbf{v} \in \mathbb{F}^n$ can be written as a linear combination

$$\mathbf{v} = \begin{pmatrix} v_1 \\ \vdots \\ v_n \end{pmatrix} = \sum_{i=1}^{n} v_i \mathbf{e}_i \,,$$

so that the standard unit vectors span $\mathbb{F}^n$. The coordinates of a vector $\mathbf{v}$ relative to the standard unit vector basis are identical to the components of $\mathbf{v}$. $\square$

Problem 7.5 *(A basis in $\mathbb{R}^3$)*

Show that the $\mathbb{R}^3$ vectors $\mathbf{v}_1 = (0, 1, 1)^T$, $\mathbf{v}_2 = (0, 1, 2)^T$, and $\mathbf{v}_3 = (1, 1, -1)^T$ form a basis of $\mathbb{R}^3$. Find the coordinates of an arbitrary vector $\mathbf{v} = (x, y, z)^T$ relative to this basis.

Solution: In Exercise 7.1 we have shown that $\mathbf{v}_1$, $\mathbf{v}_2$, $\mathbf{v}_3$ are linearly independent. We can attempt to express a general vector $\mathbf{v} = (x, y, z)^T \in \mathbb{R}^3$ as a linear combination of the vectors $\mathbf{v}_i$ by writing

$$\mathbf{v} = \begin{pmatrix} x \\ y \\ z \end{pmatrix} \stackrel{!}{=} \alpha_1 \mathbf{v}_1 + \alpha_2 \mathbf{v}_2 + \alpha_3 \mathbf{v}_3 = \begin{pmatrix} \alpha_3 \\ \alpha_1 + \alpha_2 + \alpha_3 \\ \alpha_1 + 2\alpha_2 - \alpha_3 \end{pmatrix} \,.$$

Equating the entries implies $x = \alpha_3$, $y = \alpha_1 + \alpha_2 + \alpha_3$, $z = \alpha_1 + 2\alpha_2 - \alpha_3$ and solving these equations for α_i leads to

$$\alpha_1 = -3x + 2y - z \,, \quad \alpha_2 = 2x - y + z \,, \quad \alpha_3 = x \,.$$

This result has several implications. Firstly, it shows that every vector $\mathbf{v}$ can indeed be written as a linear combination of the vectors $\mathbf{v}_i$ and, hence, that $(\mathbf{v}_1, \mathbf{v}_2, \mathbf{v}_3)$ forms a basis of $\mathbb{R}^3$. Secondly, we have explicit formulae for how to compute the coordinates α_i relative to the basis $(\mathbf{v}_1, \mathbf{v}_2, \mathbf{v}_3)$ in terms of the components x, y, and z of the vector $\mathbf{v}$. Finally, we see that both $(\mathbf{v}_1, \mathbf{v}_2, \mathbf{v}_3)$ and the standard unit vectors $(\mathbf{e}_1, \mathbf{e}_2, \mathbf{e}_3)$ form a basis of $\mathbb{R}^3$, so the basis of a vector space is by no means unique.

7.2.3 Dimension of a vector space

How should the *dimension* of a vector space be defined? Intuitively, we would say that the vector space $\mathbb{F}^n$ over $\mathbb{F}$ which consists of vector with n components should be assigned dimension n. Note that n is also the number of standard unit vectors which, as we have seen above, form a basis of $\mathbb{F}^n$ over $\mathbb{F}$. This observation suggests that, more generally, the dimension should be defined as the number of vectors in a basis.

However, for a given vector space, there are different choices of bases. Do they all have the same number of vectors? Intuitively, it seems this has to be the case but the formal proof is more difficult than expected. It comes down to the following Lemma:

Lemma 7.1 *(Exchange Lemma) Let* $(\mathbf{v}_1, \ldots, \mathbf{v}_n)$ *be a basis of* V *and* $\mathbf{w}_1, \ldots, \mathbf{w}_m \in V$ *are arbitrary vectors. If* $m > n$ *then* $\mathbf{w}_1, \ldots, \mathbf{w}_m$ *are linearly dependent.*

Proof If the vectors $\mathbf{w}_1, \ldots, \mathbf{w}_n$ are linearly dependent we are done, so assume they are not. In particular, $\mathbf{w}_1 \neq \mathbf{0}$. Since the vectors $\mathbf{v}_i$ form a basis, we can write

$$\mathbf{w}_1 = \sum_{i=1}^{n} \alpha_i \mathbf{v}_i$$

with at least one α_i (say α_1) non-zero (or else $\mathbf{w}_1$ would be zero). We can, therefore, solve this equation for $\mathbf{v}_1$ so that

$$\mathbf{v}_1 = \frac{1}{\alpha_1} \left(\mathbf{w}_1 - \sum_{i=2}^{n} \alpha_i \mathbf{v}_i \right) .$$

This shows that we can 'exchange' $\mathbf{v}_1$ for $\mathbf{w}_1$ such that $V = \text{Span}(\mathbf{w}_1, \mathbf{v}_2, \ldots, \mathbf{v}_n)$. This exchange process can be repeated. Suppose we have already exchanged $k < n$ vectors in this way so that $V = \text{Span}(\mathbf{w}_1, \ldots, \mathbf{w}_k, \mathbf{v}_{k+1}, \ldots, \mathbf{v}_n)$. Then we can write

$$\mathbf{w}_{k+1} = \sum_{i=1}^{k} \alpha_i \mathbf{w}_i + \sum_{i=k+1}^{n} \alpha_i \mathbf{v}_i .$$

If all α_i for $i > k$ are zero in this equation then $\mathbf{w}_1, \ldots, \mathbf{w}_{k+1}$ are linearly dependent and we are finished. Otherwise, we can solve the equation for one of the $\mathbf{v}_i$ with $i > k$, say $\mathbf{v}_{k+1}$, and this justifies the next step $\mathbf{v}_{k+1} \mapsto \mathbf{w}_{k+1}$ of the replacement process. We can continue in this way until all $\mathbf{v}_i$ are replaced by $\mathbf{w}_i$ and $V = \text{Span}(\mathbf{w}_1, \ldots, \mathbf{w}_n)$. Since $m > n$, there is at least one vector, $\mathbf{w}_{n+1}$, 'left over' which can be written as a linear combination:

$$\mathbf{w}_{n+1} = \sum_{i=1}^{n} \beta_i \mathbf{w}_i .$$

This shows that the vectors $\mathbf{w}_1, \ldots, \mathbf{w}_m$ are linearly dependent. $\qquad \square$

Theorem 7.1 *If* $(\mathbf{v}_1, \ldots, \mathbf{v}_n)$ *and* $(\mathbf{w}_1, \ldots, \mathbf{w}_m)$ *are bases of a vector space* V *then* $n = m$.

Proof Consider the basis $(\mathbf{v}_1, \ldots, \mathbf{v}_n)$. Since the vectors $(\mathbf{w}_1, \ldots, \mathbf{w}_m)$ also form a basis they are, in particular, linearly independent. Hence, we can apply the Exchange Lemma which implies that $m \leq n$. Repeating the argument with the roles of the two bases exchanged gives $n \leq m$ and, hence, $n = m$. □

While a vector space usually allows many choices of bases the number of basis vectors is always the same. This facilitates the definition of dimension.

Definition 7.3 *If $(\mathbf{v}_1, \ldots, \mathbf{v}_n)$ is a basis of the vector space V over $\mathbb{F}$ we call $\dim_{\mathbb{F}}(V) := n$ the dimension of V over $\mathbb{F}$. The trivial vector space $\{\mathbf{0}\}$ has an empty basis and is assigned the dimension 0.*

From what we have just seen, it does not matter which basis we use to determine the dimension. Every choice leads to the same result. Let us apply this to compute the dimension for some examples.

Example 7.5 *(Coordinate vector spaces)*

We have already established that the standard unit vectors $(\mathbf{e}_1, \ldots, \mathbf{e}_n)$ form a basis of $\mathbb{F}^n$ seen as a vector space over the field $\mathbb{F}$, so

$$\dim_{\mathbb{F}}(\mathbb{F}^n) = \dim_{\mathbb{R}}(\mathbb{R}^n) = \dim_{\mathbb{C}}(\mathbb{C}^n) = n .$$

However, $\mathbb{C}^n$ seen as a vector space over $\mathbb{R}$ has a basis $(\mathbf{e}_1, \ldots, \mathbf{e}_n, i\mathbf{e}_1, \ldots, i\mathbf{e}_n)$ and, therefore, $\dim_{\mathbb{R}}(\mathbb{C}^n) = 2n$. □

Example 7.6 *(Matrix vector spaces)*

We have seen in Example 6.3 that the space $\mathcal{M}_{n,m}(\mathbb{F})$ of $n \times m$ matrices with entries in $\mathbb{F}$ forms a vector space over $\mathbb{F}$. The standard unit matrices $E_{(ij)}$ defined in Eq. (6.6), where $i = 1, \ldots, n$ and $j = 1, \ldots, m$, clearly form a basis of this vector space. Since there are nm such matrices we have $\dim_{\mathbb{F}}(\mathcal{M}_{n,m}(\mathbb{F})) = nm$. □

Example 7.7 *(Polynomial vector spaces)*

What is the dimension of the vector space $\mathcal{P}_k(\mathbb{F})$ (where $\mathbb{F} = \mathbb{R}$ or $\mathbb{F} = \mathbb{C}$) of polynomial with degree at most k? We have already seen in Exercise 7.3 that the monomials $1, x, x^2, \ldots, x^k$ are linearly independent. Clearly, every polynomial with degree less or equal than k can be written as a linear combination of these monomials so they span the space. This means, $(1, x, x^2, \ldots, x^k)$ is a basis and $\dim_{\mathbb{F}}(\mathcal{P}_k(\mathbb{F})) = k + 1$. □

Example 7.8 *(Dimension of solution space to differential equations)*

Following up from Problem 7.4, we would like to determine the dimension of the solution vector space (of real-valued functions) for the differential equation

$$\frac{d^2 g}{dx^2} = -g .$$

The general solution is given by $g(x) = \alpha \sin(x) + \beta \cos(x)$ with arbitrary real coefficients α and β, so the solution vector space is spanned by sin and cos. We have already seen in Problem 7.4 that sin and cos are linearly independent. Hence, $(\sin, \cos)$ is a basis of the solution space and its dimension equals 2. □

7.2.4 Existence of a basis

So far, we have discussed the properties and implications of a finite basis but we have not worried about its existence.

Theorem 7.2 *Let V be a vector space spanned by vectors $\mathbf{v}_1, \ldots, \mathbf{v}_m$.*
(i) V has a basis and, hence, a well-defined dimension.
(ii)   Any linearly independent vectors $\mathbf{w}_1, \ldots, \mathbf{w}_k \in V$ can be completed to a basis.
(iii) If $n = \dim_{\mathbb{F}}(V)$, linearly independent vectors $\mathbf{w}_1, \ldots, \mathbf{w}_n \in V$ form a basis.

Proof (i) By assumption, V is spanned by the vectors $\mathbf{v}_1, \ldots, \mathbf{v}_m$. If these vectors are linearly independent, we have found a basis. If not, we know from Prop. 7.3 that one of the vectors, say $\mathbf{v}_m$, can be removed without changing the span, so that $V = \mathrm{Span}(\mathbf{v}_1, \ldots, \mathbf{v}_{m-1})$. This process can be continued until the remaining set of vectors is linearly independent and, hence, forms a basis.

(ii) If the linearly independent vectors $\mathbf{w}_1, \ldots, \mathbf{w}_k$ already span V we are finished. If not there exists a vector $\mathbf{w}_{k+1} \notin \mathrm{Span}(\mathbf{w}_1, \ldots, \mathbf{w}_k)$ and the vectors $\mathbf{w}_1, \ldots, \mathbf{w}_k, \mathbf{w}_{k+1}$ must be linearly independent (see Exercise 7.9). We can continue this process of adding new vectors for as long as the span does not equal V. It terminates when we have collected $n = \dim_{\mathbb{F}}(V)$ vectors as finding $n + 1$ linearly independent vectors would contradict the Exchange Lemma 7.1.

(iii) If $\dim_{\mathbb{F}}(V) = n$ and the linearly independent set $\mathbf{w}_1, \ldots, \mathbf{w}_n$ did not span V then, for the same reason as in the proof of (ii), we could find a vector $\mathbf{w}_{n+1} \notin \mathrm{Span}(\mathbf{w}_1, \ldots, \mathbf{w}_n)$ so that $\mathbf{w}_1, \ldots, \mathbf{w}_n, \mathbf{w}_{n+1}$ are linearly independent. However, this contradicts the exchange lemma. Hence, the vectors $\mathbf{w}_1, \ldots, \mathbf{w}_n$ must span the space and they form a basis. □

The main conclusion from this theorem is that every vector space which is spanned by a finite number of vectors has a basis and, hence, a well-defined dimension. Such vector spaces are also called *finite-dimensional*. All other vector spaces, which cannot be spanned by a finite number of vectors, are called *infinite dimensional*.

In this book, we will primarily be concerned with finite-dimensional vector spaces, although we present the occasional example which involves an infinite-dimensional space. For instance, the space of all polynomials is infinite dimensional. Indeed, any finite list of polynomials has a maximal degree and any polynomial with a degree larger than this maximum cannot be in the span. Likewise, the spaces $\mathcal{F}([a, b], \mathbb{F})$ of $\mathbb{F}$-valued functions on the interval $[a, b]$ (as well as its sub-spaces of continuous and differentiable functions) are infinite-dimensional. The systematic discussion of such infinite dimensional spaces leads into another area of mathematics, called *functional analysis*, which is beyond the scope of this text (see, for example, Rynne and Youngson 2008).

7.2.5 Properties of finite-dimensional vector spaces

What can we say about vector subspaces of finite-dimensional vector spaces? Intuitively, it seems their dimension should be bounded by the dimension of the ambient vector space, so let us proof this.

Corollary 7.1 *A vector subspace* $W \subset V$ *of a finite-dimensional vector space is finite-dimensional and* $\dim_{\mathbb{F}}(W) \leq \dim_{\mathbb{F}}(V)$. *Equality,* $W = V$, *holds iff* $\dim_{\mathbb{F}}(W) = \dim_{\mathbb{F}}(V)$.

Proof Set $n = \dim_{\mathbb{F}}(V)$. The subspace W cannot contain more than n linearly independent vectors or else there would be a contradiction with the Exchange Lemma. This shows that W is finite-dimensional and that $\dim_{\mathbb{F}}(W) \leq \dim_{\mathbb{F}}(V)$.

For the second part of the statement, clearly if $W = V$ then $\dim_{\mathbb{F}}(W) = \dim_{\mathbb{F}}(V)$. Conversely, if $\dim_{\mathbb{F}}(W) = \dim_{\mathbb{F}}(V)$, then a basis $(\mathbf{w}_1, \ldots, \mathbf{w}_k)$ of W must, from Theorem 7.2 (iii), also be a basis of V. Hence, $W = \mathrm{Span}(\mathbf{w}_1, \ldots, \mathbf{w}_k) = V$. □

This result combined with Theorem 7.2 means that every vector subspace of a finite-dimensional vector space has a basis and a dimension. We have shown earlier that every span is a vector subspace. Now we see that the opposite is also true. Every vector subspace can be written as a span, for example, as the span of its basis.

Earlier, we have mentioned the intuitive interpretation of spans as lines, planes etc. through $\mathbf{0}$. Now, we can introduce a more precise terminology which captures this intuition. We call a k-dimensional vector subspace $W \subset V$ a k-plane through $\mathbf{0}$, or k-plane for short. A 0-plane is simple the trivial vector space $\{\mathbf{0}\}$, a 1-plane is also called a *line*, a 2-plane is called a *plane* and an $(n-1)$-plane in an n-dimensional vector space V is also called a *hyperplane*. An n-dimensional vector space V contains k-planes through $\mathbf{0}$ for every $k = 0, 1, \ldots, n$. To see this, start with a basis $(\mathbf{v}_1, \ldots, \mathbf{v}_n)$ of V and note that $\mathrm{Span}(\mathbf{v}_1, \ldots, \mathbf{v}_k)$ is a k-plane through $\mathbf{0}$.

Application 7.1 *Vector spaces and magic squares*

An entertaining application of vector spaces is to magic squares. Magic squares are 3×3 (say) quadratic arrays of (rational) numbers such that all rows, all columns and both diagonals sum up to the same total. To make contact with our discussion of vector spaces, we can think of magic squares as matrices in the vector space $\mathcal{M}_{3,3}(\mathbb{Q})$ of 3×3 matrices with rational entries (seen as a vector space over the field $\mathbb{Q}$). A simple example of a magic square is

$$M = \begin{pmatrix} 4 & 9 & 2 \\ 3 & 5 & 7 \\ 8 & 1 & 6 \end{pmatrix}, \tag{7.9}$$

where every row, column, and diagonal sums up to 15. Magic squares have long held a certain fascination and an obvious problem is to find all magic squares.

In our context, the important observation is that magic squares form a vector subspace of $\mathcal{M}_{3,3}(\mathbb{Q})$. Let us agree that we add and scalar multiply magic squares in the same way as matrices (see Example 6.3), that is, entry by entry. Then, clearly, the sum of two magic squares is again a magic square, as is the scalar multiple of a magic square. Hence, from Def. 6.2, the 3×3 magic squares form a vector subspace of $\mathcal{M}_{3,3}(\mathbb{Q})$. The problem of finding all magic squares can now be phrased in the language of vector spaces. What is the dimension of the vector (sub)space of magic squares and can we write down a basis for this space?

It is relative easy to find the following three elementary examples of magic squares:

$$M_1 = \begin{pmatrix} 1 & 1 & 1 \\ 1 & 1 & 1 \\ 1 & 1 & 1 \end{pmatrix} , \quad M_2 = \begin{pmatrix} 0 & 1 & -1 \\ -1 & 0 & 1 \\ 1 & -1 & 0 \end{pmatrix} , \quad M_3 = \begin{pmatrix} -1 & 1 & 0 \\ 1 & 0 & -1 \\ 0 & -1 & 1 \end{pmatrix} . \quad (7.10)$$

It is also easy to show that these three matrices are linearly independent, using Eq. (7.5). Setting a general linear combination to zero,

$$\alpha_1 M_1 + \alpha_2 M_2 + \alpha_3 + M_3 = \begin{pmatrix} \alpha_1 - \alpha_3 & \alpha_1 + \alpha_2 + \alpha_3 & \alpha_1 - \alpha_2 \\ \alpha_1 - \alpha_2 + \alpha_3 & \alpha_1 & \alpha_1 + \alpha_2 - \alpha_3 \\ \alpha_1 + \alpha_2 & \alpha_1 - \alpha_2 - \alpha_3 & \alpha_1 + \alpha_3 \end{pmatrix} \overset{!}{=} 0 ,$$

immediately leads to $\alpha_1 = \alpha_2 = \alpha_3 = 0$. Hence, M_1, M_2, M_3 are linearly independent and

$$\mathrm{Span}(M_1, M_2, M_3) \subset \mathcal{M}_{3,3}(\mathbb{Q}) \quad (7.11)$$

is a three-dimensional vector space of magic squares. Therefore, the dimension of the magic square space is at least three. Indeed, our example (7.9) is contained in $\mathrm{Span}(M_1, M_2, M_3)$ since $M = 5M_1 + 3M_2 + M_3$. As we will see later (see Application 16.2), this is not an accident. We will show that the dimension of the magic square space equals three and, hence, that (M_1, M_2, M_3) is a basis.

Exercises

(†=challenging)

7.1 *Span of a subset*
Let $S \subset V$ be an arbitrary subset of a vector space V and define $\mathrm{Span}(S)$ as the set of all finite linear combinations of vectors in S. Show that
(a) $\mathrm{Span}(S)$ is a vector subspace.
(b) If $U \subset S$ is a vector subspace, then $\dim_{\mathbb{F}}(U) \leq \dim_{\mathbb{F}}(\mathrm{Span}(S))$.
(c) $S = \mathrm{Span}(S)$ if and only if S is a vector subspace.

7.2 *Linear dependence and independence*
Which of the following sets of vectors are linearly independent? For each linearly dependent set, identify a maximal subset of linearly independent vectors. Provide detailed reasoning in each case.
(a) The $\mathbb{R}^3$ standard unit vectors $\mathbf{e}_1, \mathbf{e}_2, \mathbf{e}_3$.

(b) The $\mathbb{R}^3$ vectors $\mathbf{v}_1 = (0, 1, 1)^T$, $\mathbf{v}_2 = (1, 1, 1)^T$ and $\mathbf{v}_3 = (0, 0, 1)^T$.
(c) The $\mathbb{R}^3$ vectors $\mathbf{v}_1 = (1, 0, 1)^T$, $\mathbf{v}_2 = (2, 3, 1)^T$ and $\mathbf{v}_3 = (1, 6, -1)$.
(d) The $\mathbb{R}^4$ vectors $\mathbf{v}_1 = (1, 2, 0, -3)^T$, $\mathbf{v}_2 = (2, 1, 1, -4)^T$ and $\mathbf{v}_3 = (-3, 6, -4, 1)^T$.

7.3 *Linear independence of functions*
(a) Show that the functions $\sin(x)$, $\sin(2x)$, and $\sin(3x)$ are linearly independent.
(b) Are the functions $\sin(x)$, $\sin(2x)$ and $\sin(x)\cos(x)$ linearly independent?

7.4 *Basis for polynomial vector spaces*
Consider the vector space $V = \mathcal{P}_3(\mathbb{F})$ of at most cubic polynomials in x.
(a) Show that the monomials $(1, x, x^2, x^3)$ form a basis of V.

(b) Show that $(1, x, (3x^2 - 1)/2, (5x^3 - 3x)/2)$ is another basis of V.

(c) Find the coordinates of a general cubic $p(x) = a_3 x^3 + a_2 x^2 + a_1 x + a_0$ relative the bases in (a) and (b).

7.5 *Basis and coordinates*
Show that the vectors $v_1 = (1, -1, 0)^T$, $v_2 = (0, 1, -1)^T$ and $v_3 = (2, 0, 1)^T$ form a basis of $\mathbb{R}^3$. Write a general vector $v = (x, y, z)^T \in \mathbb{R}^3$ as a linear combination of this basis. What are the coordinates of v relative to the basis (v_1, v_2, v_3)?

7.6 *Basis for matrix vector spaces*
Consider the vector space $\mathcal{M}_{2,2}(\mathbb{F})$ of 2×2 matrices with entries in $\mathbb{F}$.

(a) Show explicitly that the standard unit matrices $E_{(ij)}$, where $i, j = 1, 2$, form a basis of $\mathcal{M}_{2,2}(\mathbb{F})$ and, hence, that its dimension is four.

(b) The symmetric 2×2 matrices form a vector subspace of $\mathcal{M}_{2,2}(\mathbb{F})$ (see Exercise 6.7). Show that $E_{(11)}$, $E_{(22)}$, and $E_{(12)} + E_{(21)}$ form a basis of this vector subspace and, hence, that its dimension is three.

(c) Carry out a similar analysis for the vector subspace of anti-symmetric 2×2 matrices.

7.7 *Solutions to differential equation*[†]
Consider the differential equation

$$x^2 \frac{d^2 y}{dx^2} - 2x = 0$$

for real-valued functions $y \in \mathcal{C}^2((0, \infty))$.

(a) Why does the set of solutions form a vector subspace?

(b) Find a basis for this solution space, assuming that its dimension is two. (Hint: Try functions of the form $y = x^p$, for $p \in \mathbb{R}$.)

7.8 2×2 *semi-magic squares*[†]
Consider 2×2 semi-magic squares, that is, 2×2 matrices in $\mathcal{M}_{2,2}(\mathbb{Q})$ whose rows and columns sum up to the same total.

(a) Show that the 2×2 semi-magic squares form a vector subspace of $\mathcal{M}_{2,2}(\mathbb{Q})$.

(b) Show that $E_{(1,1)} + E_{(2,2)}$ and $E_{(1,2)} + E_{(2,1)}$ are semi-magic squares.

(c) Show that the matrices from (b) form a basis of the 2×2 semi-magic squares.

7.9 *Linear independence*[†]
Let $v_1, \ldots, v_k \in V$ be linearly independent vectors and $v \notin \mathrm{Span}(v_1, \ldots, v_k)$.

(a) Show that any subset of $\{v_1, \ldots, v_k\}$ is also linearly independent.

(b) Show that the vectors $v_1, \ldots, v_k, v$ are linearly independent.

(c) Show that the vectors $v_1 + v, \ldots, v_k + v$ are linearly independent. Is this statement still true if we drop the condition $v \notin \mathrm{Span}(v_1, \ldots, v_k)$?

8

Vector subspaces

In the previous chapter, we have seen that vector subspaces of an n-dimensional vector space V are also finite-dimensional vector spaces with dimension $k \leq n$. To capture the geometrical intuition, we have called such k-dimensional vector subspace k-planes through $\mathbf{0}$, with 1-planes also referred to as lines and 2-planes as planes.

It is natural to ask how vector subspaces relate to basic operations and structures on sets such as set unions, set intersections or equivalence relations. Are these consistent with the vector space structure and can they be used to create new vector subspaces from given ones? If so what happens to the dimension? The results of this chapter provide geometrical insight but will also be useful for the systematic development of the subject, in particular for the understanding of linear maps.

8.1 Intersection and sum

Summary 8.1 *For two vector subspaces U, W of a vector space V, the intersection $U \cap W$ and the sum $U + W$ are both vector subspaces. A simple formula relates the dimensions of these spaces. The two subspaces form a direct sum, $U \oplus W$, if they intersect trivially. In this case, the dimension of $U \oplus W$ is simply the sum of the dimensions of U and W.*

8.1.1 Intersection of vector subspaces

We are working with an n-dimensional vector space V over $\mathbb{F}$. An obvious question is: What happens to vector subspaces under simple set-theoretical operations? Start with two vector subspaces $U, W \subset V$ and consider their intersection $U \cap W$. By verifying the conditions in Def. 6.2, it is quite easy to show that the intersection is also a vector subspace. First, $\mathbf{0} \in U \cap W$, so the intersection is not empty. Consider two vectors $\mathbf{v}_1, \mathbf{v}_2 \in U \cap W$ in the intersection. This means both vectors must be in U and in W and since either is a vector subspace, we conclude that $\mathbf{v}_1 + \mathbf{v}_2 \in U$ and $\mathbf{v}_1 + \mathbf{v}_2 \in W$. But this means that $\mathbf{v}_1 + \mathbf{v}_2 \in U \cap W$. A similar argument shows that $U \cap W$ is closed under scalar multiplication.

8.1.2 Union and sum

Things are not so straightforward for the union $U \cup W$ of two vector subspaces. By thinking about simple examples, it should be immediately clear that the union is usually not a vector subspace. For example, consider the two subspaces $U = \mathrm{Span}(\mathbf{e}_1)$

and $W = \text{Span}(\mathbf{e}_2)$ of $\mathbb{R}^2$, that is, the two coordinate axes. Their union is merely both coordinate axis while linear combinations of $\mathbf{e}_1$ and $\mathbf{e}_2$ lead to every vector in $\mathbb{R}^2$. This example already points to a possible fix. Instead of the set-theoretical union, we should form the sum

$$U + W := \{\mathbf{u} + \mathbf{w} \mid \mathbf{u} \in U, \ \mathbf{w} \in W\}, \tag{8.1}$$

which consists of all vectors $\mathbf{u} + \mathbf{w}$, where $\mathbf{u} \in U$ and $\mathbf{w} \in W$. It is easy to see that this is a vector subspace. Consider two vectors $\mathbf{v}_1, \mathbf{v}_2 \in U + W$. By definition of $U + W$ they can be written as $\mathbf{v}_1 = \mathbf{u}_1 + \mathbf{w}_1$ and $\mathbf{v}_2 = \mathbf{u}_2 + \mathbf{w}_2$, where $\mathbf{u}_1, \mathbf{u}_2 \in U$ and $\mathbf{w}_1, \mathbf{w}_2 \in W$. Hence, $\mathbf{v}_1 + \mathbf{v}_2 = (\mathbf{u}_1 + \mathbf{u}_2) + (\mathbf{w}_1 + \mathbf{w}_2) \in U + W$ which shows that $U + W$ is closed under vector addition. Similarly, it follows that $U + W$ is closed under scalar multiplication. We summarize these results in the following Lemma:

Lemma 8.1 *For two vector subspaces $W, U \subset V$ of a vector space V, both the intersection $W \cap U$ and the sum $W + U$ are vector subspaces of V.*

Proof This follows from the arguments above. □

8.1.3 Dimension of vector space sums

What is the dimension of the sum $U + W$ of two vector subspaces W and U? The naive guess is that dimensions simply add up but this ignores a possible non-trivial intersection $U \cap W$. The correct dimension formula is stated in the following theorem.

Theorem 8.1 *For two vector subspaces U, W of a finite-dimensional vector space V over $\mathbb{F}$ we have*

$$\dim_{\mathbb{F}}(U + W) = \dim_{\mathbb{F}}(U) + \dim_{\mathbb{F}}(W) - \dim_{\mathbb{F}}(U \cap W). \tag{8.2}$$

Proof Set $p = \dim_{\mathbb{F}}(U \cap W)$, $n = \dim_{\mathbb{F}}(U)$, and $m = \dim_{\mathbb{F}}(W)$ and start with a basis $B_{U \cap W} = (\mathbf{v}_1, \dots, \mathbf{v}_p)$ of $U \cap W$. This basis can be completed to a basis $B_U = (\mathbf{v}_1, \dots, \mathbf{v}_p, \mathbf{u}_{p+1}, \dots, \mathbf{u}_n)$ of U and a to basis $B_W = (\mathbf{v}_1, \dots, \mathbf{v}_p, \mathbf{w}_{p+1}, \dots, \mathbf{w}_m)$ of W. The expectation is that $B = (\mathbf{v}_1, \dots, \mathbf{v}_p, \mathbf{u}_{p+1}, \dots, \mathbf{u}_n, \mathbf{w}_{p+1}, \dots, \mathbf{w}_m)$ is then a basis of $U + W$. This can be shown as follows. The set B contains as subsets the bases B_U and B_W, so clearly B spans $U + W$. To show linear independence of B we start with

$$\sum_{i=1}^{p} \alpha_i \mathbf{v}_i + \sum_{j=p+1}^{n} \beta_j \mathbf{u}_j + \sum_{k=p+1}^{m} \gamma_k \mathbf{w}_k \overset{!}{=} 0 \tag{8.3}$$

and define the vector $\mathbf{v}$ as the first two terms in the sum on the left-hand side, so

$$\mathbf{v} := \sum_{i=1}^{p} \alpha_i \mathbf{v}_i + \sum_{j=p+1}^{n} \beta_j \mathbf{u}_j \quad \Longrightarrow \quad \mathbf{v} = -\sum_{k=p+1}^{m} \gamma_k \mathbf{w}_k. \tag{8.4}$$

The definition of $\mathbf{v}$ means that $\mathbf{v} \in U$ and the second relation above that $\mathbf{v} \in W$, so $\mathbf{v} \in U \cap W$. This means it can be written as some linear combination

$$\mathbf{v} = \sum_{i=1}^{p} \alpha_i' \mathbf{v}_i$$

of vector in $B_{U \cap W}$. Comparing this with the first Eq. (8.4), it follows from the unique-ness of linear combinations relative to the basis B_U that all $\beta_j = 0$ (and that all $\alpha_i = \alpha_i'$). Inserting $\beta_j = 0$ into Eq. (8.3) linear independence of the basis B_W then implies that all $\alpha_i = \gamma_k = 0$. Hence, B is linearly independent and indeed a basis. The dimension of $U + W$ equals the number of basis elements in B. Counting these gives

$$\dim_{\mathbb{F}}(U + W) = |B| = p + (n - p) + (m - p)$$
$$= n + m - p = \dim_{\mathbb{F}}(U) + \dim_{\mathbb{F}}(W) - \dim_{\mathbb{F}}(U \cap W) \, .$$

$\square$

The dimension formula (8.2) has a simple interpretation. When summing up $\dim_{\mathbb{F}}(U)$ and $\dim_{\mathbb{F}}(W)$, the intersection $U \cap W$ is counted twice so its dimension has to be subtracted once for the correct overall dimension of $U + W$.

Example 8.1 *(Sum of vector subspaces)*

In $\mathbb{R}^3$, consider two-dimensional vector subspaces U and W which intersect in a line $U \cap W$.

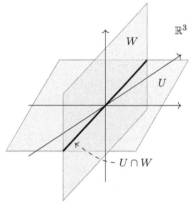

Clearly, $U + W = \mathbb{R}^3$, so that $\dim_{\mathbb{R}}(U + W) = 3$. This is matched by $\dim_{\mathbb{R}}(U) + \dim_{\mathbb{R}}(W) - \dim_{\mathbb{R}}(U \cap W) = 2 + 2 - 1 = 3$, in accordance with Eq. (8.2). $\square$

From Eq. (8.2) the dimension of the sum is always bounded from above by the sum of the dimensions,

$$\dim_{\mathbb{F}}(U + W) \leq \dim_{\mathbb{F}}(U) + \dim_{\mathbb{F}}(W) \, , \tag{8.5}$$

with equality if and only if the intersection $U \cap W$ is trivial. The dimension of $U \cap W$ can be constrained in two ways, using Cor. 7.1. First, $U \cap W$ is a vector subspace of both U and W, so its dimension must be less equal than the dimensions of U and W. Secondly, since $U + W \subset V$ it follows that $\dim_{\mathbb{F}}(U + W) \leq \dim_{\mathbb{F}}(V)$. Combining these statements with the dimension formula (8.2) leads to

$$\min(\dim_{\mathbb{F}}(U), \dim_{\mathbb{F}}(W)) \geq \dim_{\mathbb{F}}(U \cap W) \geq \dim_{\mathbb{F}}(U) + \dim_{\mathbb{F}}(W) - \dim_{\mathbb{F}}(V) \, . \tag{8.6}$$

This result implies that two vector subspaces of sufficiently large dimension, relative to the dimension of the total space, must intersect non-trivially. For example, two planes in a three-dimensional space must at least intersect in a line and two three-dimensional vector subspaces in a four-dimensional vector space must at least intersect in a plane.

8.1.4 Direct sums

If two vector subspaces $U, W \subset V$ intersect trivially, that is, if $U \cap W = \{0\}$, then the sum $U + W$ is called a *direct sum* and is written as $U \oplus W$. A direct sum has considerably nicer properties than merely a sum of two vector subspaces. For one, dimensions simply add up,

$$\dim_{\mathbb{F}}(U \oplus W) = \dim_{\mathbb{F}}(U) + \dim_{\mathbb{F}}(W) , \tag{8.7}$$

as follows immediately from Eq. (8.2). We also have the following proposition:

Proposition 8.1 *For two vector subspace $U, W \subset V$ the following statements are equivalent:*

(i) The sum $U + W$ is direct.
(ii) Every $\mathbf{v} \in U + W$ can be written uniquely as $\mathbf{v} = \mathbf{u} + \mathbf{w}$, where $\mathbf{u} \in U$ and $\mathbf{w} \in W$.

Proof (i) $\Rightarrow$ (ii): Assume that the sum $U + W$ is direct, so $U \cap W = \{0\}$. It is clear that every vector $\mathbf{v} \in U \oplus W$ can be written as in (ii) but we have to show uniqueness. Start with two decompositions $\mathbf{v} = \mathbf{u}_1 + \mathbf{w}_1 = \mathbf{u}_2 + \mathbf{w}_2$ where $\mathbf{u}_1, \mathbf{u}_2 \in U$ and $\mathbf{w}_1, \mathbf{w}_2 \in W$. It follows that $\mathbf{u}_1 - \mathbf{u}_2 = \mathbf{w}_2 - \mathbf{w}_1$ and the left-hand side of this equation is an element of U while the right-hand side is an element of W. This means that $\mathbf{u}_1 - \mathbf{u}_2, \mathbf{w}_2 - \mathbf{w}_1 \in U \cap W$ but since $U \cap W = \{0\}$ it follows that $\mathbf{u}_1 = \mathbf{u}_2$ and $\mathbf{w}_1 = \mathbf{w}_2$.

(ii) $\Rightarrow$ (i): Now assume that (ii) holds and consider a vector $\mathbf{v} \in U \cap W$. The zero vector $\mathbf{0} \in U + W$ can then be written as a sum of vectors in U and W in two ways, namely $\mathbf{0} = \mathbf{0} + \mathbf{0}$ and $\mathbf{0} = \mathbf{v} + (-\mathbf{v})$. This is only consistent with uniqueness if $\mathbf{v} = 0$. This shows that $U \cap W = \{0\}$, so that the sum is direct. □

It is now easy to argue that a direct sum $U \oplus W$ has an 'adapted' basis obtained by merging the vectors from the bases of U and W.

Corollary 8.1 *Let $U, W \subset V$ be two subspaces which form a direct sum, with bases $(\mathbf{u}_1, \ldots, \mathbf{u}_m)$ and $(\mathbf{w}_1, \ldots, \mathbf{w}_k)$, respectively, Then, $(\mathbf{u}_1, \ldots, \mathbf{u}_m, \mathbf{w}_1, \ldots, \mathbf{w}_k)$ is a basis of $U \oplus W$.*

Proof Any vector $\mathbf{v} \in U \oplus W$ can be written as $\mathbf{v} = \mathbf{u} + \mathbf{w}$, where $\mathbf{u} \in U$ and $\mathbf{w} \in W$ are unique. Further, $\mathbf{u}$ and $\mathbf{w}$ each have a unique expansion in terms of the bases on U and W. Combining these two steps, we see that every $\mathbf{v} \in U \oplus W$ can be written as a unique linear combination of $(\mathbf{u}_1, \ldots, \mathbf{u}_m, \mathbf{w}_1, \ldots, \mathbf{w}_k)$. From Cor. 7.4 this means that $(\mathbf{u}_1, \ldots, \mathbf{u}_m, \mathbf{w}_1, \ldots, \mathbf{w}_k)$ is a basis of $U \oplus W$. □

Direct sums are a very useful tool for linear algebra constructions and proofs, as they can be used to break up the vector space into smaller, often more manageable subspaces. For example, the diagonalization of linear maps and the Jordan normal form are based on direct sum decompositions, as we will see in Part VI.

8.1.5 Direct sums of vector spaces

We have studied (direct) sums $U \oplus W$ for vector spaces U and W which are both vector subspaces of an 'ambient' vector space V. But can we make sense of the sum $U \oplus W$ if U and W are not, a priori, contained in some larger vector space but are merely two abstract vector spaces over the same field $\mathbb{F}$?

In this case, we can proceed by constructing an ambient vector space which contains U and W. To do this, we observe that the Cartesian product $U \times W$ can be made into a vector space by defining vector addition and scalar multiplication as

$$(\mathbf{u}, \mathbf{w}) + (\tilde{\mathbf{u}}, \tilde{\mathbf{w}}) := (\mathbf{u} + \tilde{\mathbf{u}}, \mathbf{w} + \tilde{\mathbf{w}}) , \qquad \alpha(\mathbf{u}, \mathbf{w}) := (\alpha\mathbf{u}, \alpha\mathbf{w}) , \qquad (8.8)$$

where $(\mathbf{u}, \mathbf{w}), (\tilde{\mathbf{u}}, \tilde{\mathbf{w}}) \in U \times W$, and $\alpha \in \mathbb{F}$. This vector space has two obvious vector subspaces, $\hat{U} = \{(\mathbf{u}, \mathbf{0}) \,|\, \mathbf{u} \in U\}$ and $\hat{W} = \{(\mathbf{0}, \mathbf{w}) \,|\, \mathbf{w} \in W\}$, which can be identified with U and W, respectively. It is also clear that $U \times W = \hat{U} + \hat{W}$ and that $\hat{U} \cap \hat{W} = \{(\mathbf{0}, \mathbf{0})\}$, so we have, in fact, a direct sum $U \times W = \hat{U} \oplus \hat{W}$. Given the identifications $U \cong \hat{U}$ and $W \cong \hat{W}$ this direct sum is, by slight abuse of notation, also written as $U \oplus W$. Everything we have said about direct sums of vector subspaces in Section 8.1.4 can now be applied to this construction. In particular, the dimension formula (8.7) remains valid and we can construct a basis of $U \oplus W$ by combining bases for U and W, as stated in Cor. 8.1.

8.2 Quotient spaces*

Summary 8.2 *A vector subspace $W \subset V$ can be used to define an equivalence relation on V. The associated equivalence classes are called cosets or affine k-planes. The quotient V/W forms a vector space with dimension $\dim_{\mathbb{F}}(V) - \dim_{\mathbb{F}}(W)$.*

Quotient vector spaces are a very useful way of 'course-graining' a vector space by dividing out a vector subspace. The elements of the quotient vector space V/W are equivalence classes of vectors under an equivalence relation which declares two vectors as related if their difference is contained in the vector subspace W. The construction leads to an elegant proof of the isomorphism and rank theorems, as will be discussed in Section 14.3. However, it may seem somewhat abstract to the beginner and can be omitted at first reading.

8.2.1 Equivalence relation and cosets

Consider a vector space V and a vector subspace $W \subset V$ with dimension $k = \dim_{\mathbb{F}}(W)$. We say that two vectors in V are related if their difference is a vector in W, so

$$\mathbf{v}_1 \sim \mathbf{v}_2 \quad :\Leftrightarrow \quad \mathbf{v}_1 - \mathbf{v}_2 \in W . \qquad (8.9)$$

It is not hard to show that this defines an equivalence relation (Exercise 8.3). The associated equivalence classes are called *cosets* or *affine k-planes*, and their explicit form is $\mathbf{v} + W := \{\mathbf{v} + \mathbf{w} \,|\, \mathbf{w} \in W\}$, where $\mathbf{v} \in V$. Cosets are typically not vector subspaces as, for example, they do not need to contain the zero vector. They are obtained by 'shifting' the subspace W by vectors $\mathbf{v}$, as indicated in Fig. 8.1. The set

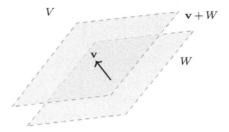

Fig. 8.1 The relationship between k-planes W and affine k-planes $\mathbf{v} + W$.

of all cosets

$$V/W := \{\mathbf{v} + W \,|\, \mathbf{v} \in V\}, \tag{8.10}$$

is called the *quotient V/W of V by W*.

Example 8.2 *(Vector space quotient in $\mathbb{R}^2$)*

In $\mathbb{R}^2$, consider a non-zero vector $\mathbf{w}$ and the one-dimensional vector subspace $W :=$ Span$(\mathbf{w})$. The equivalence classes under the relation (8.9) are then the lines parallel to W and the quotient $\mathbb{R}^2/W$ consists of all these lines. In the figure below, we have indicated some of these lines, given by $k\mathbf{v} + W$, where $\mathbf{v}$ is a fixed vector and $k = -2, -1, 0, 1, 2$.

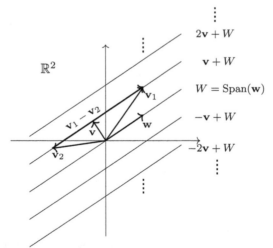

Note that the two vectors $\mathbf{v}_1$ and $\mathbf{v}_2$ indicated in the figure are related since $\mathbf{v}_1 - \mathbf{v}_2 \in W$ and are, hence, both contained in the same equivalence class, in this case the line $\mathbf{v} + W$. $\qquad\square$

8.2.2   Quotient vector space

Can the quotient space V/W be given the structure of a vector space and, if so, how should we define addition and scalar multiplication? It seems intuitive that addition of two affine k-planes should involve addition of all vectors of the first plane to all vectors of the second one. Likewise, for scalar multiplication all vectors in an affine k-plane should be multiplied by the scalar. In other words, addition and scalar multiplication on V/W should be defined as

$$(\mathbf{v}_1 + W) + (\mathbf{v}_2 + W) := (\mathbf{v}_1 + \mathbf{v}_2) + W \ , \quad \alpha(\mathbf{v} + W) := (\alpha\mathbf{v}) + W \ , \tag{8.11}$$

where $\alpha \in \mathbb{F}$ and $\mathbf{v}_1, \mathbf{v}_2, \mathbf{v} \in V$. These definitions simply rely on the vector space structure on V and, therefore, trivially satisfy all the vector space axioms in Def. 6.1. The zero vector in V/W is simply the vector subspace W.

Theorem 8.2 *Let V be a vector space over $\mathbb{F}$ and $W \subset V$ a vector subspace. Then the quotient space V/W is also a vector space over $\mathbb{F}$ with vector addition and scalar multiplication as defined in Eq. (8.11). Its dimension is*

$$\dim_{\mathbb{F}}(V/W) = \dim_{\mathbb{F}}(V) - \dim_{\mathbb{F}}(W) \ . \tag{8.12}$$

Proof   What remains to be done is to proof the dimension formula. We set $n := \dim_{\mathbb{F}}(V)$ and $k := \dim_{\mathbb{F}}(W)$ and start by introducing a basis $B_W := (\mathbf{w}_1, \dots, \mathbf{w}_k)$ of W. This basis can be completed to a basis $B_V := (\mathbf{w}_1, \dots, \mathbf{w}_k, \mathbf{v}_1, \dots, \mathbf{v}_{n-k})$ of V, using Theorem 7.2 (ii). We claim that $B := (\mathbf{v}_1 + W, \dots, \mathbf{v}_{n-k} + W)$ is a basis of the quotient V/W.

First, we show that B spans the quotient V/W. Start with an arbitrary vector $\mathbf{v} + W \in V/W$ and write $\mathbf{v}$ as a linear combination

$$\mathbf{v} = \sum_{i=1}^{k} \alpha_i \mathbf{w}_i + \sum_{i=1}^{n-k} \beta_i \mathbf{v}_i$$

of the basis B_V. Then we have

$$\mathbf{v} + W = \sum_{i=1}^{k} \alpha_i \mathbf{w}_i + \sum_{i=1}^{n-k} \beta_i \mathbf{v}_i + W = \sum_{i=1}^{n-k} \beta_i \mathbf{v}_i + W = \sum_{i=1}^{n-k} \beta_i (\mathbf{v}_i + W) \ ,$$

which shows that B spans V/W.

For linear independence, start with

$$\sum_{i=1}^{n-k} \alpha_i (\mathbf{v}_i + W) = \mathbf{0} \quad \Rightarrow \quad \sum_{i=1}^{n-k} \alpha_i \mathbf{v}_i \in W$$

where the second statement follows because the zero 'vector' in V/W is really the entire vector subspace W. We need to show that all $\alpha_i = 0$. Since B_W is a basis of W this means we can find scalars β_i such that

$$\sum_{i=1}^{n-k} \alpha_i \mathbf{v}_i = -\sum_{i=1}^{k} \beta_i \mathbf{w}_i \ ,$$

and linear independence of B_V then implies that all $\alpha_i = 0$.

In conclusion, B is a basis of V/W and

$$\dim_{\mathbb{F}}(V/W) = |B| = n - k = \dim_{\mathbb{F}}(V) - \dim_{\mathbb{F}}(W) \, .$$

$\square$

Note that the dimension formula (8.12) is in line with the intuitive idea of dividing by a vector subspace W. Taking the quotient 'removes' W (the entirety of W becomes the zero vector of V/W) and, hence, in passing from V to V/W, the dimension reduces by $\dim_{\mathbb{F}}(W)$.

Exercises

(†=challenging)

8.1 *Dimension formula for an example*
For $V = \mathbb{R}^3$ the vector subspace U is spanned by $\mathbf{u}_1 = \mathbf{i} + 2\mathbf{j}$, $\mathbf{u}_2 = \mathbf{k}$ and the vector subspace W is spanned by $\mathbf{w}_1 = \mathbf{j} + \mathbf{k}$, $\mathbf{w}_2 = -\mathbf{i} + 2\mathbf{j}$. Explicitly verify the dimension formula (8.2) for this example.

8.2 *Intersections*
Consider two 3-planes in $\mathbb{R}^4$. What are the possible dimensions for their intersections? Provide an explicit example for each case.

8.3 Show that Eq. (9.9) defines an equivalence relation and find the associated equivalence classes.

8.4 *Quotients*
In $V = \mathbb{R}^3$ consider the vector subspace W spanned by $\mathbf{i}$ and $\mathbf{j}$ ('the x–y plane'). What are the cosets in V/W and what is the dimension of V/W?

8.5 *Quotients in polynomial vector spaces*
Consider the polynomial vector space $V = \mathcal{P}_3(\mathbb{R})$ and its vector subspace $W = \{ax + b \,|\, a, b \in \mathbb{R}\} \subset V$. Work out the dimensions of V, W, and V/W

and use the result to verify the dimension formula (8.12). Describe the cosets in V/W.

8.6 *General sums of vector spaces*
Let $W_1, \ldots, W_k \subset V$ be vector subspaces and define the sum $W := W_1 + \cdots + W_k = \{\mathbf{w}_1 + \cdots + \mathbf{w}_k \,|\, \mathbf{w}_i \in W_i, \ i = 1, \ldots, k\}$. Show that
(a) $W = \mathrm{Span}(W_1 \cup \ldots \cup W_k)$. (See Exercise 7.1.)
(b) W is a vector subspace.
(c) $\dim_{\mathbb{F}}(W) \leq \sum_{i=1}^{k} \dim_{\mathbb{F}}(W_i)$.

8.7 *Generalizing directs sums*†
Vector subspaces $W_1, \ldots, W_k \subset V$ are said to form a direct sum $W = W_1 \oplus \cdots \oplus W_k$ if
(i) $W = W_1 + \cdots + W_k$.
(ii) $\mathbf{w}_1 + \cdots + \mathbf{w}_k = \mathbf{0}$ for $\mathbf{w}_i \in W_i$ implies that all $\mathbf{w}_i = \mathbf{0}$.
If $W = W_1 \oplus \cdots \oplus W_k$ show that
(a) every $\mathbf{w} \in W$ can be uniquely written as $\mathbf{w} = \mathbf{w}_1 + \cdots + \mathbf{w}_k$, where $\mathbf{w}_i \in W_i$.
(b) combining bases of $W_1, \ldots, W_k$ into a single list gives a basis for W.
(c) $\dim_{\mathbb{F}}(W) = \sum_{i=1}^{k} \dim_{\mathbb{F}}(W_i)$.

Part III

Basic geometry

In this part, we pause the systematic development of the subject, and turn to a number of more practical topics, related to elementary geometry. One of the dilemmas of presenting a linear algebra course for scientists is that some of the practical methods needed early on in science, such as dot and cross products, only appear relatively late in the systematic mathematical development of the subject. The present part intends to address this problem. We focus on coordinate vector in $\mathbb{R}^n$ (and, for some parts, in $\mathbb{R}^3$) and introduce the dot and cross products in a somewhat informal manner, focusing on techniques for calculation and applications to the geometry of lines and planes.

This is done for a number of reasons. For one, the reader has a chance to engage with some of the practical methods used in science early on. The material developed in this part is also a good source of examples to illustrate some of the more abstract ideas which follow. Learning about new structures, such as the scalar product, in a special and more familiar setting first may help getting to grips with the axiomatic approach taken later on. As we go along, we will present powerful methods for calculation with indices. These techniques are extremely useful for calculations but are rarely covered in linear algebra textbooks.

School mathematics sometimes talks about vectors as objects with 'length and direction'. Such a statement lacks the rigour required for a mathematical definition but, worse, it is also seriously misleading. Vector spaces and vectors have been defined in Def. 6.1 and the words 'length' and 'direction' have not even been mentioned. Vectors are elements of vector spaces, objects which can be added and scalar multiplied, subject to a number of rules. Length and direction play no role at this level. Of course, we can still talk about length and direction of vectors but we do need more structure — in addition to the vector space structure — to do this. The required structure is that of a scalar product on a vector space. Its simplest incarnation, the dot product on $\mathbb{R}^n$, will be introduced in the next chapter. It allows us to introduce the length and the direction of vectors as well as angles between vectors and the notion of orthogonality.

In Chapter 10 we introduce the cross product in $\mathbb{R}^3$. From a mathematical point of view, the cross product is a somewhat exotic operation whose natural home is in advanced (multi-) linear algebra, and we will return to the subject in our discussion of tensors in Chapter 27. However, the cross product is widely used in scientific applications and should be discussed early on. On a geometrical level, the cross product is a method to obtain orthogonal vectors and it will be introduced with this motivation in mind. Combining the dot and cross products leads to the triple product on $\mathbb{R}^3$ which is, in fact, the same as the determinant in three dimensions. This provides an opportunity to develop some properties of the determinant in a special case, before general determinants are introduced in Chapter 18.

In the final Chapter 11 of this part, we apply some of the new tools to elementary geometry, mainly the geometry of lines and planes in $\mathbb{R}^2$ and $\mathbb{R}^3$.

9

The dot product

Geometry often requires the notion of a length of a vector and an angle between vectors. As we have emphasized above, a vector space by itself does not provide for these notions, so we require additional structure.

9.1 Basic properties

Summary 9.1 *The dot product* $\cdot : \mathbb{R}^n \times \mathbb{R}^n \to \mathbb{R}^n$ *is a bi-linear, symmetric, and positive map. For two vectors* $\mathbf{v}, \mathbf{w} \in \mathbb{R}^n$ *it is defined by* $\mathbf{v} \cdot \mathbf{w} = v_1 w_1 + \cdots + v_n w_n$.

What should we require for a structure on $\mathbb{R}^n$ which can provide us with an angle between two vectors? Above all, since the angle is a scalar quantity, we require a map $\cdot : \mathbb{R}^n \times \mathbb{R}^n \to \mathbb{R}$ which assigns to a pair of vectors $(\mathbf{v}, \mathbf{w})$ a scalar, which we denote by $\mathbf{v} \cdot \mathbf{w}$. The angle should not depend on the ordering of the two vectors so we should demand that $\mathbf{v} \cdot \mathbf{w} = \mathbf{w} \cdot \mathbf{v}$. Since linearity is a key feature of vector spaces it also makes sense to demand that the map $(\mathbf{v}, \mathbf{w}) \to \mathbf{v} \cdot \mathbf{w}$ is linear, in the same sense as a linear map (see Def. 6.3), in each of its two arguments. Finally, for a notion of length we need to impose a positivity condition.

9.1.1 Definition of dot product

It turns out that these simple requirements are satisfied by the *dot product* on $\mathbb{R}^n$ which is defined as

$$\left. \begin{array}{c} \cdot : \mathbb{R}^n \times \mathbb{R}^n \to \quad \mathbb{R} \\ (\mathbf{v}, \mathbf{w}) \quad \mapsto \mathbf{v} \cdot \mathbf{w} \end{array} \right\} \qquad \mathbf{v} \cdot \mathbf{w} = v_1 w_1 + \cdots + v_n w_n = \sum_{i=1}^{n} v_i w_i . \qquad (9.1)$$

It is customary to omit the sum symbol in this definition and simply write

$$\mathbf{v} \cdot \mathbf{w} = v_i w_i , \qquad (9.2)$$

adopting the convention that an index which appears twice in a given term (such as the index i in the present case) is summed over. This is also referred to as the *Einstein summation convention*. This convention is routinely used in Einstein's general theory of relativity which suffers from a proliferation of indices but it also facilitates a simplified notation and more effective computations in many other contexts. We will soon see explicit examples.

Problem 9.1 *(Dot product)*

Work out the dot product of the $\mathbb{R}^3$ vectors $\mathbf{v} = (1, 3, -2)^T$, $\mathbf{w} = (5, 2, 4)^T$, and of the $\mathbb{R}^4$ vectors $\mathbf{r} = (1, 3, 2, -1)^T$, $\mathbf{s} = (0, -4, 7, 5)^T$.

Solution:

$$\mathbf{v} \cdot \mathbf{w} = \begin{pmatrix} 1 \\ 3 \\ -2 \end{pmatrix} \cdot \begin{pmatrix} 5 \\ 2 \\ 4 \end{pmatrix} = 1 \cdot 5 + 3 \cdot 2 + (-2) \cdot 4 = 3$$

$$\mathbf{r} \cdot \mathbf{s} = \begin{pmatrix} 1 \\ 3 \\ 2 \\ -1 \end{pmatrix} \cdot \begin{pmatrix} 0 \\ -4 \\ 7 \\ 5 \end{pmatrix} = 1 \cdot 0 + 3 \cdot (-4) + 2 \cdot 7 + (-1) \cdot 5 = -3$$

9.1.2   Properties of the dot product

The following proposition shows that the dot product does indeed satisfy the requirements of linearity, symmetry, and positivity, discussed above.

Proposition 9.1 *The dot product on $\mathbb{R}^n$ satisfies the following properties for all* $\mathbf{v}, \mathbf{w}, \mathbf{u} \in \mathbb{R}^n$ *and all* $\alpha \in \mathbb{R}$.

(D1) $\mathbf{v} \cdot (\mathbf{w} + \mathbf{u}) = \mathbf{v} \cdot \mathbf{w} + \mathbf{v} \cdot \mathbf{u}$ *and* $\mathbf{v} \cdot (\alpha\mathbf{w}) = \alpha(\mathbf{v} \cdot \mathbf{w})$	*(linearity)*
(D2) $\mathbf{v} \cdot \mathbf{w} = \mathbf{w} \cdot \mathbf{v}$	*(symmetry)*
(D3) $\mathbf{v} \cdot \mathbf{v} > 0$ *for all* $\mathbf{v} \neq \mathbf{0}$	*(positivity)*

Proof   This is our first opportunity to compute with indices, using the Einstein summation convention 9.2.

(D1) $\mathbf{v} \cdot (\mathbf{w} + \mathbf{u}) = v_i(\mathbf{w} + \mathbf{u})_i = v_i(w_i + u_i) = v_i w_i + v_i u_i = \mathbf{v} \cdot \mathbf{w} + \mathbf{v} \cdot \mathbf{u}$
   $\mathbf{v} \cdot (\alpha\mathbf{w}) = v_i(\alpha\mathbf{w})_i = \alpha v_i w_i = \alpha(\mathbf{v} \cdot \mathbf{w})$

(D2) $\mathbf{v} \cdot \mathbf{w} = v_i w_i = w_i v_i = \mathbf{w} \cdot \mathbf{v}$

(D3) $(\mathbf{v}, \mathbf{v}) = \sum_i v_i^2 > 0$ for $\mathbf{v} \neq \mathbf{0}$.

Note that the components v_i, w_i, u_i are just numbers, not vectors, so all the rules for calculating in a field can be applied. This feature is one of the strengths of index calculations (which comes at the price of having to deal with indexed objects). This is the reason we are allowed to use the distributive law in the proof for (D1) or reverse the order, $v_i w_i = w_i v_i$, in the proof for (D2).   □

The condition (D1) is indeed a linearity condition, similar to the one for linear maps in Def. 6.3, on the second argument of the dot product. Since the dot product is symmetric, linearity in the second argument immediately translates into linearity in the first argument.

$$(\mathbf{w} + \mathbf{u}) \cdot \mathbf{v} \overset{(D2)}{=} \mathbf{v} \cdot (\mathbf{w} + \mathbf{u}) \overset{(D1)}{=} \mathbf{v} \cdot \mathbf{w} + \mathbf{v} \cdot \mathbf{u} \overset{(D2)}{=} \mathbf{w} \cdot \mathbf{v} + \mathbf{u} \cdot \mathbf{v}$$

$$(\alpha\mathbf{w}) \cdot \mathbf{v} \overset{(D2)}{=} \mathbf{v} \cdot (\alpha\mathbf{w}) \overset{(D1)}{=} \alpha(\mathbf{v} \cdot \mathbf{w}) \overset{(D2)}{=} \alpha(\mathbf{w} \cdot \mathbf{v})$$

Hence, the dot product is *bi-linear* and this can be applied to arbitrary linear combinations, so that

$$\left(\sum_i \alpha_i \mathbf{v}_i\right) \cdot \left(\sum_j \beta_j \mathbf{w}_j\right) = \sum_{i,j} \alpha_i \beta_j (\mathbf{v}_i \cdot \mathbf{w}_j) . \tag{9.3}$$

The properties in Prop. 9.1 will be used later to define general scalar products axiomatically, much as the rules for calculating with coordinate vectors inspire the general definition of vector spaces.

9.2 Length and angle

Summary 9.2 *The Euklidean norm* $|\cdot| : \mathbb{R}^n \to \mathbb{R}^{\geq 0}$ *is defined in terms of the dot product. It is positive, it scales under scalar multiplication and, as a result of the Cauchy–Schwarz inequality, it satisfied the triangle inequality. The Cauchy–Schwarz inequality also facilitates the definition of an angle between two vectors, in terms of the dot product and the Euklidean norm.*

Our original motivation for introducing the dot product was to facilitate the notions of length and angle. We begin by explaining how the dot product can be used to define length.

9.2.1 Definition of length

The (Euklidean) *length* (or *norm*) on $\mathbb{R}^n$ is defined as

$$\left.\begin{array}{c} |\cdot| : \mathbb{R}^n \to \mathbb{R}^{\geq 0} \\ \mathbf{v} \mapsto |\mathbf{v}| \end{array}\right\} \qquad |\mathbf{v}| = \sqrt{\mathbf{v} \cdot \mathbf{v}} = \left(\sum_{i=1}^n v_i^2\right)^{1/2} . \tag{9.4}$$

The square root in this definition makes sense because of the positivity property (D3) of the dot product. The length is strictly positive, except for the zero vector which has length zero.

But this by itself is not enough to convince us that the above is a sensible notion of length. For example, what happens to the length under scalar multiplication of a vector?

$$|\alpha \mathbf{v}| = \sqrt{(\alpha \mathbf{v}) \cdot (\alpha \mathbf{v})} = \sqrt{\alpha^2\, \mathbf{v} \cdot \mathbf{v}} = |\alpha|\sqrt{\mathbf{v} \cdot \mathbf{v}} = |\alpha|\,|\mathbf{v}| . \tag{9.5}$$

Evidently, if a vector is multiplied with a scalar α its length scales with the modulus $|\alpha|$ [1]. This certainly makes intuitive sense and explains why the square root has been included in Eq. (9.4). (Otherwise, the length would scale with the square of the scalar.)

---

[1] The notation $|\cdot|$ indicates the length of a vector whenever the argument is vectorial and the (real or complex) modulus whenever the argument is a scalar.

The property (9.5) allows us to define, for any non-zero vector $\mathbf{v} \in \mathbb{R}^n$, an associated vector $\mathbf{n}$ with unit length given by

$$\mathbf{n} = |\mathbf{v}|^{-1}\mathbf{v} \quad \Rightarrow \quad |\mathbf{n}| = ||\mathbf{v}|^{-1}\mathbf{v}| \overset{(9.5)}{=} |\mathbf{v}|^{-1}|\mathbf{v}| = 1 \; . \tag{9.6}$$

We can think about this vector $\mathbf{n}$ as the *direction* of $\mathbf{v}$. Vectors with unit length are also called *unit vectors* for short.

Problem 9.2 *(Euklidean length)*

Work out the length of the vectors $\mathbf{v} = (3,4)^T$, $\mathbf{w} = (-1,1,2)^T$, and $\mathbf{u} = (-3,2,-1,1)^T$. What is the unit vector associated to $\mathbf{v}$?

Solution:

$$|\mathbf{v}| = \sqrt{3^2 + 4^2} = 5 \; , \qquad |\mathbf{w}| = \sqrt{(-1)^2 + 1^2 + 2^2} = \sqrt{6}$$
$$|\mathbf{u}| = \sqrt{(-3)^2 + 2^2 + (-1)^2 + 1^2} = \sqrt{15}$$

The unit vector associated to $\mathbf{v}$ is $\mathbf{n} = \mathbf{v}/|\mathbf{v}| = (3,4)^T/5$.

9.2.2 The Cauchy–Schwarz inequality

The remaining geometrical notion we still need to define is the angle between two vectors. To this end, we need to prove an important and famous inequality which relates the dot product and the length.

Theorem 9.1 *(Cauchy–Schwarz inequality) For any two vectors* $\mathbf{v}, \mathbf{w} \in \mathbb{R}^n$ *we have*

$$|\mathbf{v} \cdot \mathbf{w}| \leq |\mathbf{v}|\,|\mathbf{w}| \; . \tag{9.7}$$

Equality holds if and only if $\mathbf{v}$ *and* $\mathbf{w}$ *are multiples of each other.*

Proof The proof is a bit tricky. We start by considering two unit length vectors $\mathbf{a}, \mathbf{b} \in \mathbb{R}^n$, so $|\mathbf{a}| = |\mathbf{b}| = 1$. A quick calculation shows that

$$0 \leq |\mathbf{a} \pm \mathbf{b}|^2 \overset{(9.4)}{=} (\mathbf{a} \pm \mathbf{b}) \cdot (\mathbf{a} \pm \mathbf{b}) \overset{(9.3)}{=} |\mathbf{a}|^2 \pm 2\mathbf{a} \cdot \mathbf{b} + |\mathbf{b}|^2 = 2(1 \pm \mathbf{a} \cdot \mathbf{b}) \; , \tag{9.8}$$

and, hence, that $|\mathbf{a} \cdot \mathbf{b}| \leq 1$. Now consider arbitrary vectors $\mathbf{v}$ and $\mathbf{w}$. If one of these vector is zero then (9.7) is satisfied since both sides equal zero. We can therefore assume that both $\mathbf{v}$ and $\mathbf{w}$ are non-zero. In this case we can introduce the unit vectors $\mathbf{a} := |\mathbf{v}|^{-1}\mathbf{v}$ and $\mathbf{b} := |\mathbf{w}|^{-1}\mathbf{w}$. The earlier result $|\langle \mathbf{a}, \mathbf{b} \rangle| \leq 1$ translates into

$$|\mathbf{v} \cdot \mathbf{w}| = |(|\mathbf{v}|\,\mathbf{a}) \cdot (|\mathbf{w}|\,\mathbf{b})| \overset{(9.3)}{=} |\mathbf{v}|\,|\mathbf{w}|\,|\mathbf{a} \cdot \mathbf{b}| \leq |\mathbf{v}|\,|\mathbf{w}| \; ,$$

which proves the Cauchy–Schwarz inequality.

It remains to prove the second statement about equality. If $\mathbf{v}$ and $\mathbf{w}$ are multiples of each other, for example $\mathbf{w} = \alpha \mathbf{v}$, then

$$\mathbf{v} \cdot \mathbf{w} = \mathbf{v} \cdot (\alpha \mathbf{v}) \overset{(D1)}{=} \alpha \mathbf{v} \cdot \mathbf{v} \overset{(9.4)}{=} \alpha |\mathbf{v}|^2 = |\mathbf{v}|(\alpha|\mathbf{v}|) \overset{(9.5)}{=} |\mathbf{v}|\,|\alpha \mathbf{v}| = |\mathbf{v}|\,|\mathbf{w}| \; ,$$

so that (9.7) is indeed satisfied with an equality.

Conversely, assume that $|\mathbf{v} \cdot \mathbf{w}| = |\mathbf{v}| \, |\mathbf{w}|$ and we want to show that $\mathbf{v}$ and $\mathbf{w}$ are multiples of each other. If one of the vectors is zero the statement holds trivially (as the zero vector is a multiple of any vector with 0), so we can again assume both $\mathbf{v}$ and $\mathbf{w}$ are non-zero. Their directions $\mathbf{a} = |\mathbf{v}|^{-1}\mathbf{v}$ and $\mathbf{b} = |\mathbf{w}|^{-1}\mathbf{w}$ then satisfy $|\mathbf{a} \cdot \mathbf{b}| = 1$ which shows that the right-hand side of Eq. (9.8) vanishes for one choice of sign. Therefore, $|\mathbf{a} \pm \mathbf{b}| = 0$ and, from (D3), $\mathbf{b} = \mp\mathbf{a}$ for one of the signs. It follows that $\mathbf{w} = |\mathbf{w}|\mathbf{b} = \mp|\mathbf{w}|\mathbf{a} = \mp|\mathbf{w}| \, |\mathbf{v}|^{-1}\mathbf{v}$ and, hence, that $\mathbf{w}$ is a multiple of $\mathbf{v}$. $\qquad\square$

9.2.3 Properties of the length

The Cauchy–Schwarz inequality implies another important inequality for the length, the *triangle inequality* $|\mathbf{v} + \mathbf{w}| \leq |\mathbf{v}| + |\mathbf{w}|$. (For its geometrical interpretation see Fig. 9.1.) It follows from the short calculation

$$|\mathbf{v} + \mathbf{w}|^2 \overset{(9.4)}{=} (\mathbf{v} + \mathbf{w})^2 \overset{(9.3)}{=} |\mathbf{v}|^2 + |\mathbf{w}|^2 + 2\,\mathbf{v} \cdot \mathbf{w} \leq |\mathbf{v}^2| + |\mathbf{w}|^2 + 2\,|\mathbf{v} \cdot \mathbf{w}|$$

$$\overset{(9.7)}{\leq} |\mathbf{v}|^2 + |\mathbf{w}|^2 + 2\,|\mathbf{v}| \, |\mathbf{w}| = (|\mathbf{v}| + |\mathbf{w}|)^2 \, . \tag{9.9}$$

The following proposition summarizes the properties of the Euklidean length on $\mathbb{R}^n$.

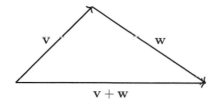

Fig. 9.1 Geometrical interpretation of the triangle inequality. The length $|\mathbf{v} + \mathbf{w}|$ of one side of the triangle is always less equal than the sum $|\mathbf{v}| + |\mathbf{w}|$ for the other two sides.

Proposition 9.2 *The Euklidean length (norm) $|\cdot| : \mathbb{R}^n \to \mathbb{R}^{\geq 0}$ defined in Eq. (9.4) has the following properties, for all $\mathbf{v}, \mathbf{w} \in \mathbb{R}^n$ and all $\alpha \in \mathbb{R}$.*

   (N1) $|\mathbf{v}| > 0$ *for* $\mathbf{v} \neq \mathbf{0}$ *(positivity)*
   (N2) $|\alpha\mathbf{v}| = |\alpha| \, |\mathbf{v}|$ *(scaling)*
   (N3) $|\mathbf{v} + \mathbf{w}| \leq |\mathbf{v}| + |\mathbf{w}|$ *(triangle inequality)*

Proof (N1) is immediately evident from the definition (9.4). (N2) and (N3) have been shown in Eqs. (9.5) and (9.9), respectively. $\qquad\square$

The properties in this proposition are what one would intuitively require from a sensible notion of length and they justify the definition in Eq. 9.4. Later, in Chapter 22, we will use these properties to define the general notion of norms on vector spaces axiomatically.

9.2.4 The angle between vectors

The Cauchy–Schwarz inequality (9.7) implies for any two non-zero vectors $\mathbf{v}, \mathbf{w} \in \mathbb{R}^n$ that

$$-1 \leq \frac{\mathbf{v} \cdot \mathbf{w}}{|\mathbf{v}| \, |\mathbf{w}|} \leq 1 \,. \tag{9.10}$$

This means there is a unique $\theta \in [0, \pi]$ such that

$$\cos(\theta) = \frac{\mathbf{v} \cdot \mathbf{w}}{|\mathbf{v}| \, |\mathbf{w}|} \,. \tag{9.11}$$

This quantity θ is called the *angle between the two vectors* $\mathbf{v}$ and $\mathbf{w}$ and is also denoted by $\sphericalangle(\mathbf{v}, \mathbf{w})$. By rearranging Eq. (9.11), we can write the dot product as

$$\mathbf{v} \cdot \mathbf{w} = |\mathbf{v}| \, |\mathbf{w}| \, \cos(\sphericalangle(\mathbf{v}, \mathbf{w})) \,, \tag{9.12}$$

and this equation suggests the geometrical interpretation of the dot product indicated in Fig. 9.2.

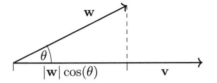

Fig. 9.2 Geometrical interpretation of the dot product which is obtained by multiplying $|\mathbf{v}|$ with $|\mathbf{w}| \cos(\theta)$, the length of the projection of $\mathbf{w}$ into the direction of $\mathbf{v}$.

Problem 9.3 *(Angle between vectors)*

Is the above definition of the angle between two vectors sensible and does it match our geometrical intuition? Find arguments that this is indeed the case.

Solution: A non-zero vector $\mathbf{v}$ should form an angle 0 with itself. Since $\cos \sphericalangle(\mathbf{v}, \mathbf{v}) = \mathbf{v} \cdot \mathbf{v}/|\mathbf{v}|^2 = 1$ this is indeed the case (and can be seen as the motivation for using the cos, rather than the sin, in the definition). We also have $\cos \sphericalangle(\mathbf{v}, -\mathbf{v}) = (\mathbf{v} \cdot (-\mathbf{v}))/|\mathbf{v}|^2 = -1$, so that $\sphericalangle(\mathbf{v}, -\mathbf{v}) = \pi$, the expected result for the angle between a vector and its negative.

We can also check that the definition of the angle is consistent with the geometrical definition of the cosine function. Think of Fig. 9.2 in $\mathbb{R}^2$ with the vector $\mathbf{v} = |\mathbf{v}| \mathbf{e}_1$ along the x-axis and $\mathbf{w} = (w_1, w_2)^T$. Then, the geometrical definition of the cosine function says that $\cos(\theta) = w_1/|\mathbf{w}|$. On the other hand, the angle from Eq. (9.11) is

$$\cos(\theta) = \frac{\mathbf{v} \cdot \mathbf{w}}{|\mathbf{v}| \, |\mathbf{w}|} = \frac{|\mathbf{v}| \, (\mathbf{e}_1 \cdot \mathbf{w})}{|\mathbf{v}| \, |\mathbf{w}|} = \frac{w_1}{|\mathbf{w}|} \,,$$

which does indeed give the same result.

Problem 9.4 *(Calculating the angle)*

Calculate the angle $\sphericalangle(\mathbf{v}, \mathbf{w})$ between the vectors $\mathbf{v} = (2, 4, -2)^T$ and $\mathbf{w} = (2, 1, 1)$. Do the same for the vectors $\mathbf{r} = (1, 0, 1, 1)^T$ and $\mathbf{s} = (2, 1, 0, 1)^T$.

Solution: With $\mathbf{v} \cdot \mathbf{w} = 6$, $|\mathbf{v}| = 2\sqrt{6}$, and $|\mathbf{w}| = \sqrt{6}$, we have from Eq. (9.11)

$$\cos \sphericalangle(\mathbf{v}, \mathbf{w}) = \frac{6}{2\sqrt{6} \cdot \sqrt{6}} = \frac{1}{2} \quad \Rightarrow \quad \sphericalangle(\mathbf{v}, \mathbf{w}) = \frac{\pi}{3}.$$

From $\mathbf{r} \cdot \mathbf{s} = 3$, $|\mathbf{r}| = \sqrt{3}$, and $\mathbf{s} = \sqrt{6}$, it follows that

$$\cos \sphericalangle(\mathbf{r}, \mathbf{s}) = \frac{3}{\sqrt{3} \cdot \sqrt{6}} = \frac{1}{\sqrt{2}} \quad \Rightarrow \quad \sphericalangle(\mathbf{r}, \mathbf{s}) = \frac{\pi}{4}.$$

Note that this last example in four dimensions is difficult to visualize. Yet, there is no problem computing the angle — having a precise definition pays off!

9.3   Orthogonality

> **Summary 9.3** *Two vectors in* $\mathbb{R}^n$ *are defined to be orthogonal if their dot product vanishes. Two non-zero vectors are orthogonality iff they form an angle* $\pi/2$. *The dot product can be expressed in terms of the Kronecker delta symbol, which is a useful tool for calculating with indices. A basis of* $\mathbb{R}^n$ *is called orthonormal if it consists of pairwise orthogonal unit vectors.*

9.3.1   Definition of orthogonality

Two vectors $\mathbf{v}, \mathbf{w} \in \mathbb{R}^n$ are called *orthogonal* or *perpendicular*, also written as $\mathbf{v} \perp \mathbf{w}$, if $\mathbf{v} \cdot \mathbf{w} = 0$. This definition means that every vector is orthogonal to the zero vector but, more importantly, for two non-zero vectors $\mathbf{v}, \mathbf{w} \in \mathbb{R}^n$ the angle formula (9.12) for the dot product implies that

$$\mathbf{v} \perp \mathbf{w} \quad \Leftrightarrow \quad \sphericalangle(\mathbf{v}, \mathbf{w}) = \frac{\pi}{2}. \tag{9.13}$$

Our definition makes sense: two non-zero vectors are orthogonal if and only if they form an angle $\pi/2$.

Problem 9.5 *(Orthogonality)*

Are the vectors $\mathbf{v} = (1, 3, -1, 2)^T$ and $\mathbf{w} = (0, 1/2, 1, -1/4)^T$ orthogonal? Find a vector orthogonal to $\mathbf{r} = (3, 2)^T$.

Solution: Since $\mathbf{v} \cdot \mathbf{w} = 1 \cdot 0 + 3 \cdot (1/2) + (-1) \cdot 1 + 2 \cdot (-1/4) = 0$ the vectors are indeed orthogonal.

A vector perpendicular to $\mathbf{r}$ can be obtained by exchanging its two components and multiplying one of them with -1, leading, for example, to $\mathbf{s} = (-2, 3)$. Indeed, $\mathbf{r} \cdot \mathbf{s} = 3 \cdot (-2) + 2 \cdot 3 = 0$.

The notion of orthogonality is a very useful tool for geometry, as the following example illustrates.

Example 9.1 *(Projections along a vector)*

Consider a unit vector $\mathbf{n} \in \mathbb{R}^n$ and define a map $p_{\mathbf{n}} : \mathbb{R}^n \to \mathbb{R}^n$ by $p_{\mathbf{n}}(\mathbf{w}) = (\mathbf{n} \cdot \mathbf{w})\mathbf{n}$. Comparison with Fig. 9.2 (setting $\mathbf{v} = \mathbf{n}$ in the figure) shows that $p_{\mathbf{n}}(\mathbf{w})$ should be thought of as the projection of $\mathbf{w}$ in the direction of $\mathbf{n}$. It is clear that $p_{\mathbf{n}}$ is a linear map, due to the linearity of the scalar product. Moreover, using linearly of $p_{\mathbf{n}}$ and $\mathbf{n} \cdot \mathbf{n} = 1$, it follows that

$$p_{\mathbf{n}}(p_{\mathbf{n}}(\mathbf{w})) = p_{\mathbf{n}}((\mathbf{n} \cdot \mathbf{w})\mathbf{n}) = (\mathbf{n} \cdot \mathbf{w}) \, p_{\mathbf{n}}(\mathbf{n}) = (\mathbf{n} \cdot \mathbf{w}) \, (\mathbf{n} \cdot \mathbf{n})\mathbf{n} = (\mathbf{n} \cdot \mathbf{w})\mathbf{n} = p_{\mathbf{n}}(\mathbf{w}) \, .$$

This means that $p_{\mathbf{n}} \circ p_{\mathbf{n}} = p_{\mathbf{n}}$, a property characteristic for projectors: applying the projection a second time does not have an effect. Using the above projection we can

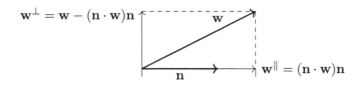

Fig. 9.3 Decomposing a vector $\mathbf{w}$ into a component $\mathbf{w}^{\|}$ along a unit vector $\mathbf{n}$ and a component $\mathbf{w}^{\perp}$ orthogonal to $\mathbf{n}$.

show that every vector $\mathbf{w} \in \mathbb{R}^n$ can be written as a unique sum $\mathbf{w} = \mathbf{w}^{\|} + \mathbf{w}^{\perp}$, where $\mathbf{w}^{\|}$ is a multiple of the vector $\mathbf{n}$ and $\mathbf{w}^{\perp}$ is orthogonal to $\mathbf{n}$. Indeed, if we write $\mathbf{w}^{\|} = \alpha\mathbf{n}$, then the scalar α is determined by

$$\mathbf{n} \cdot \mathbf{w} = \mathbf{n} \cdot (\alpha\mathbf{n} + \mathbf{w}^{\perp}) = \alpha \, ,$$

and, hence, $\mathbf{w}^{\|} = (\mathbf{n} \cdot \mathbf{w})\mathbf{n} = p_{\mathbf{n}}(\mathbf{w})$ is uniquely fixed. This fixes $\mathbf{w}^{\perp}$ uniquely to $\mathbf{w}^{\perp} = \mathbf{w} - \mathbf{w}^{\|}$ and since

$$\mathbf{n} \cdot \mathbf{w}^{\perp} = \mathbf{n} \cdot (\mathbf{w} - \mathbf{w}^{\|}) = \mathbf{n} \cdot \mathbf{w} - (\mathbf{n} \cdot \mathbf{w}) \, (\mathbf{n} \cdot \mathbf{n}) = 0$$

it is indeed orthogonal to $\mathbf{w}^{\|}$. In summary, for every vector $\mathbf{w} \in \mathbb{R}^n$ we have the unique decomposition

$$\mathbf{w} = \mathbf{w}^{\|} + \mathbf{w}^{\perp} \, , \quad \mathbf{w}^{\|} = p_{\mathbf{n}}(\mathbf{w}) = (\mathbf{n} \cdot \mathbf{w})\mathbf{n} \, , \quad \mathbf{w}^{\perp} = \mathbf{w} - \mathbf{w}^{\|} = \mathbf{w} - (\mathbf{n} \cdot \mathbf{w})\mathbf{n} \quad (9.14)$$

into a component $\mathbf{w}^{\|}$ along the unit vector $\mathbf{n}$ and a component $\mathbf{w}^{\perp}$ orthogonal to $\mathbf{n}$. This is illustrated in Fig. 9.3. □

9.3.2 The Kronecker delta symbol

The vector space $\mathbb{R}^n$ has a canonical basis of standard unit vectors $(\mathbf{e}_1, \ldots, \mathbf{e}_n)$. Recall, that the ith component of $\mathbf{e}_i$ equals one, while all other components are zero. This implies that all standard unit vectors have length one, $|\mathbf{e}_i| = 1$, and that they are mutually orthogonal, $\mathbf{e}_i \cdot \mathbf{e}_j = 0$ for $i \neq j$. These two properties are often written more concisely as

$$\mathbf{e}_i \cdot \mathbf{e}_j = \delta_{ij} \quad \text{where} \quad \delta_{ij} = \begin{cases} 1 \text{ for } i = j \\ 0 \text{ for } i \neq j \end{cases} . \tag{9.15}$$

The symbol δ_{ij} is called the *Kronecker delta symbol*. Its value depends on the value of its two indices and is equal to 1 if the indices are equal and 0 if they are not.

The Kronecker delta is a very useful tool for calculations with indices. At the most basic level it acts as an 'index replacer', meaning that

$$\delta_{ij} v_j = v_i \tag{9.16}$$

for a vector $\mathbf{v}$ with components v_i. In Eq. (9.16) we have used the Einstein summation convention: the index j appears twice in the term on the left and is, hence, summed over (a 'summation index') while the index i labels different components of the equation (a 'free index'). Given these conventions it is easy to argue that Eq. (9.16) is correct. The only term from the sum over j which contributes is the one for $j = i$ (since $\delta_{ij} = 0$ for $i \neq j$), giving v_i.

Another useful property of the Kronecker delta is

$$\delta_{ii} = n . \tag{9.17}$$

Here, the index i appears twice so it is summed over. It runs over the values $i = 1, \ldots, n$, each of which contributes 1 to the sum, for a total of n.

The dot product can also be expressed in terms of the Kronecker delta as

$$\mathbf{v} \cdot \mathbf{w} = v_i w_i = \delta_{ij} v_i w_j , \tag{9.18}$$

where the second equality follows from Eq. (9.16).

9.3.3   Orthonormal basis

Orthogonality and linear independence relate in an interesting way.

Proposition 9.3 *Non-zero vectors* $\mathbf{v}_1, \ldots, \mathbf{v}_k \in \mathbb{R}^n$ *which are pairwise orthogonal, so* $\mathbf{v}_i \cdot \mathbf{v}_j = 0$ *for all* $i \neq j$, *are linearly independent.*

Proof Consider the equation $\sum_j \alpha_j \mathbf{v}_j = \mathbf{0}$. For linear independence, we need to show that it is only solved if all $\alpha_j = 0$. If we take the dot product of this equation with $\mathbf{v}_i$ we find, given that $\mathbf{v}_i \cdot \mathbf{v}_j = 0$ for $j \neq i$, that $|\mathbf{v}_i|^2 \alpha_i = 0$. Dividing this by $|\mathbf{v}_i|^2$ (which is non-zero since the vectors $\mathbf{v}_i$ are non-zero) we have $\alpha_i = 0$. This holds for all $i = 1, \ldots, n$, so linear independence follows. $\square$

This result can be paraphrased by saying that 'orthogonality implies linear independence' and it motivates defining the concept of a basis consisting of mutually orthogonal (unit) vectors.

Definition 9.1 *A basis* $(\boldsymbol{\epsilon}_1, \ldots, \boldsymbol{\epsilon}_n)$ *of* $\mathbb{R}^n$ *is called an orthonormal basis if* $\boldsymbol{\epsilon}_i \cdot \boldsymbol{\epsilon}_j = \delta_{ij}$ *for all* $i, j = 1, \ldots, n$.

Orthonormal bases have considerably nicer properties than general bases. For example, it is easy to compute the coordinates of a vector relative to an orthonormal basis since

$$\mathbf{v} = \sum_{j=1}^{n} \alpha_j \epsilon_j \quad \Leftrightarrow \quad \alpha_i = \epsilon_i \cdot \mathbf{v} \,. \tag{9.19}$$

This follows simply by taking the dot product of the equation on the left with ϵ_i which gives

$$\epsilon_i \cdot \mathbf{v} = \sum_{j=1}^{n} \alpha_j \, \epsilon_i \cdot \epsilon_j = \sum_{j=1}^{n} \alpha_j \delta_{ij} = \alpha_i \,.$$

By virtue of Eq. (9.15) the standard unit vectors form an ortho-normal basis of $\mathbb{R}^n$ and the coordinates of a vector $\mathbf{v}$ relative to this basis are the components $\mathbf{e}_i \cdot \mathbf{v} = v_i$. But there are other, less simple orthonormal bases, as the following exercise shows.

Problem 9.6 *(Orthonormal basis)*

Find an orthonormal basis (ϵ_1, ϵ_2) of $\mathbb{R}^2$ which is different from the standard unit vector basis. Find the coordinates of a vector $\mathbf{v} = (v_1, v_2)^T$ relative to this basis, first by explicitly solving the equations, similar to Exercise 7.5, and then by using Eq. (22.15).

Solution: A possible choice for an orthonormal basis (ϵ_1, ϵ_2) on $\mathbb{R}^2$ is

$$\epsilon_1 = \frac{1}{\sqrt{2}} \begin{pmatrix} 1 \\ 1 \end{pmatrix}, \quad \epsilon_2 = \frac{1}{\sqrt{2}} \begin{pmatrix} 1 \\ -1 \end{pmatrix}.$$

Indeed, $|\epsilon_1| = |\epsilon_2| = 1$ and $\epsilon_1 \cdot \epsilon_2 = 0$.
   We start with the pedestrian way to find the coordinates. Writing an arbitrary vector as $\mathbf{v} = (v_1, v_2)^T = \alpha_1 \epsilon_1 + \alpha_2 \epsilon_2$ and inserting (ϵ_1, ϵ_2) from above gives

$$\begin{pmatrix} v_1 \\ v_2 \end{pmatrix} = \frac{1}{\sqrt{2}} \begin{pmatrix} \alpha_1 + \alpha_2 \\ \alpha_1 - \alpha_2 \end{pmatrix}.$$

Splitting up into the two components, $v_1 = (\alpha_1 + \alpha_2)/\sqrt{2}$ and $v_2 = (\alpha_1 - \alpha_2)\sqrt{2}$, and solving for α_i then leads to the desired result, $\alpha_1 = (v_1 + v_2)/\sqrt{2}$ and $\alpha_2 = (v_1 - v_2)/\sqrt{2}$, for the coordinates.
   On the other hand, we can use Eq. (22.15) to find the same result

$$\alpha_1 = \epsilon_1 \cdot \mathbf{v} = \frac{1}{\sqrt{2}} \begin{pmatrix} 1 \\ 1 \end{pmatrix} \cdot \begin{pmatrix} v_1 \\ v_2 \end{pmatrix} = \frac{1}{\sqrt{2}} (v_1 + v_2)$$

$$\alpha_2 = \epsilon_2 \cdot \mathbf{v} = \frac{1}{\sqrt{2}} \begin{pmatrix} 1 \\ -1 \end{pmatrix} \cdot \begin{pmatrix} v_1 \\ v_2 \end{pmatrix} = \frac{1}{\sqrt{2}} (v_1 - v_2)$$

somewhat more directly and efficiently.

Orthonormal bases are of great importance for vector spaces with scalar products and we will return to the subject in Chapter 22.

Application 9.1 *(Projections and graphical representation)*

In Example 9.1 we have seen how to project a vector $\mathbf{v} \in \mathbb{R}^3$ into the direction of a unit vector $\mathbf{n} \in \mathbb{R}^3$. We would like to use this method to graphically represent three-dimensional objects, defined by a set of vectors in $\mathbb{R}^3$. One way to tackle this problem is to define a linear map $P : \mathbb{R}^3 \to \mathbb{R}^2$ whose images provide the points which need to be plotted in two dimensions.

We start by introducing spherical polar coordinates (r, θ, φ), where $r \in [0, \infty)$, $\theta \in [0, \pi]$ and $\varphi \in [0, 2\pi)$, such that any vector $\mathbf{x} \in \mathbb{R}^3$ can be written as

$$\mathbf{x} = r(\sin\theta\cos\varphi, \sin\theta\cos\varphi, \cos\theta)^T . \tag{9.20}$$

Associated to these coordinates we introduce the three vectors

$$\begin{aligned} \mathbf{e}_r &= (\sin\theta\cos\varphi, \sin\theta\cos\varphi, \cos\theta)^T \\ \mathbf{e}_\theta &= (\cos\theta\cos\varphi, \cos\theta\sin\varphi, -\sin\theta)^T \\ \mathbf{e}_\varphi &= (-\sin\varphi, \cos\varphi, 0)^T \end{aligned} \tag{9.21}$$

which are easily checked to form an ortho-normal basis of $\mathbb{R}^3$ for any value of θ and φ. The geometrical interpretation of these vectors and of the angles θ and φ is indicated in the figure below.

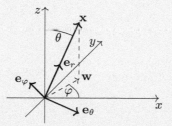

The angle $\theta = \sphericalangle(\mathbf{x}w, \mathbf{e}_3)$ is the angle $\mathbf{x}$ forms with the z-axis and the angle $\varphi = \sphericalangle(\mathbf{w}, \mathbf{e}_1)$ is the angle between the x-axis and the projection $\mathbf{w}$ of $\mathbf{x}$ onto the x–y plane.

For our purposes we would like to think of $\mathbf{e}_r$ as the 'direction of viewing' which can be adjusted by changing the angles θ and φ. To obtain two-dimensional vectors we use the coordinates of a vector $\mathbf{v}$ in the directions $\mathbf{e}_\theta$ and $\mathbf{e}_\varphi$, so we define the map P as

$$P_{\theta,\varphi}(\mathbf{v}) = (\mathbf{v} \cdot \mathbf{e}_\theta, \mathbf{v} \cdot \mathbf{e}_\varphi)^T . \tag{9.22}$$

Let us apply this to the simple example of a tetrahedron with the four vertices

$$\mathbf{v}_1 = (1,1,1)^T , \quad \mathbf{v}_2 = (1,-1,-1)^T , \quad \mathbf{v}_3 = (-1,1,-1)^T , \quad \mathbf{v}_4 = (-1,-1,1)^T ,$$

and edges l_{ij} connecting $\mathbf{v}_i$ and $\mathbf{v}_j$ given by $l_{ij} = \{\mathbf{v}_i + t(\mathbf{v}_j - \mathbf{v}_i) \,|\, t \in [0,1]\}$. The result of drawing the line segments $P_{\theta,\varphi}(l_{ij})$ for values $(\theta, \varphi) = (\pi/9, \pi/6) + k(\pi, \pi)/18$, where $k = 0, \ldots, 5$, is shown in the figure below.

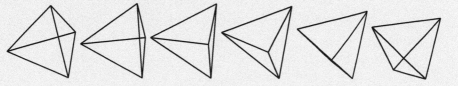

Exercises

(†=challenging)

9.1 *Length, dot product, and angle*
Find the lengths of the following vectors and the dot products and the (cosines of the) angles between them.
(a) $\mathbf{v} = (1, -1)^T$, $\mathbf{u} = (2, 1)^T$, $\mathbf{w} = (-3, 1)^T$
(b) $\mathbf{v} = (1, 0, 1)^T$, $\mathbf{u} = (1, 2, -1)^T$, $\mathbf{w} = (3, 1, -2)^T$
(c) $\mathbf{v} = (1, 1, 1, 1)^T$, $\mathbf{u} = (1, -1, 1, -1)$, $\mathbf{w} = (1, 2, 0, -1)^T$

9.2 For vectors $\mathbf{v} = (1, -1, 2)^T$ and $\mathbf{u} = (1, 1, 1)^T$ find the scalar multiple $\alpha\mathbf{u}$ for which the 'distance' $|\mathbf{v} - \alpha\mathbf{u}|$ is minimal.

9.3 Find a vector of the form $\mathbf{w} = (1, \alpha, \beta)^T$, where $\alpha, \beta \in \mathbb{R}$, which is orthogonal to $\mathbf{v} = (1, 3, 0)^T$ and $\mathbf{u} = (-1, 2, -1)$.

9.4 For two non-zero vectors $\mathbf{v}, \mathbf{w} \in \mathbb{R}^n$, show that $\sphericalangle(\mathbf{v}, \mathbf{w}) = \pi$ if and only if $\mathbf{w} = \alpha\mathbf{v}$ for $\alpha < 0$.

9.5 For $z, \zeta \in \mathbb{C}$, show that $\Re(z\bar{\zeta}) \leq |z|\,|\zeta|$.

9.6 *Triangle inequality*†
(a) For vectors $\mathbf{v}, \mathbf{w} \in \mathbb{R}^n$ show that $|\mathbf{v} - \mathbf{w}| \geq ||\mathbf{v}| - |\mathbf{w}||$.
(b) For vectors $\mathbf{v}_1, \ldots, \mathbf{v}_k \in \mathbb{R}^n$, show that $|\mathbf{v}_1 + \cdots + \mathbf{v}_k| \leq |\mathbf{v}_1| + \cdots + |\mathbf{v}_k|$.

9.7 *Index notation*
(a) Convert the expressions $\mathbf{v} \cdot \mathbf{w}$, $|\mathbf{v}|^2$, $(\mathbf{v} \cdot \mathbf{w})(\mathbf{u} \cdot \mathbf{x})$, and $(\mathbf{v} \cdot \mathbf{w})\mathbf{u} - \mathbf{v}$ into index notation.
(b) Convert the expressions $\delta_{ij}v_iw_j$, $v_iw_iu_ju_jx_k$, and $v_i\delta_{ij}\delta_{jk}w_k$ into vector notation.

9.8 *Orthonormal basis*
(a) Show that $(\boldsymbol{\epsilon}_1, \boldsymbol{\epsilon}_2, \boldsymbol{\epsilon}_3)$ with $\boldsymbol{\epsilon}_1 = (1, 1, 0)^T/\sqrt{2}$, $\boldsymbol{\epsilon}_2 = (1, -1, 1)^T/\sqrt{2}$ and $\boldsymbol{\epsilon}_3 = (1, -1, -2)^T/\sqrt{6}$ is an ortho-normal basis of $\mathbb{R}^3$. Find the coordinates of the vector $\mathbf{v} = (5, -4, 2)^T$ relative to this basis.
(b) Show the $\mathbb{R}^4$ vectors $\boldsymbol{\epsilon}_1 = (1, 1, 1, 1)^T/2$ and $\boldsymbol{\epsilon}_2 = (1, 1, -1, -1)^T/2$ are ortho-normal. Find two further vectors $\boldsymbol{\epsilon}_3, \boldsymbol{\epsilon}_4$ such that $(\boldsymbol{\epsilon}_1, \boldsymbol{\epsilon}_2, \boldsymbol{\epsilon}_3\boldsymbol{\epsilon}_4)$ is an ortho-normal basis of $\mathbb{R}^4$. Find the coordinates of a general vector $\mathbf{v} = (v_1, v_2, v_3, v_4)^T \in \mathbb{R}^4$ relative to this basis.

9.9 *Projectors from ortho-normal vectors*
Let $\mathbf{v}_1, \ldots, \mathbf{v}_k \in \mathbb{R}^n$ be a set of ortho-normal vectors. Show that the map $p : \mathbb{R}^n \to \mathbb{R}^n$ defined by $p(\mathbf{v}) = \sum_{i=1}^{k}(\mathbf{v} \cdot \mathbf{v}_i)\mathbf{v}_i$ is linear and that it satisfies the projector condition $p \circ p = p$.

9.10 *Polar coordinates*†
In $\mathbb{R}^2$ with vectors $\mathbf{x} = x\mathbf{e}_1 + y\mathbf{e}_2$ introduce polar coordinates $r \in [0, \infty)$ and $\varphi \in [0, 2\pi)$ by $x = r\cos\varphi$ and $y = r\sin\varphi$.
(a) With $\mathbf{e}_r := (\cos\varphi, \sin\varphi)^T$ and $\mathbf{e}_\varphi := (-\sin\varphi, \cos\varphi)^T$, show that $(\mathbf{e}_r, \mathbf{e}_\varphi)$ is an ortho-normal basis of $\mathbb{R}^2$. Write the vector $\mathbf{x}$ as a linear combination of this basis.
(b) Assume that x, y, r, and φ are functions of t and work out the t derivatives of $\mathbf{e}_r$, $\mathbf{e}_\varphi$, and $\mathbf{x}$.
(c) Solve the simultaneous differential equations $\dot{x} = -by$, $\dot{y} = bx$, (where the dot denotes the t derivative) using polar coordinates. Interpret the resulting solution.

9.11 *Projections and graphics*†
Apply the discussion in Application 9.1 to a cube in $\mathbb{R}^3$.

10
Vector and triple product

We have seen that two vectors in $\mathbb{R}^n$ are, by definition, orthogonal if their dot product vanishes. Orthogonality is easy to check for any given two vectors since the dot product can be readily carried out. But how can we find a vector orthogonal to another, given vector or to a list of vectors? Trying to answer this question in three dimensions leads to the vector or cross product in $\mathbb{R}^3$.

Before we tackle the three-dimensional case, we start with vectors in $\mathbb{R}^2$. In this case, finding a vector perpendicular to a given one is rather simple but the discussion will provide guidance for how to deal with the three-dimensional case. In three dimensions, the cross product allows us to compute a vector which is perpendicular to two given vectors. We will see how the cross product can be elegantly expressed in index notation, using the Levi-Civita symbol, and how vector identities with cross and dot product can be efficiently derived with index techniques.

The triple product in $\mathbb{R}^3$ is formed by combining the cross and dot products. It is, in fact, nothing but the three-dimensional determinant, so this is a good opportunity to get used to determinants, before we develop their general theory in Chapter 18.

10.1 The cross product

Summary 10.1 *The cross product $\times : \mathbb{R}^3 \times \mathbb{R}^3 \to \mathbb{R}^3$ is an anti-symmetric, bilinear map which produces orthogonal vectors. The cross product can be expressed in term of the Levi-Civita symbol which is an efficient tool for index calculations. Geometrically, the cross product gives a vector orthogonal to both of its arguments whose length equals the area of the parallelogram defined by the arguments.*

10.1.1 Orthogonality in two dimensions

Suppose we would like to construct a linear map $\mathbb{R}^2 \to \mathbb{R}^2$ which maps vectors $\mathbf{v}$ into orthogonal vectors $\mathbf{v}^\times$ with the same length. Since the standard unit vectors are mutually orthogonal a reasonable starting point is to demand that $\mathbf{e}_1^\times = -\mathbf{e}_2$ and $\mathbf{e}_2^\times = \mathbf{e}_1$. These conditions together with linearity already fix the map. To see this consider an arbitrary vector $\mathbf{v} = v_1\mathbf{e}_1 + v_2\mathbf{e}_2$ and use linearity.

$$\mathbf{v}^\times = (v_1\mathbf{e}_1 + v_2\mathbf{e}_2)^\times = v_1\mathbf{e}_1^\times + v_2\mathbf{e}_2^\times = -v_1\mathbf{e}_2 + v_2\mathbf{e}_1 \quad \Rightarrow \quad \begin{pmatrix} v_1 \\ v_2 \end{pmatrix}^\times = \begin{pmatrix} v_2 \\ -v_1 \end{pmatrix} \quad (10.1)$$

(We have already encountered this method to generate an orthogonal vector in Exercise 9.5.) We can verify that $\mathbf{v}^{\times}$ is indeed orthogonal to $\mathbf{v}$ by calculating

$$\mathbf{v} \cdot \mathbf{v}^{\times} = (v_1 \mathbf{e}_1 + v_2 \mathbf{e}_2) \cdot (-v_1 \mathbf{e}_2 + v_2 \mathbf{e}_1) = v_1 v_2 - v_1 v_2 = 0 \, ,$$

where we have used bi-linearity of the dot product and Eq. (9.15). It is also easily checked that all vectors orthogonal to $\mathbf{v}$ are multiples of $\mathbf{v}^{\times}$, that the map $\times$ preserves the length, so $|\mathbf{v}^{\times}| = |\mathbf{v}|$, and that $(\mathbf{v}^{\times})^{\times} = -\mathbf{v}$ (see Exercise 10.7).

From Section 1.2.4 we know that a linear map between two-dimensional coordinate vector can be described by a 2×2 matrix which we can find by acting on the standard unit vectors. Since $\mathbf{e}_1^{\times} = -\mathbf{e}_2$ and $\mathbf{e}_2^{\times} = \mathbf{e}_1$ comparison with the results from Section 1.2.4 shows that this matrix is given by

$$\epsilon = \begin{pmatrix} 0 & 1 \\ -1 & 0 \end{pmatrix} \quad \Rightarrow \quad \epsilon_{ij} = \begin{cases} 1 & \text{for } (i,j) = (1,2) \\ -1 & \text{for } (i,j) = (2,1) \\ 0 & \text{otherwise} \end{cases} . \tag{10.2}$$

With this notation, the perpendicular vector can be computed from

$$(\mathbf{v}^{\times})_i = \epsilon_{ij} v_j \, , \tag{10.3}$$

where a sum over j is implied by the Einstein convention. The two-index object ϵ_{ij} is called the *Levi-Civita symbol* in two dimensions (also see Exercise 10.6).

10.1.2 Definition of cross product in $\mathbb{R}^3$

In three dimensions it makes sense to ask for a (bi-) linear map which assigns to two vectors a third which is orthogonal to either. This map $\times : \mathbb{R}^2 \times \mathbb{R}^3 \to \mathbb{R}^3$ is called the *cross product* or *vector product* and is written as $(\mathbf{v}, \mathbf{w}) \mapsto \mathbf{v} \times \mathbf{w}$. Apart from linearity in each of its two arguments we would like the cross product to be anti-symmetric, that is, $\mathbf{v} \times \mathbf{w} = -\mathbf{w} \times \mathbf{v}$, so that the orthogonal vector points into the opposite direction when the order of the arguments is changed. Finally, given that the standard unit vectors are mutually orthogonal it makes sense to demand that the cross product of two standard unit vectors gives the third. Altogether this motivates the following definition.

Definition 10.1 *(Cross product) A map $\times : \mathbb{R}^2 \times \mathbb{R}^3 \to \mathbb{R}^3$ is called a cross product if is satisfied the following conditions for all vectors $\mathbf{v}, \mathbf{w}, \mathbf{u} \in \mathbb{R}^3$ and all scalars $\alpha, \beta \in \mathbb{R}$.*

(C1) $\quad \mathbf{v} \times \mathbf{w} = -\mathbf{w} \times \mathbf{v}$ $\hspace{3cm}$ *(anti-symmetry)*

(C2) $\quad \mathbf{v} \times (\mathbf{w} + \mathbf{u}) = \mathbf{v} \times \mathbf{w} + \mathbf{v} \times \mathbf{u}$ $\hspace{1.5cm}$ *(linearity)*

$\hspace{1.1cm} \mathbf{v} \times (\alpha \mathbf{w}) = \alpha(\mathbf{v} \times \mathbf{w})$

(C3) $\quad \mathbf{e}_1 \times \mathbf{e}_2 = \mathbf{e}_3, \; \mathbf{e}_2 \times \mathbf{e}_3 = \mathbf{e}_1, \; \mathbf{e}_3 \times \mathbf{e}_1 = \mathbf{e}_2$ $\hspace{0.5cm}$ *(orthogonality)*

While (C2) demands linearity in the second argument it is clear, by combining (C2) with (C1), that the cross product is also linear in the first argument and, hence, that it is bi-linear. This means that $\mathbb{R}^3$ with the cross product forms an algebra, in the sense of Def. 6.4. Anti-symmetry implies that $\mathbf{v} \times \mathbf{v} = -\mathbf{v} \times \mathbf{v}$, so the cross product of any vector with itself vanishes, that is,

$$\mathbf{v} \times \mathbf{v} = \mathbf{0} \quad \text{for all } \mathbf{v} \in \mathbb{R}^3 \, . \tag{10.4}$$

10.1.3 Existence and uniqueness of the cross product

Def. 10.1 fixes the cross product uniquely. To see this, we start with two vectors $\mathbf{v} = \sum_{i=1}^{3} v_i \mathbf{e}_i$ and $\mathbf{w} = \sum_{j=1}^{3} w_j \mathbf{e}_j$ and use bi-linearity of the cross product to get

$$\mathbf{v} \times \mathbf{w} = \left(\sum_{i=1}^{3} v_i \mathbf{e}_i \right) \times \left(\sum_{j=1}^{3} w_j \mathbf{e}_j \right) \overset{(C2)}{=} \sum_{i,j=1}^{3} v_i w_j \mathbf{e}_i \times \mathbf{e}_j$$

$$= (v_2 w_3 - v_3 w_2)\mathbf{e}_1 + (v_3 w_1 - v_1 w_3)\mathbf{e}_2 + (v_1 w_2 - v_2 w_1)\mathbf{e}_3 \qquad (10.5)$$

In the last step, we have removed the terms proportional to $\mathbf{e}_i \times \mathbf{e}_i$, as they vanish from Eq. (10.4), and have worked out the remaining terms proportional to $\mathbf{e}_i \times \mathbf{e}_j$ using the rules (C3) together with anti-symmetry (C1) (so that, for example, $\mathbf{e}_1 \times \mathbf{e}_2 = \mathbf{e}_3$ and $\mathbf{e}_2 \times \mathbf{e}_1 = -\mathbf{e}_3$). To summarize, the unique expression for the cross product which follows from Def. 10.1 is given by

$$\mathbf{v} \times \mathbf{w} = \begin{pmatrix} v_1 \\ v_2 \\ v_3 \end{pmatrix} \begin{matrix} \\ \times \\ \end{matrix} \begin{pmatrix} w_1 \\ w_2 \\ w_3 \end{pmatrix} = \begin{pmatrix} v_2 w_3 - v_3 w_2 \\ v_3 w_1 - v_1 w_3 \\ v_1 w_2 - v_2 w_1 \end{pmatrix} . \qquad (10.6)$$

It is not hard to verify that this expression does indeed satisfy all the axioms in Def. 10.1 and that the vector product is orthogonal to its two arguments. We defer this for now and will come back to it in a moment once we have introduced more efficient notation.

The formula (10.6) is easy to remember. The third component of the cross product is computed by ignoring the third entries of $\mathbf{v}$ and $\mathbf{w}$ and by multiplying the remaining entries as indicated by the thick lines. The first and second component are obtained analogously, with multiplications indicated by the thin lines.

Problem 10.1 *(Cross product)*

Work out the cross product of the vectors $\mathbf{v} = (2, 4, 3)^T$ and $\mathbf{w} = (-2, 1, 5)^T$.

Solution: Using Eq. (10.6), and multiplying as indicated by the lines in this equation, the cross product is

$$\mathbf{v} \times \mathbf{w} = \begin{pmatrix} 2 \\ 4 \\ 3 \end{pmatrix} \times \begin{pmatrix} -2 \\ 1 \\ 5 \end{pmatrix} = \begin{pmatrix} 4 \cdot 5 - 3 \cdot 1 \\ 3 \cdot (-2) - 2 \cdot 5 \\ 2 \cdot 1 - 4 \cdot (-2) \end{pmatrix} = \begin{pmatrix} 17 \\ -16 \\ 10 \end{pmatrix} .$$

10.1.4 The Levi-Civita symbol in $\mathbb{R}^3$

While Eq. (10.6) is convenient for working out the cross product of explicit numerical vectors, as in the previous problem, it becomes quite cumbersome when the vectors contain symbolic entries and when multiple products are involved. A more efficient way of writing the cross product is facilitated by the Levi-Cicita symbol ϵ_{ijk} in three dimensions, a generalization of the two-dimensional Levi-Civita symbol (10.2). It is defined by

$$
\epsilon_{ijk} = \begin{cases} +1 \text{ if } (i,j,k) = (1,2,3),(2,3,1),(3,1,2) & \text{(cyclic permutations)} \\ -1 \text{ if } (i,j,k) = (2,1,3),(3,2,1),(1,3,2) & \text{(anti-cyclic permutations)} \\ 0 \quad \text{otherwise} & \text{(two same indices)} \end{cases} \quad (10.7)
$$

Using this symbol, the cross product can be written as

$$
(\mathbf{v} \times \mathbf{w})_i = \epsilon_{ijk} v_j w_k \, , \tag{10.8}
$$

generalizing the two-dimensional formula (10.3). We recall that the summation convention is assumed and implies a summation over the indices j and k on the right-hand side of Eq. (10.8). To check that this expression is correct work out its first component

$$
(\mathbf{v} \times \mathbf{w})_1 = \epsilon_{123} v_2 w_3 + \epsilon_{132} v_3 w_2 = v_2 w_3 - v_3 w_2 \, , \tag{10.9}
$$

and note that this does indeed coincide with the first entry in Eq. (10.6). Similarly, one can verify that the second and third components are correctly reproduced. A low-key way of thinking about the Levi-Civita symbol is as a convenient short-hand for the factors of ± 1 and 0 which appear in the cross product. To work with the Levi-Civita symbol efficiently we need to understand its properties, which are collected in the following proposition.

Proposition 10.1 *The Levi-Civita symbol* (10.7) *has the following properties:*

(LC1) It remains unchanged under cyclic index permutation, for example $\epsilon_{ijk} = \epsilon_{jki}$.
(LC2) It changes sign under anti-cyclic index permutation, for example $\epsilon_{ijk} = -\epsilon_{ikj}$.
(LC3) It vanishes if two indices coincide, for example $\epsilon_{ijj} = 0$.
(LC4) $\epsilon_{ijk}\epsilon_{ilm} = \delta_{jl}\delta_{km} - \delta_{jm}\delta_{kl}$.
(LC5) $\epsilon_{ijk}\epsilon_{ijm} = 2\delta_{km}$.
(LC6) $\epsilon_{ijk}\epsilon_{ijk} = 6$.
(LC7) $\epsilon_{ijk} v_j v_k = 0$ for any vector $\mathbf{v} \in \mathbb{R}^3$.

Proof (LC1), (LC2), (LC3) These properties follow directly from the definition of the Levi-Civita symbol.
(LC4) This can be reasoned out as follows. If the index pair (j,k) is different from (l,m) (in any order) then clearly both sides of (LC4) are zero. On the other hand, if the two index pairs equal each other they can do so in the same or the opposite ordering and these two possibilities correspond to the two terms on the right-hand side of (LC4).
(LC5) If we multiply (LC4) by δ_{jl}, using the index replacing property (9.16) of the Kronecker delta, we obtain

$$
\epsilon_{ijk}\epsilon_{ijm} = (\delta_{jl}\delta_{km} - \delta_{jm}\delta_{kl})\delta_{jl} = 3\delta_{km} - \delta_{km} = 2\delta_{km}
$$

and this is the desired result.
(LC6) Further, multiplying (LC5) with δ_{km} we have

$$\epsilon_{ijk}\epsilon_{ijk} = 2\delta_{km}\delta_{km} = 2\delta_{kk} = 6 \ .$$

(LC7) From (LC2) we have $\epsilon_{ijk}v_jv_k = -\epsilon_{ikj}v_kv_j = -\epsilon_{ijk}v_jv_k$, where the summation indices j and k have been swapped in the last step, and, hence, $2\epsilon_{ijk}v_jv_k = 0$. $\qquad\square$

10.1.5 Properties of the cross product

With these techniques available it is now quite easy to verify that the cross product, as defined by Eq. (10.1) or Eq. (10.8), does indeed satisfy the axioms in Def. 10.1. The property (C3) is easily checked by explicit computation, applying the formula (10.6) to the standard unit vectors. Axioms (C1) and (C2) are verified by

$$(\mathbf{v}\times\mathbf{w})_i \overset{(10.8)}{=} \epsilon_{ijk}v_jw_k = \epsilon_{ijk}w_kv_j \overset{(LC2)}{=} -\epsilon_{ikj}w_kv_j \overset{(10.8)}{=} -(\mathbf{w}\times\mathbf{v})_i$$

$$(\mathbf{v}\times(\mathbf{w}+\mathbf{u}))_i \overset{(10.8)}{=} \epsilon_{ijk}v_j(\mathbf{w}+\mathbf{u})_k = \epsilon_{ijk}v_j(w_k+u_k) = \epsilon_{ijk}v_jw_k + \epsilon_{ijk}v_ju_k$$

$$\overset{(10.8)}{=} (\mathbf{v}\times\mathbf{w})_i + (\mathbf{v}\times\mathbf{u})_i$$

$$(\mathbf{v}\times(\alpha\mathbf{w})) \overset{(10.8)}{=} \epsilon_{ijk}v_j(\alpha w)_k = \alpha\,\epsilon_{ijk}v_jw_k \overset{(10.8)}{=} \alpha\,\mathbf{v}\times\mathbf{w}$$

Note that all the quantities in indexed expressions are numbers so we can calculate using the standard rules in a field. The cross product is indeed orthogonal to its two arguments since

$$\mathbf{v}\cdot(\mathbf{v}\times\mathbf{w}) \overset{(9.2)}{=} v_i(\mathbf{v}\times\mathbf{w})_i \overset{(10.8)}{=} \epsilon_{ijk}v_iv_jw_k \overset{(LC7)}{=} 0 \ , \qquad (10.10)$$

and similarly for $\mathbf{w}\cdot(\mathbf{v}\times\mathbf{w})$. There are a few more complicated relations which involve double cross products and combinations of cross and dot products, which can be very useful for explicit computations.

Proposition 10.2 *The dot and cross product in $\mathbb{R}^3$ satisfy the following relations for all vectors $\mathbf{a},\mathbf{b},\mathbf{c},\mathbf{d}\in\mathbb{R}^3$.*

(a) $\mathbf{a}\times(\mathbf{b}\times\mathbf{c}) = (\mathbf{a}\cdot\mathbf{c})\mathbf{b} - (\mathbf{a}\cdot\mathbf{b})\mathbf{c}$
(b) $(\mathbf{a}\times\mathbf{b})\cdot(\mathbf{c}\times\mathbf{d}) = (\mathbf{a}\cdot\mathbf{c})(\mathbf{b}\cdot\mathbf{d}) - (\mathbf{a}\cdot\mathbf{d})(\mathbf{b}\cdot\mathbf{c})$ *(Lagrange identity)*
(c) $|\mathbf{a}\times\mathbf{b}|^2 = |\mathbf{a}|^2|\mathbf{b}|^2 - (\mathbf{a}\cdot\mathbf{b})^2$

Proof Some of these proofs are very tedious when carried out in vector notation, using Eq. (10.6) to evaluate the cross product. Index notation is much more efficient.

(a) Using the symmetry properties of the Levi-Civita symbol, in particular (LC1), and the identity (LC4), we have

$$(\mathbf{a}\times(\mathbf{b}\times\mathbf{c}))_i = \epsilon_{ijk}a_j(\mathbf{b}\times\mathbf{c})_k = \epsilon_{ijk}\epsilon_{kmn}a_jb_mc_n = \epsilon_{kij}\epsilon_{kmn}a_jb_mc_n$$

$$= (\delta_{im}\delta_{jn} - \delta_{in}\delta_{jm})a_jb_mc_n = a_jc_jb_i - a_jb_jc_i$$

$$= \mathbf{a}\cdot\mathbf{c}\,b_i - \mathbf{a}\cdot\mathbf{b}\,c_i = ((\mathbf{a}\cdot\mathbf{c})\mathbf{b} - (\mathbf{a}\cdot\mathbf{b})\mathbf{c})_i \ .$$

(b) Proving the Lagrange identity from (LC4) is even easier.

$$(\mathbf{a}\times\mathbf{b})\cdot(\mathbf{c}\times\mathbf{d}) = \epsilon_{ijk}\epsilon_{imn}a_jb_kc_md_n = (\delta_{jm}\delta_{kn} - \delta_{jn}\delta_{km})a_jb_kc_md_n$$

$$= (\mathbf{a}\cdot\mathbf{c})(\mathbf{b}\cdot\mathbf{d}) - (\mathbf{a}\cdot\mathbf{d})(\mathbf{b}\cdot\mathbf{c})$$

(c) This identity follows by setting $\mathbf{c}=\mathbf{a}$ and $\mathbf{d}=\mathbf{b}$ in the Lagrange identity. $\qquad\square$

The cross product of any vector with itself vanishes (Eq. (10.4)) and the following Corollary is a generalization of this statement.

Corollary 10.1 *Vectors* $\mathbf{v}, \mathbf{w} \in \mathbb{R}^3$ *are multiples of each other if and only if* $\mathbf{v} \times \mathbf{w} = \mathbf{0}$.

Proof '$\Rightarrow$': This is the easier direction. Assume that $\mathbf{v}$ and $\mathbf{w}$ are multiples of each other, for example $\mathbf{w} = \alpha \mathbf{v}$. Then $\mathbf{v} \times \mathbf{w} = \mathbf{v} \times (\alpha \mathbf{v}) = \alpha \mathbf{v} \times \mathbf{v} = \mathbf{0}$, where we have used linearity of the cross product and, in the final step, Eq. (10.4).

'$\Leftarrow$': Now assume that $\mathbf{v} \times \mathbf{w} = \mathbf{0}$. Using Prop. 10.2 (c) this implies $0 = |\mathbf{v} \times \mathbf{w}|^2 = |\mathbf{v}|^2 |\mathbf{w}|^2 - (\mathbf{v} \cdot \mathbf{w})^2$ and, hence, that $\mathbf{v}$ and $\mathbf{w}$ satisfy Cauchy–Schwarz with an equality. From Theorem 9.7 this means that $\mathbf{v}$ and $\mathbf{w}$ are multiples of each other. $\square$

We can now show that we can always construct an $\mathbb{R}^3$ basis of mutually orthogonal vectors, starting from a given non-zero vector.

Corollary 10.2 *For any non-zero vector* $\mathbf{w}_1 \in \mathbb{R}^3$ *there exists a basis* $(\mathbf{w}_1, \mathbf{w}_2, \mathbf{w}_3)$ *of* $\mathbb{R}^3$ *of mutually orthogonal vectors.*

Proof We need to construct two further vectors which are orthogonal to $\mathbf{w}_1$ (and to each other). Since $\mathbf{w}_1$ is non-zero it has at least one non-zero component, say $w_{11} \neq 0$. Hence, $\mathbf{w}_1$ and $\mathbf{e}_3$ are not multiples of each other, so from Cor. 10.1, $\mathbf{w}_2 := \mathbf{w}_1 \times \mathbf{e}_3$ is non-zero. If we further define $\mathbf{w}_3 = \mathbf{w}_1 \times \mathbf{w}_2$ then the vectors $(\mathbf{w}_1, \mathbf{w}_2, \mathbf{w}_3)$ are mutually orthogonal and, hence, from Prop. 9.3 they are linearly independent. Since we are in a three-dimensional space, they must form a basis. $\square$

Of course we can normalize the vectors $\mathbf{w}_i$ in the above corollary to unit length to obtain an orthonormal basis.

10.1.6   Geometrical interpretation of the cross product

For a geometrical interpretation we first recall that the cross product is orthogonal to both its arguments. In order to find an interpretation for its length we use Prop. 10.2 (c) which leads to

$$|\mathbf{v} \times \mathbf{w}| = (|\mathbf{v}|^2 |\mathbf{w}|^2 - (\mathbf{v} \cdot \mathbf{w}))^{\frac{1}{2}} = |\mathbf{v}| \, |\mathbf{w}| \left(1 - \frac{(\mathbf{v} \cdot \mathbf{w})^2}{|\mathbf{v}|^2 |\mathbf{w}|^2} \right)^{\frac{1}{2}}$$

$$= |\mathbf{v}| \, |\mathbf{w}| \sin(\sphericalangle(\mathbf{v}, \mathbf{w})) \tag{10.11}$$

This formula together with Fig. 10.1 implies that the length $|\mathbf{v} \times \mathbf{w}|$ of the cross product equals the area of the parallelogram, that is

$$\text{Par}(\mathbf{v}, \mathbf{w}) := \{\alpha \mathbf{v} + \beta \mathbf{w} \,|\, \alpha, \beta \in [0, 1]\}\,, \qquad \text{Area}(\text{Par}(\mathbf{v}, \mathbf{w})) = |\mathbf{v} \times \mathbf{w}|\,. \tag{10.12}$$

While this statement is intuitively clear, a formal proof requires defining the notion of area first. This is really part of another subject, called *measure theory*, which will not be covered here. However, we can sharpen the argument somewhat if we accept two plausible properties of the area:

(*i*) $\text{Area}(\text{Par}(\mathbf{v}, \mathbf{w})) = |\mathbf{v}| \, |\mathbf{w}| \quad$ if $\mathbf{v} \perp \mathbf{w}$;

(*ii*) $\text{Area}(\text{Par}(\mathbf{v}, \mathbf{w})) = \text{Area}(\text{Par}(\mathbf{v}, \mathbf{w} + \alpha \mathbf{v})) \quad \forall \mathbf{v}, \mathbf{w} \in \mathbb{R}^3, \, \forall \alpha \in \mathbb{R}$.

The first statement is simply the formula for the area of a rectangle, the second one says that the area is invariant under a *shear*. (An example of this is the parallelogram

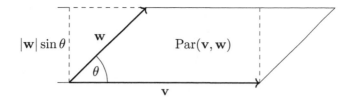

Fig. 10.1 Geometrical interpretation of the length of the cross product $|\mathbf{v} \times \mathbf{w}|$ which equals the area of the parallelogram defined by $\mathbf{v}$ and $\mathbf{w}$.

and the dashed rectangle in Fig. 10.1 which are related by a shear and have the same area.) For $\mathbf{v}, \mathbf{w} \in \mathbb{R}^3$, we can define $\mathbf{w}' = \mathbf{w} + \alpha\mathbf{v}$ with $\alpha = -(\mathbf{v} \cdot \mathbf{w})/|\mathbf{v}|^2$, so that $\mathbf{w}' \perp \mathbf{v}$. Then the area formula (10.12) can be shown as follows:

$$|\mathbf{v} \times \mathbf{w}| = |\mathbf{v} \times (\mathbf{w}' - \alpha\mathbf{v}| \overset{(10.4)}{=} |\mathbf{v} \times \mathbf{w}'|$$

$$\overset{(10.11)}{=} |\mathbf{v}|\,|\mathbf{w}'| \overset{(i)}{=} \text{Area}(\text{Par}(\mathbf{v}, \mathbf{w}')) \overset{(ii)}{=} \text{Area}(\text{Par}(\mathbf{v}, \mathbf{w})) \ .$$

Problem 10.2 (*Parallelogram area from the cross product*)

Work out the cross product of $\mathbf{a} = (1, -2, 0)^T$ and $\mathbf{b} = (3, 0, -1)^T$. Verify that $\mathbf{a} \times \mathbf{b}$ is perpendicular to $\mathbf{a}$ and $\mathbf{b}$ and calculate the area of the parallelogram $\text{Par}(\mathbf{a}, \mathbf{b})$.

Solution: From Eq. (10.6), the cross product is:

$$\mathbf{c} := \mathbf{a} \times \mathbf{b} = \begin{pmatrix} 1 \\ -2 \\ 0 \end{pmatrix} \times \begin{pmatrix} 3 \\ 0 \\ -1 \end{pmatrix} = \begin{pmatrix} 2 \\ 1 \\ 6 \end{pmatrix} \ . \tag{10.13}$$

It follows immediately that $\mathbf{c} \cdot \mathbf{a} = 1 \cdot 2 + (-2) \cdot 1 = 0$ and $\mathbf{c} \cdot \mathbf{b} = 2 \cdot 3 + (-1) \cdot 6 = 0$, so that the cross product $\mathbf{a} \times \mathbf{b}$ is indeed perpendicular to $\mathbf{a}$ and $\mathbf{b}$. Using Eq. (10.12), the area of the parallelogram defined by $\mathbf{a}$ and $\mathbf{b}$ is given by $\text{Par}(\mathbf{a}, \mathbf{b}) = |\mathbf{a} \times \mathbf{b}| = \sqrt{2^2 + 1^2 + 6^2} = \sqrt{41}$. The area of the triangle defined by $\mathbf{a}$ and $\mathbf{b}$ is half the area of the parallelogram, that is $\sqrt{41}/2$ for the example.

Application 10.1 *Kinetic energy of a rotating rigid body*

We would like to discuss an application of some of the above identities and index techniques in the context of classical mechanics. Specifically, the task is to work out a formula for the kinetic energy of a rigid body which rotates around the origin O, as in the figure below.

To avoid having to carry out integrals we will think of the rigid body as consisting of (a possibly large number of) mass points, labelled by an index α, each with mass m_α, position vector $\mathbf{r}_\alpha$ and velocity $\mathbf{v}_\alpha$. The total kinetic energy of this body is of course obtained by summing over the kinetic energy of all mass points, that is, $E_{\text{kin}} = \frac{1}{2} \sum_\alpha m_\alpha \mathbf{v}_\alpha^2$.

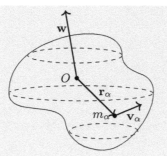

Since the body is rigid, the velocities of the mass points are not independent but are related to their positions by $\mathbf{v}_\alpha = \boldsymbol{\omega} \times \mathbf{r}_\alpha$, where $\boldsymbol{\omega}$ is the angular velocity of the body. (The length $|\boldsymbol{\omega}|$ is the angular speed and the direction of $\boldsymbol{\omega}$ indicates the axis of rotation.) The kinetic energy of the rotating body can then be written as

$$E_{\text{kin}} = \frac{1}{2}\sum_\alpha m_\alpha \mathbf{v}_\alpha^2 = \frac{1}{2}\sum_a m_\alpha \mid \boldsymbol{\omega} \times \mathbf{r}_\alpha \mid^2 = \frac{1}{2}\sum_\alpha m_\alpha (\mid \boldsymbol{\omega} \mid^2 \mid \mathbf{r}_\alpha \mid^2 - (\boldsymbol{\omega} \cdot \mathbf{r}_\alpha)^2)$$

$$= \frac{1}{2}\sum_\alpha m_\alpha (\omega_i \omega_j \delta_{ij} \mid \mathbf{r}_\alpha \mid^2 - w_i r_{\alpha i} w_j r_{\alpha j}) = \frac{1}{2}\omega_i \underbrace{\left[\sum_\alpha m_\alpha (\mid \mathbf{r}_\alpha \mid^2 \delta_{ij} - r_{\alpha i} r_{\alpha j})\right]}_{=:\, I_{ij}} \omega_j \ ,$$

where we have used Prop. (10.2) (c) in the third step.

The object in the square bracket, denoted by I_{ij}, is called the *moment of inertia tensor*, a characteristic quantity of the rigid body. It plays a role in rotational motion analogous to that of regular mass in linear motion. In terms of the moment of inertia tensor, the total kinetic energy of the rigid body can be written as

$$E_{\text{kin}} = \frac{1}{2}I_{ij}\omega_i\omega_j \ . \tag{10.14}$$

This relation is of fundamental importance for the mechanics of rigid bodies, in particular the motion of tops. (For more on the mechanics of rigid bodies see, for example, Goldstein 2013; Landau and Lifshitz 1982.)

10.2   The triple product

Summary 10.2 *The triple product is a map $\mathbb{R}^3 \times \mathbb{R}^3 \times \mathbb{R}^3 \to \mathbb{R}$ obtained by combining the cross and dot products. It is identical to the three-dimensional determinant. Three vectors in $\mathbb{R}^3$ form a basis if and only if their triple product is non-vanishing. The absolute value of the triple product equals the volume of the parallelepiped defined by its arguments.*

10.2.1   Definition of triple product

The dot and cross products in $\mathbb{R}^3$ can be combined to a new product, the *triple product* or *determinant*, a map $\det : \mathbb{R}^3 \times \mathbb{R}^3 \times \mathbb{R}^3 \to \mathbb{R}$ defined by

$$\det(\mathbf{a}, \mathbf{b}, \mathbf{c}) := \mathbf{a} \cdot (\mathbf{b} \times \mathbf{c}) = \epsilon_{ijk} a_i b_j c_k \; . \tag{10.15}$$

The index expression on the right-hand side follows from $\mathbf{a} \cdot (\mathbf{b} \times \mathbf{c}) = a_i (\mathbf{b} \times \mathbf{c})_i = \epsilon_{ijk} a_i b_j c_k$. The determinant of a 3×3 matrix $A = (\mathbf{a}, \mathbf{b}, \mathbf{c})$ with columns $\mathbf{a}$, $\mathbf{b}$ and $\mathbf{c}$ is defined as $\det(A) = \det(\mathbf{a}, \mathbf{b}, \mathbf{c})$. The basic properties of the determinant are summarized in the following proposition.

Proposition 10.3 *The determinant* (10.15) *satisfies the following properties.*

(T1) It is linear in each of its three arguments, for example $\det(\alpha \mathbf{a} + \beta \mathbf{b}, \mathbf{c}, \mathbf{d}) = \alpha \det(\mathbf{a}, \mathbf{c}, \mathbf{d}) + \beta \det(\mathbf{b}, \mathbf{c}, \mathbf{d})$ *and analogously for the second and third argument.*

(T2) Exchanging two arguments flips the sign, for example $\det(\mathbf{a}, \mathbf{b}, \mathbf{c}) = -\det(\mathbf{b}, \mathbf{a}, \mathbf{c})$.

(T3) $\det(\mathbf{e}_1, \mathbf{e}_2, \mathbf{e}_3) = 1$.

Proof (T1) This property follows directly from Eq. (10.15) and the bi-linearity of the dot and cross product. We can also proceed more explicitly and use the index version expression in Eq. (10.15):

$$\det(\alpha \mathbf{a} + \beta \mathbf{b}, \mathbf{c}, \mathbf{d}) = \epsilon_{ijk}(\alpha a_i + \beta b_i) c_j d_k = \alpha \epsilon_{ijk} a_i c_j b_k + \beta \epsilon_{ijk} b_i c_j d_k$$
$$= \alpha \det(\mathbf{a}, \mathbf{c}, \mathbf{d}) + \beta \det(\mathbf{b}, \mathbf{c}, \mathbf{d}) \; .$$

The proofs for the second and third argument work complete analogously.

(T2) This is a direct consequence of the symmetry properties (LC1) and (LC2) of the Levi-Civita symbol in Lemma 10.1. For example,

$$\det(\mathbf{a}, \mathbf{b}, \mathbf{c}) = \epsilon_{ijk} a_i b_j c_k = -\epsilon_{jik} b_j a_i c_k = -\det(\mathbf{b}, \mathbf{a}, \mathbf{c}) \; .$$

(T3) From Def. 10.1 (C3) and Eq. (10.15) we have

$$\det(\mathbf{e}_1, \mathbf{e}_2, \mathbf{e}_3) = \mathbf{e}_1 \cdot (\mathbf{e}_2 \times \mathbf{e}_3) = \mathbf{e}_1 \cdot \mathbf{e}_1 = 1 \; .$$

$\square$

This proposition can be summarized by saying that the determinant is multi-linear (T1), that is it totally anti-symmetric (T2) and that it is normalized to one (T3). We note that (T2) implies that any determinant with two same arguments vanishes, for example

$$\det(\mathbf{a}, \mathbf{a}, \mathbf{b}) = 0 \; . \tag{10.16}$$

In Chapter 18 we will discuss the determinant more generally and in arbitrary dimensions and we will use precisely those properties for its axiomatic definition. For now, we focus on the three-dimensions case.

10.2.2 Calculation of the determinant

One way to calculate three-dimensional determinants is to carry out the cross and dot product from definition (10.15). Alternatively, we can write out the six terms which appear on the right-hand side of Eq. (10.15) explicitly by using the definition (10.7) of the Levi-Civita symbol. This leads to

$$\det(\mathbf{a}, \mathbf{b}, \mathbf{c}) = \det \begin{pmatrix} a_1 & b_1 & c_1 & a_1 & b_1 \\ a_2 & b_2 & c_2 & a_2 & b_2 \\ a_3 & b_3 & c_3 & a_3 & b_3 \end{pmatrix} = \begin{cases} +a_1 b_2 c_3 + a_2 b_3 c_1 + a_3 b_1 c_2 \\ -a_1 b_3 c_2 - a_2 b_1 c_3 - a_3 b_2 c_1 \end{cases}.$$

$$(10.17)$$

We have arranged the components of the three vectors into a 3×3 matrix and have repeated the first and second column after the vertical bar on the right in order to simplify notation. The components encircled by solid lines are each multiplied together to form the first three terms with a positive sign in Eq. (10.17) while the components encircled by dashed lines lead to the last three terms, with a negative sign.

Problem 10.3 *(Calculating the three-dimensional determinant)*

Calculate the determinant $\det(\mathbf{a}, \mathbf{b}, \mathbf{c}) = \mathbf{a} \cdot (\mathbf{b} \times \mathbf{c})$ for the three vectors $\mathbf{a} = (-1, 2, -3)^T$, $\mathbf{b} = (-2, 5, 1)^T$, and $\mathbf{c} = (4, -6, 3)^T$.

Solution: One method is to first work out the cross product between $\mathbf{b}$ and $\mathbf{c}$ and then dot the result with $\mathbf{a}$:

$$\mathbf{b} \times \mathbf{c} = \begin{pmatrix} -2 \\ 5 \\ 1 \end{pmatrix} \times \begin{pmatrix} 4 \\ -6 \\ 3 \end{pmatrix} = \begin{pmatrix} 21 \\ 10 \\ -8 \end{pmatrix} \quad \Rightarrow \quad \mathbf{a} \cdot (\mathbf{b} \times \mathbf{c}) = \begin{pmatrix} -1 \\ 2 \\ -3 \end{pmatrix} \cdot \begin{pmatrix} 21 \\ 10 \\ -8 \end{pmatrix} = 23 .$$

Alternatively and equivalently, we can use the rule (10.17) which gives

$$\det(\mathbf{a}, \mathbf{b}, \mathbf{c}) = \det \begin{pmatrix} -1 & -2 & 4 \\ 2 & 5 & -6 \\ -3 & 1 & 3 \end{pmatrix} = \begin{aligned} &(-1) \cdot 5 \cdot 3 + (-2) \cdot (-6) \cdot (-3) + 4 \cdot 2 \cdot 1 \\ &- (-1) \cdot (-6) \cdot 1 - (-2) \cdot 2 \cdot 3 - 4 \cdot 5 \cdot (-3) \end{aligned}$$

$$= -15 - 36 + 8 - 6 + 12 + 60 = 23 .$$

The determinant has many more properties and applications which we will explore in more depth later. For now, we prove the following criterion for a basis of $\mathbb{R}^3$.

Theorem 10.1 $(\mathbf{v}_1, \mathbf{v}_2, \mathbf{v}_3)$ *is a basis of* $\mathbb{R}^3$ *iff* $\det(\mathbf{v}_1, \mathbf{v}_2, \mathbf{v}_3) \neq 0$.

Proof '$\Rightarrow$': If $(\mathbf{v}_1, \mathbf{v}_2, \mathbf{v}_3)$ is a basis of $\mathbb{R}^3$ then the standard unit vectors can be written as $\mathbf{e}_i = \alpha_{ij} \mathbf{v}_j$ (sum over j implied) for some $\alpha_{ij} \in \mathbb{R}$. Using linearity of the determinant we find

$$1 = \det(\mathbf{e}_1, \mathbf{e}_2, \mathbf{e}_3) = \det(\alpha_{1j} \mathbf{v}_j, \alpha_{2k} \mathbf{v}_k, \alpha_{3l} \mathbf{v}_l) = \alpha_{1j} \alpha_{2k} \alpha_{3l} \det(\mathbf{v}_j, \mathbf{v}_k, \mathbf{v}_l) , \quad (10.18)$$

where sums over j, k, l are implied. If any two of the indices (i, j, k) are the same then $\det(\mathbf{v}_j, \mathbf{v}_k, \mathbf{v}_l) = 0$. Hence, only terms with $\{j, k, l\} = \{1, 2, 3\}$ contribute to the sum on the right. Due to the anti-symmetry of the determinant each of these terms is proportional to $\det(\mathbf{v}_1, \mathbf{v}_2, \mathbf{v}_3)$, so if this determinant is zero then so is the right-hand side of Eq. (10.18). This is of course a contradiction and, hence, $\det(\mathbf{v}_1, \mathbf{v}_2, \mathbf{v}_3) \neq 0$.

'$\Leftarrow$': We proof this indirectly so assume that $(\mathbf{v}_1, \mathbf{v}_2, \mathbf{v}_3)$ is not a basis. This means the vector are not linearly independent and one of them, say $\mathbf{v}_3$, can be written as a linear combination of the other two, $\mathbf{v}_3 = \alpha_1 \mathbf{v}_1 + \alpha_2 \mathbf{v}_3$. From linearity of the determinant it follows that

$$\det(\mathbf{v}_1, \mathbf{v}_2, \mathbf{v}_3) = \det(\mathbf{v}_1, \mathbf{v}_2, \alpha_1 \mathbf{v}_1 + \alpha_2 \mathbf{v}_2)$$
$$= \alpha_1 \det(\mathbf{v}_1, \mathbf{v}_2, \mathbf{v}_1) + \alpha_2 \det(\mathbf{v}_1, \mathbf{v}_2, \mathbf{v}_2) \overset{(10.4)}{=} 0 .$$

$\square$

The above theorem allows us to derive an explicit formula for the coordinates relative to a basis $(\mathbf{v}_1, \mathbf{v}_2, \mathbf{v}_3)$ of $\mathbb{R}^3$. Write an arbitrary vector $\mathbf{v} \in \mathbb{R}^3$ as a linear combination $\mathbf{v} = \alpha_1 \mathbf{v}_1 + \alpha_2 \mathbf{v}_2 + \alpha_3 \mathbf{v}_3$. Forming the dot product of this expression with the cross product $\mathbf{v}_2 \times \mathbf{v}_3$ (and remembering that both $\mathbf{v}_2$ and $\mathbf{v}_3$ are orthogonal to the cross product) gives the formula $\det(\mathbf{v}, \mathbf{v}_2, \mathbf{v}_3) = \alpha_1 \det(\mathbf{v}_1, \mathbf{v}_2, \mathbf{v}_3)$. Since $\det(\mathbf{v}_1, \mathbf{v}_2, \mathbf{v}_3) \neq 0$ from the theorem we can divide and obtain a formula for α_1. The other two coordinates are obtained analogously (by dotting with $\mathbf{v}_3 \times \mathbf{v}_1$ for α_2 and $\mathbf{v}_1 \times \mathbf{v}_3$ for α_3). As a result, a vector $\mathbf{v} = \alpha_1 \mathbf{v}_1 + \alpha_2 \mathbf{v}_2 + \alpha_3 \mathbf{v}_3$ has the coordinates

$$\alpha_1 = \frac{\det(\mathbf{v}, \mathbf{v}_2, \mathbf{v}_3)}{\det(\mathbf{v}_1, \mathbf{v}_2, \mathbf{v}_3)} , \quad \alpha_2 = \frac{\det(\mathbf{v}_1, \mathbf{v}, \mathbf{v}_3)}{\det(\mathbf{v}_1, \mathbf{v}_2, \mathbf{v}_3)} , \quad \alpha_3 = \frac{\det(\mathbf{v}_1, \mathbf{v}_2, \mathbf{v})}{\det(\mathbf{v}_1, \mathbf{v}_2, \mathbf{v}_3)} , \quad (10.19)$$

relative to the basis $(\mathbf{v}_1, \mathbf{v}_2, \mathbf{v}_3)$.

Problem 10.4 *(Coordinates relative to a basis in $\mathbb{R}^3$)*

Find the coordinates of a general vector $\mathbf{v} = (x, y, z)^T$ relative to the basis $(\mathbf{v}_1, \mathbf{v}_2, \mathbf{v}_3)$, where $\mathbf{v}_1 = (1, 1, 0)^T$, $\mathbf{v}_2 = (1, 0, -1)^T$, and $\mathbf{v}_3 = (0, 1, 2)^T$.

Solution: We can use the 'pedestrian' methods and write down the equation

$$\mathbf{v} = \begin{pmatrix} x \\ y \\ z \end{pmatrix} = \alpha_1 \mathbf{v}_1 + \alpha_2 \mathbf{v}_2 + \alpha_3 \mathbf{v}_3 = \begin{pmatrix} \alpha_1 + \alpha_2 \\ \alpha_1 + \alpha_3 \\ -\alpha_2 + 2\alpha_3 \end{pmatrix} .$$

Splitting up into components and solving the resulting three equations, $x = \alpha_1 + \alpha_2$, $y = \alpha_1 + \alpha_3$ and $z = -\alpha_2 + 2\alpha_3$ for x, y, z, gives the coordinates

$$\alpha_1 = -x + 2y - z , \quad \alpha_2 = 2x - 2y + z , \quad \alpha_3 = x - y + z .$$

Alternatively, we can use the formulae (10.19). First, since $\det(\mathbf{v}_1, \mathbf{v}_2, \mathbf{v}_3) = -1$, we can conclude from Theorem 10.1 that $(\mathbf{v}_1, \mathbf{v}_2, \mathbf{v}_3)$ is indeed a basis of $\mathbb{R}^3$. For the determinants in the numerators of (10.19) we find

$$\det(\mathbf{v}, \mathbf{v}_2, \mathbf{v}_3) = x - 2y + z , \quad \det(\mathbf{v}, \mathbf{v}_3, \mathbf{v}_1) = -2x + 2y - z , \quad \det(\mathbf{v}, \mathbf{v}_1, \mathbf{v}_2) = -x + y - z .$$

Dividing by $\det(\mathbf{v}_1, \mathbf{v}_2, \mathbf{v}_3) = -1$ then leads to the same results for the coordinates.

10.2.3   Interpretation of the triple product

We have argued earlier that the length of the cross product equals the area of the parallelogram defined by its two arguments. We would expect the (absolute value of the) triple product is equal to the volume of the parallelepiped (see Fig. 10.2)

$$\mathrm{Par}(\mathbf{v}_1, \mathbf{v}_2, \mathbf{v}_2) = \left\{ \sum_{i=1}^{3} \alpha_i \mathbf{v}_i \,|\, \alpha_i \in [0,1] \right\} , \qquad (10.20)$$

that is,

$$\mathrm{Vol}(\mathrm{Par}(\mathbf{v}_1, \mathbf{v}_2, \mathbf{v}_2)) = |\det(\mathbf{v}_1, \mathbf{v}_2, \mathbf{v}_2)| . \qquad (10.21)$$

To see this we can use arguments similar to the ones we have used for the cross product, based on invariance under shears. We assume (i) the volume formula of a

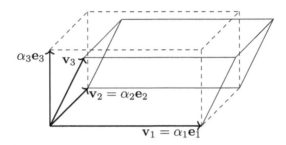

Fig. 10.2 Cuboid with edges $(\alpha_1 \mathbf{e}_1, \alpha_2 \mathbf{e}_2, \alpha_3 \mathbf{e}_3)$, sheared to a parallelepiped with the same volume.

cuboid, so that, for mutually orthogonal vector $\mathbf{v}_i$, we have $\mathrm{Vol}(\mathrm{Par}(\mathbf{v}_1, \mathbf{v}_2, \mathbf{v}_3)) = |\mathbf{v}_1| |\mathbf{v}_2| |\mathbf{v}_3|$ and (ii) the invariance of the volume under shear, so that, for example, $\mathrm{Vol}(\mathrm{Par}(\mathbf{v}_1, \mathbf{v}_2, \mathbf{v}_3)) = \mathrm{Vol}(\mathrm{Par}(\mathbf{v}_1, \mathbf{v}_2, \mathbf{v}_3 + \alpha \mathbf{v}_1))$ (and similar for the other arguments). If we start with mutually orthogonal vectors $\mathbf{v}_i = \alpha_i \mathbf{e}_i$, proportional to the standard unit vectors, which form a cuboid we have

$$|\det(\mathbf{v}_1, \mathbf{v}_2, \mathbf{v}_3)| = |\alpha_1| |\alpha_2| \alpha_3| = \mathrm{Vol}(\mathrm{Par}(\mathbf{v}_1, \mathbf{v}_2, \mathbf{v}_3)) ,$$

where we have used multi-linearity of the determinant in the first step. The key observation is that the determinant is invariant under shears as well since, for example

$$\det(\mathbf{v}_1, \mathbf{v}_2, \mathbf{v}_3 + \alpha \mathbf{v}_1) = \det(\mathbf{v}_1, \mathbf{v}_2, \mathbf{v}_3) + \alpha \det(\mathbf{v}_1, \mathbf{v}_2, \mathbf{v}_1) \overset{(18.1)}{=} \det(\mathbf{v}_1, \mathbf{v}_2, \mathbf{v}_3) .$$

Every parallelepiped can be obtained from a cuboid by shears. Given that the values of the determinant and the volume agree for cuboids and both quantities are unchanged under shears the argument is complete.

Problem 10.5 (*Volume of a parallelepiped from the triple product*)

Find the volume of the parallelepiped defined by the vectors $\mathbf{v}_1 = (1, 1, 0)^T$, $\mathbf{v}_2 = (1, 0, -1)^T$, and $\mathbf{v}_3 = (0, 1, 2)^T$.

Solution: A quick calculation shows that $\det(\mathbf{v}_1, \mathbf{v}_2, \mathbf{v}_3) = -1$, so $\mathrm{Vol}(\mathrm{Par}(\mathbf{v}_1, \mathbf{v}_2, \mathbf{v}_3)) = 1$.

Exercises

(†=challenging)

10.1 (a) For $\mathbf{a} = (2, -3, 1)^T$, $\mathbf{b} = (1, 0, -5)^T$, and $\mathbf{c} = (-1, 1, 2)^T$, compute the cross products $\mathbf{a} \times \mathbf{b}$, $\mathbf{a} \times \mathbf{c}$ and $\mathbf{b} \times \mathbf{c}$ and the triple product $\mathbf{a} \cdot (\mathbf{b} \times \mathbf{c})$.
(b) Work out the determinants of the matrices

$$A = \begin{pmatrix} 1 & 0 & -2 \\ 0 & 3 & -1 \\ 2 & 2 & -5 \end{pmatrix}, \quad B = \begin{pmatrix} \frac{1}{2} & 0 & -\frac{1}{6} \\ \frac{2}{3} & \frac{1}{3} & 1 \\ 0 & -\frac{1}{6} & -\frac{1}{2} \end{pmatrix}.$$

10.2 For vectors $\mathbf{a}, \mathbf{b}, \mathbf{c} \in \mathbb{R}^3$, show the following:
(a) $\mathbf{a} \times \mathbf{b} = \mathbf{a} - \mathbf{b}$ implies that $\mathbf{a} = \mathbf{b}$;
(b) $\mathbf{c} = \lambda \mathbf{a} + \mu \mathbf{b}$ implies that $(\mathbf{a} \times \mathbf{b}) \cdot \mathbf{c} = 0$;
(c) $(\mathbf{a} \times \mathbf{b}) \times (\mathbf{c} \times \mathbf{b}) = \mathbf{b}[\mathbf{b} \cdot (\mathbf{a} \times \mathbf{c})]$;
(d) If $\mathbf{c} = \mathbf{a} + \mathbf{b}$ then $|\mathbf{a} \times \mathbf{b}| = |\mathbf{b} \times \mathbf{c}| = |\mathbf{c} \times \mathbf{a}|$.

10.3 Show that the cross product is not associative.

10.4 Use index notation to re-write the following expressions in a simpler form.
(a) $\mathbf{a} \cdot (\mathbf{b} \times (\mathbf{a} \times \mathbf{c}))$
(b) $\mathbf{a} \times (\mathbf{b} \times (\mathbf{c} \times \mathbf{a}))$

10.5 For two linearly independent vectors $\mathbf{u}, \mathbf{w} \in \mathbb{R}$, define the lines $U = \mathrm{Span}(\mathbf{u})$ and $W = \mathrm{Span}(\mathbf{w})$ and select non-zero vectors $\mathbf{u}' \in U$ and $\mathbf{w}' \in W$, such that $\mathbf{u}' - \mathbf{w}'$ is a multiple of $\mathbf{u} - \mathbf{w}$.

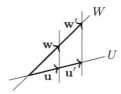

(a) Show that $|\mathbf{u}'|/|\mathbf{u}| = |\mathbf{w}'|/|\mathbf{w}|$.
(b) Show that $|\mathbf{u}' - \mathbf{w}'|/|\mathbf{u} - \mathbf{w}| = |\mathbf{u}'|/|\mathbf{u}|$.

10.6 *Two-dimensional Levi-Civita symbol*
For the two-dimensional Levi-Civita symbol ϵ_{ij} show the following identities:
(a) $\epsilon_{ij} = -\epsilon_{ji}$, $\epsilon_{ii} = 0$;
(b) $\epsilon_{ij} \epsilon_{kl} = \delta_{ik} \delta_{jl} - \delta_{il} \delta_{jk}$;
(c) $\epsilon_{ij} \epsilon_{il} = \delta_{jl}$;
(d) $\epsilon_{ij} \epsilon_{ij} = 2$.

10.7 For $\mathbf{a}, \mathbf{b} \in \mathbb{R}^2$ use index notation to show that:
(a) $(\mathbf{a}^\times)^\times = -\mathbf{a}$;
(b) $\mathbf{a}^\times \cdot \mathbf{b}^\times = \mathbf{a} \cdot \mathbf{b}$;
(c) $\mathbf{a}^\times \cdot \mathbf{b} = -\mathbf{a} \cdot \mathbf{b}^\times$.

10.8 For which values of $a, b \in \mathbb{R}$ is $(\mathbf{v}_1, \mathbf{v}_2, \mathbf{v}_3)$, with $\mathbf{v}_1 = (1, a, 2)^T$, $\mathbf{v}_2 = (0, 3, -1)^T$ and $\mathbf{v}_3 = (1, 0, b)^T$, not a basis of $\mathbb{R}^3$?

10.9 *Reciprocal vectors*[†]
Consider a basis $(\mathbf{v}_1, \mathbf{v}_2, \mathbf{v}_3)$ of $\mathbb{R}^3$ and the triple product $V := \mathbf{v}_1 \cdot (\mathbf{v}_2 \times \mathbf{v}_3)$. Define the *reciprocal vectors* $\mathbf{v}'_i$ by $\mathbf{v}'_1 = \frac{1}{V} \mathbf{v}_2 \times \mathbf{v}_3$, $\mathbf{v}'_2 = \frac{1}{V} \mathbf{v}_3 \times \mathbf{v}_1$ and $\mathbf{v}'_3 = \frac{1}{V} \mathbf{v}_1 \times \mathbf{v}_2$. Show that
(a) $\mathbf{v}_i \cdot \mathbf{v}'_j = \delta_{ij}$;
(b) $(\mathbf{v}'_1, \mathbf{v}'_2, \mathbf{v}'_3)$ is a basis of $\mathbb{R}^3$;
(c) the coordinates of $\mathbf{w} \in \mathbb{R}^3$ relative to the basis $(\mathbf{v}_1, \mathbf{v}_2, \mathbf{v}_3)$ are given by $\mathbf{w} \cdot \mathbf{v}'_i$;
(d) $V' = 1/V$, where $V' := \mathbf{v}'_1 \cdot (\mathbf{v}'_2 \times \mathbf{v}'_3)$;
(e) $\frac{1}{V'} \mathbf{v}'_2 \times \mathbf{v}'_3 = \mathbf{v}_1$, $\frac{1}{V'} \mathbf{v}'_3 \times \mathbf{v}'_1 = \mathbf{v}_2$ and $\frac{1}{V'} \mathbf{v}'_1 \times \mathbf{v}'_2 = \mathbf{v}_3$.

10.10 *Reciprocal vectors and linear equations*
(a) For a basis $(\mathbf{v}_1, \mathbf{v}_2, \mathbf{v}_3)$ of $\mathbb{R}^3$ and $\mathbf{x} \in \mathbb{R}^3$ consider the equations

$$\mathbf{v}_1 \cdot \mathbf{x} = b_1, \quad \mathbf{v}_2 \cdot \mathbf{x} = b_2, \quad \mathbf{v}_3 \cdot \mathbf{x} = b_3,$$

where $b_i \in \mathbb{R}$. Show that these equations have a unique solution for $\mathbf{x}$ and find this solution in terms of reciprocal vectors.

(b) Find the reciprocal vectors for $\mathbf{v}_1 = (1, 1, 1)^T$, $\mathbf{v}_2 = (2, 0, -1)^T$, and $\mathbf{v}_3 = (-2, 1, 1)^T$. For $b_1 = 2$, $b_2 = -3$, and $b_3 = 1$, find the solution to the system of equations in (a). Check your results.

10.11 (a) For two linearly independent vectors $\mathbf{w}_1, \mathbf{w}_2 \in \mathbb{R}^3$, and $\mathbf{w}_3 = \mathbf{w}_1 \times \mathbf{w}_2$

show that $(\mathbf{w}_1, \mathbf{w}_2, \mathbf{w}_3)$ is a basis of $\mathbb{R}^3$.

(b) For a two-dimensional vector subspace $W \subset \mathbb{R}^3$ show that there are precisely two unit length vectors which are orthogonal to all vectors in W.

10.12 *Angular momentum of a rigid body*[†]
Following the approach in Application 10.1, compute the total angular momentum of a rigid body in terms of the moment of inertia tensor.

11
Lines and planes

In this chapter we discuss some aspects of elementary geometry in $\mathbb{R}^n$, focusing on the dimensions $n = 2, 3$. These topics are not really part of linear algebra but of a related area of mathematics called affine geometry which we cannot possibly do justice to in our brief account (see, for example, Bennett 2011). However, what we do cover should be enough to appreciate the geometrical ideas underlying and motivating linear algebra and help the reader as we move forward with the formal development of the subject in the next part. Of course, geometrical applications of linear algebra are also important in science and need to be covered for this reason alone.

Affine geometry usually starts by thinking about $\mathbb{R}^n$ in two roles, as a space of points — the affine space — and as a vector space, with vectors acting on points by translation. We will refrain from making this distinction as it would add little substance to our discussion. Instead, we work with $\mathbb{R}^n$ as a vector space and think of vectors and points as being identified. By the distance between two vectors $\mathbf{v}, \mathbf{w} \in \mathbb{R}^n$, we simply mean the length $|\mathbf{v} - \mathbf{w}|$.

Our main objects of interest are affine k-planes in $\mathbb{R}^n$, that is, subsets $\mathbf{p} + W \subset \mathbb{R}^n$, where $\mathbf{p} \in \mathbb{R}^n$ is a vector and W is a k-dimensional vector subspace of $\mathbb{R}^n$. In this chapter, we will focus on the cases of lines ($k = 1$) and planes ($k = 2$) in two and three dimensions. As we will see, the dot and cross products are very useful tools for dealing with those objects. A more general discussion of affine k-planes is most easily carried out once we have developed the theory of linear systems and is, hence, postponed until Chapter 17.

In the next section, we begin with the simplest case of lines in $\mathbb{R}^2$ and their properties. They can be described in parametric or Cartesian form. In Section 11.2 we generalize this discussion to lines and planes in $\mathbb{R}^3$.

11.1 Lines in $\mathbb{R}^2$

Summary 11.1 *All lines in $\mathbb{R}^2$ can be described in parametric and in Cartesian form. The intersection of two lines in $\mathbb{R}^2$ is described by a system of two linear equations in two variables.*

11.1.1 Parametric and Cartesian form

Lines in $\mathbb{R}^2$ are subsets of the form $L = \mathbf{p} + W$, where $\mathbf{p} \in \mathbb{R}^2$ and W is a one-dimensional vector subspace of $\mathbb{R}^2$. They can be explicitly described in *parametric* or

Cartesian form.

Theorem 11.1 *The following sets are lines in $\mathbb{R}^2$.*

(i) $L_p = \{\mathbf{p} + t\mathbf{w} \,|\, t \in \mathbb{R}\}$ *with* $\mathbf{p}, \mathbf{w} \in \mathbb{R}^2$, $\mathbf{w} \neq \mathbf{0}$ *(parametric form)*

(ii) $L_c = \{\mathbf{x} \in \mathbb{R}^2 \,|\, \mathbf{n} \cdot \mathbf{x} = b\}$ *with* $\mathbf{n} \in \mathbb{R}^2$, $b \in \mathbb{R}$, $\mathbf{n} \neq \mathbf{0}$ *(Cartesian form)*

Every line can be written in the form (i) or (ii). If $\mathbf{n} \perp \mathbf{w}$ and $b = \mathbf{n} \cdot \mathbf{p}$ then $L_p = L_c$.

Proof First, we show the equality, $L_p = L_c$, under the stated conditions.

$L_p \subset L_c$: Start with an element $\mathbf{x} = \mathbf{p} + t\mathbf{w} \in L_p$. Multiply this equation with $\mathbf{n}$ (keeping in mind that $\mathbf{n} \cdot \mathbf{w} = 0$) leads to $\mathbf{n} \cdot \mathbf{x} = \mathbf{n} \cdot \mathbf{p} = b$ which shows that $\mathbf{x} \in L_c$.

$L_c \subset L_p$: Conversely, consider an element $\mathbf{x} \in L_c$ which, by definition, satisfies $\mathbf{n} \cdot \mathbf{x} = b$. Given that $b = \mathbf{n} \cdot \mathbf{p}$, this equation can be re-written as $\mathbf{n} \cdot (\mathbf{x} - \mathbf{p}) = \mathbf{0}$. Every vector $\mathbf{x} - \mathbf{p}$ can be represented as a linear combination $\mathbf{x} - \mathbf{p} = \alpha\mathbf{n} + t\mathbf{w}$ (since $(\mathbf{n}, \mathbf{w})$, as two non-zero orthogonal vectors, form a basis of $\mathbb{R}^2$). From $0 = \mathbf{n} \cdot (\mathbf{x} - \mathbf{p}) = \alpha|\mathbf{n}|^2$ we conclude that $\alpha = 0$ and, hence, $\mathbf{x} = \mathbf{p} + t\mathbf{w} \in L_p$.

It is clear that every line $\mathbf{p} + W$ can be written in parametric form by choosing a non-zero vector $\mathbf{w} \in \mathbb{R}^2$ such that $W = \mathrm{Span}(\mathbf{w})$. Conversely, every parametric form defines a line by setting $W = \mathrm{Span}(\mathbf{w})$.

To complete the proof, we show that every line in parametric form can be converted into Cartesian form and vice versa. To see the former, start with vectors $\mathbf{p}, \mathbf{w}$ defining a parametric line L_p. Setting $\mathbf{n} = \mathbf{w}^\times$ (so that $\mathbf{n} \perp \mathbf{w}$) and $b = \mathbf{p} \cdot \mathbf{n}$ defines a line in Cartesian form with $L_p = L_c$. If we start with the Cartesian line L_c, specified by $\mathbf{n}$ and b, then setting $\mathbf{w} = \mathbf{n}^\times$ and choosing $\mathbf{p}$ to be any solution of $\mathbf{n} \cdot \mathbf{p} = b$, we get a parametric line L_p with $L_p = L_c$. $\square$

The geometric interpretation of the various vectors which enter the parametric and Cartesian form is indicated in Fig. 11.1. For the parametric form, $\mathbf{x}(t) = \mathbf{p} + t\mathbf{w}$, the

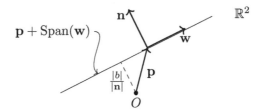

Fig. 11.1 Line in $\mathbb{R}^2$ with parametric form $\mathbf{x}(t) = \mathbf{p} + t\mathbf{w}$ and Cartesian form $\mathbf{n} \cdot \mathbf{x} = b$.

vector $\mathbf{p}$ is a vector 'to the line' and the vector $\mathbf{w}$ a vector 'along the line'. For the Cartesian form, $\mathbf{n} \cdot \mathbf{x} = b$, the vector $\mathbf{n}$ is a vector orthogonal to the line while $|b|/|\mathbf{n}|$ is the minimal distance of the line from the origin. To verify the last statement, write the parametric form of a line as

$$|\mathbf{n}|\,|\mathbf{x}|\,\cos \sphericalangle(\mathbf{n}, \mathbf{x}) = b \,. \tag{11.1}$$

The value of $|\mathbf{x}|$ is minimal when $|\cos \sphericalangle(\mathbf{n}, \mathbf{w})|$ takes its maximal value which is, of course, $|\cos \sphericalangle(\mathbf{n}, \mathbf{w})| = 1$. It follows that the minimal value of $|\mathbf{x}|$ is indeed $|b|/|\mathbf{n}|$.

We should stress that neither the parametric nor the Cartesian form of a line is unique, that is, one and the same line can be described by different equations of this form. For example, in the parametric form we can use instead of $\mathbf{w}$ any non-zero multiple $\beta\mathbf{w}$ or, instead of $\mathbf{p}$ any other vector $\mathbf{p} + \alpha\mathbf{w}$ to the line. For the Cartesian form, using $\alpha\mathbf{n}$ and αb instead of $\mathbf{n}$ and b describes the same line. Theorem 11.1 indicates a method to convert a line in parametric form to Cartesian form and vice versa. It is worth practising this by looking at examples.

Problem 11.1 *(Converting a line from parametric into Cartesian form)*

Convert the parametric form

$$\mathbf{x}(t) = \begin{pmatrix} 1 \\ 2 \end{pmatrix} + t \begin{pmatrix} 1 \\ -3 \end{pmatrix}$$

of a line into Cartesian form. Find the minimal distance of the line from the origin.

Solution: With $\mathbf{p} = (1,2)^T$, $\mathbf{w} = (1,-3)^T$, and $\mathbf{x} = (x,y)^T$ we have $\mathbf{n} = \mathbf{w}^\times = (3,1)^T$ and $b = \mathbf{n} \cdot \mathbf{p} = 5$. Hence, the Cartesian form of the line is

$$\begin{pmatrix} 3 \\ 1 \end{pmatrix} \cdot \mathbf{x} = 5 \quad \text{or} \quad 3x + y = 5 \,.$$

It minimal distance from the origin is $|b|/|\mathbf{n}| = 5/\sqrt{3^2 + 1^2} = \sqrt{5/2}$.

Problem 11.2 *(Converting a line from Cartesian into parametric form)*

A line in Cartesian form is given by $2x - 5y = 7$. Find its parametric from.

Solution: We can write this equation as $\mathbf{n} \cdot \mathbf{x} = b$ with $\mathbf{n} = (2,-5)^T$, $\mathbf{x} = (x,y)^T$, and $b = 7$. A vector $\mathbf{w}$ along the line is found by $\mathbf{w} = \mathbf{n}^\times = (5,2)^T$. To find a vector $\mathbf{p}$ to the line we can use any solution to the Cartesian equation, for example, $\mathbf{p} = (1,-1)$. Hence, a parametric form of the line is

$$\mathbf{x}(t) = \begin{pmatrix} 1 \\ -1 \end{pmatrix} + t \begin{pmatrix} 5 \\ 2 \end{pmatrix} \,.$$

11.1.2   Intersection of two lines

The intersection of two lines in $\mathbb{R}^2$ can be discussed using either parametric or Cartesian forms. We opt for the latter and write the equations for the two lines as

$$L_1 = \{\mathbf{x} \in \mathbb{R}^2 \,|\, \mathbf{n}_1 \cdot \mathbf{x} = b_1\} \,, \qquad L_2 = \{\mathbf{x} \in \mathbb{R}^2 \,|\, \mathbf{n}_2 \cdot \mathbf{x} = b_2\} \,,$$

with intersection

$$L_1 \cap L_2 = \{\mathbf{x} \in \mathbb{R}^2 \,|\, \mathbf{n}_1 \cdot \mathbf{x} = b_1 \wedge \mathbf{n}_2 \cdot \mathbf{x} = b_2\} \,.$$

Evidently, the intersection $L_1 \cap L_2$ is determined by the solution to two simultaneous linear equations in two variables. With $\mathbf{n}_1 = (n_{11}, n_{12})^T$, $\mathbf{n}_2 = (n_{21}, n_{22})^T$, and $\mathbf{x} = (x,y)^T$ these can also be written as

$$n_{11}x + n_{12}y = b_1 \,, \qquad n_{21}x + n_{22}y = b_2 \,,$$

or, in matrix-vector form, as $N\mathbf{x} = \mathbf{b}$, where N is a 2×2 matrix with entries n_{ij} and $\mathbf{b} = (b_1, b_2)^T$. This example illustrates how the geometry of lines (and planes, as we will see) is related to systems of linear equations.

We already know from the introduction that there are several possibilities for the solutions to a system of two linear equations in two variables. The solution can be unique, but we can also have an entire line of solutions or there may be no solution at all, depending on the coefficients in the linear equations. In geometrical terms, these three cases correspond to the two lines intersecting in a point, being identical and being parallel (see Fig. 11.2).

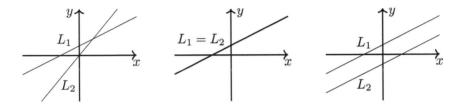

Fig. 11.2 Two lines in $\mathbb{R}^2$ can intersect in a point, in a line or do not intersect at all.

Problem 11.3 *(Intersection of two lines)*

Find the intersection of the two lines in Cartesian form with equations

$$\begin{pmatrix} 1 \\ a \end{pmatrix} \cdot \mathbf{x} = 1 \,, \qquad \begin{pmatrix} 2 \\ -3 \end{pmatrix} \cdot \mathbf{x} = b \,,$$

for all possible values of the parameters $a, b \in \mathbb{R}$.

Solution: For $\mathbf{x} = (x, y)^T$ the linear system which corresponds to the intersection of these two lines is

$$\left.\begin{array}{r} x + ay = 1 \\ 2x - 3y = b \end{array}\right\} \qquad \Rightarrow \qquad (2a + 3)y = 2 - b \,, \tag{11.2}$$

where the result on the right follows by subtracting the second equation from twice the first. We have to distinguish the following cases.

(1) $a \neq -3/2$: We can divide by $2a + 3$ and find the unique solution

$$x = \frac{ab + 3}{2a + 3} \,, \qquad y = \frac{2 - b}{2a + 3} \,.$$

This is the unique intersection point of the two lines.

(2a) $a = -3/2$ and $b = 2$: The equation on the right in (11.2) is trivial or, equivalently, the two linear equations become multiples of each other. This means the two lines are identical so that the intersection consists of the entire line with Cartesian form $2x - 3y = 2$.

(2b) $a = -3/2$ and $b \neq 2$: The equation on the right in (11.2) leads to a contradiction so there is no solution. This corresponds to the two lines being parallel (but not identical).

11.2   Lines and planes in $\mathbb{R}^3$

Summary 11.2 *Line and planes in* $\mathbb{R}^3$ *can be described in parametric or Cartesian form. Formulae for minimal distances, such as between lines and points or planes and points, can be derived using cross and dot products. The intersection of two (three) planes is described by a system of two (three) linear equations in three variables.*

Next, we discuss lines and planes in $\mathbb{R}^3$, their intersections and other related properties.

11.2.1   Parametric and Cartesian form for planes

Planes are sets of the form $P = \mathbf{p} + W \subset \mathbb{R}^3$, where $\mathbf{p} \in \mathbb{R}^3$ and $W \subset \mathbb{R}^3$ is a two-dimensional vector subspace.

Theorem 11.2 *The following sets are planes in* $\mathbb{R}^3$.

(i) $P_p = \{\mathbf{p} + t_1\mathbf{w}_1 + t_2\mathbf{w}_2 \,|\, t_1, t_2 \in \mathbb{R}\}$
      with $\mathbf{p} \in \mathbb{R}^3$ *and* $\mathbf{w}_1, \mathbf{w}_2 \in \mathbb{R}^3$ *linearly independent*   *(parametric form)*

(ii) $P_c = \{\mathbf{x} \in \mathbb{R}^3 \,|\, \mathbf{n} \cdot \mathbf{x} = b\}$ *with* $\mathbf{n} \in \mathbb{R}^3$ *nonzero,* $b \in \mathbb{R}$ *(Cartesian form)*

Every plane can be written as in (i) or (ii). If $\mathbf{n} \perp \mathbf{w}_1, \mathbf{w}_2$ *and* $b = \mathbf{p} \cdot \mathbf{n}$ *then* $P_p = P_c$.

Proof   As in theorem 11.1, we begin by showing $P_p = P_c$ under the stated conditions.

$P_p \subset P_c$: For a vector $\mathbf{x} = \mathbf{p} + t_1\mathbf{w}_1 + t_2\mathbf{w}_2 \in P_p$, take the dot product with $\mathbf{n}$, keeping in mind that $\mathbf{n} \cdot \mathbf{w}_1 = \mathbf{n} \cdot \mathbf{w}_2 = 0$. This leads to $\mathbf{n} \cdot \mathbf{x} = \mathbf{n} \cdot \mathbf{p} = b$, so that $\mathbf{x} \in P_c$.

$P_c \subset P_p$: Start with a vector $\mathbf{x} \in P_c$, so that $\mathbf{n} \cdot \mathbf{x} = b$ or, equivalently, $\mathbf{n} \cdot (\mathbf{x} - \mathbf{p}) = 0$. To solve this last equation we first note that $(\mathbf{n}, \mathbf{w}_1, \mathbf{w}_2)$ forms a basis of $\mathbb{R}^3$. This means we can write every vector $\mathbf{x} - \mathbf{p}$ as a linear combination $\mathbf{x} - \mathbf{p} = \alpha\mathbf{n} + t_1\mathbf{w}_1 + t_2\mathbf{w}_2$. From $0 = \mathbf{n} \cdot (\mathbf{x} - \mathbf{p}) = \alpha|\mathbf{n}|^2$ it follows that $\alpha = 0$ and, hence, $\mathbf{x} = \mathbf{p} + t_1\mathbf{w}_1 + t_2\mathbf{w}_2 \in P_p$.

Every plane $\mathbf{p} + W$ can be written in parametric form by choosing a basis $(\mathbf{w}_1, \mathbf{w}_2)$ of W. Conversely, a parametric form with $\mathbf{w}_1, \mathbf{w}_2$ defines a plane by setting $W = \mathrm{Span}(\mathbf{w}_1, \mathbf{w}_2)$.

Every plane in parametric form can be converted into Cartesian form. Start with vectors $\mathbf{p}, \mathbf{w}_1, \mathbf{w}_2$, specifying a parametric plane P_p. Setting $\mathbf{n} = \mathbf{w}_1 \times \mathbf{w}_2$ and $b = \mathbf{p} \cdot \mathbf{n}$ defines a Cartesian plane P_c with $P_c = P_p$.

On the other hand, for a Cartesian plane P_c, specified by $\mathbf{n}$ and p, we can find, from Cor. 10.2, mutually orthogonal vectors $(\mathbf{n}, \mathbf{w}_1, \mathbf{w}_2)$. Then, $\mathbf{w}_1, \mathbf{w}_2$ together with any solution of $\mathbf{p}$ of $\mathbf{n} \cdot \mathbf{p} = b$ defines a parametric plane P_p with $P_p = P_c$.   □

In the parametric form, $\mathbf{p}$ is a vector 'to the plane' while $\mathbf{w}_1$ and $\mathbf{w}_2$ are vectors 'along the plane'. For the Cartesian form, $\mathbf{n}$ is a vector orthogonal to every vector in the plane and $|b|/|\mathbf{n}|$ is the minimal distance of the plane from the origin. (The last statement follows from Eq. (11.1) in the same way as for lines in $\mathbb{R}^2$.) The geometrical interpretation of the various vectors is illustrated in Fig. 11.3. Just as for lines, the parametric and Cartesian forms of planes are not unique — different choices of vectors can describe the same plane. For example, in the parametric from we can choose any basis $(\mathbf{w}_1, \mathbf{w}_2)$ for the two-dimensional subspace W.

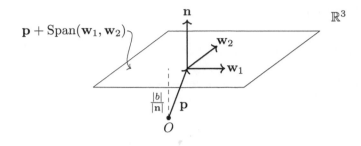

Fig. 11.3 Plane with parametric form $\mathbf{x}(t_1, t_2) = \mathbf{p} + t_1\mathbf{w}_1 + t_2\mathbf{w}_2$ and Cartesian form $\mathbf{n} \cdot \mathbf{x} = b$.

The method for converting between the parametric and Cartesian form of a plane, implicit in the proof of Theorem 11.2, is worth practising.

Problem 11.4 *(Converting a plane from parametric to Cartesian form)*

Convert the plane with parametric form

$$\mathbf{x}(t_1, t_2) = \begin{pmatrix} -1 \\ 2 \\ -1 \end{pmatrix} + t_1 \begin{pmatrix} 2 \\ 0 \\ 3 \end{pmatrix} + t_2 \begin{pmatrix} 1 \\ -2 \\ 0 \end{pmatrix}$$

into Cartesian form. Find the minimal distance of the plane from the origin.

Solution: With $\mathbf{p} = (-1, 2, -1)^T$, $\mathbf{w}_1 = (2, 0, 3)^T$, and $\mathbf{w}_2 = (1, -2, 0)^T$ we have

$$\mathbf{n} = \mathbf{w}_1 \times \mathbf{w}_2 = \begin{pmatrix} 2 \\ 0 \\ 3 \end{pmatrix} \times \begin{pmatrix} 1 \\ -2 \\ 0 \end{pmatrix} = \begin{pmatrix} 6 \\ 3 \\ -4 \end{pmatrix}, \qquad b = \mathbf{n} \cdot \mathbf{p} = 4.$$

Hence, with $\mathbf{x} = (x, y, z)^T$, a Cartesian form of the pane is

$$\begin{pmatrix} 6 \\ 3 \\ -4 \end{pmatrix} \cdot \mathbf{x} = 4 \quad \text{or} \quad 6x + 3y - 4z = 4.$$

The minimal distance from the origin is $|b|/|\mathbf{n}| = 4/\sqrt{6^2 + 3^2 + (-4)^2} = 4/\sqrt{61}$.

Problem 11.5 *(Converting a plane from Cartesian into parametric form)*

Convert the plane with Cartesian form $2x - 3y + z = 5$ into parametric form.

Solution: With $\mathbf{n} = (2, -3, 1)^T$, $\mathbf{x} = (x, y, z)^T$, and $b = 5$ the plane can be written in the Cartesian standard form $\mathbf{n} \cdot \mathbf{x} = b$. As a vector $\mathbf{p}$ 'to the plane' we can use any vector which satisfies $\mathbf{n} \cdot \mathbf{p} = b$; for example, $\mathbf{p} = (1, -1, 0)^T$. To get two vectors $\mathbf{w}_1$, $\mathbf{w}_2$ 'in the plane' we need two linearly independent solutions to $\mathbf{n} \cdot \mathbf{x} = 0$; for example, $\mathbf{w}_1 = (3, 2, 0)^T$ and $\mathbf{w}_2 = (0, 1, 3)^T$. Hence, a parametric form of the plane is

$$\mathbf{x}(t_1, t_2) = \mathbf{p} + t_1\mathbf{w}_1 + t_2\mathbf{w}_2 = \begin{pmatrix} 1 \\ -1 \\ 0 \end{pmatrix} + t_1 \begin{pmatrix} 3 \\ 2 \\ 0 \end{pmatrix} + t_2 \begin{pmatrix} 0 \\ 1 \\ 3 \end{pmatrix}.$$

11.2.2   Parametric and Cartesian form for lines

Lines in $\mathbb{R}^3$ are subsets of the form $\mathbf{p} + W$, where $\mathbf{p} \in \mathbb{R}^3$ and $W \subset \mathbb{R}^3$ is a one-dimensional vector subspace.

Theorem 11.3 *A following sets are lines in* $\mathbb{R}^3$.

(i) $L_p = \{\mathbf{p} + t\mathbf{w} \,|\, t \in \mathbb{R}\}$ *with* $\mathbf{p}, \mathbf{w} \in \mathbb{R}^3$ *and* $\mathbf{w} \neq \mathbf{0}$ (*parametric form*)

(ii) $L_c = \{\mathbf{x} \in \mathbb{R}^3 \,|\, \mathbf{n}_1 \cdot \mathbf{x} = b_1, \, \mathbf{n}_2 \cdot \mathbf{x} = b_2\}$
    with $\mathbf{n}_1, \mathbf{n}_2$ *linearly independent*, $b_1, b_2 \in \mathbb{R}$ (*Cartesian form*)

All lines in $\mathbb{R}^3$ *can be written in the form (i) and (ii). If* $(\mathbf{w}, \mathbf{n}_1, \mathbf{n}_2)$ *is a basis of mutually orthogonal vectors and* $b_1 = \mathbf{n}_1 \cdot \mathbf{p}$, $b_2 = \mathbf{n}_2 \cdot \mathbf{p}$ *then* $L_p = L_c$.

Proof   We begin by showing the equality of the two sets.

$L_p \subset L_c$: Start with $\mathbf{x} = \mathbf{p} + t\mathbf{w} \in L_p$. Multiplying this equation with $\mathbf{n}_1$ and $\mathbf{n}_2$ gives $\mathbf{n}_i \cdot \mathbf{x} = \mathbf{n}_i \cdot \mathbf{p} = b_i$ for $i = 1, 2$ and, therefore, $\mathbf{x} \in L_c$.

$L_c \subset L_p$: A vector $\mathbf{x} \in L_c$ satisfies $\mathbf{n}_i \cdot \mathbf{x} = b_i$ or, equivalently, $\mathbf{n}_i \cdot (\mathbf{x} - \mathbf{p}) = 0$ for $i = 1, 2$. We can write the vector $\mathbf{x} - \mathbf{p}$ as a linear combination of the basis $(\mathbf{w}, \mathbf{n}_1, \mathbf{n}_2)$, so $\mathbf{x} - \mathbf{p} = t\mathbf{w} + \alpha_1\mathbf{n}_1 + \alpha_2\mathbf{n}_2$. It follows that $0 = \mathbf{n}_i \cdot (\mathbf{x} - \mathbf{p}) = \mathbf{n}_i \cdot (t\mathbf{w} + \alpha_1\mathbf{n}_1 + \alpha_2\mathbf{n}_2) = \alpha_i |\mathbf{n}_i|^2$ and, hence, $\alpha_i = 0$ for $i = 1, 2$. Therefore, $\mathbf{x} = \mathbf{p} + t\mathbf{w} \in L_p$.

Clearly, every line $\mathbf{p} + W$ can be written in parametric form by choosing a vector $\mathbf{w}$ with $W = \mathrm{Span}(\mathbf{w})$; conversely, every parametric form given a line with $W = \mathrm{Span}(\mathbf{w})$.

Given a parametric line L_p, specified by vectors $\mathbf{p}$, $\mathbf{w}$, Cor. 10.2 tells us we can find mutually orthogonal vectors $(\mathbf{w}, \mathbf{n}_1, \mathbf{n}_2)$. Then $\mathbf{n}_1, \mathbf{n}_2$ along with $b_1 = \mathbf{n}_1 \cdot \mathbf{p}$ and $b_2 = \mathbf{n}_2 \cdot \mathbf{p}$ define a Cartesian line L_c with $L_c = L_p$.

Conversely, let L_c be a Cartesian line specified by $\mathbf{n}_1, \mathbf{n}_2$ and b_1, b_2. Define $\mathbf{w} = \mathbf{n}_1 \times \mathbf{n}_2$. We can always find a solution $\mathbf{p}$ to the equations $\mathbf{n}_1 \cdot \mathbf{p} = b_1$ and $\mathbf{n}_2 \cdot \mathbf{p} = b_2$. Then $\mathbf{w}$ and $\mathbf{p}$ define a parametric line L_p with $L_p = L_c$.   □

The interpretation of the various vectors in the theorem is illustrated in Fig. 11.4. The vector $\mathbf{p}$ is a vector 'to the line', the vector $\mathbf{w}$ is 'along the line' and $\mathbf{n}_1$, $\mathbf{n}_2$ are orthogonal to it. The two Cartesian equations for a line, $\mathbf{n}_1 \cdot \mathbf{x} = b_1$ and $\mathbf{n}_2 \cdot \mathbf{x} = b_2$, can also be written as a linear system $N\mathbf{x} = \mathbf{b}$ with two equations and three variables, where N is the matrix with entries n_{ij} and $\mathbf{b} = (b_1, b_2)^T$. From this point of view, the line is the solution space of the linear system. This is another illustration of the close relation between affine k-planes and linear systems.

Note that a plane in $\mathbb{R}^3$ requires two parameters for its parametric form and one equation for its Cartesian form. For a line in $\mathbb{R}^3$ this is reversed and we need one parameter and two equations. This suggests that, more generally, an affine k-plane in $\mathbb{R}^n$ needs k parameters for its parametric form and $n - k$ equations for its Cartesian form. We will show this in Section 17.3.

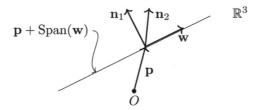

Fig. 11.4 Line in $\mathbb{R}^3$ with parametric form $\mathbf{x}(t) = \mathbf{p} + t\mathbf{w}$ and Cartesian form $\mathbf{n}_i \cdot \mathbf{x} = b_i$, $i = 1, 2$.

As before, we should practice converting between parametric and Cartesian form.

Problem 11.6 *(Converting a line from parametric to Cartesian form)*

Convert the line in parametric form given by

$$\mathbf{x}(t) = \begin{pmatrix} 1 \\ 0 \\ -1 \end{pmatrix} + t \begin{pmatrix} -2 \\ 1 \\ -1 \end{pmatrix}$$

into Cartesian form.

Solution: With $\mathbf{p} = (1, 0, -1)^T$ and $\mathbf{w} = (-2, 1, -1)^T$ we can construct the two orthogonal vectors as $\mathbf{n}_1 = \mathbf{w} \times \mathbf{e}_3 = (1, 2, 0)^T$ and $\mathbf{n}_2 = \mathbf{n}_1 \times \mathbf{w} = (-2, 1, 5)^T$. Since $b_1 = \mathbf{n}_1 \cdot \mathbf{p} = 1$ and $b_2 = \mathbf{n}_2 \cdot \mathbf{p} = -7$ the two equations of the Cartesian form (using $\mathbf{x} = (x, y, z)^T$) read

$$\begin{pmatrix} 1 \\ 2 \\ 0 \end{pmatrix} \cdot \mathbf{x} = 1 \,, \quad \begin{pmatrix} -2 \\ 1 \\ 5 \end{pmatrix} \cdot \mathbf{x} = -7 \quad \text{or} \quad x + 2y = 1 \,, \quad -2x + y + 5z = -7 \,.$$

Problem 11.7 *(Converting a line from Cartesian to parametric form)*

Convert the line in Cartesian form specified by the equations $x - y + 3z = 8$ and $2x + y - z = 2$ into parametric form.

Solution: Since $\mathbf{n}_1 = (1, -1, 3)^T$ and $\mathbf{n}_2 = (2, 1, -1)^T$ are the two vectors orthogonal to the line, $\mathbf{w} = \mathbf{n}_1 \times \mathbf{n}_2 = (-2, 7, 3)^T$ is 'along the line'. A vector $\mathbf{p}$ 'to the line' is obtained by finding a solution to the two equations, for example $\mathbf{p} = (2, 0, 2)^T$. Hence, a parametric form is

$$\mathbf{x}(t) = \mathbf{p} + t\mathbf{w} = \begin{pmatrix} 2 \\ 0 \\ 2 \end{pmatrix} + t \begin{pmatrix} -2 \\ 7 \\ 3 \end{pmatrix} \,.$$

11.2.3   Minimal distances

As an application of some of the techniques based on dot and cross products let us determine some minimal distances, starting with the minimal distance of a line in $\mathbb{R}^3$ from a given point.

Proposition 11.1 *Let* $L = \{\mathbf{p} + t\mathbf{w} \,|\, t \in \mathbb{R}\} \subset \mathbb{R}^3$ *be a line in parametric form and* $\mathbf{p}_0 \in \mathbb{R}^3$. *The minimal distance of* $\mathbf{p}_0$ *from the line arises at* $t_{\min} = -\mathbf{d} \cdot \mathbf{w}/|\mathbf{w}|^2$, *where* $\mathbf{d} = \mathbf{p} - \mathbf{p}_0$. *The minimal distance is given by* $d_{\min} = |\mathbf{d} \times \mathbf{w}|/|\mathbf{w}|$.

Proof We simply work out the distance square $d^2(t) := |\mathbf{x}(t) - \mathbf{p}_0|^2$ of an arbitrary point $\mathbf{x}(t) = \mathbf{p} + t\mathbf{w}$ on the line from the point $\mathbf{p}_0$. This leads to

$$d^2(t) = |\mathbf{d} + t\mathbf{w}|^2 = |\mathbf{w}|^2 t^2 + 2(\mathbf{d} \cdot \mathbf{w})\,t + |\mathbf{d}|^2 = \left[|\mathbf{w}|\,t + \frac{\mathbf{d} \cdot \mathbf{w}}{|\mathbf{w}|} \right]^2 + |\mathbf{d}|^2 - \frac{(\mathbf{d} \cdot \mathbf{w})^2}{|\mathbf{w}|^2}.$$

This is minimal when the expression inside the square bracket vanishes which happens for $t = t_{\min} = -\mathbf{d} \cdot \mathbf{w}/|\mathbf{w}|^2$. This proves the first part of the claim. For the second part we simply compute the distance at $t_{\min}$ which gives

$$d^2_{\min} := d^2(t_{\min}) = \frac{1}{|\mathbf{w}|^2} \left(|\mathbf{d}|^2 |\mathbf{w}|^2 - (\mathbf{d} \cdot \mathbf{w})^2 \right) \overset{(10.2)(c)}{=} \frac{|\mathbf{d} \times \mathbf{w}|^2}{|\mathbf{w}|^2}.$$

$\square$

Problem 11.8 *(Minimal distance of a line from a point)*

Find the minimal distance of the line $\mathbf{x}(t) = \mathbf{p} + t\mathbf{w}$ from the point $\mathbf{p}_0$, where $\mathbf{p} = (2, -1, 4)^T$, $\mathbf{w} = (3, -5, 2)^T$ and $\mathbf{p}_0 = (1, 1, 1)^T$.

Solution: Using the notation and results from the previous proposition we have

$$\mathbf{d} = \mathbf{p} - \mathbf{p}_0 = \begin{pmatrix} 1 \\ -2 \\ 3 \end{pmatrix}, \qquad \mathbf{d} \times \mathbf{w} = \begin{pmatrix} 1 \\ -2 \\ 3 \end{pmatrix} \times \begin{pmatrix} 3 \\ -5 \\ 2 \end{pmatrix} = \begin{pmatrix} 11 \\ 7 \\ 1 \end{pmatrix}. \tag{11.3}$$

Hence, $|\mathbf{d} \times \mathbf{w}| = 3\sqrt{19}$ and $|\mathbf{w}| = \sqrt{38}$ so that $d_{\min} = |\mathbf{d} \times \mathbf{w}|/|\mathbf{w}| = 3/\sqrt{2}$.

For the minimal distance of a plane from a point, we have the following statement:

Proposition 11.2 *A plane is described in Cartesian form by* $\mathbf{n} \cdot \mathbf{x} = b$ *and in parametric form by* $\mathbf{x}(t_1, t_2) = \mathbf{p} + t_1 \mathbf{w}_1 + t_2 \mathbf{w}_2$. *The minimal distance of the plane from* $\mathbf{p}_0 \in \mathbb{R}^3$ *is given by*

$$d_{\min} = \frac{|b - \mathbf{n} \cdot \mathbf{p}_0|}{|\mathbf{n}|} = \frac{|\mathbf{d} \cdot (\mathbf{w}_1 \times \mathbf{w}_2)|}{|\mathbf{w}_1 \times \mathbf{w}_2|}, \qquad \mathbf{d} = \mathbf{p} - \mathbf{p}_0. \tag{11.4}$$

Proof Start with the Cartesian form $\mathbf{n} \cdot \mathbf{x} = b$ and subtract $\mathbf{n} \cdot \mathbf{p}_0$, so that $\mathbf{n} \cdot (\mathbf{x} - \mathbf{p}_0) = b - \mathbf{n} \cdot \mathbf{p}_0$ or, equivalently

$$|\mathbf{n}|\,|\mathbf{x} - \mathbf{p}_0|\cos \sphericalangle(\mathbf{n}, \mathbf{x} - \mathbf{p}_0) = b - \mathbf{n} \cdot \mathbf{p}_0.$$

The value of $|\mathbf{x} - \mathbf{p}_0|$ is maximal when the (absolute) value of the cosine is 1, so that $d_{\min} = |b - \mathbf{n} \cdot \mathbf{p}_0|/|\mathbf{n}|$. This proves the first part of Eq. (11.4). For the second part, simple insert the relations $\mathbf{n} = \mathbf{w}_1 \times \mathbf{w}_2$ and $b = \mathbf{p} \cdot \mathbf{n}$.

$\square$

Problem 11.9 *(Minimal distance of a point from a plane)*

Find the minimal distance of the plane $x - 2y + z = 4$ from $\mathbf{p}_0 = (1, 0, 1)^T$.

Solution: The plane is given in Cartesian form, $\mathbf{n} \cdot \mathbf{x} = b$, with $\mathbf{n} = (1, -2, 1)$ and $b = 4$. Inserting $\mathbf{n} \cdot \mathbf{p}_0 = 2$, $|\mathbf{n}| = \sqrt{6}$ and $b = 4$ into the formula (11.4) gives $d_{\min} = \sqrt{2/3}$.

11.2.4　Intersection of two planes

Consider two planes in Cartesian form

$$P_1 = \{\mathbf{x} \in \mathbb{R}^3 \,|\, \mathbf{n}_1 \cdot \mathbf{x} = b_1\}, \qquad P_2 = \{\mathbf{x} \in \mathbb{R}^3 \,|\, \mathbf{n}_2 \cdot \mathbf{x} = b_2\},$$

and their intersection

$$L = P_1 \cap P_2 = \{\mathbf{x} \in \mathbb{R}^3 \,|\, \mathbf{n}_1 \cdot \mathbf{x} = b_1 \wedge \mathbf{n}_2 \cdot \mathbf{x} = b_2\}.$$

If the two vectors $\mathbf{n}_1$ and $\mathbf{n}_2$ are linearly independent then L is a line, written down in Cartesian form, as comparison with Theorem 11.3 shows. If $\mathbf{n}_1$ and $\mathbf{n}_2$ are linearly dependent, $\mathbf{n}_2 = \alpha \mathbf{n}_1$ for a non-zero $\alpha \in \mathbb{R}$, then the equations for the two planes turn into $\mathbf{n}_1 \cdot \mathbf{x} = b_1$ and $\mathbf{n}_1 \cdot \mathbf{x} = b_2/\alpha$. If $b_1 = b_2/\alpha$ the two planes are identical and their intersection is the entire plane, so $L = P_1 = P_2$. On the other hand, if $b_1 \neq b_2/\alpha$, then the planes are parallel and the intersection is empty, $L = \{\}$ (see Fig. 11.5).

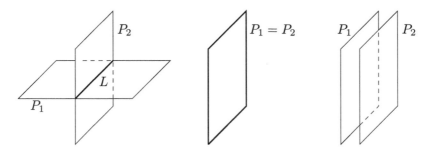

Fig. 11.5 Two planes in $\mathbb{R}^3$ can intersect in a line, in a plane or have an empty intersection.

Problem 11.10 *(Intersection of two planes)*

Find the intersection of the two planes $-x - y + z = 2$ and $ax - 2y + 2z = b$ for all values of the parameters $a, b \in \mathbb{R}$. If the intersection is a line find its parametric form.

Solution: We have $\mathbf{n}_1 = (-1, -1, 1)^T$, $b_1 = 2$ and $\mathbf{n}_2 = (a, -2, 2)$, $b_2 = b$.

(1) $a \neq -2$: In this case $\mathbf{n}_1$ and $\mathbf{n}_2$ are linearly independent and the intersection must be a line. With $\mathbf{w} = \mathbf{n}_1 \times \mathbf{n}_2 = (0, 2+a, 2+a)^T$ and a special solution $\mathbf{p} = (b-4, -b-2a, 0)^T/(2+a)$ to the two equations the parametric form of the intersection line is

$$\mathbf{x}(t) = \frac{-1}{2+a} \begin{pmatrix} 4 - b \\ 2a + b \\ 0 \end{pmatrix} + t \begin{pmatrix} 0 \\ 1 \\ 1 \end{pmatrix}.$$

(2a) $a = -2$ and $b \neq 4$: In this case the two planes are parallel so the intersection is empty.

(2b) $a = -2$ and $b = 4$: The two planes are identical.

11.2.5   Intersection of line and plane

The intersection of a line and a plane in $\mathbb{R}^3$ is easiest discussed with the line in parametric and the plane in Cartesian form, that is

$$L = \{\mathbf{p} + t\mathbf{w} \,|\, t \in \mathbb{R}\}\,, \qquad P = \{\mathbf{x} \in \mathbb{R}^3 \,|\, \mathbf{n} \cdot \mathbf{x} = b\}\,.$$

By inserting the parametrization of the line into the equation for the plane we find for the intersection

$$L \cap P = \{\mathbf{p} + t\mathbf{w} \,|\, (\mathbf{n} \cdot \mathbf{w})t = b - \mathbf{n} \cdot \mathbf{p}\}\,.$$

If $\mathbf{n} \cdot \mathbf{w} \neq 0$ then we can solve for $t = t_0 = (b - \mathbf{n} \cdot \mathbf{p})/(\mathbf{n} \cdot \mathbf{w})$ and the intersection consists of a single point, $L \cap P = \{\mathbf{p} + t_0\mathbf{w}\}$. On the other hand, if $\mathbf{n} \cdot \mathbf{w} = 0$ and $b \neq \mathbf{n} \cdot \mathbf{p}$ there is no solution for t, so the intersection is empty, $L \cap P = \{\}$. For $\mathbf{n} \cdot \mathbf{w} = 0$ and $b = \mathbf{n} \cdot \mathbf{p}$ every $t \in \mathbb{R}$ is a solution so that $L \cap P = L$ — the line is a subset of the plane (see Fig. 11.6).

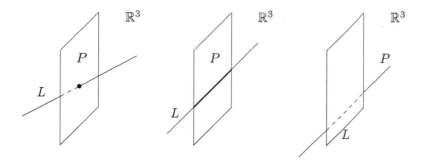

Fig. 11.6 A line and a plane in $\mathbb{R}^3$ can intersect in a point, in a line or have an empty intersection.

Problem 11.11 *(Intersection of line and plane)*

Find the intersection of the plane with Cartesian equation $2x - 5y + z = 7$ with the line in parametric form given by

$$\mathbf{x}(t) = \begin{pmatrix} 0 \\ -2 \\ 3 \end{pmatrix} + t \begin{pmatrix} 2 \\ 1 \\ -1 \end{pmatrix}\,.$$

Solution: With $\mathbf{x} = (x, y, z)^T$, we can split the parametric form of the line into its components, $x(t) = 2t$, $y(t) = -2 + t$ and $z(t) = 3 - t$, and insert these into the equation for the plane. This gives

$$4t - 5(-2 + t) + 3 - t = 7 \qquad \Rightarrow \qquad t = 3\,.$$

Hence, the intersection point is $\mathbf{x}(3) = (6, 1, 0)^T$. As a check, we note that this point does indeed satisfy the equation for the plane.

11.2.6   Intersection of three planes

Suppose that we are interested in the intersection $P_1 \cap P_2 \cap P_3$ of the three planes in $\mathbb{R}^3$ in Cartesian form:

$$P_i = \{\mathbf{x} \in \mathbb{R}^2 \,|\, \mathbf{n}_i \cdot \mathbf{x} = b_i\}\,, \quad \text{where} \quad i = 1, 2, 3\,.$$

This intersection is clearly the same as the solution to the system of the three linear equations in three variables given by $\mathbf{n}_i \cdot \mathbf{x} = b_i$ for $i = 1, 2, 3$, another example of the close relationship between the geometry of affine k-planes and linear systems. Of course, the intersection can be found by solving the linear system explicitly. However, the qualitatively different cases which can arise can be easily reasoned out from what we have discussed so far.

Start by considering the intersection $P_1 \cap P_2$ of the first two planes. This intersection may be a line, a plane or it may be empty. If $L = P_1 \cap P_2$ is a line then the triple intersection $P_1 \cap P_2 \cap P_3 = L \cap P_3$ can be a point, a line or be empty. This already covers all the cases but one, which is the case $P_1 = P_2 = P_3$, so that the intersection is a plane. These four cases are illustrated in Fig. 11.7.

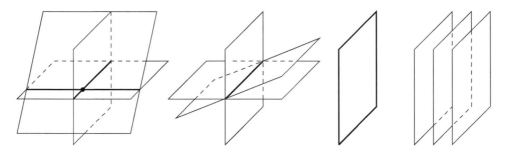

Fig. 11.7 Three planes in $\mathbb{R}^3$ can intersect in a point, a line, a plane or have an empty intersection.

Problem 11.12 *(Intersection of three planes)*

Find the intersection of the three planes with Cartesian equations $x-y+3z = 5$, $2x-2y+z = 0$ and $x + 3y - 8z = 1$.

Solution: Just add suitable multiples of the three equations to eliminate variables but do so in an organized manner (labelling equations helps!).

$$
\begin{array}{ll}
(E1) & x - y + 3z = 5 \\
(E2) & 2x - 2y + z = 0 \\
(E3) & x + 3y - 8z = 1
\end{array}\Bigg\}
\qquad
\begin{array}{ll}
2(E1) - (E2) & 10z = 5 \\
(E2) - 2(E3) & -8y + 17z = -2
\end{array}\Bigg\}
\qquad
\begin{array}{l}
z = 2 \\
y = 9/2
\end{array}\Bigg\}
$$

Inserting the results for z and y into, say, the first equation gives $x = 7/2$, so that the intersection point is $(x, y, z) = (7/2, 9/2, 2)$.

Application 11.1 *The perceptron — a simple artificial neural networks*

Artificial neural networks are motivated by the structure of the human brain and they play an important role in modern computing. Many of the operating principles of artificial neural networks can be formulated and understood in terms of linear algebra. Here, we would like to introduce one of the basic building blocks of artificial neural networks — the *perceptron*.

The structure of the perceptron is schematically illustrated in the figure below.

$$\mathbb{R}^n \ni \mathbf{x} \longrightarrow \boxed{z = \mathbf{w} \cdot \mathbf{x} - b} \xrightarrow{\;z \in \mathbb{R}\;} \boxed{y = \sigma(z)} \longrightarrow y \in \mathbb{R}$$

The perceptron receives an input vector $\mathbf{x} = (x_1, \ldots, x_n)^T \in \mathbb{R}^n$ which it converts into a real output $y \in \mathbb{R}$ in two steps. In the first step, it transforms $\mathbf{x}$ as

$$\mathbf{x} \mapsto z = \mathbf{w} \cdot \mathbf{x} - b \,, \tag{11.5}$$

where $\mathbf{w} \in \mathbb{R}$ is called the *weight vector* of the perceptron and $b \in \mathbb{R}$ is called the *bias*. The weight vector and the bias represent the internal state of the perceptron and, for now, we think of them as given quantities. In the second step, the output z from the first step is transformed as

$$z \mapsto y = \sigma(z) \,. \tag{11.6}$$

Here, σ is called the *activation function* and there are several possible choices for this function. A common choice which we adopt here is called the *logistic sigmoid* function:

$$\sigma(z) := \frac{1}{1 + \exp(-z)} \,. \tag{11.7}$$

Its graph is shown in the figure below.

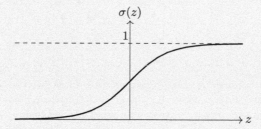

Clearly, the logistic sigmoid has two asymptotes, namely $\sigma(z) \to 0$ for $z \to -\infty$ and $\sigma(z) \to 1$ for $z \to \infty$. Its presence makes the overall action of the perceptron non-linear.

Given this set-up, the functioning of the perceptron can be phrased in geometrical terms. To do this, consider the hyperplane in $\mathbb{R}^n$ (which is a line for $n = 2$ and a plane for $n = 3$) in Cartesian form

$$\mathbf{w} \cdot \mathbf{x} = b \,, \tag{11.8}$$

which is determined by the weight vector $\mathbf{w}$ and the bias b of the perceptron. If a point $\mathbf{x} \in \mathbb{R}^n$ is 'above' this hyperplane, so that $z = \mathbf{w} \cdot \mathbf{x} - b > 0$, then, from Eqs. (11.5), (11.6) and the asymptotic behaviour of the logistic sigmoid, the output of the perceptron is close to 1. On the other hand, for a point $\mathbf{x} \in \mathbb{R}^n$ below this hyperplane, so that $z = \mathbf{w} \cdot \mathbf{x} - b < 0$, the perceptron's output is close to 0. In other words, the purpose of the perceptron is to 'decide' whether a given input vector $\mathbf{x}$ is above or below the hyperplane (11.8).

So far this does not seem to hold much interest — all we have done is to re-formulate a sequence of simple mathematical operations related to the Cartesian form of a (hyper)plane in a different language. The point is that the internal state of the perceptron, that is the choice of hyperplane specified by the weight vector $\mathbf{w}$ and the bias b, is not inserted 'by hand' but rather determined by a learning process. This works as follows. Imagine a certain quantity, y, rapidly changes from 0 to 1 across a certain hyperplane in $\mathbb{R}^n$ whose location is not a priori known. Let us perform m measurements of y at locations $\mathbf{x}^{(1)}, \ldots, \mathbf{x}^{(m)} \in \mathbb{R}^n$ resulting in measured values $y^{(1)}, \ldots, y^{(m)} \in \{0, 1\}$. These measurements can then be used to train the perceptron. Starting from random values $\mathbf{w}^{(1)}$ and $b^{(1)}$ of the weight vector and the bias we can iteratively improve those values by carrying out the operations

$$\mathbf{w}^{(a+1)} = \mathbf{w}^{(a)} + \lambda(y^{(a)} - y)\,\mathbf{x}^{(a)}\,, \qquad b^{(a+1)} = b^{(a)} - \lambda(y^{(a)} - y)\,. \tag{11.9}$$

Here, y is the output value produced by the perceptron given the input vector $\mathbf{x}^{(a)}$ and λ is a real value, typically chosen in the interval $[0, 1]$, called the learning rate of the perceptron. Evidently, if the value y produced by the perceptron differs from the true, measured value $y^{(a)}$, the weight vector and the bias of the perceptron are adjusted according to Eqs. (11.9). This training process continues until all measurements are used up and the final values $\mathbf{w} = \mathbf{w}^{(m+1)}$ and $b = b^{(m+1)}$ have been obtained. In this state the perceptron can then be used to 'predict' the value of y for new input vectors $\mathbf{x}$. Essentially, the perceptron has 'learned' about the location of the hyperplane via the training process and is now able to decide whether a given point is located above or below.

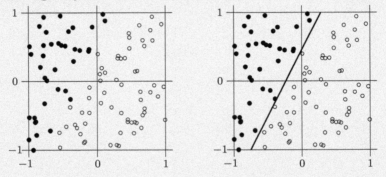

For a simple two-dimensional case, $n = 2$, this process is illustrated in the above figure. The plot on the left shows the data points $\mathbf{x}^{(a)}$, where filled dots have a value $y^{(a)} = 1$ and open dots $y^{(a)} = 0$. After training the perceptron with this data set, using Eq. (11.9), the values of weights and biases are $\mathbf{w} \simeq (-2, 1)^T$ and $b \simeq 1/2$. The resulting line, Eq. (11.8), is shown in the plot on the right.

In the context of artificial neural networks, the perceptron corresponds to a single neuron. More complicated neural networks can be constructed by combining several perceptrons (and, frequently, other building blocks). The learning process for such larger networks is similar to the one for the perceptron described above and it underlies many applications, for example to pattern recognition. We will return to some aspects of this in Application 13.3.

Exercises

(†=challenging, ††=difficult, wide-ranging)

11.1 Consider two lines $L_i = \{\mathbf{p}_i + t\mathbf{w}_i \mid t \in \mathbb{R}\} \subset \mathbb{R}^2$, where $i = 1, 2$, with $\mathbf{p}_1 = (1, b)^T$, $\mathbf{w}_1 = (-1, 1)^T$, $\mathbf{p}_2 = (0, -1)^T$ and $\mathbf{w}_2 = (1, a)^T$ and $a, b \in \mathbb{R}$. Find the intersection $L_1 \cap L_2$ for all values of a, b.

11.2 In $\mathbb{R}^3$, write down the vector equation of the line

$$\frac{(x - 2)}{4} = \frac{(y - 1)}{3} = \frac{(z - 5)}{2}$$

and find the minimum distance of this line from the origin.

11.3 For $\mathbf{p}_1, \mathbf{p}_2 \in \mathbb{R}^2$ or $\mathbb{R}^3$ and $\mathbf{p}_1 \neq \mathbf{p}_2$ show that there is precisely one line L with $\mathbf{p}_1, \mathbf{p}_2 \in L$.

11.4 Show that two lines $L_i = \{\mathbf{p}_i + t\mathbf{w}_i \mid t \in \mathbb{R}\} \subset \mathbb{R}^3$, where $i = 1, 2$ and $\mathbf{w}_1, \mathbf{w}_2$ are linearly independent, have a non-trivial intersection if and only if $(\mathbf{p}_1 - \mathbf{p}_2) \cdot (\mathbf{w}_1 \times \mathbf{w}_2) = 0$.

11.5 Find the Cartesian equation for the plane in $\mathbb{R}^3$ which contains $\mathbf{p}_1 = (2, -1, 1)$, $\mathbf{p}_2 = (3, 2, -1)$, and $\mathbf{p}_3 = (-1, 3, 2)$.

11.6 Find the Cartesian and parametric form of the plane in $\mathbb{R}^3$ which contains $\mathbf{p}_1 = 3\mathbf{i} + 2\mathbf{j} + \mathbf{k}$, $\mathbf{p}_2 = -\mathbf{i} + 3\mathbf{j} - 2\mathbf{k}$, and $\mathbf{p}_3 = 2\mathbf{i} + 2\mathbf{j} + 2\mathbf{k}$.

11.7 In $\mathbb{R}^3$, find the intersection of the line which contains $\mathbf{0}$ and $\mathbf{p} = (1, 1, 1)^T$ and the plane which contains $\mathbf{p}_1 = (-1, 1, -2)^T$, $\mathbf{p}_2 = (1, 5, -5)^T$, and $\mathbf{p}_3 = (0, 2, -3)^T$.

11.8 *Smallest distance of trajectories*[†]
In $\mathbb{R}^3$, consider the equations $\mathbf{x}_1(t) = \mathbf{p}_1 + t\mathbf{w}_1$ and $\mathbf{x}_2(t) = \mathbf{p}_2 + t\mathbf{w}_2$, where $\mathbf{w}_1$ and $\mathbf{w}_2$ are linearly independent and $t \in \mathbb{R}$.
(a) For which value of t is the distance of $\mathbf{x}_1(t)$ and $\mathbf{x}_2(t)$ minimal and what is this minimal distance?
(b) Consider two planes whose trajectories are described by the above equations with $\mathbf{p}_1 = (10, 10, 0)^T$, $\mathbf{w}_1 = (-1, -1, 1)^T$, $\mathbf{p}_2 = (-10, -10, 0)^T$, and $\mathbf{w}_2 = (1, 1/2, 1)^T$. For which time t are the planes closest and what is their distance at this time? Do they collide?

11.9 *The perceptron*[††]
Write code in your favourite programming language which realizes the perceptron in Application 11.1 and which trains the perceptron from a given data set using Eqs. (11.9). For two-dimensional cases, check that your realization is capable of identifying the separating line between the two sets of points.

Part IV

Linear maps and matrices

After our interlude on geometry in the previous part, it is now time to come back to the main narrative. We have already developed the main properties of vector spaces — linear independence, basis, dimension — and now we need to have a closer look at its morphisms, the linear maps. Analysing the morphisms of an algebraic structure is often key to a deep understanding of a mathematical area and linear algebra is no different in this regard.

In the next chapter, we cover the basics of linear maps: their existence and construction; the vector space $\mathrm{Hom}(V, W)$ of linear maps $V \to W$; composition of linear maps and their inverse and the *general linear group* $\mathrm{GL}(V)$ of invertible linear maps $V \to V$. We also present a number of interesting examples, including *coordinate maps* and differential operators.

Earlier, we have encountered matrices in their role as constituents of vector spaces (see Section 6.2.2). We also know from Section 1.2.4 that linear maps $\mathbb{R}^2 \to \mathbb{R}^2$ can be identified with 2×2 matrices with real entries. In Chapter 13 we will generalize this statement and show that, thanks to the existence of the standard unit vector basis, linear maps $\mathbb{F}^n \to \mathbb{F}^m$ can be identified with $m \times n$ matrices with entries in $\mathbb{F}$. This leaves us with an obvious task: the features of linear maps need to be translated into the language of matrices. Indeed, this is how the main properties of matrices emerge. As we will see, the action of a linear map on a vector turns into matrix-vector multiplication, the composition of linear maps corresponds to matrix multiplication and the map inverse becomes the matrix inverse. We will also introduce transposition and Hermitian conjugation for matrices as well as the matrices invariant under these operations, the symmetric and Hermitian matrices. These are basic and useful operations for matrices whose introduction cannot be deferred, although their mathematical meaning will only become clear later, in the context of inner product vector spaces in Chapter 23 and duality in Chapter 26.

In Chapter 14 we introduce and prove the central statement about the structure of linear maps: the *rank theorem*[1]. There are two natural vector subspaces associated to a linear map $f : V \to W$, namely the kernel, $\mathrm{Ker}(f) \subset V$, which consists of all vectors mapped to $\mathbf{0}$, and the image, $\mathrm{Im}(f) \subset W$. The linear map's domain is a vec-

---

[1]In the literature this is often referred to as the 'rank-nullity theorem'. We will avoid this somewhat cumbersome terminology.

tor space with dimension $\dim_{\mathbb{F}}(V)$ but $\dim_{\mathbb{F}}(\text{Ker}(f))$ dimensions are 'lost' under the action of f, since the entire kernel is mapped to the zero vector. This suggests that $\dim_{\mathbb{F}}(V) - \dim_{\mathbb{F}}(\text{Ker}(f))$ dimensions are available for the image and this is precisely the statement of the rank theorem.

Finally, in Chapter 15 we will see that all linear maps can be described by matrices, just as all vectors can be described by coordinate vectors, once bases have been chosen. Computing the matrix associated to a linear map is one of the key tasks in linear algebra and it is of great importance in linear algebra applications. A prominent example is the relation between Scrödinger's formulation of quantum mechanics in terms of differential operators and Heisenberg's in terms of matrices (see Example 15.2). We also need to understand how the matrix which describes a linear map changes under a change of basis. In other words, we will derive the transformation law for matrices under a change of the 'coordinate system', another key aspect of linear algebra with many applications.

12

Introduction to linear maps

This chapter begins with the theoretical foundations for the understanding of linear maps. We show that linearity is a 'nice' property in that is it preserved under basic map operations, including addition and scalar multiplication of maps as well as map composition and map inversion. This implies that the linear maps $V \to W$ form a vector space $\mathrm{Hom}(V, W)$, called the vector space of *homomorphisms* from V to W and that the invertible linear maps $V \to V$ form a group $\mathrm{GL}(V)$, called the *general linear group* of V.

We finish the discussion with examples of linear maps. One of them are *coordinate maps* which relate vectors to their coordinate vectors and are a very useful tool to describe the relationship between linear maps and matrices. We also discuss linear differential operators.

12.1   First properties of linear maps

Summary 12.1 *Linear maps are the maps between vector spaces which are consistent with vector addition and scalar multiplication. Given a basis of the domain, there exists a unique linear map for every choice of images for the basis vectors. Addition and scalar multiplication of maps preserves linearity. This means that the space $\mathrm{Hom}(V, W)$ of all linear maps $V \to W$ forms a vector space of functions. Linearity is preserved under map composition and under carrying out the inverse. As a result, the set $\mathrm{GL}(V)$ of all invertible linear maps $V \to V$ forms a group, called the general linear group of V. Invertible linear maps $V \to W$ are called (vector space) isomorphisms and two vector spaces related by an isomorphism are called isomorphic.*

12.1.1   Reminder of definition

Let us recall from Def. 6.3 that linear maps $f : V \to W$ between two vector spaces V and W over the same field $\mathbb{F}$ are maps which are consistent with vector addition and scalar multiplication, in the sense that

$$\left. \begin{array}{c} f(\mathbf{v}_1 + \mathbf{v}_2) = f(\mathbf{v}_1) + f(\mathbf{v}_2) \\ f(\alpha \mathbf{v}) = \alpha f(\mathbf{v}) \end{array} \right\} \quad \Leftrightarrow \quad f(\alpha_1 \mathbf{v}_1 + \alpha_2 \mathbf{v}_2) = \alpha f(\mathbf{v}_1) + \alpha_2 f(\mathbf{v}_2) \quad (12.1)$$

for all $\mathbf{v}, \mathbf{v}_1, \mathbf{v}_2 \in V$ and all $\alpha, \alpha_1, \alpha_2 \in \mathbb{F}$. These linearity conditions can be combined and generalized to arbitrary linear combinations, so that

$$f\left(\sum_{i=1}^{k}\alpha_i\mathbf{v}_i\right) = \sum_{i=1}^{k}\alpha_i f(\mathbf{v}_i)\,, \tag{12.2}$$

for all $\mathbf{v}_i \in V$ and all $\alpha_i \in \mathbb{F}$. In other words, forming a linear combination and applying a linear map are two operations which 'commute'.

Another simple but important property of linear maps is that they map the zero vector of the domain vector space to the zero vector of the co-domain, so

$$f(\mathbf{0}) = \mathbf{0}\,. \tag{12.3}$$

This is easily seen if we recall from Prop 6.1 (ii) that $0\,\mathbf{v} = \mathbf{0}$ for all $\mathbf{v} \in V$ and, hence, $f(\mathbf{0}) = f(0\,\mathbf{0}) = 0f(\mathbf{0}) = \mathbf{0}$.

We also note the simple fact that the restriction $f|_U$ of a linear map $f : V \to W$ to a sub vector space $U \subset V$ is also linear. Indeed, since f satisfies the linearity conditions (12.1) for all vectors in V, they are also satisfied for vectors in $U \subset V$.

The identity map $\mathrm{id}_V : V \to V$ (defined by $\mathrm{id}_V(\mathbf{v}) = \mathbf{v}$ for all $\mathbf{v} \in V$) is a trivial example of a linear map. Another, slightly more interesting example is the map $f_\alpha : V \to V$ which multiplies vectors with a fixed scalar $\alpha \in \mathbb{F}$, so $f_\alpha(\mathbf{v}) = \alpha\mathbf{v}$. This map is indeed linear (Exercise 12.1) but is still a rather special example of a linear maps.

12.1.2 Existence and construction of linear maps

The full scope of linear maps is described by the following theorem which also provides us with a practical construction method.

Theorem 12.1 *For two vector spaces V, W over the same field $\mathbb{F}$, let $(\mathbf{v}_1,\ldots,\mathbf{v}_n)$ be a basis of V and $\mathbf{w}_1,\ldots,\mathbf{w}_n \in W$ arbitrary vectors. Then there exists a unique linear map $f : V \to W$ with $f(\mathbf{v}_i) = \mathbf{w}_i$, for $i = 1,\ldots,n$.*

Proof *Existence:* Since $(\mathbf{v}_1,\ldots,\mathbf{v}_n)$ is a basis of V every vector $\mathbf{v} \in V$ can be written as a linear combination $\mathbf{v} = \sum_{i=1}^{n}\alpha_i\mathbf{v}_i$ with unique coordinates α_i. Let $\mathbf{w}_i \in W$ be the intended images of the basis vector. Then we define the map $f : V \to W$ by

$$f(\mathbf{v}) = \sum_{i=1}^{n}\alpha_i\mathbf{w}_i \quad \text{for} \quad \mathbf{v} = \sum_{i=1}^{n}\alpha_i\mathbf{v}_i\,. \tag{12.4}$$

Clearly, f is well-defined and $f(\mathbf{v}_i) = \mathbf{w}_i$. It remains to be shown that f is linear. For a second vector $\mathbf{u} = \sum_{i=1}^{n}\beta_i\mathbf{v}_i$ we have

$$f(\mathbf{v}+\mathbf{u}) = f\left(\sum_{i=1}^{n}(\alpha_i+\beta_i)\mathbf{v}_i\right) = \sum_{i=1}^{n}(\alpha_i+\beta_i)\mathbf{w}_i = \sum_{i=1}^{n}\alpha_i\mathbf{w}_i + \sum_{i=1}^{n}\beta_i\mathbf{v}_i = f(\mathbf{v})+f(\mathbf{u})$$

which shows that the first Eq. (12.1) is satisfied. To check the second Eq. (12.1) we start with a scalar $\alpha \in \mathbb{F}$ and work out

$$f(\alpha\mathbf{v}) = f\left(\sum_{i=1}^{n}\alpha\alpha_i\mathbf{v}_i\right) = \sum_{i=1}^{n}\alpha\alpha_i\mathbf{w}_i = \alpha\sum_{i=1}^{n}\alpha_i\mathbf{w}_i = \alpha f(\mathbf{v})\,.$$

This completes the existence proof.

Uniqueness: For a linear map $f : V \to W$ with $f(\mathbf{v}_i) = \mathbf{w}_i$ and a linear combination $\mathbf{v} = \sum_{i=1}^{n} \alpha_i \mathbf{v}_i$ it follows from generalized linearity (12.2) that

$$f(\mathbf{v}) = f\left(\sum_{i=1}^{n} \alpha_i \mathbf{v}_i\right) = \sum_{i=1}^{n} \alpha_i f(\mathbf{v}_i) = \sum_{i=1}^{n} \alpha_i \mathbf{w}_i .$$

Hence, f is already determined for all $\mathbf{v} \in V$ and is, therefore, unique. $\qquad \square$

In short, we can construct a linear map by choosing a basis $(\mathbf{v}_1, \dots, \mathbf{v}_n)$ on the domain vector space and by selecting an arbitrary image vector $\mathbf{w}_i = f(\mathbf{v}_i)$ for each basis vector. From Eq. (12.4), this fixes the linear map uniquely and clearly every linear map (with a finite-dimensional domain vector space) can be obtained in this way. Let us illustrate this result with a simple example.

Problem 12.1 *(Constructing linear maps)*

Construct the linear map $f : \mathbb{R}^2 \to \mathbb{R}^2$ with $f(\mathbf{e}_1) = \mathbf{w}_1$ and $f(\mathbf{e}_2) = \mathbf{w}_2$, where $\mathbf{w}_1 = (1, -2)^T$ and $\mathbf{w}_2 = (-3, 6)^T$.

Solution: Simply write a general vector $\mathbf{v} = (v_1, v_2)^T \in \mathbb{R}^2$ as a linear combination $\mathbf{v} = v_1 \mathbf{v}_1 + v_2 \mathbf{e}_2$ and use linearity of f.

$$f(\mathbf{v}) = f(v_1 \mathbf{e}_1 + v_2 \mathbf{e}_2) = v_1 f(\mathbf{e}_1) + v_2 f(\mathbf{e}_2) = v_1 \mathbf{w}_1 + v_2 \mathbf{w}_2 = \begin{pmatrix} v_1 - 3v_2 \\ -2v_1 + 6v_2 \end{pmatrix} . \qquad (12.5)$$

12.1.3   Addition and scalar multiplication of linear maps

How does linearity relate to basic operations that can be performed with maps? We begin with the addition and scalar multiplication of maps. We already know from Section 6.2.3 that the space $\mathcal{F}(V, W)$ of all function $f : V \to W$ between two vector spaces V and W over $\mathbb{F}$ can be made into a vector space over the same field. Vector addition and multiplication of functions are defined 'point-wise',

$$(f + g)(\mathbf{v}) = f(\mathbf{v}) + g(\mathbf{w}) , \qquad (\alpha f)(\mathbf{v}) = \alpha f(\mathbf{v}) , \qquad (12.6)$$

where $f, g \in \mathcal{F}(V, W)$, $\mathbf{v} \in V$ and $\alpha \in \mathbb{F}$ (see Eq, (6.9)). Is the property of linearity preserved under addition and scalar multiplication of functions?

Proposition 12.1 *Let $f, g : V \to W$ be two linear maps between vector spaces V and W over $\mathbb{F}$ and $\alpha \in \mathbb{F}$ a scalar. Then the sum $f + g$ and the scalar multiple αf, as defined in Eq. (12.6), are linear.*

Proof We need to check the linearity condition (12.1) for $f + g$ and αf, given it is satisfied for f and g.

$$
\begin{aligned}
(f + g)(\alpha_1 \mathbf{v}_1 + \alpha_2 \mathbf{v}_2) &\stackrel{(12.6)}{=} f(\alpha_1 \mathbf{v}_1 + \alpha_2 \mathbf{v}_2) + g(\alpha_1 \mathbf{v}_1 + \alpha_2 \mathbf{v}_2) \\
&\stackrel{(12.1)}{=} \alpha_1 f(\mathbf{v}_1) + \alpha_2 f(\mathbf{v}_2) + \alpha_1 g(\mathbf{v}_1) + \alpha_2 g(\mathbf{v}_2) \\
&= \alpha_1 (f(\mathbf{v}_1) + g(\mathbf{v}_1)) + \alpha_2 (f(\mathbf{v}_2) + g(\mathbf{v}_2)) \\
&\stackrel{(12.6)}{=} \alpha_1 (f + g)(\mathbf{v}_1) + \alpha_2 (f + g)(\mathbf{v}_2)
\end{aligned}
$$

The proof for αf works analogously and is left as Exercise 12.1. $\qquad\qquad\square$

We can re-formulate this result more abstractly by saying that the set of linear maps $f :$ $V \rightarrow W$ forms a vector subspace of $\mathcal{F}(V, W)$ (see Def. 6.2). This vector space of linear maps from V to W is also denoted by $\mathrm{Hom}(V, W)$, which stands for *homomorphisms* from V to W. For the special case $V = W$, linear maps $V \rightarrow V$ are also called *(vector space) endomorphisms* and the vector space of endomorphisms is denoted by $\mathrm{End}(V)$.

Theorem 12.2 *The space* $\mathrm{Hom}(V, W)$ *of linear maps* $V \rightarrow W$, *with addition and scalar multiplication as defined in Eq. (12.6), forms a vector spaces over the same field as V and W. If V and W are finite-dimensional its dimension is given by*

$$\dim_{\mathbb{F}}(\mathrm{Hom}(V, W)) = \dim_{\mathbb{F}}(V) \dim_{\mathbb{F}}(W) . \qquad (12.7)$$

Proof It remains to proof the dimension formula. We choose bases $(\mathbf{v}_1, \ldots, \mathbf{v}_n)$ and $(\mathbf{w}_1, \ldots, \mathbf{w}_m)$ of V and W and define the linear maps $f_{ij} \in \mathrm{Hom}(V, W)$ for $i = 1, \ldots, n$ and $j = 1, \ldots, m$ by

$$f_{ij}(\mathbf{v}_k) = \begin{cases} \mathbf{w}_j & \text{for } k = i \\ \mathbf{0} & \text{for } k \neq i \end{cases}$$

We want to show that these linear maps form a basis of $\mathrm{Hom}(V, W)$. For linear independence, start with the equation $\sum_{ij} \lambda_{ij} f_{ij} = 0$ and act on the vector $\mathbf{v}_k$ which results in $\sum_j \lambda_{kj} \mathbf{w}_j = \mathbf{0}$. Since $(\mathbf{w}_1, \ldots, \mathbf{w}_m)$ forms a basis, this implies that all $\lambda_{kj} = 0$. Hence, the f_{ij} are linearly independent.

Next, consider the function $f = \sum_{i,j} a_{ij} f_{ij}$, for $a_{ij} \in \mathbb{F}$. Since $f(\mathbf{v}_k) = \sum_j a_{kj} \mathbf{w}_j$ and the $\mathbf{w}_j$ are a basis, it follows that any image vectors $f(\mathbf{v}_k)$ can be obtained for suitable choices of the a_{ij}. From Theorem 12.1 this means the f_{ij} span $\mathrm{Hom}(V, W)$. Since the number of these functions is $n\,m$, Eq. (12.7) follows. $\qquad\qquad\square$

12.1.4 Map composition and inverse

Linearity is also preserved under map composition and inversion, as the following proposition shows.

Proposition 12.2 *Let* $f, f_1, f_2 : V \rightarrow W$ *and* $g, g_1, g_2 : W \rightarrow U$ *be linear maps and* $\alpha_1, \alpha_2 \in \mathbb{F}$.
  (i)   *The composition* $g \circ f : V \rightarrow U$ *is linear.*
  (ii) $g \circ (\alpha_1 f_1 + \alpha_2 f_2) = \alpha_1(g \circ f_1) + \alpha_2(g \circ f_2)$.
  (iii) $(\alpha_1 g_1 + \alpha_2 g_2) \circ f = \alpha_1(g_1 \circ f) + \alpha_2(g_2 \circ f)$.
  (iv) *If* $f^{-1} : W \rightarrow V$ *exists it is linear.*

Proof (i) All we need to do is check the linearity condition (12.1) for $g \circ f$ given it is satisfied for f and g.

$$\begin{aligned}(g \circ f)(\alpha_1 \mathbf{v}_1 + \alpha_2 \mathbf{v}_2) &= g(f(\alpha_1 \mathbf{v}_1 + \alpha_2 \mathbf{v}_2)) = g(\alpha_1 f(\mathbf{v}_1) + \alpha_2 f(\mathbf{v}_2)) \\ &= \alpha_1 g(f(\mathbf{v}_1)) + \alpha_2 g(f(\mathbf{v}_2)) = \alpha_1 (g \circ f)(\mathbf{v}_1) + \alpha_2 (g \circ f)(\mathbf{v}_2) .\end{aligned}$$

(ii) For $\mathbf{v} \in V$ we have, from the definition of map composition and linearity, that

$$(g \circ (\alpha_1 f_1 + \alpha_2 f_2))(\mathbf{v}) = g((\alpha_1 f_1 + \alpha_2 f_2)(\mathbf{v})) = g(\alpha_1 f_1(\mathbf{v}) + \alpha_2 f_2(\mathbf{v}))$$
$$= \alpha_1 (g \circ f_1)(\mathbf{v}) + \alpha_2 (g \circ f_2)(\mathbf{v}) = (\alpha_1 (g \circ f_1) + \alpha_2 (g \circ f_2))(\mathbf{v}) \,.$$

(iii) This works exactly like the proof for part (ii).

(iv) Let $f : V \to W$ be an invertible linear maps with inverse $f^{-1} : W \to V$. Consider two vectors $\mathbf{w}_1, \mathbf{w}_2 \in W$. Since f is surjective they can be written as $\mathbf{w}_1 = f(\mathbf{v}_1)$ and $\mathbf{w}_2 = f(\mathbf{v}_2)$ for two vectors $\mathbf{v}_1, \mathbf{v}_2 \in V$, so that $\mathbf{v}_1 = f^{-1}(\mathbf{w}_1)$ and $\mathbf{v}_2 = f^{-1}(\mathbf{w}_2)$. The linearity condition (12.1) for f^{-1} is verified by

$$f^{-1}(\alpha_1 \mathbf{w}_1 + \alpha_2 \mathbf{w}_2) = f^{-1}(\alpha_1 f(\mathbf{v}_1) + \alpha_2 f(\mathbf{v}_2)) = f^{-1}(f(\alpha_1 \mathbf{v}_1 + \alpha_2 \mathbf{v}_2))$$
$$= \alpha_1 \mathbf{v}_1 + \alpha_2 \mathbf{v}_2 = \alpha_1 f^{-1}(\mathbf{w}_1) + \alpha_2 f^{-1}(\mathbf{w}_2) \,.$$

$\square$

Not only does map composition preserve linearity, it is also a bi-linear operation, from (ii) and (iii) of the proposition. In particular, this means that $\text{End}(V)$ forms an algebra, with map composition as multiplication, as comparison with Def. 6.4 shows. Since map composition is associative and has a unit element, id_V, this is an associative algebra with unit.

While map composition is associative it is, in general, not commutative. For two linear maps $f, g \in \text{End}(V)$ we can introduce the commutator $[f, g] := f \circ g - g \circ f$, a linear map in $\text{End}(V)$ which vanishes (equals the zero map) if and only if f and g commute. The commutator has a number of interesting properties. It is anti-symmetric, linear in each of its arguments and it satisfies an equation referred to as *Jacobi identity* (see Exercise 12.8).

12.1.5   Isomorphisms and general linear groups

Bijective linear maps are of particular importance since they can be used to identify two vector spaces and this motivates introducing the following terminology.

Definition 12.1 *A bijective linear map $f : V \to W$ is called a vector space isomorphism, or isomorphism for short, from V to W. If such an isomorphism from V to W exists then V and W are called isomorphic, written as $V \cong W$.*

Isomorphic vector spaces should be regarded as identical with regard to their vector space structure. In other words, it does not matter in which of the two spaces calculations are carried out — the isomorphism (and its inverse) can always be used to translate to the other space in a way that is consistent with addition and scalar multiplication.

Note that the notion of vector spaces being isomorphic is an equivalence relation. Indeed, every vector space is isomorphic to itself, $V \cong V$ (since the identity map id_V is linear and bijective), so the relation is reflexive. If $V \cong W$ then, by definition, there is a bijective linear map $f : V \to W$ and, from Prop. 12.2 (ii) we know that its inverse $f^{-1} : W \to V$ is also linear. Therefore, $V \cong W$ implies that $W \cong V$, so the relation is symmetric. Finally, we need to show that it is also transitive. Consider three vector spaces V, W, U with $V \cong W$ and $W \cong U$, so that there are bijective linear maps $f : V \to W$ and $g : W \to U$. Then, the map $g \circ f : V \to U$ is linear from Prop. 12.2 (i) and bijective from Prop. 2.4, so that $V \cong U$. In conclusion, being isomorphic is an equivalence relation. As a result, the vector spaces over a fixed field fall into disjoint equivalence classes each of which consists of all vector spaces isomorphic to each other. We will soon learn how to characterize these equivalence classes.

From Example 3.1 we know that the set of bijective maps $V \to V$ on a vector space V forms a group, $\mathrm{Bij}(V)$. Recall that the group multiplication is composition of maps, the group inverse is the inverse map and the group identity is the identity map. The invertible linear maps $V \to V$ are called *(vector space) automorphisms* and they form a subset of $\mathrm{Bij}(V)$, which is denoted by $\mathrm{Aut}(V)$. The interesting observation is that $\mathrm{Aut}(V)$ forms a sub-group of $\mathrm{Bij}(V)$, as can be verified by checking the conditions in Def. 3.2. Indeed, $\mathrm{Aut}(V)$ contains the identity map and, from Prop. 12.2 it is closed under map composition and under taking the map inverse. This sub-group is also called the *general linear group* of V and is sometimes denoted by $\mathrm{GL}(V) = \mathrm{Aut}(V)$.

Theorem 12.3 *The set* $\mathrm{Aut}(V) = \mathrm{GL}(V)$ *of invertible linear maps* $V \to V$ *forms a group under map composition.*

General linear groups are quite important. For example, they are a key ingredient in the theory of linear group representations, a more advanced subject which studies the interaction between groups and vector spaces (see, for example, Fulton and Harris 2013) and which has many applications in modern physics (see, for example, Cornwell 1997). General linear groups also have many interesting sub-groups some of which arise and will be discussed in the context of scalar products (see Chapter 23).

12.2 Examples of linear maps

Summary 12.2 *Coordinate maps* $\mathbb{F}^n \to V$ *map coordinate vectors relative to a basis into the associated vectors. They are vector spaces isomorphisms which implies that every n-dimensional vector space V over $\mathbb{F}$ is isomorphic to $\mathbb{F}^n$. Linear differential operators can be viewed as linear maps on suitable vector spaces of functions.*

12.2.1 Coordinate maps

There is a simple but extremely useful way to formalize the relationship between vectors and their coordinates relative to a basis. Consider a vector space V over $\mathbb{F}$ with basis $(\mathbf{v}_1, \ldots, \mathbf{v}_n)$. From Theorem 12.1 we know there exists a unique linear map $\varphi : \mathbb{F}^n \to V$ with

$$\varphi(\mathbf{e}_i) = \mathbf{v}_i \tag{12.8}$$

for all $i = 1, \ldots, n$. It maps a coordinate vector $\boldsymbol{\alpha} = (\alpha_1, \ldots, \alpha_n)^T \in \mathbb{F}^n$ to

$$\varphi(\boldsymbol{\alpha}) = \sum_{i=1}^{n} \alpha_i \mathbf{v}_i \,, \tag{12.9}$$

that is, to the associated vector. For this reason, φ is called a *coordinate map*. Since the coordinates of a vector are unique, this map is injective and since $(\mathbf{v}_1, \ldots, \mathbf{v}_n)$ spans V it is surjective. Hence, coordinate maps are bijective. (This also follows from Exercise 12.5.)

In conclusion, coordinate maps are vector space isomorphisms and their existence shows that an n-dimensional vector space V over $\mathbb{F}$ is isomorphic to the coordinate vector space $\mathbb{F}^n$ over $\mathbb{F}$, so $V \cong \mathbb{F}^n$. In this sense, we can think of any vector space with a basis as a coordinate vector space. However, it is important to remember that this identification of vectors with coordinate vectors is not 'canonical' but depends on the choice of basis. In other words, the same vector is represented by different coordinate vectors for different choices of basis. Coordinate maps will be very useful later when we examine the relationship between linear maps and matrices.

Problem 12.2 *(Coordinate maps)*

Find the coordinate map $\varphi : \mathbb{R}^2 \to \mathbb{R}^2$ associated to the basis $(\mathbf{v}_1, \mathbf{v}_2)$ of $\mathbb{R}^2$, where $\mathbf{v}_1 = (1, 1)^T$ and $\mathbf{v}_2 = (1, -1)^T$.

Solution: We write an arbitrary vector $\mathbf{w} \in \mathbb{R}^2$ as a linear combination $\mathbf{w} = \alpha_1 \mathbf{v}_1 + \alpha_2 \mathbf{v}_2$ of the basis. With $\boldsymbol{\alpha} = (\alpha_1, \alpha_2)^T$, the coordinate map associated to this basis is

$$\varphi(\boldsymbol{\alpha}) = \alpha_1 \mathbf{v}_1 + \alpha_2 \mathbf{v}_2 - \begin{pmatrix} \alpha_1 + \alpha_2 \\ \alpha_1 - \alpha_2 \end{pmatrix}.$$

Problem 12.3 *(Coordinate map for a polynomial vector space)*

Show that (p_0, p_1, p_2, p_3) with $p_0(x) = 1$, $p_1(x) = x$, $p_2(x) = 3x^2 - 1$ and $p_3(x) = 5x^3 - 3x$ is a basis of $\mathcal{P}_3(\mathbb{R})$ and find the coordinate map associated to this basis.

Solution: To show linear independence we consider the equation

$$0 = \sum_{i=0}^{3} \alpha_i p_i(x) = (\alpha_0 - \alpha_2) + (\alpha_1 - 3\alpha_3)x + 3\alpha_2 x^2 + 5\alpha_3 x^3 .$$

Linear independence of the monomials $(1, x, x^2, x^3)$ implies that $\alpha_0 - \alpha_2 = 0$, $\alpha_1 - 3\alpha_3 = 0$, $3\alpha_2 = 0$ and $5\alpha_3 = 0$. It follows that all $\alpha_i = 0$ so the p_i are linearly independent. Since they are four polynomials in a four-dimensional space they must form a basis. The associated coordinate map is

$$\varphi(\boldsymbol{\alpha})(x) = \sum_{i=0}^{3} \alpha_i p_i(x) = (\alpha_0 - \alpha_2) + (\alpha_1 - 3\alpha_3)x + 3\alpha_2 x^2 + 5\alpha_3 x^3 .$$

12.2.2 Differential operators

We recall that $\mathcal{C}^\infty([a, b], \mathbb{R})$ is the vector space of (real-valued) infinitely many times differentiable functions on the interval $[a, b] \subset \mathbb{R}$. This is of course an infinite-dimensional vector space, so it is somewhat outside the main topic of this book. Nevertheless, linear maps on this and other function spaces are so important for many applications that we have to mention them. An important class of such maps are linear differential operators of order n which are of the form

$$L = \sum_{k=0}^{n} p_k \frac{d^k}{dx^k} : \mathcal{C}^\infty([a, b], \mathbb{R}) \to \mathcal{C}^\infty([a, b], \mathbb{R}) , \tag{12.10}$$

where $p_k \in \mathcal{C}^\infty([a, b], \mathbb{R})$ are fixed functions. Differentiation is a linear operation since

$$\frac{d}{dx}(\alpha g + \beta g)(x) = \alpha \frac{dg}{dx}(x) + \beta \frac{dh}{dx}(x) \tag{12.11}$$

for functions $g, h \in \mathcal{C}^\infty([a, b], \mathbb{R})$ and all scalars $\alpha, \beta \in \mathbb{R}$. Likewise, multiplication with a fixed function $p \in \mathcal{C}^\infty([a, b], \mathbb{R})$ is linear since

$$(p(\alpha g + \beta h))(x) = \alpha p(x)g(x) + \beta p(x)h(x) = (\alpha(pg) + \beta(ph))(x) . \tag{12.12}$$

Since the differential operator L is built up from compositions and sums of differentiations and multiplications with fixed functions we know from Props. 12.2 and 12.1 that it must be linear as well. Of course, we can also verify this explicitly.

$$L(\alpha g + \beta h) = \sum_{k=0}^{n} p_k \frac{d^k}{dx^k}(\alpha g + \beta g) = \alpha \sum_{k=0}^{n} p_k \frac{d^k g}{dx^k} + \beta \sum_{k=0}^{n} p_k \frac{d^k h}{dx^k} = \alpha L(g) + \beta L(h)$$

All results for linear maps which do not assume finite dimensionality can be directly applied to such linear differential operators.

Problem 12.4 *(Playing with differential operators)*

Define two linear maps $D, X : \mathcal{C}^\infty([a, b]) \to \mathcal{C}^\infty([a, b])$ by $D(g)(x) = g'(x)$ (single derivative operator) and $X(g)(x) = xg(g)$ (map which multiplies with x). Are these maps injective or surjective? Do they commute?

Solution: The derivative operator D maps any constant function to the zero function so it cannot be injective. On the other hand, every function h is an image $h = Dg$ of a function g (take g to be an indefinite integral of h) so that D is surjective.

As for X, $Xg = Xh$ implies that $xg(x) = xh(x)$ for all $x \in [a, b]$. Dividing by x leads to $g = h$ so that X is injective. For surjectivity of X the discussion is a bit more subtle. We need to check if, for every $h \in \mathcal{C}^\infty([a, b])$ there exists a $g \in \mathcal{C}^\infty([a, b])$ with $h = Xg$ or, equivalently, $h(x) = xg(x)$ for all $x \in [a, b]$. The obvious (and only possible) choice is to take $g(x) = h(x)/x$ but this is only an element of $\mathcal{C}^\infty([a, b])$ if $0 \notin [a, b]$. In conclusion, X is surjective if and only if $0 \notin [a, b]$.

For the final part, we work out the commutator $[D, X]$:

$$[D, X](g)(x) = \frac{d}{dx}(xg(x)) - xg'(x) = g(x) = \mathrm{id}(g)(x) \quad \Rightarrow \quad [D, X] = \mathrm{id} .$$

Hence, D and X do not commute, a result which is of profound importance for quantum mechanics.

Exercises

(†=challenging)

12.1 For a linear map $f : V \to W$ and a scalar $\alpha \in \mathbb{F}$, show that the function αf, as defined in Eq. (12.6), is linear.

12.2 Find the linear map $f : \mathbb{R}^2 \to \mathbb{R}^2$ for which $f(\mathbf{e}_1) = (1, -2)^T$ and $f(\mathbf{e}_2) = (-1, 1)^T$ and find the 2×2 matrix which describes f.

12.3 Two maps $f, g : \mathbb{R}^3 \to \mathbb{R}^3$ are defined by $f(\mathbf{x}) = (\mathbf{n} \cdot \mathbf{x})\mathbf{n}$ and $g(\mathbf{x}) = \mathbf{v} \times \mathbf{x}$, where $\mathbf{n}, \mathbf{v} \in \mathbb{R}^3$ are given non-zero vectors.
(a) Show that f and g are linear.
(b) Show that f and g are neither surjective nor injective.
(c) Work out the composite maps $f \circ f$, $g \circ f$ and $g \circ g$.

12.4 Consider the vector space $V = \mathcal{P}_2(\mathbb{R})$ of quadratic polynomials with monomial basis $(1, x, x^2)$.
(a) Find the unique linear map $F : V \to V$ which maps the basis vectors as $F(1) = 1$, $F(x) = 2 + x$, and $F(x^2) = 2x + x^2$.
(b) Show that the map from (a) can be written as a linear differential operator of the form (12.10).
(c) Can every linear map $F : V \to V$ be expressed as a linear differential operator?

12.5 For a vector space V with basis $(\mathbf{v}_1, \ldots, \mathbf{v}_n)$ and vectors $\mathbf{w}_1, \ldots, \mathbf{w}_m \in W$, let $f : V \to W$ be the unique linear map with $f(\mathbf{v}_i) = \mathbf{w}_i$, for $i = 1, \ldots, n$.
(a) Show that f is surjective iff the vectors $\mathbf{w}_1, \ldots, \mathbf{w}_n$ span W.
(b) Show that f is injective iff the vectors $\mathbf{w}_1, \ldots, \mathbf{w}_n$ are linearly independent.
(c) Show that f is an isomorphism iff $(\mathbf{w}_1, \ldots, \mathbf{w}_n)$ is a basis of W.
(d) If f is an isomorphism, show that $\dim_{\mathbb{F}}(V) = \dim_{\mathbb{F}}(W)$.

12.6 *Bases and general linear group*†
Use the results from Exercise 12.5 to show that there is a bijective map between the bases of a vector space V and the general linear group $\mathrm{GL}(V)$.

12.7 For two vector spaces V, W and a vector subspace $U \subset V$, we have a linear map $f : U \to W$. Show that there exists a linear map $F : V \to W$ with $\mathrm{Im}(F) = \mathrm{Im}(f)$.

12.8 *Commutator properties*
Let $f, g, h : V \to V$ be linear maps.
(a) Show that the commutator is anti-symmetric, so $[f, g] = -[g, f]$.
(b) Show that the commutator is bi-linear, by verifying that $[f, \alpha g + \beta h] = \alpha[f, g] + \beta[f, h]$.
(c) Show that the commutator satisfies $[f, [g, h]] + [g, [h, f]] + [h, [f, g]] = 0$ (Jacobi identity).

12.9 *Calculating with commutators*†
(a) For linear maps $f, g, h : V \to V$ show that the commutator satisfies $[f \circ g, h] = f \circ [g, h] + [f, h] \circ g$.
(b) For a vector space V of differentiable function, use the formula from (a) to work out the commutators $[xD, x]$, $[D^2, x]$ and $[D^3, x]$, where $D = d/dx$ is the derivative operator and x denotes the linear map which multiplies functions by x.

13
Matrices

Probably the most important class of linear maps are linear maps between coordinate vector spaces. As we will see, they can be identified with matrices. Under this identification, basic map operations are turned into matrix operations: the action of a linear map on a vector becomes matrix-vector multiplication, the composition of maps turns into matrix multiplication and the map inverse corresponds to the matrix inverse. In this way, matrix properties which may, at first, seem contrived become perfectly natural — they reflect elementary map operations.

We will also use the opportunity to introduce the basic matrix operations of *transposition* and *Hermitian conjugation*. The matrices invariant under these operations are called *symmetric* and *Hermitian matrices*, respectively. The mathematical meaning of transposition and Hermitian conjugation will become clear in Chapter 22.

13.1 Matrices as linear maps

Summary 13.1 *Linear maps $f : \mathbb{F}^n \to \mathbb{F}^m$ between coordinate vector spaces are identified with $m \times n$ matrices with entries in $\mathbb{F}$. Under this correspondence, the action of linear maps on vectors turns into matrix-vector multiplication, map composition becomes matrix multiplication and the inverse map is described by the inverse matrix.*

In the introduction we have verified that linear maps between two-coordinate vectors can be identified with 2×2 matrices. We are now ready to consider the generalization of this statement to an arbitrary number of components.

13.1.1 Linear maps between coordinate vectors

Suppose, we are interested in a linear map $f : \mathbb{F}^n \to \mathbb{F}^m$ from n-dimensional to m-dimensional coordinate vectors. First, we introduce the standard unit vector bases $(\mathbf{e}_1, \dots \mathbf{e}_n)$ and $(\tilde{\mathbf{e}}_1, \dots, \tilde{\mathbf{e}}_m)$ on $\mathbb{F}^n$ and $\mathbb{F}^m$, respectively[1]. We can write the images $f(\mathbf{e}_j)$ as a linear combination of the standard unit vectors $\tilde{\mathbf{e}}_i$ of the co-domain, so

$$f(\mathbf{e}_j) = \begin{pmatrix} A_{1j} \\ \vdots \\ A_{mj} \end{pmatrix} = \sum_{i=1}^{m} A_{ij} \tilde{\mathbf{e}}_i , \tag{13.1}$$

---

[1]Since n and m are allowed to be different the standard unit vectors on either space can have different numbers of components and the tilde notation has been used to indicate this.

for suitable numbers $A_{ij} \in \mathbb{F}$. For general vectors $\mathbf{v} = \sum_{j=1}^n v_j \mathbf{e}_j \in \mathbb{F}^n$, this implies

$$f(\mathbf{v}) = f\left(\sum_{j=1}^n v_j \mathbf{e}_j\right) \overset{(12.2)}{=} \sum_{j=1}^n v_j f(\mathbf{e}_j) \overset{(13.1)}{=} \sum_{i=1}^m \sum_{j=1}^n A_{ij} v_j \tilde{\mathbf{e}}_i \qquad (13.2)$$

so that the ith component of $f(\mathbf{v})$ is given by

$$(f(\mathbf{v}))_i = \sum_{j=1}^n A_{ij} v_j \ . \qquad (13.3)$$

This result shows the linear map is described by the numbers A_{ij} and the action of the linear map corresponds to carrying out the sum on the right-hand side of Eq. (13.3).

13.1.2 Matrix-vector multiplication

Given that the numbers A_{ij} in Eq. (13.3) are labelled by two indices in the range $i = 1, \ldots, m$ and $j = 1, \ldots, n$ it is natural to arrange them into a $m \times n$ matrix

$$A = \begin{pmatrix} A_{11} & \cdots & A_{1n} \\ \vdots & & \vdots \\ A_{m1} & \cdots & A_{mn} \end{pmatrix} . \qquad (13.4)$$

We will frequently need to refer to the *row vectors* and *column vectors* of such a matrix for which we introduce the following notation:

$$\mathbf{A}_i = (A_{i1}, \ldots, A_{in})^T \ , \qquad \mathbf{A}^j = (A_{1j}, \ldots, A_{mj})^T \ . \qquad (13.5)$$

Hence, $\mathbf{A}_i$ is an n-dimensional column vector which contains the entries in the ith row of A and $\mathbf{A}^j$ is an m-dimensional column vector which contains the entries in the jth column of A. In terms of its row and column vectors, we sometimes write a matrix as

$$A = \begin{pmatrix} \mathbf{A}_1^T \\ \vdots \\ \mathbf{A}_m^T \end{pmatrix} = (\mathbf{A}^1, \ldots, \mathbf{A}^n) \ . \qquad (13.6)$$

Having introduced the relevant notation we now get to the important point: matrix-vector multiplication. The product of an $m \times n$ matrix A with an n-dimensional coordinate vector $\mathbf{v} \in \mathbb{F}^n$ it is defined by

$$(A\mathbf{v})_i := \sum_{j=1}^n A_{ij} v_j \ . \qquad (13.7)$$

and it leads to an m-dimensional coordinate vector $A\mathbf{v} \in \mathbb{F}^m$. Of course, this definition is motivated by the action of a linear map in Eq. (13.3) and can, therefore, be seen as a natural consequence of linearity. Using the above notation for row vectors matrix-vector multiplication can also be written as

$$Av = \begin{pmatrix} \mathbf{A}_1^T \\ \vdots \\ \mathbf{A}_m^T \end{pmatrix} \mathbf{v} = \begin{pmatrix} \mathbf{A}_1 \cdot \mathbf{v} \\ \vdots \\ \mathbf{A}_m \cdot \mathbf{v} \end{pmatrix} . \tag{13.8}$$

The dot on the right-hand side denotes the dot product between two vectors which, as in Chapter 9, is defined as

$$\mathbf{v} \cdot \mathbf{w} := \sum_{i=1}^{n} v_i w_i , \tag{13.9}$$

where $\mathbf{v}, \mathbf{w} \in \mathbb{F}^n$. In other words, we can say that a matrix and a vector are multiplied by performing the dot product between the vector and the row vectors of the matrix. Note that, for this process to make sense, the vector needs to have as many components as the matrix has columns. A useful observation is that the action of a matrix on the standard unit vectors gives the column vectors,

$$A\mathbf{e}_i = \mathbf{A}^i , \tag{13.10}$$

as Eq. (13.8) shows. The above results can be summarized in the following theorem.

Theorem 13.1 *For a linear map $f : \mathbb{F}^n \to \mathbb{F}^m$ the $m \times n$ matrix $A = (f(\mathbf{e}_1), \dots, f(\mathbf{e}_n))$, whose columns are the images of the standard unit vectors $\mathbf{e}_i$, satisfies $f(\mathbf{v}) = A\mathbf{v}$ for all $\mathbf{v} \in \mathbb{F}^n$.*

Combining this statement with Theorem 12.1, we conclude that linear maps between coordinate vectors can be identified with matrices. Moreover, the action of such linear maps corresponds to matrix-vector multiplication. From now on we will take this identification for granted and freely switch between linear maps $f : \mathbb{F}^n \to \mathbb{F}^m$ and their associated matrices $A = (f(\mathbf{e}_1), \dots, f(\mathbf{e}_n))$. This identification can be viewed as 'canonical' in the sense that it arises from a preferred basis — the standard unit vector basis.

We will see later that linear maps between abstract vector spaces can also be described by matrices. However, abstract vector spaces do not have a preferred basis, so the general relationship between linear maps and matrices will not be canonical.

Problem 13.1 *(Multiplication of matrices and vectors)*

Consider a linear map $f : \mathbb{R}^3 \to \mathbb{R}^4$, with $f(\mathbf{v}) = A\mathbf{v}$ and the 4×3 matrix A given by

$$A = \begin{pmatrix} 1 & 0 & -1 \\ 2 & 1 & 3 \\ -2 & 1 & 1 \\ 0 & 0 & 4 \end{pmatrix} .$$

Work out the image of the vector $\mathbf{v} = (1, -2, 3)$ under this linear map.

Solution: The image of vectors $\mathbf{v}$ under f is found by computing the matrix-vector product $A\mathbf{v}$, which gives

$$A\mathbf{v} = \begin{pmatrix} 1 & 0 & -1 \\ 2 & 1 & 3 \\ -2 & 1 & 1 \\ 0 & 0 & 4 \end{pmatrix} \begin{pmatrix} 1 \\ -2 \\ 3 \end{pmatrix} = \begin{pmatrix} -2 \\ 9 \\ -1 \\ 12 \end{pmatrix} =: \mathbf{w} \ .$$

Note that the image vector $\mathbf{w} \in \mathbb{R}^4$ has been computed by carrying out the dot product between the vector $\mathbf{v}$ and the rows of A.

Problem 13.2 *(The matrix associated to a linear map between coordinate vectors)*

Consider the linear map $f : \mathbb{R}^2 \rightarrow \mathbb{R}^2$ defined by $f(\mathbf{e}_1) = \mathbf{w}_1$ and $f(\mathbf{e}_2) = \mathbf{w}_2$, where $\mathbf{w}_1 = (2,3)^T$ and $\mathbf{w}_2 = (-2,4)^T$. Find the 2×2 matrix A which describes this linear map.

Solution: From Theorem 13.1 the columns of this matrix A are precisely the images $\mathbf{w}_1, \mathbf{w}_2$ of the standard unit vectors, so that

$$A = (\mathbf{w}_1, \mathbf{w}_2) = \begin{pmatrix} 2 & -2 \\ 3 & 4 \end{pmatrix} \ .$$

It is easy to check that $A\mathbf{e}_1 = \mathbf{w}_1$ and $A\mathbf{e}_2 = \mathbf{w}_2$, as should be the case.

13.1.3   The two faces of matrices

Theorem 13.1 can be used to translate features of (linear) maps into features of matrices. The most elementary example of this is linearity of the map itself which translates into linearity of matrix-vector multiplication. This means for every $m \times n$ matrix $A \in \mathcal{M}_{m,n}(\mathbb{F})$, vectors $\mathbf{v}_1, \mathbf{v}_2 \in \mathbb{F}^n$ and scalars $\alpha, \alpha_2 \in \mathbb{F}$ we have

$$A(\alpha_1 \mathbf{v}_1 + \alpha_2 \mathbf{v}_2) = \alpha_1 A\mathbf{v}_1 + \alpha_2 A\mathbf{v}_2 \ . \tag{13.11}$$

Of course this can also be verified directly from the definition, Eq. (13.7), of matrix-vector multiplication (Exercise 13.1).

In Prop. 12.1 we have seen that addition and scalar multiplication of maps preserves linearity, so that the homomorphisms $\mathrm{Hom}(V, W)$ form a vector space. As we have just seen, the homomorphisms $\mathrm{Hom}(\mathbb{F}^n, \mathbb{F}^m)$ can be identified with matrices, so that

$$\mathrm{Hom}(\mathbb{F}^n, \mathbb{F}^m) \cong \mathcal{M}_{m,n}(\mathbb{F}) \ , \qquad \mathrm{End}(\mathbb{F}^n) \cong \mathcal{M}_{n,n}(\mathbb{F}) \ . \tag{13.12}$$

These relations have a somewhat abstract flavour: they talk about a bijective linear map between linear maps and their associated matrices. In practice this is not so hard to understand. The map which underlies the identifications (13.12) is the map $f \mapsto A$ which assigns to a linear map $f \in \mathrm{Hom}(\mathbb{F}^n, \mathbb{F}^m)$ the associated matrix $A = (f(\mathbf{e}_1), \ldots f(\mathbf{e}_n))$, as in Theorem 13.1. This map is bijective, as follows from Theorem 12.1. It is also linear. Indeed, if two maps $f, g \in \mathrm{Hom}(\mathbb{F}^n, \mathbb{F}^m)$ are identified with matrices $A, B \in \mathcal{M}_{m,n}(\mathbb{F})$ then the linear map $\alpha f + \beta g$ is identified with $\alpha A + \beta B$ (Exercise 13.4). In this way, addition and scalar multiplication of linear maps between coordinate vectors translate into addition and scalar multiplication of matrices. This explains the dual role of matrices as elements of the matrix vector spaces $\mathcal{M}_{m,n}(\mathbb{F})$ and as linear maps in $\mathrm{Hom}(\mathbb{F}^n, \mathbb{F}^m)$.

13.1.4 Square and diagonal matrices

Under the identification (13.12), special properties of a linear map between coordinate vectors translate into special properties of the associated matrix. An endomorphism $f \in \mathrm{End}(\mathbb{F}^n)$ is described by an $n \times n$ matrix $A \in \mathcal{M}_{n,n}(\mathbb{F})$ with as many rows as columns. Such matrices are called *square matrices*. The entries A_{ii} of a square matrix A are called the *diagonal entries* and all other entries A_{ij} with $i \neq j$ are called *off-diagonal entries*.

The identity map $\mathrm{id}_{\mathbb{F}^n} \in \mathrm{End}(\mathbb{F}^n)$ is associated to a square matrix called *unit or identity matrix* and denoted by $\mathbb{1}_n$. Since $\mathrm{id}_{\mathbb{F}^n}(\mathbf{e}_i) = \mathbf{e}_i$, for $i = 1, \ldots, n$, we know from Theorem 13.1 that

$$\mathbb{1}_n = (\mathbf{e}_1, \ldots, \mathbf{e}_n) = \begin{pmatrix} 1 & & 0 \\ & \ddots & \\ 0 & & 1 \end{pmatrix} . \tag{13.13}$$

It is clear that $\mathbb{1}_n \mathbf{v} = \mathbf{v}$ for all $\mathbf{v} \in \mathbb{F}^n$ but this can also be explicitly verified from matrix-vector multiplication. In fact, this is easy to do if we note that the entries of the unit matrix, $(\mathbb{1}_n)_{ij} = \delta_{ij}$, are precisely given by the Kronecker delta symbol, so that $(\mathbb{1}_n \mathbf{v})_i = (\mathbb{1}_n)_{ij} v_j = \delta_{ij} v_j = v_i$. (The last step is just using the index replacing property of the Kronecker delta.)

Slightly generalizing from the identity map, another simple class of linear maps $f \in \mathrm{End}(\mathbb{F}^n)$ are those which only scale the standard unit vectors, that is, $f(\mathbf{e}_i) = \lambda_i \mathbf{e}_i$ for some $\lambda_i \in \mathbb{F}$. From Theorem 13.1, the matrices associated to such linear maps have column vectors $\lambda_i \mathbf{e}_i$ and are, hence, of the form

$$D = (\lambda_1 \mathbf{e}_1, \ldots, \lambda_n \mathbf{e}_n) = \begin{pmatrix} \lambda_1 & & 0 \\ & \ddots & \\ 0 & & \lambda_n \end{pmatrix} =: \mathrm{diag}(\lambda_1, \ldots, \lambda_n) . \tag{13.14}$$

Such matrices are called *diagonal* and are also denoted by $\mathrm{diag}(\lambda_1, \ldots, \lambda_n)$. Generalizing even further, it is sometimes convenient to talk about *block-diagonal matrices*

$$A = \begin{pmatrix} A_1 & & 0 \\ & \ddots & \\ 0 & & A_n \end{pmatrix} =: \mathrm{diag}(A_1, \ldots, A_n) , \tag{13.15}$$

which are built up from square matrices A_i (of possibly different sizes) arranged along the diagonal.

13.2 Matrix multiplication

Suppose we have two matrices, each describing a linear map between coordinate vector spaces. Since the composition of these maps is again linear (see Prop. 12.2) and it is also a map between coordinate vector spaces, it must be described by a matrix as well. It turns out this matrix is obtained by matrix multiplication, as we now explain.

13.2.1 Matrix multiplication from map composition

Consider the following composition of linear maps.

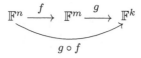

We know from Theorem 13.1 that f corresponds to an $m \times n$ matrix A and g to a $k \times m$ matrix B. But from Prop. 12.2 the composition $g \circ f$ is also linear and, hence, corresponds to a $k \times n$ matrix C. How does this matrix C relate to A and B?

From Eq. (13.3), the matrix C satisfies $((g \circ f)\mathbf{v}))_i = \sum_{k=1}^{n} C_{ik} v_k$. On the other hand, be applying f and g sequentially, we find

$$((g \circ f)(\mathbf{v}))_i = (g(f(\mathbf{v})))_i = \sum_{j=1}^{m} B_{ij}(A\mathbf{v})_j = \sum_{j=1}^{m} \sum_{k=1}^{n} B_{ij} A_{jk} v_k \ . \tag{13.16}$$

A comparison of these two results shows that

$$C_{ik} = \sum_{j=1}^{m} B_{ij} A_{jk} \ , \tag{13.17}$$

and this is the desired relationship. Eq. (13.17) defines *matrix multiplication* and is, in matrix notation, also written as $C = BA$. The sum over the adjacent index j in Eq. (13.17) means that the product matrix C is obtained by performing all possible dot product between the rows of B and the columns of A. Using our notation for row and column vectors this can also be written as

$$BA = \begin{pmatrix} \mathbf{B}_1^T \\ \vdots \\ \mathbf{B}_k^T \end{pmatrix} (\mathbf{A}^1, \dots, \mathbf{A}^n) = \begin{pmatrix} \mathbf{B}_1 \cdot \mathbf{A}^1 \cdots \mathbf{B}_1 \cdot \mathbf{A}^n \\ \vdots \qquad \vdots \\ \mathbf{B}_k \cdot \mathbf{A}^1 \cdots \mathbf{B}_k \cdot \mathbf{A}^n \end{pmatrix} \ . \tag{13.18}$$

$$\underset{k \times m}{} \qquad \underset{m \times n}{} \qquad \rightarrow \qquad \underset{k \times n}{}$$

The size of the various matrices is indicated underneath. Note that matrix multiplication only makes sense if the first matrix has as many columns as the second one has rows — only then are the dot products well-defined. We emphasize that there is nothing strange or unnatural about matrix multiplication. As we have seen, it is simply the way composition of linear maps is carried out when they are represented by matrices.

Problem 13.3 *(Matrix multiplication)*

Consider the 3×3 matrix $A : \mathbb{R}^3 \to \mathbb{R}^3$ and the 2×3 matrix $B : \mathbb{R}^3 \to \mathbb{R}^2$ given by

$$B = \begin{pmatrix} 1 & 0 & -1 \\ 2 & 3 & -2 \end{pmatrix} , \qquad A = \begin{pmatrix} 0 & 1 & 1 \\ 2 & 0 & 1 \\ 1 & -1 & 1 \end{pmatrix} .$$

Compute all well-defined products of these matrices (including with themselves).

Solution: We can compute the 2×3 product matrix $BA : \mathbb{R}^3 \to \mathbb{R}^2$ and the $3\times$ product matrix $A^2 : \mathbb{R}^3 \to \mathbb{R}^3$ by performing the dot products between the column vectors the second matrix and the row vectors of the first matrix. This leads to

$$BA = \begin{pmatrix} 1 & 0 & -1 \\ 2 & 3 & -2 \end{pmatrix} \begin{pmatrix} 0 & 1 & 1 \\ 2 & 0 & 1 \\ 1 & -1 & 1 \end{pmatrix} = \begin{pmatrix} -1 & 2 & 0 \\ 4 & 4 & 3 \end{pmatrix}$$

$$A^2 = \begin{pmatrix} 0 & 1 & 1 \\ 2 & 0 & 1 \\ 1 & -1 & 1 \end{pmatrix} \begin{pmatrix} 0 & 1 & 1 \\ 2 & 0 & 1 \\ 1 & -1 & 1 \end{pmatrix} = \begin{pmatrix} 3 & -1 & 2 \\ 1 & 1 & 3 \\ -1 & 0 & 1 \end{pmatrix} .$$

The products AB and B^2 do not make sense — the number of rows of the second matrix does not match the number of columns of the first.

Application 13.1 *Matrices in graph theory*

Graphs are objects which consist of a certain number of vertices, $V_1, \ldots, V_n$, and links connecting these vertices. A simple example with five vertices is shown below.

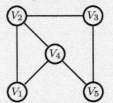

Here we focus on undirected graphs for which the links have no direction, but our considerations can easily be generalized to directed graphs. Graphs can be related to linear algebra via the *adjacency matrix* which is defined by

$$M_{ij} = \begin{cases} 1 & \text{if } V_i \text{ and } V_j \text{ are linked} \\ 0 & \text{otherwise} \end{cases} .$$

For example, for the above graph the adjacency matrix is given by

$$M = \begin{pmatrix} 0 & 1 & 0 & 1 & 0 \\ 1 & 0 & 1 & 1 & 0 \\ 0 & 1 & 0 & 0 & 1 \\ 1 & 1 & 0 & 0 & 1 \\ 0 & 0 & 1 & 1 & 0 \end{pmatrix} . \tag{13.19}$$

The following fact (which we will not try to prove here) makes the adjacency matrix a useful object.

Fact: The number of possible walks from vertex V_i to vertex V_j over precisely n links in a graph is given by $(M^n)_{ij}$, where M is the adjacency matrix of the graph.

To illustrate this, compute the low powers of the adjacency matrix M in Eq. (13.19).

$$M^2 = \begin{pmatrix} 2 & 1 & 1 & 1 & 1 \\ 1 & 3 & 0 & 1 & 2 \\ 1 & 0 & 2 & 2 & 0 \\ 1 & 1 & 2 & 3 & 0 \\ 1 & 2 & 0 & 0 & 2 \end{pmatrix} , \quad M^3 = \begin{pmatrix} 2 & 4 & 2 & 4 & 2 \\ 4 & 2 & 5 & 6 & 1 \\ 2 & 5 & 0 & 1 & 4 \\ 4 & 6 & 1 & 2 & 5 \\ 2 & 1 & 4 & 5 & 0 \end{pmatrix} .$$

For example, the number of possible walks from V_1 to V_3 over three links is given by $(M^3)_{13} = 2$. By inspecting the above figure it can be seen that these two walks correspond to $V_1 \to V_4 \to V_5 \to V_3$ and $V_1 \to V_4 \to V_2 \to V_3$.

13.2.2 Rules for matrix multiplication

The rules for matrix multiplication follow immediately from the rules for composition of linear maps. We know that composition of maps is associative so the same must be true for matrix multiplications. Hence, for three matrices A, B, and C (with suitable size so that multiplication makes sense) we have

$$A(BC) = (AB)C \ . \tag{13.20}$$

This can also be verified using Eq. (13.17) and working with index notation (and the Einstein summation convention).

$$(A(BC))_{il} = A_{ij}(BC)_{jl} = A_{ij}B_{jk}C_{kl} = (AB)_{ik}C_{kl} = ((AB)C)_{il} \ .$$

The identity map acts as the neutral element of map composition, so the unit matrix must be the neutral element of matrix multiplication. This means for an $m \times n$ matrix A we have

$$A\mathbb{1}_n = \mathbb{1}_m A = A \ . \tag{13.21}$$

Again this is easily verified explicitly using index notation, for example

$$(A\mathbb{1}_n)_{ik} = A_{ij}(\mathbb{1}_n)_{jk} = A_{ij}\delta_{jk} \stackrel{(9.16)}{=} A_{ik} \ .$$

If A is a square matrix it can be multiplied with itself an arbitrary number of times and this is also written as $A^k := A\,A \cdots A$ for $k \in \mathbb{Z}^{>0}$. It is also useful to define $A^0 := \mathbb{1}_n$.

The relation between matrix multiplication on the one hand and matrix addition and scalar multiplication on the other hand is governed by simple distributive and associative rules. Specifically, we have

$$\begin{aligned} (A + B)C &= AC + BC \\ C(A + B) &= CA + CB \\ A(\alpha B) &= (\alpha A)B = \alpha(AB) \end{aligned} \ , \tag{13.22}$$

where $\alpha \in \mathbb{F}$ and the sizes of A, B, and C should be such that all operations make sense. These rules follow immediately from the corresponding ones for linear maps in Prop. 12.2 but they can also be shown explicitly. For example, for the first of these equations, simply perform the index computation

$$((A + B)C)_{ik} \stackrel{(13.17)}{=} \sum_j (A + B)_{ij}C_{jk} \stackrel{(6.5)}{=} \sum_j (A_{ij}C_{jk} + B_{ij}C_{jk})$$

$$\stackrel{(13.17)}{=} (AC)_{ik} + (BC)_{ik} \stackrel{(6.5)}{=} (AC + BC)_{ik} \ .$$

Another way to state the rules in Eq. (13.22) is to say that matrix multiplication is bi-linear. In particular, this means that the vector space of square matrices $\mathcal{M}_{n,n}(\mathbb{F}) \cong$

$\text{End}(\mathbb{F}^n)$ with matrix multiplication forms an algebra, in the sense of Def. 6.4. Since matrix multiplication is associative and has a unit element, the unit matrix $\mathbb{1}_n$, this is an associative algebra with a unit.

While matrix multiplication is associative it is, in general, not commutative, so, typically, $AB \neq BA$ for two square matrices A and B. The 'degree of non-commutativity' of two matrices is often measured by the commutator defined as

$$[A, B] := AB - BA . \tag{13.23}$$

Evidently, the matrices A, B commute if and only if $[A, B] = 0$. The matrix commutator has the same properties as the commutator of linear maps (see Exercise 12.8), namely it is anti-symmetric, bi-linear, and it satisfies the Jacobi identity.

Problem 13.4 *(Non-commutativity of matrix multiplication)*

For the matrices

$$A = \begin{pmatrix} 1 & 2 \\ -1 & 0 \end{pmatrix} , \quad B = \begin{pmatrix} 3 & -1 \\ 0 & 2 \end{pmatrix} , \quad C = \begin{pmatrix} a & b \\ b & a \end{pmatrix} , \quad D = \begin{pmatrix} c & d \\ d & c \end{pmatrix} ,$$

where $a, b, c, d \in \mathbb{R}$, show that A and B do not commute and that $[C, D] = 0$.

Solution: By straightforward computation we have

$$AB = \begin{pmatrix} 1 & 2 \\ -1 & 0 \end{pmatrix} \begin{pmatrix} 3 & -1 \\ 0 & 2 \end{pmatrix} = \begin{pmatrix} 3 & 3 \\ -3 & 1 \end{pmatrix} , \quad BA = \begin{pmatrix} 3 & -1 \\ 0 & 2 \end{pmatrix} \begin{pmatrix} 1 & 2 \\ -1 & 0 \end{pmatrix} = \begin{pmatrix} 4 & 6 \\ -2 & 0 \end{pmatrix} ,$$

so that indeed $AB \neq BA$.

For the second part, we work out the commutator

$$[C, D] = \begin{pmatrix} a & b \\ b & a \end{pmatrix} \begin{pmatrix} c & d \\ d & c \end{pmatrix} - \begin{pmatrix} c & d \\ d & c \end{pmatrix} \begin{pmatrix} a & b \\ b & a \end{pmatrix} = \begin{pmatrix} 0 & 0 \\ 0 & 0 \end{pmatrix} .$$

This shows that matrices with a specific structure can still commute.

13.2.3   Matrix inverse and general linear group

A bijective linear map $f \in \text{End}(\mathbb{F}^n)$ has a unique linear inverse f^{-1} (see Prop. 12.2) and, from Theorem 13.1, both f and f^{-1} are described by an $n \times n$ matrix. If A is the $n \times n$ matrix which describes f, then the matrix which corresponds to f^{-1} is called the *inverse matrix* of A, and it is denoted by A^{-1}. Since map composition becomes matrix multiplication, the defining relations $f \circ f^{-1} = f^{-1} \circ f = \text{id}$ for the inverse map translate into the defining relations

$$AA^{-1} = A^{-1}A = \mathbb{1}_n . \tag{13.24}$$

for the inverse matrix. Conversely, a general $n \times n$ matrix A is called *invertible* if it corresponds to an invertible linear map or, equivalently, if an $n \times n$ matrix A^{-1} satisfying Eq. (13.24) can be found. If an inverse does not exist, the matrix A is called *singular.*

Powers of an inverse matrix A^{-1} are also written as $A^{-k} := (A^{-1})^k$ for $k \in \mathbb{N}$. Therefore, for invertible matrices A the powers A^k are meaningful for all integers k and we have $A^k A^l = A^{k+l}$ for all $k, l \in \mathbb{Z}$.

In Section 12.1.5 we have introduced the general linear group $GL(V)$ of invertible linear maps $V \to V$ and this may have seemed a somewhat abstract object. For $V = \mathbb{F}^n$ we can be more concrete and state that the general linear group $GL(\mathbb{F}^n)$ consists of all invertible $n \times n$ matrices with entries in $\mathbb{F}$. The group multiplication is matrix multiplication (since it corresponds to map composition) and we have already seen that this is associative. The group inverse is the matrix inverse and, from Eq. (13.21), the identity matrix is the neutral element.

We recall from Prop. 3.1 that the left inverse in a group is always the right inverse as well. Hence, once we know that a matrix A is invertible its inverse is already determined by one of the relations (13.24). We will soon discuss systematic methods to calculate the inverse matrix but for now we note that this can be done by solving Eq. (13.24).

Problem 13.5 *(Matrix inverse the pedestrian way)*

Using Eq. (13.24), compute the inverse of the matrix:

$$A = \begin{pmatrix} 1 & 2 \\ 3 & -1 \end{pmatrix} .$$

Solution: We start with a general Ansatz for A^{-1} and impose Eq. (13.24).

$$A^{-1} = \begin{pmatrix} a & b \\ c & d \end{pmatrix} \quad \Rightarrow \quad AA^{-1} = \begin{pmatrix} 1 & 2 \\ 3 & -1 \end{pmatrix} \begin{pmatrix} a & b \\ c & d \end{pmatrix} = \begin{pmatrix} a+2c & b+2d \\ 3a-c & 3b-d \end{pmatrix} \overset{!}{=} \begin{pmatrix} 1 & 0 \\ 0 & 1 \end{pmatrix} .$$

Splitting the last equation into components gives a linear system $a + 2c = 1$, $b + 2d = 0$, $3a - c = 0$ and $3b - d = 1$ of four equations in four variables. Its solution is easily found to be $a = 1/7$, $b = 2/7$, $c = 3/7$ and $d = -1/7$. Inserting this into the Ansatz for A^{-1} gives

$$A^{-1} = \frac{1}{7} \begin{pmatrix} 1 & 2 \\ 3 & -1 \end{pmatrix} .$$

After such a computation it is always worth checking the result by verifying Eq. (13.24).

Application 13.2 *Matrices in cryptography*

Matrices can be used for encryption. Here is a basic example for how this works. Suppose we would like to encrypt the text: '`linear␣algebra␣`'. First, we translate this text into numerical form using the simple code $␣ \mapsto 0, a \mapsto 1, b \mapsto 2, \cdots$ and then we partition the resulting list of numbers into sub-lists of the same size. Here we use sub-lists of size three for definiteness. Next, we arrange these numbers into a matrix, with each sub-list forming a column of the matrix. For our sample text this results in

$$T = \begin{pmatrix} 12 & 5 & 0 & 7 & 18 \\ 9 & 1 & 15 & 1 \\ 14 & 18 & 12 & 2 & 0 \end{pmatrix} \qquad \text{for} \qquad \begin{array}{l} \texttt{l e ␣ g r} \\ \texttt{i a a e a .} \\ \texttt{n r l b ␣} \end{array}$$

So far, this is relatively easy to decode, even if we had decided to permute the assignment of letters to numbers. As long as same letters are represented by same numbers, the code can be deciphered by a frequency analysis, at least for a sufficiently long text. To do this, the relative frequency of each number is determined and compared with the typical frequency with which letters appear in the English language. Matching similar frequencies leads to the key.

For a more sophisticated encryption, define a square 'encoding' matrix whose size equals the length of the sub-lists, so a 3×3 matrix for our case. Basically, the only other restriction on this matrix is that it should be invertible. For our example, let us choose

$$A = \begin{pmatrix} -1 & -1 & 1 \\ 2 & 0 & -1 \\ -2 & 1 & 1 \end{pmatrix}.$$

To encode the text, carry out the matrix multiplication

$$T_{\mathrm{enc}} = AT = \begin{pmatrix} -1 & -1 & 1 \\ 2 & 0 & -1 \\ -2 & 1 & 1 \end{pmatrix} \begin{pmatrix} 12 & 5 & 0 & 7 & 18 \\ 9 & 1 & 1 & 5 & 1 \\ 14 & 18 & 12 & 2 & 0 \end{pmatrix} = \begin{pmatrix} -7 & 12 & 11 & -10 & -19 \\ 10 & -8 & -12 & 12 & 36 \\ -1 & 9 & 13 & -7 & -35 \end{pmatrix}.$$

Note there is no longer an identification of numbers with letters in T_{enc}. For example, the letter 'a' appears three times and corresponds to the three 1's in T. However, there is no corresponding repetition of numbers in T_{enc}. Without knowledge of the encoding matrix A it is quite difficult to de-cypher T_{enc}, particularly for large block sizes. The legitimate receiver of the text should be provided with the inverse A^{-1} of the encoding matrix. For our example, it is given by

$$A^{-1} = \begin{pmatrix} 1 & 2 & 1 \\ 0 & 1 & 1 \\ 2 & 3 & 2 \end{pmatrix},$$

as can be checked by verifying that $A^{-1}A = \mathbb{1}_3$. The receiver can then recover the message by the simply matrix multiplication

$$T = A^{-1}T_{\mathrm{enc}}.$$

13.3 Transpose and Hermitian conjugate

Summary 13.2 *For a matrix A the transpose A^T is obtained by exchanging rows and columns. Square matrices invariant under transposition are called symmetric. If they change by an overall sign they are called anti-symmetric. The Hermitian conjugate $A^\dagger$ of a matrix A with complex entries is a combination of complex conjugation and transposition. Hermitian matrices are those that are invariant under Hermitian conjugation, anti-Hermitian matrices change by a sign.*

Transposition and Hermitian conjugation are basic matrix operations whose mathematical meaning will only emerge later (see Chapter 22). However, since they are widely used in matrix calculations it makes sense to introduce them somewhat ahead of their proper mathematical context.

13.3.1 The transpose of a matrix

For an $n \times m$ matrix $A \in \mathcal{M}_{n,m}(\mathbb{F})$ the transpose matrix $A^T \in \mathcal{M}_{m,n}(\mathbb{F})$ is an $m \times n$ matrix obtained by exchanging the rows and columns of A. Using index notation the transpose can be defined by

$$(A^T)_{ij} := A_{ji} \,. \tag{13.25}$$

Problem 13.6 *(Transposition)*

Write down 2×2, 2×3 and 3×3 matrices with general entries together with their transpose.

Solution: Exchange rows and columns to obtain the transpose.

$$
\begin{aligned}
A &= \begin{pmatrix} a & b \\ c & d \end{pmatrix} & A &= \begin{pmatrix} a & b & c \\ d & e & f \end{pmatrix} & A &= \begin{pmatrix} a & b & c \\ d & e & f \\ g & h & i \end{pmatrix} \\
A^T &= \begin{pmatrix} a & c \\ b & d \end{pmatrix} & A^T &= \begin{pmatrix} a & d \\ b & e \\ c & f \end{pmatrix} & A^T &= \begin{pmatrix} a & d & g \\ b & e & h \\ c & f & i \end{pmatrix}
\end{aligned}
\tag{13.26}
$$

13.3.2 Symmetric and anti-symmetric matrices

It makes sense to single out matrices which remain unchanged (or nearly unchanged) under transposition. Since transposition for non-square matrices changes the size of the matrix (from $n \times m$ to $m \times n$), this can of course only happen for square matrices. A square matrix $A \in \mathcal{M}_{n,n}(\mathbb{F})$ is called *symmetric* if it is unchanged under transposition, so if $A = A^T$. In view of Eq. (13.25), this translates into the condition $A_{ij} = A_{ji}$ for all $i, j = 1, \ldots n$. On the diagonal (for $i = j$) this condition becomes trivial, $A_{ii} = A_{ii}$, so the diagonal entries of symmetric matrices are arbitrary. The entries above the diagonal have to equal their counterparts below the diagonal. In particular, this means every diagonal matrix is symmetric.

A square matrix $A \in \mathcal{M}_{n,n}(\mathbb{F})$ is called *anti-symmetric* if it changes the overall sign under transposition, so if $A = -A^T$ or, in index notation, if $A_{ij} = -A_{ji}$ for all $i, j = 1, \ldots, n$. For the diagonal entries this implies $A_{ii} = -A_{ii}$ so that $(1+1)A_{ij} = 0$. So if $1+1 \neq 0$ in $\mathbb{F}$ (true unless the field is $\mathbb{F}_2$) the diagonal entries of an anti-symmetric matrix vanish. In addition, the entries above the diagonal have to be the negatives of their counterparts below the diagonal.

Problem 13.7 *(Symmetric and anti-symmetric matrices)*

Write down the most general symmetric and anti-symmetric 2×2 and 3×3 matrices.

Solution: This can be done by equating the entries of the matrices and their transpose in Eq. (13.26). This leads to the general symmetric 2×2 and 3×3 matrices:

$$A = \begin{pmatrix} a & b \\ b & d \end{pmatrix}, \qquad A = \begin{pmatrix} a & b & c \\ b & e & f \\ c & f & i \end{pmatrix}.$$

Equating the entries of A to the negatives of the entries of A^T in Eq. (13.26) gives the most general anti-symmetric 2×2 and 3×3 matrices:

$$A = \begin{pmatrix} 0 & b \\ -b & 0 \end{pmatrix}, \qquad A = \begin{pmatrix} 0 & b & c \\ -b & 0 & f \\ -c & -f & 0 \end{pmatrix}.$$

13.3.3 Properties of transposition

We should think about how transposition relates to addition, scalar multiplication and multiplication of matrices.

Proposition 13.1 *For matrices A, B of suitable sizes with entries in $\mathbb{F}$ and $\alpha \in \mathbb{F}$ matrix transposition satisfies the following rules:*

(i) $(A + B)^T = A^T + B^T$
(ii) $(\alpha A)^T = \alpha A^T$
(iii) $(AB)^T = B^T A^T$
(iv) If A is invertible then so is A^T and $(A^T)^{-1} = (A^{-1})^T$.

Proof The proofs are most easily carried out in index notation, using Eq. (13.25).

(i) $((A + B)^T)_{ij} = (A + B)_{ji} = A_{ji} + B_{ji} = (A^T + B^T)_{ij}$
(ii) $((\alpha A)^T)_{ij} = (\alpha A)_{ji} = \alpha A_{ji} = (\alpha A^T)_{ij}$
(iii) $((AB)^T)_{ij} = (AB)_{ji} = A_{jk}B_{ki} = B_{ki}A_{jk} = (B^T)_{ik}(A^T)_{kj} = (B^T A^T)_{ij}$
(iv) Taking the transpose of $AA^{-1} = \mathbb{1}$ and applying (iii) gives $(A^{-1})^T A^T = \mathbb{1}$. This means that $(A^{-1})^T$ is the inverse of A^T, so $(A^T)^{-1} = (A^{-1})^T$. $\qquad\square$

From property (iii), matrix multiplication and transposition relate in a well-defined way but note the change of the order in the multiplication! Property (iv) is also very useful: the operations of taking the inverse and the transpose commute! Finally, properties (i) and (ii) mean that transposition is a linear map $\mathcal{M}_{n,m} \to \mathcal{M}_{m,n}$. This has immediate implications for symmetric and anti-symmetric matrices.

Example 13.1 (*Vector spaces of symmetric and anti-symmetric matrices*)

By $\mathcal{S}_n(\mathbb{F})$ and $\mathcal{A}_n(\mathbb{F})$ we denote the sets of symmetric and anti-symmetric $n \times n$ matrices with entries in $\mathbb{F} \neq \mathbb{F}_2$. These sets are closed under addition and scalar multiplication. To see this, start with two (anti-) symmetric matrices A, B, so $A = \pm A^T$ and $B = \pm B^T$ and use Prop (13.1).

$$(\alpha A + \beta B)^T = \alpha A^T + \beta B^T = \pm(\alpha A + \beta B)$$

This means that (anti-) symmetry is preserved under addition and scalar multiplication of matrices and, hence, that $\mathcal{S}_n(\mathbb{F})$ and $\mathcal{A}_n(\mathbb{F})$ are vector subspaces of $\mathcal{M}_{n,n}(\mathbb{F})$. What are the dimensions of these spaces? It is easy to construct bases by starting with the elementary unit matrices $E_{(ij)}$ (see Eq. (6.6)). For the symmetric matrices, a basis is given by the matrices $E_{(ij)} + E_{(ji)}$, where $i, j = 1, \ldots, n$ and $i \leq j$. For the anti-symmetric matrices, we have a basis $E_{(ij)} - E_{(ji)}$, where $i, j = 1, \ldots, n$ and $i < j$. It follows that

$$\dim_{\mathbb{F}}(\mathcal{S}_n(\mathbb{F})) = \frac{1}{2}n(n+1) , \qquad \dim_{\mathbb{F}}(\mathcal{A}_n(\mathbb{F})) = \frac{1}{2}n(n-1) . \tag{13.27}$$

For any matrix $A \in \mathcal{M}_{n,n}(\mathbb{F})$ we can introduce its symmetric part A_+ and anti-symmetric part A_- by

$$A_{\pm} := \frac{1}{2}(A \pm A^T) \qquad \Rightarrow \qquad A = A_+ + A_- . \tag{13.28}$$

Evidently, a matrix is the sum of its symmetric and anti-symmetric parts. Furthermore, this decomposition is unique. Indeed, if $A_+ + A_- = \tilde{A}_+ + \tilde{A}_-$ taking the transpose of this equation gives $A_+ - A_- = \tilde{A}_+ - \tilde{A}_-$. Adding and subtracting these two equations immediately implies that $\tilde{A}_{\pm} = A_{\pm}$. More formally, the existence and uniqueness of the decomposition (13.28) is expressed by the equation

$$\mathcal{M}_{n,n}(\mathbb{F}) = \mathcal{S}_n(\mathbb{F}) \oplus \mathcal{A}_n(\mathbb{F}) .$$

In other words, the vector space of $n \times n$ matrices is a direct sum of the subspaces of symmetric and anti-symmetric matrices (see Prop. 8.1). $\qquad \square$

Problem 13.8 *(Basis for 2×2 symmetric and anti-symmetric matrices)*

Write down a basis for the space $\mathcal{S}_2(\mathbb{F})$ of symmetric 2×2 matrices and a basis for the space $\mathcal{A}_2(\mathbb{F})$ of anti-symmetric 2×2 matrices.

Solution: We can specialize the bases constructed in Exercise 13.1 giving, for the symmetric case, the basis $(E_{(11)}, E_{(22)}, E_{(12)} + E_{(21)})$ and, for the anti-symmetric case, the basis $(E_{(12)} - E_{(21)})$. More explicitly, these are

$$\mathcal{S}_2(\mathbb{F}) = \mathrm{Span}\left(\begin{pmatrix} 1 & 0 \\ 0 & 0 \end{pmatrix}, \begin{pmatrix} 0 & 0 \\ 0 & 1 \end{pmatrix}, \begin{pmatrix} 0 & 1 \\ 1 & 0 \end{pmatrix} \right) , \qquad \mathcal{A}_2(\mathbb{F}) = \mathrm{Span}\left(\begin{pmatrix} 0 & 1 \\ -1 & 0 \end{pmatrix} \right) .$$

Problem 13.9 *(Symmetric and anti-symmetric parts of a matrix)*

Write the following matrix as a sum of its symmetric and anti-symmetric parts.

$$A = \begin{pmatrix} 1 & 0 & -2 \\ -2 & 1 & 6 \\ 2 & 4 & 0 \end{pmatrix} .$$

Solution: We use Eq. (13.28).

$$A^T = \begin{pmatrix} 1 & -2 & 2 \\ 0 & 1 & 4 \\ -2 & 6 & 0 \end{pmatrix}, \quad A_+ = \frac{1}{2}(A + A^T) = \begin{pmatrix} 1 & -1 & 0 \\ -1 & 1 & 5 \\ 0 & 5 & 0 \end{pmatrix}, \quad A_- = \frac{1}{2}(A - A^T) = \begin{pmatrix} 0 & 1 & -2 \\ -1 & 0 & 1 \\ 2 & -1 & 0 \end{pmatrix}$$

Note that A_+ is symmetric and A_- is anti-symmetric. It is easy to verify that $A = A_+ + A_-$.

13.3.4 The Hermitian conjugate of a matrix

Another, somewhat ad-hoc but widely used matrix operation is *Hermitian conjugation*. Its mathematical interpretation will emerge in the context of scalar products (see Chapter 22). For the purpose of this discussion, we will be working with matrices $A \in \mathcal{M}_{n,m}(\mathbb{C})$ with complex entries. For such a matrix A, we can define the *complex conjugate* matrix $\bar{A}$ whose entries $(\bar{A})_{ij} = \overline{A_{ij}}$ are obtained from those of A by complex conjugation.

The *Hermitian conjugate* of an $n \times m$ matrix $A \in \mathcal{M}_{n,m}(\mathbb{C})$ is an $m \times n$ matrix denoted by $A^\dagger \in \mathcal{M}_{m,n}(\mathbb{C})$. It is obtained from A by combining complex conjugation and transposition, so that $A^\dagger = \bar{A}^T$, or, using index notation

$$(A^\dagger)_{ij} = \overline{A_{ji}}. \tag{13.29}$$

Problem 13.10 *(Hermitian conjugate)*

Work out the Hermitian conjugate of the matrix

$$A = \begin{pmatrix} a & b \\ c & d \end{pmatrix}, \quad B = \begin{pmatrix} i & 1 & 2-i \\ 2 & 3 & -3i \\ 1-i & 4 & 2+i \end{pmatrix}.$$

Solution: In addition to transposition, carried out by exchanging rows and columns, all entries are complex conjugated, so that

$$A^\dagger = \begin{pmatrix} \bar{a} & \bar{c} \\ \bar{b} & \bar{d} \end{pmatrix}, \quad B^\dagger = \begin{pmatrix} -i & 2 & 1+i \\ 1 & 3 & 4 \\ 2+i & 3i & 2-i \end{pmatrix}.$$

13.3.5 Hermitian and anti-Hermitian matrices

A square matrix $A \in \mathcal{M}_{n,n}(\mathbb{C})$ is called *Hermitian* if it is invariant under Hermitian conjugation, so if $A = A^\dagger$ or, in index notation, if $A_{ij} = \overline{A_{ji}}$ for all $i, j = 1, \ldots, n$. The diagonal entries of a Hermitian matrix A satisfy $A_{ii} = \overline{A_{ii}}$, so they are real. The entries above the diagonal are the complex conjugates of their counterparts below the diagonal.

A square matrix $A \in \mathcal{M}_{n,n}(\mathbb{C})$ is called *anti-Hermitian* if $A = -A^\dagger$ or, in index notation, if $A_{ij} = -\overline{A_{ji}}$ for all $i, j = 1, \ldots, n$. Anti-symmetric matrices have purely imaginary diagonals and the entries above the diagonal are the negative complex conjugates of their counterparts below the diagonal.

Problem 13.11 *(Hermitian and anti-Hermitian matrices)*

Write down the most general Hermitian and anti-Hermitian 2×2 matrices.

Solution: By imposing $A = \pm A^\dagger$ on the matrices A and $A^\dagger$ from Exercise 13.10, we find

$$A_{\text{herm}} = \begin{pmatrix} a & b \\ \bar{b} & d \end{pmatrix} , \quad a, d \in \mathbb{R} , \quad b \in \mathbb{C}$$

$$A_{\text{anti-herm}} = \begin{pmatrix} a & b \\ -\bar{b} & d \end{pmatrix} , \quad a, d \in i\mathbb{R} , \quad b \in \mathbb{C} .$$

13.3.6 Properties of Hermitian conjugation

Just as for transposition, we should think about how Hermitian conjugation relates to matrix addition, scalar multiplication, and matrix multiplication.

Proposition 13.2 *For matrices A, B of suitable sizes with entries in $\mathbb{C}$ and $\alpha \in \mathbb{C}$ Hermitian conjugation satisfies the following rules.*
(i) $(A + B)^\dagger = A^\dagger + B^\dagger$
(ii) $(\alpha A)^\dagger - \bar{\alpha} A^\dagger$
(iii) $(AB)^\dagger = B^\dagger A^\dagger$
(iv) If A is invertible then so is $A^\dagger$ and $(A^\dagger)^{-1} = (A^{-1})^\dagger$.

Proof These rules follow directly by including complex conjugating in the corresponding proofs for transposition in Prop. 13.1. $\qquad \square$

These rules are similar to those for transposition but there is one crucial difference: in (ii) the scalar is extracted with a complex conjugation. This seemingly innocent modification means that Hermitian conjugation is not a linear map on $\mathcal{M}_{n,m}(\mathbb{C})$ when viewed as a vector space over $\mathbb{C}$ but only when it is viewed as a vector space over $\mathbb{R}$. Rule (ii) also implies that multiplication with $\pm i$ converts between Hermitian and anti-Hermitian matrices, so if $B = \pm iA$ then

$$A = A^\dagger \quad \Leftrightarrow \quad B = -B^\dagger . \tag{13.30}$$

Example 13.2 *(Vector spaces of Hermitian and anti-Hermitian matrices)*

Prop. (13.2) (i) implies that the sum of two (anti-) Hermitian matrices is again (anti-) Hermitian. From Prop. (13.2) (ii) the same is true for scalar multiplication only if we restrict to real scalars. The conclusion is that the sets $\mathcal{H}_n$ and $\mathcal{A}_n$ of Hermitian and anti-Hermitian $n \times n$ matrices form a vector space over $\mathbb{R}$ (but not over $\mathbb{C}$).

Every Hermitian matrix can be written as a sum of its real part, which is symmetric, and its imaginary part, which is anti-symmetric. Hence, combining the basis $E_{(ij)} + E_{(ji)}$, where $i, j = 1, \ldots n$ and $i \leq j$ for symmetric matrices and the basis $i(E_{(ij)} - E_{(ji)})$, where $i, j = 1, \ldots, n$ and $i < j$ for anti-symmetric matrices from Example 13.1

gives a basis for $\mathcal{H}_n$. From Eq. (13.30), multiplying all these basis matrices with i gives a basis for $\mathcal{A}_n$. As a result,

$$\dim_{\mathbb{R}}(\mathcal{H}_n) = \dim_{\mathbb{R}}(\mathcal{A}_n) = \frac{1}{2}n(n+1) + \frac{1}{2}n(n-1) = n^2 \; .$$

Every matrix $A \in \mathcal{M}_{n,n}(\mathbb{C})$ can be written as a sum $A = A_+ + A_-$, where $A_\pm = \frac{1}{2}(A \pm A^\dagger)$ are the Hermitian and anti-Hermitian parts of A. As was the case for transposition, this decomposition is unique, so we can write

$$\mathcal{M}_{n,n}(\mathbb{C}) = \mathcal{H}_n \oplus \mathcal{A}_n \; ,$$

and this should be understood as a relationship between vector spaces over $\mathbb{R}$. $\square$

Problem 13.12 *(Hermitian and anti-Hermitian 2×2 matrices)*

Write down an explicit basis for the vector space $\mathcal{H}_2$ of 2×2 Hermitian matrices and a basis for the vector space $\mathcal{A}_2$ of 2×2 anti-Hermitian matrices.

Solution: We can specialize the general construction in Example 13.2, keeping in mind that we should only use real scalars to form linear combinations. For the Hermitian case a basis is given by $(E_{(11)}, E_{(22)}, E_{(12)} + E_{(21)}, i(E_{(12)} - E_{(21)}))$ or, taking some linear combinations, by

$$\mathcal{H}_2 = \mathrm{Span}\left(\mathbb{1}_2 = \begin{pmatrix} 1 & 0 \\ 0 & 1 \end{pmatrix}, \sigma_1 := \begin{pmatrix} 0 & 1 \\ 1 & 0 \end{pmatrix}, \sigma_2 := \begin{pmatrix} 0 & -i \\ i & 0 \end{pmatrix}, \sigma_3 := \begin{pmatrix} 1 & 0 \\ 0 & -1 \end{pmatrix} \right) . \quad (13.31)$$

The matrices σ_i are called the *Pauli matrices* and they play an important role in quantum mechanics. A basis for $\mathcal{A}_2$ is obtained by multiplying with i, so $(i\mathbb{1}_2, i\sigma_1, i\sigma_2, i\sigma_3)$.

Application 13.3 *More on neural networks*

In Application 11.1 we have introduced the simplest building block of neural networks, the perceptron. Now we would like to take things further and set up a more complicated neural network, built up from perceptrons. Recall that a single perceptron realizes a map $\mathbb{R}^n \ni \mathbf{x} \mapsto y \in \mathbb{R}$, with

$$y = \sigma(\mathbf{w} \cdot \mathbf{x} + b) \qquad \sim \qquad \xrightarrow{\;\mathbb{R}^n\;} \boxed{\mathbf{w}, b} \xrightarrow{\;\mathbb{R}\;}$$

where $\mathbf{w} \in \mathbb{R}^n$ is the weight vector, $b \in \mathbb{R}$ is the bias, and an example for the activation function σ has been given in Eq. (11.7).

As a first step towards a *multi-layer perceptron* we arrange k perceptrons in parallel. Their weight vectors $\mathbf{w}_1, \ldots, \mathbf{w}_k$ can be assembled into an $k \times n$ weight matrix $W = (\mathbf{w}_1, \ldots, \mathbf{w}_k)^T$ and their biases $b_1, \ldots, b_k$ form a bias vector $\mathbf{b} = (b_1, \ldots, b_k)^T$. The network realizes a map $\mathbb{R}^n \ni \mathbf{x} \mapsto \mathbf{y} \in \mathbb{R}^k$ defined by

$$\mathbf{y} = \sigma(W\mathbf{x} + \mathbf{b}) \qquad \sim \qquad \xrightarrow{\;\mathbb{R}^n\;} \begin{array}{c} \boxed{\mathbf{w}_1, b_1} \\ \vdots \\ \boxed{\mathbf{w}_i, b_i} \\ \vdots \\ \boxed{\mathbf{w}_k, b_k} \end{array} \xrightarrow{\;\mathbb{R}^k\;} \qquad = \qquad \xrightarrow{\;\mathbb{R}^n\;} \boxed{W, \mathbf{b}} \xrightarrow{\;\mathbb{R}^k\;}$$

where it is understood that the activation function σ acts on each component of its argument vector. The final step is to arrange d such layers to act one after the other with the output of one layer providing the input for the next one.

$$\mathbf{x} = \mathbf{x}^{(0)} \xrightarrow[\mathbf{x}^{(0)}]{\mathbb{R}^{n_0}} \boxed{W_1, \mathbf{b}_1} \xrightarrow[\mathbf{x}^{(1)}]{\mathbb{R}^{n_1}} \boxed{W_2, \mathbf{b}_2} \xrightarrow[\mathbf{x}^{(2)}]{\mathbb{R}^{n_2}} \cdots\cdots \xrightarrow[\mathbf{x}^{(d-1)}]{\mathbb{R}^{n_{d-1}}} \boxed{W_d, \mathbf{b}_d} \xrightarrow[\mathbf{x}^{(d)}]{\mathbb{R}^{n_d}} \mathbf{x}^{(d)} = \mathbf{y}$$

For the dimensions of input and output to match from one layer to the next each weight matrix W_i must have size $n_{i-1} \times n_i$ and each bias vector $\mathbf{b}_i$ must have dimension n_i. The action of this multi-layer perceptron represents a complicated function, obtained by the iteration

$$\mathbf{x}^{(i+1)} = \sigma(W_i \mathbf{x}^{(i)} + \mathbf{b}_i) \quad \text{where} \quad i = 0, \ldots, d-1 .$$

For ease of notation, we call this function $p_\theta : \mathbb{R}^{n_0} \to \mathbb{R}^{n_d}$, where $\theta = (W_1, \mathbf{b}_1, \ldots, W_d, \mathbf{b}_d)$ denotes all the weighs and biases. To train the multi-layer perceptron we require a training set $\{\mathbf{x}_a, \mathbf{y}_a\}$ of input values $\mathbf{x}_a$ and 'desirable' outputs $\mathbf{y}_a$. Training goal is to adjust the weights and biases θ such that the square deviation

$$\sum_a |p_\theta(\mathbf{x}_a) - \mathbf{y}_a|^2$$

of the network's output from the desired values $\mathbf{y}_a$ is minimized. This is usually accomplished by an algorithm referred to as *stochastic gradient descent* and which involves formulae for the weight and bias adjustment similar to Eq. (11.9). Rather than going into the details of this algorithm which is beyond our scope, we would like to conclude with an explicit example which illustrates an application of the multi-layer perceptron. (For a more comprehensive introduction to neural networks see, for example, da Silva 2017; Goodfellow *et al.* 2016.)

Consider a relatively simple network with a two-dimensional input, a one-dimensional output and two layers, the first with k perceptrons and the second one with a single perceptron.

$$\mathbf{x} \xrightarrow{\mathbb{R}^2} \boxed{W, \mathbf{b}} \xrightarrow{\mathbb{R}^k} \boxed{\tilde{W}, \tilde{\mathbf{b}}} \xrightarrow{\mathbb{R}} y$$

The training set $\{(\mathbf{x}_a, y_a)\}$ for this network is shown in the figure on the left-hand side below.

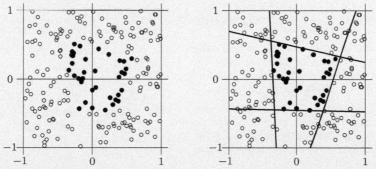

The points indicate the coordinates $\mathbf{x}_a \in \mathbb{R}^2$ and the desired outcome is $y_a = 1$ for a filled circle and $y_a = 0$ for an open circle. The idea is that the points marked with filled circles trace out a certain region in $\mathbb{R}^2$ which we would like the network to 'learn'.

After training the network for $k = 4$ perceptrons in the first layer, we can read out the 4×2 weight matrix W and the four-dimensional bias vector $\mathbf{b}$ from the first layer. These quantities correspond to four lines in $\mathbb{R}^2$, defined by the equations $\mathbf{W}_i \cdot \mathbf{x} + b_i = 0$, where $i = 1, 2, 3, 4$ and $\mathbf{W}_i$ are the row vectors of W. These four lines have been plotted in the above figure on the right-hand side. Evidently, they form the boundary between filled and open points. The first layer 'decides' on which side of any of these lines a given input point lies, the second layer effectively performs a logical and-operation. In this way, the trained network can now decide if a given input point is inside or outside the region. An input point $\mathbf{x} \in \mathbb{R}$ is judged to be inside if its output is $y \simeq 1$ and outside if its output is $y \simeq 0$.

Exercises

(†=challenging, ††=difficult, wide-ranging)

13.1 Show that matrix-vector multiplication is linear, starting with its definition, Eq. (13.7).

13.2 Two maps $f : \mathbb{R}^3 \to \mathbb{R}^3$ and $g : \mathbb{R}^3 \to \mathbb{R}^2$ are defined by $f(\mathbf{v}) = (v_1 - v_2, v_2 + 2v_3, -2v_1)^T$ and $g(\mathbf{v}) = (v_1 + 2v_2 - v_3, v_3)^T$.
(a) Why are these maps linear?
(b) Find the matrices A and B which describe f and g.
(c) Work out the map $g \circ f$, its associated matrix C, and show explicitly that $C = BA$.

13.3 A linear map $f : \mathbb{F}^2 \to \mathbb{F}^2$ is defined by $f(\mathbf{e}_1) = (a, 0)^T$ and $f(\mathbf{e}_2) = (1, -1)^T$, where $a \in \mathbb{F}$.
(a) Find the matrix A which describes f.
(b) Find the matrices which describe $f \circ f$ and $f \circ f \circ f$.
(c) Based on the results in (b), write down a guess for the matrix which represents $f^k = f \circ f \circ \cdots \circ f$ and prove this guess by induction.

13.4 Let $f, g : \mathbb{F}^n \to \mathbb{F}^m$ be two linear maps described by $m \times n$ matrices A and B. Show that the linear maps $\alpha f + \beta g$, where $\alpha, \beta \in \mathbb{F}$ is represented by the matrix $\alpha A + \beta B$.

13.5 † Consider the linear maps $M_1, M_2 : \mathbb{R}^2 \to \mathbb{R}^2$ defined by the matrices

$$M_1 = \begin{pmatrix} -1 & 0 \\ m_1 & 1 \end{pmatrix}, \quad M_2 = \begin{pmatrix} 1 & m_2 \\ 0 & -1 \end{pmatrix},$$

where m_1, m_2 are positive integers, and $M = M_1 M_2$. Show that
(a) $M_1^2 = M_2^2 = \mathbb{1}_2$ and M_1, M_2 do not commute.
(b) the set which consist of $\mathbb{1}_2$ and all 'words' $\cdots M_1 M_2 M_1 M_2 \cdots$ is a subgroup $G \subset \mathrm{GL}(\mathbb{R}^2)$.
(c) the set $H = \{M^k \,|\, k \in \mathbb{Z}\}$ is a sub-group of G which is isomorphic to either a cyclic group $\mathbb{Z}_n$ or to $\mathbb{Z}$. For $m_1 = m_2 = 3$ find all images $g(P)$, where $g \in G$ and $P = \{(x, y)^T \in \mathbb{R}^2 \,|\, x > 0 \cap y > 0\}$ is the positive quadrant. (The picture on the title page originates from a generalization of this structure to 4×4 matrices.)

13.6 *Block matrices*
Consider the block matrices $M, \tilde{M} \in \mathcal{M}_{n,n}(\mathbb{F})$ given by

$$M = \begin{pmatrix} A & B \\ C & D \end{pmatrix}, \quad \tilde{M} = \begin{pmatrix} \tilde{A} & \tilde{B} \\ \tilde{C} & \tilde{D} \end{pmatrix}$$

which are made up from the constituent matrices $A, \tilde{A} \in \mathcal{M}_{k,k}(\mathbb{F})$, $D, \tilde{D} \in \mathcal{M}_{n-k,n-k}(\mathbb{F})$, $B, \tilde{B} \in \mathcal{M}_{k,n-k}(\mathbb{F})$ and $C, \tilde{C} \in \mathcal{M}_{n-k,k}(\mathbb{F})$. Show that their matrix product is given by

$$M\tilde{M} = \begin{pmatrix} A\tilde{A} + B\tilde{C} & A\tilde{B} + B\tilde{D} \\ C\tilde{A} + D\tilde{C} & C\tilde{B} + D\tilde{D} \end{pmatrix}.$$

This means multiplication of block matrices follows the same pattern as normal matrix multiplication, but the products of the entries are replaced by matrix products of the blocks.

13.7 *Inverse for block matrices*
A matrix $M \in \mathcal{M}_{n+1,n+1}(\mathbb{F})$ has the block structure

$$M = \begin{pmatrix} a & \mathbf{b}^T \\ \mathbf{0} & A \end{pmatrix} ,$$

where $a \in \mathbb{F}$ is non-zero, $\mathbf{b} \in \mathbb{F}^n$ and $A \in \mathcal{M}_{n,n}(\mathbb{F})$ is invertible. Show that M is invertible by finding an explicit expression for M^{-1} in terms of A^{-1}.

13.8 *Matrix commutators*
Let $A, B \in \mathcal{M}_{n,n}(\mathbb{F})$ be two square matrices.
(a) If A and B are invertible and commute show that A^{-1} and B^{-1} also commute.
(b) Under the assumptions in (a) show that A^k and B^l commute for all $k, l \in \mathbb{Z}$.
(c) If A and B are symmetric show that AB is symmetric iff $[A, B] = 0$.

13.9 *Anti-symmetric matrices*
Define the 3×3 matrices $T_1 = E_{(32)} - E_{(23)}$, $T_2 = E_{(13)} - E_{(31)}$ and $T_3 = E_{(21)} - E_{(12)}$.
(a) Show that (T_1, T_2, T_3) is a basis of the space of anti-symmetric matrices $\mathcal{A}_3(\mathbb{R})$.
(b) Show that $[T_i, T_j] = \epsilon_{ijk} T_k$.
(c) For two matrices $A, B \in \mathcal{A}_3(\mathbb{R})$ with $A = a_i T_i$ and $B = b_i T_i$ show that $[A, B] = (\mathbf{a} \times \mathbf{b})_i T_i$.

13.10 *Adjacency matrix*
Consider the following simple graph

with three vertices.
(a) Write down the adjacency matrix M for this graph.
(b) Work out the matrix powers M^2 and M^3 and interpret their entries in terms of walks in the graph.

13.11 *Encoding with matrices*[†]
(a) Following Application 13.2, a text has been encoded with the matrix

$$A = \begin{pmatrix} 3 & -1 \\ 5 & -2 \end{pmatrix}$$

leading to the encoded matrix

$$T_{\text{enc}} = \begin{pmatrix} 52 & 8 & -23 & -16 & -5 & -16 \\ 84 & 7 & -46 & -33 & -10 & -33 \end{pmatrix} .$$

Decipher the matrix and find the underlying text.
(b) Another message is encoded in

$$\begin{pmatrix} 4 & 35 & 30 & 53 & 28 & 59 & 17 \\ -2 & 25 & 22 & 34 & -14 & -40 & -14 \end{pmatrix}$$

but the encoding matrix is not known. Try to decipher it.

13.12 *(Multi-layer perceptron*[††]*)*
Write code in your favourite programming language which realizes a simple two-layer perceptron such as the one described in Application 13.3. Read up on the training algorithm, implement it computationally, and apply your code to simple examples, such as the two-dimensional data sets described in Application 13.3.

14

The structure of linear maps

With a basic understanding of linear maps under our belt and examples readily available we can now move the discussion to a more profound level. Central to this are two vector subspaces associated to any linear map: the *image* and the *kernel*. The dimension of the image is called the *rank*, an important characteristic of a linear map. The rank theorem, which is one of the central results of linear algebra, states that the rank of a linear map equals the dimension of the domain vector space minus the dimension of the kernel. As we will see in the next part, the rank theorem is key to a qualitative understanding of the solution structure for linear systems.

We have seen in the last chapter that linear maps between coordinate vectors can be canonically identified with matrices. What about linear maps between abstract vector spaces? It turns out, such linear maps can be *described* by matrices, relative to a choice of bases on the domain and co-domain. However, this relationship between linear maps and matrices is not canonical — it depends on the choice of bases. In the final part of this chapter, we explain this general correspondence between linear maps and matrices and derive the formula for basis transformations of matrices.

14.1 Image and kernel

Summary 14.1 *The image of a linear map is a vector subspace of the co-domain. Another set associated to a linear map is the kernel which consists of all vectors in the domain which are mapped to the zero vector. It is a vector subspace of the domain. The dimension of the image is called the rank of the linear map. A linear map is surjective iff its rank equals the dimension of the co-domain and it is injective iff the dimension of its kernel is zero.*

14.1.1 Definition of image and kernel

As for any map, we can consider the image of a linear map which is a subset of the co-domain vector space. Since vector spaces have a special element, the zero vector, there is another set, the *kernel*, which can be associated to a linear map. The kernel is a subset of the domain vector space and consists of all vectors whose image is the zero vector of the co-domain. For a linear map $f : V \to W$, these two spaces are formally defined as

$$\text{Im}(f) = f(V) = \{f(\mathbf{v}) \,|\, \mathbf{v} \in V\} \subset W \,, \quad \text{Ker}(f) = \{\mathbf{v} \in V \,|\, f(\mathbf{v}) = \mathbf{0}\} \subset V \,. \quad (14.1)$$

Note that both sets are non-empty, the image trivially so, and the kernel since it contains at least the zero vector of V, as a consequence of Eq. (12.3). If the kernel only contains the zero vector it is called *trivial*, otherwise *non-trivial*.

Since both sets are associated to a morphism of vector spaces, it is natural to expect that they carry more structure. In fact, it turns out they are vector subspaces as shown in the following proposition:

Proposition 14.1 *(Properties of kernel and image) Let $f : V \to W$ be a linear map between two vector spaces V and W over a field $\mathbb{F}$.*
(i) The kernel of f is a sub vector space of V.
(ii) The image of f is a sub vector space of W.

Proof   (i) We need to check the conditions in Def. 6.2. Since $\mathbf{0} \in \mathrm{Ker}(f)$, the kernel is not empty. If $\mathbf{v}_1, \mathbf{v}_2 \in \mathrm{Ker}(f)$ then, by definition of the kernel, $f(\mathbf{v}_1) = f(\mathbf{v}_2) = \mathbf{0}$. It follows that $f(\mathbf{v}_1 + \mathbf{v}_1) = f(\mathbf{v}_1) + f(\mathbf{v}_2) = \mathbf{0}$ so that $\mathbf{v}_1 + \mathbf{v}_2 \in \mathrm{Ker}(f)$. Similarly, if $\mathbf{v} \in \mathrm{Ker}(f)$, so that $f(\mathbf{v}) = 0$ it follows that $f(\alpha\mathbf{v}) = \alpha f(\mathbf{v}) = \mathbf{0}$, hence, $\alpha\mathbf{v} \in \mathrm{Ker}(f)$.

(ii) The image is obviously not empty. To show closure under addition start with two vectors $\mathbf{w}_1, \mathbf{w}_2 \in \mathrm{Im}(f)$. By definition of the image, there exist vectors $\mathbf{v}_1, \mathbf{v}_2 \in V$ such that $\mathbf{w}_1 = f(\mathbf{v}_1)$ and $\mathbf{w}_2 = f(\mathbf{v}_2)$. It follows that $\mathbf{w}_1 + \mathbf{w}_2 = f(\mathbf{v}_1) + f(\mathbf{v}_2) = f(\mathbf{v}_1 + \mathbf{v}_2)$ and, hence, $\mathbf{w}_1 + \mathbf{w}_2 \in \mathrm{Im}(f)$. For closure under scalar multiplication, consider a vector $\mathbf{w} \subset \mathrm{Im}(f)$, which can be written as $\mathbf{w} = f(\mathbf{v})$ for a $\mathbf{v} \in V$. Then, for a scalar $\alpha \in \mathbb{F}$, we have $\alpha\mathbf{w} = \alpha f(\mathbf{v}) = f(\alpha\mathbf{v})$ and, hence, $\alpha\mathbf{w} \in \mathrm{Im}(f)$. $\qquad\square$

14.1.2   Rank of a linear map

Since both image and kernel of a linear map are vector subspaces, they can be assigned dimensions. Clearly, these dimensions are characteristic properties of the underlying linear map and, as we will see, they play an important role in analysing its structure. The dimension of the image is of particular relevance and is given a special name.

Definition 14.1 *The dimension of the image of a linear map f is called the rank of f, in symbols $\mathrm{rk}(f) := \dim_{\mathbb{F}}(\mathrm{Im}(f))$.*

The image and the rank can be expressed more explicitly in terms of a basis $(\mathbf{v}_1, \dots, \mathbf{v}_n)$ of V. The image of a vector $\mathbf{v} = \sum_i \alpha_i \mathbf{v}_i$ is $f(\mathbf{v}) = \sum_i \alpha_i f(\mathbf{v}_i)$, so the image

$$\mathrm{Im}(f) = \left\{ \sum_i \alpha_i f(\mathbf{v}_i) \,\middle|\, \alpha_i \in \mathbb{F} \right\} = \mathrm{Span}(f(\mathbf{v}_1), \dots, f(\mathbf{v}_n)) \tag{14.2}$$

is spanned by the images of the basis vectors. Therefore, the rank equals

$$\mathrm{rk}(f) = \dim_{\mathbb{F}}(\mathrm{Span}(f(\mathbf{v}_1), \dots, f(\mathbf{v}_n))) = \begin{cases} \text{maximal number of linearly} \\ \text{independent images } f(\mathbf{v}_i) \end{cases} . \tag{14.3}$$

Example 14.1 *(The image and the rank of a matrix)*
An $m \times n$ matrix A with entries in $\mathbb{F}$ defines a linear map $A : \mathbb{F}^n \to \mathbb{F}^m$. We can,

therefore, talk about the rank of a matrix. To determine the image of a matrix we can apply Eq. (14.2) to the standard unit vector basis, remembering that the images of the standard unit vectors, $A\mathbf{e}_i = \mathbf{A}^i$, are the columns of the matrix. Hence, the image

$$\text{Im}(A) = \text{Span}(\mathbf{A}^1, \dots, \mathbf{A}^n),$$

of a matrix is spanned by its column vectors. For the rank of A, this implies

$$\text{rk}(A) = \dim_{\mathbb{F}}(\text{Span}(\mathbf{A}^1, \dots, \mathbf{A}^n)) = \begin{cases} \text{maximal number of linearly} \\ \text{independent column vectors} \end{cases}. \qquad (14.4)$$

All we need to do is look at the column vectors of the matrix and find out how many of them are linearly independent — this number equals the rank of the matrix. To do this we can use, for now, the standard methods for checking linear independence.

The rank of a matrix as defined above is also called the *column rank*. We can also define the *row rank* of a matrix as the maximal number of linearly independent row vectors. We will see later that these two ranks are, in fact, always equal! □

Problem 14.1 *(Image, rank, and kernel of a matrix 'by inspection')*

Find the images, kernels, and ranks of the matrices:

$$A = \begin{pmatrix} 2 & -1 \\ 1 & 0 \end{pmatrix}, \qquad B = \begin{pmatrix} -1 & 4 & 3 \\ 2 & -3 & -1 \\ 3 & 2 & 5 \end{pmatrix}. \qquad (14.5)$$

Solution: The matrix A defines a linear map $A : \mathbb{R}^2 \to \mathbb{R}^2$. Since its two columns are linearly independent (they are not multiples of each other) they form a basis of $\mathbb{R}^2$ and, therefore, $\text{Im}(A) = \mathbb{R}^2$ and the rank is maximal, $\text{rk}(A) = 2$. In order to find the kernel we have to solve the equation $A\mathbf{v} = \mathbf{0}$ which, split up into components, reads $2v_1 - v_2 = 0$ and $v_1 = 0$. Clearly, the only solution is $v_1 = v_2 = 0$ so that the kernel is trivial, $\text{Ker}(A) = \{\mathbf{0}\}$.

The first two columns, $\mathbf{B}^1$ and $\mathbf{B}^2$, of B are linearly independent (they are not each other's multiples) while the third column is the sum of the first two. Hence, the image, $\text{Im}(B) = \text{Span}(\mathbf{B}^1, \mathbf{B}^2)$, is two-dimensional and the rank, $\text{rk}(B) = 2$, is not maximal. To find the kernel we have to solve

$$\mathbf{0} = B\mathbf{v} = v_1\mathbf{B}^1 + v_2\mathbf{B}^2 + v_3\mathbf{B}^3 = (v_1 + v_3)\mathbf{B}^1 + (v_2 + v_3)\mathbf{B}^2,$$

where the last step follows from $\mathbf{B}^3 = \mathbf{B}^1 + \mathbf{B}^2$. Since $\mathbf{B}^1$ and $\mathbf{B}^2$ are linearly independent the solution is $v_1 + v_3 = v_2 + v_3 = 0$. This leads to the one-dimensional kernel $\text{Ker}(B) = \text{Span}((1, 1, -1)^T)$.

Example 14.2 *(Kernel of a differential operator)*

In Eq. (12.10) we have introduced linear differential operators $L : \mathcal{C}^\infty([a, b]) \to \mathcal{C}^\infty([a, b])$. These differential operators define homogeneous, linear differential equations of the form $Lg = 0$. The simple but important observation is that the space of solutions to such a differential equation is, in fact, the kernel, $\text{Ker}(L)$ of the differential operator. □

Problem 14.2 *(Kernel of a linear differential operators)*

Find the kernel of the linear differential operator

$$L = \frac{d^2}{dx^2} + 4\frac{d}{dx} - 5 \; : \; \mathcal{C}^\infty(\mathbb{R}) \to \mathcal{C}^\infty(\mathbb{R}) \; ,$$

and determine its dimension.

Solution: The kernel of L consists of all solutions to the homogeneous differential equation $L(g) = 0$. This equation is solved by $g_1(x) = \exp(x)$ and $g_2(x) = \exp(-5x)$ and the most general solution is a linear combination of g_1 and g_2. Moreover, g_1 and g_2 are linearly independent. Indeed, evaluating the equation $\alpha_1 g_1 + \alpha_2 g_2 = 0$ at $x = 0$ gives $\alpha_1 + \alpha_2 = 0$ and evaluating it at $x = 1$ gives $\alpha_1 e + \alpha_2 e^{-5} = 0$. These two equations are only solved simultaneously for $\alpha_1 = \alpha_2 = 0$. In conclusion $\text{Ker}(L) = \text{Span}(g_1, g_2)$ and the kernel is two-dimensional.

Problem 14.3 *(Image, rank, and kernel of a differential operator)*

The space $\mathcal{C}^\infty(\mathbb{R})$ of infinitely many times differentiable functions is infinite-dimensional which makes it difficult to discuss the image. For a simple example which gets around this problem, we restrict to the vector space $V = \mathcal{P}_3(\mathbb{R})$ of at most cubic polynomials. On this space we introduce the first-order differential operator

$$L = x\frac{d}{dx} - 1 \; : \; V \to V \; . \tag{14.6}$$

Determine the image, the rank, and the kernel of this operator.

Solution: Recall that the monomials $(1, x, x^2, x^3)$ form a basis of V. A general cubic and its image under L are given by

$$p(x) = a_3 x^3 + a_2 x^2 + a_1 x + a_0 \quad \Rightarrow \quad L(p)(x) = 2a_3 x^3 + a_2 x^2 - a_0 \; . \tag{14.7}$$

It follows immediately that $\text{Im}(L) = \text{Span}(1, x^2, x^3)$ and $\text{rk}(L) = 3$. On the other hand, Eq. (14.7) shows that the polynomials p for which $L(p) = 0$ are precisely those of the form $p(x) = a_1 x$. This means that $\text{Ker}(L) = \text{Span}(x)$ and $\dim_\mathbb{R}(\text{Ker}(L)) = 1$.

14.1.3   Injective and surjective linear maps

It might be difficult to check if a map is surjective or injective, using the definitions of these properties (see Def. 2.4). For linear maps, simple criteria can be formulated in terms of the image and the kernel and their dimensions.

Proposition 14.2 *(Criteria for surjectivity and injectivity) For a linear map $f : V \to W$ we have the following statements:*

(i)   *f surjective* $\Leftrightarrow$ $\text{Im}(f) = W$ $\Leftrightarrow$ $\text{rk}(f) = \dim_\mathbb{F}(W)$
(ii) *f injective* $\Leftrightarrow$ $\text{Ker}(f) = \{\mathbf{0}\}$ $\Leftrightarrow$ $\dim_\mathbb{F}(\text{Ker}(f)) = 0$ *(kernel trivial)*

(The dimension statements on the right apply to finite-dimensional V and W.)

Proof   (i) The first equivalence, f surjective $\Leftrightarrow \text{Im}(f) = W$, is clear by the definitions of surjective and the image. Turning to the second equivalence, if $\text{Im}(f) = W$, then both spaces have the same dimension. Conversely, from Lemma 7.2, two vector spaces with the same dimension and one contained in the other (here $\text{Im}(f) \subset W$) must be identical.

(ii) Suppose f is injective and consider a vector $\mathbf{v} \in \text{Ker}(f)$. Then $f(\mathbf{v}) = \mathbf{0} = f(\mathbf{0})$, which implies that $\mathbf{v} = \mathbf{0}$ from injectivity and, hence, $\text{Ker}(f) = \{\mathbf{0}\}$. Conversely, assume that $\text{Ker}(f) = \{\mathbf{0}\}$. Then, from linearity, $f(\mathbf{v}_1) = f(\mathbf{v}_2)$ implies that $f(\mathbf{v}_1 - \mathbf{v}_2) = \mathbf{0}$ so that $\mathbf{v}_1 - \mathbf{v}_2 \in \text{Ker}(f) = \{\mathbf{0}\}$. Hence, $\mathbf{v}_1 - \mathbf{v}_2 = \mathbf{0}$ and f is injective. This proves the first equivalence. The second equivalence is evident as the trivial vector space, $\{\mathbf{0}\}$, is the only one with dimension zero. $\square$

Problem 14.4 (*Criteria for surjective and injective maps*)

Determine if the linear maps defined by the matrices A and B in Eq. (14.5) and the matrix $C : \mathbb{R}^3 \to \mathbb{R}^2$ with

$$C = \begin{pmatrix} 2 & 0 & -1 \\ 0 & 3 & 1 \end{pmatrix}$$

are injective or surjective. Are these matrices invertible? Is the differential operator in Eq. (14.6) injective or surjective?

Solution: In Exercise 14.1 we have found that the rank of A is maximal, and its kernel is trivial. From Prop. (14.2) this means that A is bijective and, hence, invertible.

For the 3×3 matrix B in Eq. (14.5) we know from Exercise 14.1 that $\text{rk}(B) = 2$ and $\dim_{\mathbb{R}}(\text{Ker}(B)) = 1$, so B is neither surjective nor injective and, hence, not invertible.

The first two columns of C are linearly independent so that $\text{Im}(C) = \mathbb{R}^2$, $\text{rk}(C) = 2$ and C is surjective. On the other hand, it is easy to see that the vector $\mathbf{v} = (3, -2, 6)^T$ satisfies $C\mathbf{v} = \mathbf{0}$. Hence, $\mathbf{v} \in \text{Ker}(C)$ so that the kernel is non-trivial and C is neither injective nor invertible.

From Exercise 14.3, the differential operator L in Eq. (14.6) has a non-maximal rank and a non-trivial kernel, so it is neither surjective nor injective.

14.2   The rank theorem

Summary 14.2 *The rank theorem states that the rank of a linear map equals the difference of the domain dimension and the dimension of the kernel. It can be used to show that two vector spaces are isomorphic if and only if they have the same dimension. In particular, invertible linear maps only exist between vector spaces of the same dimension. A linear map between same-dimensional vector spaces is invertible iff its kernel is trivial or iff its rank is maximal. The rank theorem is implied by the isomorphism theorem which states that for any linear map $f : V \to W$ the quotient $V/\text{Ker}(f)$ is isomorphic to $\text{Im}(f)$.*

14.2.1 Motivation

To develop a better intuition we recall our interpretation of vector subspaces as lines, planes, and their higher-dimensional analogues through **0**. We should think of both the kernel and the image of a linear map in this way, the former residing in the domain vector space, the latter in the co-domain.

Consider a linear map $f : V \to W$. The entire kernel, $\mathrm{Ker}(f)$, is mapped to the zero vector, so it does not at all contribute to creating a non-trivial image. What is more, the image of an affine plane $\mathbf{v} + \mathrm{Ker}(f)$ under f consists of a single vector. To see this consider two vectors $\mathbf{v}_1 = \mathbf{v} + \mathbf{w}_1$ and $\mathbf{v}_2 = \mathbf{v} + \mathbf{w}_2$, where $\mathbf{w}_1, \mathbf{w}_2 \in \mathrm{Ker}(f)$. Then, since $f(\mathbf{w}_1) = f(\mathbf{w}_2) = \mathbf{0}$, it follows

$$f(\mathbf{v}_1) = f(\mathbf{v} + \mathbf{w}_1) = f(\mathbf{v}) + f(\mathbf{w}_1) = f(\mathbf{v}) = f(\mathbf{v}) + f(\mathbf{w}_2) = f(\mathbf{v} + \mathbf{w}_2) = f(\mathbf{v}_2) \,.$$

This suggest that the dimensions associated to the kernel are lost under the action of the map and that the remaining $\dim_{\mathbb{F}}(V) - \dim_{\mathbb{F}}(\mathrm{Ker}(f))$ dimensions are available to form the image of f. This is precisely the content of the rank theorem.

14.2.2 The theorem

Theorem 14.1 *For a linear map $f : V \to W$ between (finite-dimensional) vector space V and W we have*

$$\dim_{\mathbb{F}}(\mathrm{Ker}(f)) + \mathrm{rk}(f) = \dim_{\mathbb{F}}(V) \,. \tag{14.8}$$

Proof To simplify notation, set $k = \dim_{\mathbb{F}}(\mathrm{Ker}(f))$ and $n = \dim_{\mathbb{F}}(V)$. Let $(\mathbf{v}_1, \cdots , \mathbf{v}_k)$ be a basis of $\mathrm{Ker}(f)$ which we complete to a basis $(\mathbf{v}_1, \ldots , \mathbf{v}_k, \mathbf{v}_{k+1}, \ldots , \mathbf{v}_n)$ of V. (This is indeed possible from Theorem 7.2 (ii).) We will show that $f(\mathbf{v}_{k+1}), \ldots , f(\mathbf{v}_n)$ forms a basis of $\mathrm{Im}(f)$. To do this we need to check the two conditions in Definition 7.2.

(B1) First we need to show that $\mathrm{Im}(f)$ is spanned by $f(\mathbf{v}_{k+1}), \ldots , f(\mathbf{v}_n)$. We begin with an arbitrary vector $\mathbf{w} \in \mathrm{Im}(f)$. This vector must be the image of a $\mathbf{v} \in V$, so that $\mathbf{w} = f(\mathbf{v})$. We can expand $\mathbf{v}$ as a linear combination

$$\mathbf{v} = \sum_{i=1}^{n} \alpha_i \mathbf{v}_i$$

of the basis in V. Acting on this equation with f and using linearity we find

$$\mathbf{w} = f(\mathbf{v}) = f\left(\sum_{i=1}^{n} \alpha_i \mathbf{v}_i\right) = \sum_{i=1}^{n} \alpha_i f(\mathbf{v}_i) = \sum_{i=k+1}^{n} \alpha_i f(\mathbf{v}_i) \,,$$

where the last step follows since the vectors $\mathbf{v}_i$ for $i = 1, \ldots , k$ are in the kernel so that $f(\mathbf{v}_i) = \mathbf{0}$. Hence, we have written $\mathbf{w}$ as a linear combination of the vectors $f(\mathbf{v}_{k+1}), \ldots , f(\mathbf{v}_n)$ which, therefore, span the image of f.

(B2) For the second step, we have to show that the vectors $f(\mathbf{v_{k+1}}), \ldots, f(\mathbf{v_n})$ are linearly independent. As usual, we start with the equation

$$\sum_{i=k+1}^{n} \alpha_i f(\mathbf{v}_i) = \mathbf{0} \quad \Rightarrow \quad f\left(\sum_{i=k+1}^{n} \alpha_i \mathbf{v}_i \right) = \mathbf{0} .$$

The second of these equations means that the vector $\sum_{i=k+1}^{n} \alpha_i \mathbf{v}_i$ is in the kernel of f and, given that $\mathbf{v}_1, \ldots, \mathbf{v}_k$ form a basis of the kernel, there are coefficients $\alpha_1, \ldots, \alpha_k$ such that

$$\sum_{i=k+1}^{n} \alpha_i \mathbf{v}_i = - \sum_{i=1}^{k} \alpha_i \mathbf{v}_i \quad \Rightarrow \quad \sum_{i=1}^{n} \alpha_i \mathbf{v}_i = \mathbf{0} .$$

Since $(\mathbf{v}_1, \ldots, \mathbf{v}_n)$ is a basis of V it follows that all $\alpha_i = 0$ and, hence, $f(\mathbf{v_{k+1}}), \ldots, f(\mathbf{v_n})$ are linearly independent.

In summary, $(f(\mathbf{v_{k+1}}), \cdots, f(\mathbf{v_n}))$ forms a basis of $\mathrm{Im}(f)$. Hence, by counting the number of basis elements, we have $\dim_{\mathbb{F}}(\mathrm{Im}(f)) = n - k = \dim_{\mathbb{F}}(V) - \dim_{\mathbb{F}}(\mathrm{Ker}(f))$.
□

We emphasize again that this theorem has a simple and intuitive interpretation. We have $\dim_{\mathbb{F}}(V)$ dimensions of the domain vector space available but the $\dim_{\mathbb{F}}(\mathrm{Ker}(f))$ dimensions of the kernel are removed since the entire kernel is mapped to zero. Hence, the difference of these two dimensions is available to account for the dimension of the image.

Example 14.3 *(Structure of a linear map $\mathbb{R}^3 \to \mathbb{R}^2$)*

Consider a linear map $f : \mathbb{R}^3 \to \mathbb{R}^2$ with a two-dimensional kernel, $\dim_{\mathbb{R}}(\mathrm{Ker}(f)) = 2$. In this case the dimension formula (14.13) implies $\mathrm{rk}(f) = \dim_{\mathbb{R}}(\mathbb{R}^3) - \dim_{\mathbb{R}}(\mathrm{Ker}(f)) = 3 - 2 = 1$, so the image of f is one-dimensional. In other words, f has removed the two kernel dimensions by mapping them to the zero vector so that one dimension remains available to create the image. This is schematically illustrated in the figure below.

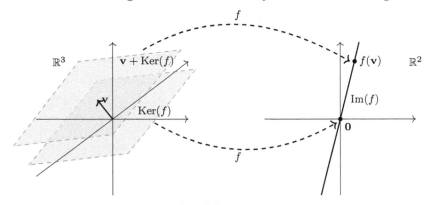

The entire two-dimensional kernel $\mathrm{Ker}(f)$ is mapped to the zero vector, while any affine plane $\mathbf{v} + \mathrm{Ker}(f)$ is mapped to the single vector $f(\mathbf{v})$. In this way, a one-dimensional image, $\mathrm{Im}(f)$, is created, and the image becomes isomorphic to the set of all affine planes $\mathbf{v} + \mathrm{Ker}(f)$, a statement that will be made more general and precise in Theorem 14.3. □

Problem 14.5 *(Checking the rank theorem)*

Check the rank theorem for the matrices A and B in Eq. (14.5) and for the differential operator L in Eq. (14.6).

Solution: From Exercise 14.1 the 2×2 matrix $A : \mathbb{R}^2 \to \mathbb{R}^2$ has $\mathrm{rk}(A) = 2$ and $\dim_{\mathbb{R}}(\mathrm{Ker}(A)) = 0$. Hence, we have $\dim_{\mathbb{R}}(\mathrm{Ker}(A)) + \mathrm{rk}(A) = 0 + 2 = \dim_{\mathbb{R}}(\mathbb{R}^2)$.

For the 3×3 matrix $B : \mathbb{R}^3 \to \mathbb{R}^3$ we have found in Exercise 14.1 that $\mathrm{rk}(B) = 2$ and $\dim_{\mathbb{R}}(\mathrm{Ker}(B)) = 1$ so that $\dim_{\mathbb{R}}(\mathrm{Ker}(B)) + \mathrm{rk}(B) = 1 + 2 = 3 = \dim_{\mathbb{R}}(\mathbb{R}^3)$.

Finally, the operator $L : V \to V$ on the space V of at most cubic polynomials in Eq. (14.6) has $\mathrm{rk}(L) = 3$ and $\dim_{\mathbb{R}}(\mathrm{Ker}(L)) = 1$, as was shown in Exercise 14.3. This means $\dim_{\mathbb{R}}(\mathrm{Ker}(L)) + \mathrm{rk}(L) = 1 + 3 = 4 = \dim_{\mathbb{R}}(V)$.

14.2.3 Easy conclusions from the rank theorem

We can use the rank theorem to derive an upper bound on the rank of a linear map $f : V \to W$. Since the image of f is a vector subspace of W it is clear that $\mathrm{rk}(f) \leq \dim_{\mathbb{F}}(W)$. On the other hand, the rank theorem implies that $\dim_{\mathbb{F}}(V) - \mathrm{rk}(f) = \dim_{\mathbb{F}}(\mathrm{Ker}(f)) \geq 0$, so that $\mathrm{rk}(f) \leq \dim_{\mathbb{F}}(V)$. Together, this leads to

$$\mathrm{rk}(f) \leq \min\left(\dim_{\mathbb{F}}(V), \dim_{\mathbb{F}}(W)\right) . \tag{14.9}$$

If this bound is saturated the rank of f is called *maximal*. For an $m \times n$ matrix $A : \mathbb{F}^n \to \mathbb{F}^m$ Eq. (14.9) reads

$$\mathrm{rk}(A) \leq \min(n, m) , \tag{14.10}$$

so the rank of a matrix cannot exceed the number of its rows and columns.

Another simple and useful conclusion from the rank theorem is that injective linear maps preserve dimension.

Corollary 14.1 *Let $f : V \to W$ be an injective linear map and $U \subset V$ a (finite-dimensional) vector subspace. Then $\dim_{\mathbb{F}}(U) = \dim_{\mathbb{F}}(f(U))$.*

Proof Define the restricted map $g = f|_U$, so that $f(U) = \mathrm{Im}(g)$. Since f is injective, so is its restriction g and from Prop. 14.2 (ii) this means that $\dim_{\mathbb{F}}(\mathrm{Ker}(g)) = 0$. Hence, we have $\dim_{\mathbb{F}}(f(U)) = \dim_{\mathbb{F}}(\mathrm{Im}(g)) = \mathrm{rk}(g) = \dim_{\mathbb{F}}(U)$, where the rank formula (14.8) applied to g has been used in the last step. □

14.2.4 Isomorphisms

Recall that a (vector space) isomorphism $f : V \to W$ is a bijective linear map and that two vector spaces V and W are called isomorphic, denoted as $V \cong W$, if an isomorphism $f : V \to W$ exists. We have seen in Section 12.2.1 that an n-dimensional vector space over $\mathbb{F}$ is isomorphic to the coordinate vector space $\mathbb{F}^n$. We can now generalize this statement and use the rank theorem to obtain a simple criterion for two vector spaces to be isomorphic.

Theorem 14.2 *Two (finite-dimensional) vector spaces are isomorphic if and only if they have the same dimension.*

Proof '$\Rightarrow$': If the vector spaces V and W are isomorphic there exists an isomorphism $f : V \rightarrow W$. Since f is bijective it follows from Prop. 14.2 that $\mathrm{rk}(f) = \dim_\mathbb{F}(W)$ and $\dim_\mathbb{F}(\mathrm{Ker}(f)) = 0$. Inserting this into the rank formula (14.13) leads to $\dim_\mathbb{F}(V) = \dim_\mathbb{F}(W)$.

'$\Leftarrow$': Suppose the two vector spaces V and W have the same dimension $n = \dim_\mathbb{F}(V) = \dim_\mathbb{F}(W)$. First, choose bases $(\mathbf{v}_1, \ldots, \mathbf{v}_n)$ for V and $(\mathbf{w}_1, \ldots, \mathbf{w}_n)$ for W. Theorem 12.1 tells us there exists a unique linear map $f : V \rightarrow W$ with $f(\mathbf{v}_i) = \mathbf{w}_i$ for $i = 1, \ldots, n$. For a vector $\mathbf{v} = \sum_i \alpha_i \mathbf{v}_i$ we have, from linearity, that $f(\mathbf{v}) = \sum_i \alpha_i \mathbf{w}_i$ and since the $\mathbf{w}_i$ form a basis of W this shows that f is surjective. Prop. 14.2 then implies that $\mathrm{rk}(f) = \dim_\mathbb{F}(W) = n$. Inserting this into the rank formula (14.8) implies $\dim_\mathbb{F}(\mathrm{Ker}(f)) = 0$, so, from Prop. 14.2, f is also injective. Hence, f is an isomorphism and V and W are isomorphic. $\qquad\square$

We have seen earlier that being isomorphic is an equivalence relation on the (finite-dimensional) vector spaces over a given field $\mathbb{F}$. The associated equivalence classes consist of all vector spaces isomorphic to each other. We can now easily describe these equivalence classes: they contain all vector spaces over $\mathbb{F}$ with the same dimension.

14.2.5 The inverse of a linear map

It is clear from Theorem 14.2 that a linear map $f : V \rightarrow W$ can only have an inverse if V and W have the same dimension (or else we would have an isomorphism between vector spaces of different dimensions which the theorem excludes). Of course a given linear map between vector spaces with the same dimension does not have to be an isomorphism but we can use the rank theorem to derive simple criteria for when this is the case.

Corollary 14.2 *Let $f : V \rightarrow W$ be a linear map between two (finite-dimensional) vector spaces with the same dimension. Then the following statements are equivalent.*

  (i)   *f is an isomorphism (has an inverse)*
  (ii) *$\dim_\mathbb{F}(\mathrm{Ker}(f)) = 0$*   *(kernel trivial)*
  (iii) *$\mathrm{rk}(f) = \dim_\mathbb{F}(V)$*   *(rank maximal)*

Proof Proving the equivalence of three statements seems a lot of work. A common trick which simplifies matters is to show the 'cyclic' implications (i) $\Rightarrow$ (ii) $\Rightarrow$ (iii) $\Rightarrow$ (i) from which all the others follow.

'(i) $\Rightarrow$ (ii)': If f is an isomorphism then it is injective and, hence, from Prop. 14.2 (ii), we have $\dim_\mathbb{F}(\mathrm{Ker}(f)) = 0$.

'(ii) $\Rightarrow$ (iii)': If $\dim - \mathbb{F}(\mathrm{Ker}(f)) = 0$ then the dimension formula (14.13) implies that $\mathrm{rk}(f) = \dim_\mathbb{F}(V)$.

'(iii) $\Rightarrow$ (i)': If $\mathrm{rk}(f) = \dim_\mathbb{F}(V) = \dim_\mathbb{F}(W)$ then Prop. 14.2 (i) implies that f is surjective. Further, inserting $\mathrm{rk}(f) = \dim_\mathbb{F}(V)$ into the dimension formula (14.13) leads to $\dim_\mathbb{F}(\mathrm{Ker}(f)) = 0$ which, from Prop. 14.2 (ii), means that f is injective. $\qquad\square$

So, a linear map $f : V \to W$ between two vector spaces of the same dimension is an isomorphism iff its kernel is trivial or iff its rank is maximal.

Applied to matrices, Theorem (14.2) means that only square matrices can have an inverse. From Cor. 14.2, such a matrix is invertible iff its rank is maximal and this is the case iff its column vectors are linearly independent.

Problem 14.6 (*Checking if a matrix is invertible*)

Are the following matrices invertible?

$$
A = \begin{pmatrix} -3 & 2 & 0 \\ 1 & -1 & -2 \\ 5 & -8 & 0 \end{pmatrix}, \qquad B = \begin{pmatrix} 1 & -3 & 9 \\ -1 & 0 & -3 \\ 2 & 4 & -2 \end{pmatrix}
$$

Solution: Since the equation

$$
0 \stackrel{!}{=} \alpha_1 \mathbf{A}^1 + \alpha_2 \mathbf{A}^2 + \alpha_3 \mathbf{A}_3 = \begin{pmatrix} -3\alpha_1 + 2\alpha_2 \\ \alpha_1 - \alpha_2 - 2\alpha_3 \\ 5\alpha_1 - 8\alpha_2 \end{pmatrix} \quad \Rightarrow \quad \alpha_1 = \alpha_2 = \alpha_3 = 0
$$

only has the trivial solution, $\alpha_i = 0$, the columns of A are linearly independent and the rank is maximal, $\mathrm{rk}(A) = 3$. Hence, A is invertible.

It is easy to see that $B(-3, 2, 1)^T = \mathbf{0}$, so the kernel of B is non-trivial and the rank non-maximal. Therefore, B is not invertible.

14.3   Another proof of the rank theorem*

Our proof the rank theorem in Section 14.2.2 has been elementary. We have explicitly constructed bases for the various vector spaces to determine dimensions in order to verify the dimension formula. There is a slightly more abstract approach which provides more insight into the structure of linear maps and which relies on the results for vector subspaces and quotients in Chapter 8. The reader who has skipped this material at first reading should move on to the next chapter.

Consider a linear map $f : V \to W$ with kernel $\mathrm{Ker}(f)$. We know that f maps the entire kernel to the zero vector and also that the affine planes $\mathbf{v} + \mathrm{Ker}(f)$ are each mapped to the single vector $f(\mathbf{v})$. This structure is reminiscent of the quotient vector space construction we have discussed in Section 8.2. Recall that the quotient $V/\mathrm{Ker}(f)$ consists precisely of the affine planes $\mathbf{v} + \mathrm{Ker}(f)$ and on each these the map f only has a single value. This means we can write down a well-defined map

$$
\hat{f} : V/\mathrm{Ker}(f) \to \mathrm{Im}(f) , \qquad \hat{f}(\mathbf{v} + \mathrm{Ker}(f)) := f(\mathbf{v}) , \tag{14.11}
$$

whose domain is the quotient $V/\mathrm{Ker}(f)$. Clearly, this map is linear as it inherits the linearity properties from f. What is more, since we have divided by $\mathrm{Ker}(f)$, there is no obstruction to injectivity. In fact, the kernel $\mathrm{Ker}(f)$ is the zero vector of the quotient $V/\mathrm{Ker}(f)$. Moreover, since we have replaced the co-domain by $\mathrm{Im}(f)$ the map $\hat{f}$ is surjective. It appears that $\hat{f}$ is an isomorphism and this is shown in the following theorem.

Theorem 14.3 *(Isomorphism theorem) The map $\hat{f}$ defined in Eq. (14.11) is a vector space isormorphism. Hence, for a linear map $f : V \to W$, we have*

$$V/\mathrm{Ker}(f) \cong \mathrm{Im}(f) .\tag{14.12}$$

Proof The linearity of $\hat{f}$ in Eq. (14.11) follows directly from the linearity of f. It is obviously surjective since the co-domain is $\mathrm{Im}(f)$. To show injectivity, set $U = \mathrm{Ker}(f)$ and start with an element $\mathbf{v} + U \in \mathrm{Ker}(\hat{f})$ in the kernel. It follows from the definition, Eq. (14.11), of $\hat{f}$ that $\mathbf{0} = \hat{f}(\mathbf{v} + U) = f(\mathbf{v})$, which implies $\mathbf{v} \in U$. This means that $\mathbf{v} + U = U$ which is the zero vector in V/U. Hence, the kernel of $\hat{f}$ is trivial and, from Prop. 14.2 (ii), this means $\hat{f}$ is injective. $\square$

The rank theorem now follows very easily from this isomorphism statement and the dimension formula for quotient spaces.

Theorem 14.4 *For a linear map $f : V \to W$ between (finite-dimensional) vector spaces V and W we have*

$$\dim_{\mathbb{F}}(\mathrm{Ker}(f)) + \mathrm{rk}(f) = \dim_{\mathbb{F}}(V) .\tag{14.13}$$

Proof From Theorem 14.2 we know that the isomorphism (14.12) implies equality of dimensions, so that $\dim_{\mathbb{F}}(V/\mathrm{Ker}(f)) = \dim_{\mathbb{F}}(\mathrm{Im}(f)) = \mathrm{rk}(f)$. On the other hand, Theorem 8.2 implies that $\dim_{\mathbb{F}}(V/\mathrm{Ker}(f)) = \dim_{\mathbb{F}}(V) - \dim_{\mathbb{F}}(\mathrm{Ker}(f))$. Combining these two equations gives the rank theorem. $\square$

Application 14.1 *Coding theory*

Coding theory deals with the problem of errors in information such as they may arise when information is transmitted in the presence of noise. Whenever information may be faulty, methods are required for both error detection and error correction. A simple but potentially inefficient method is to transmit the information repeatedly. Here, we would like to discuss a more sophisticated method, referred to as *Hamming code*, which is based on some of the linear algebra methods we have explored.

Information is conveniently described in binary form, that is, as a sequence of bits, $\beta_1, \ldots, \beta_n \in \{0, 1\}$. Mathematically, a bit can be seen as an element of the finite field $\mathbb{F}_2 = \{0, 1\}$ which we have introduced in Example 4.3 and information encoded by n bits can be seen as an element of the n-dimensional vector space $V = \mathbb{F}_2^n$ over the field $\mathbb{F}_2$. In other words, we can think of the above bit sequence as a column vector $(\beta_1, \ldots, \beta_n)^T \in \mathbb{F}_2^n$. Through this simple re-interpretation all the tools of linear algebra are now available to deal with information.

To be specific we focus on the case of four bits, $\boldsymbol{\beta} = (\beta_1, \beta_2, \beta_3, \beta_4)^T$, but the method can be generalized to arbitrary dimensions. We begin by writing down the matrix

$$H = (\mathbf{H}^1, \ldots, \mathbf{H}^7) = \begin{pmatrix} 0 & 0 & 0 & 1 & 1 & 1 & 1 \\ 0 & 1 & 1 & 0 & 0 & 1 & 1 \\ 1 & 0 & 1 & 0 & 1 & 0 & 1 \end{pmatrix} ,\tag{14.14}$$

whose columns consist of all non-zero vectors of $\mathbb{Z}_2^3$. Clearly, $\mathrm{rk}(H) = 3$ (since $\mathbf{H}^1$, $\mathbf{H}^2$, $\mathbf{H}^4$ are linearly independent) and, from Eq. (14.13), its kernel has dimension $\dim_{\mathbb{F}_2}(\mathrm{Ker}(H)) =$

$7 - 3 = 4$. It is easy to see that this four-dimensional kernel has a basis consisting of the vectors

$$\mathbf{k}_1 = (1,0,0,0,0,1,1)^T \qquad \mathbf{k}_2 = (0,1,0,0,1,0,1)^T$$
$$\mathbf{k}_3 = (0,0,1,0,1,1,0)^T \qquad \mathbf{k}_4 = (0,0,0,1,1,1,1)^T \ .$$

The key idea is now to encode the information stored in $\beta_1, \ldots, \beta_4$ by forming the linear combination of these numbers with the above vectors $\mathbf{k}_i$. In other words, we encode the information in the following seven-dimensional vector

$$\mathbf{v} = \sum_{i=1}^{4} \beta_i \mathbf{k}_i \ .$$

Note that, given the choice of the vectors $\mathbf{k}_i$, the first four bits in $\mathbf{v}$ coincide with the actual information $\beta_1, \ldots, \beta_4$. By construction, the vector $\mathbf{v}$ is an element of $\mathrm{Ker}(H)$.

Now suppose that the transmission of $\mathbf{v}$ has resulted in a vector $\mathbf{w}$ which may have an error in at most one bit. How do we detect whether such an error has occurred? We note that the seven-dimensional standard unit vectors $\mathbf{e}_1, \ldots, \mathbf{e}_7$ are not in the kernel of H. Given that $\mathbf{v} \in \mathrm{Ker}(H)$, it follows that none of the vectors $\mathbf{w} = \mathbf{v} + \mathbf{e}_i$ is in $\mathrm{Ker}(H)$. This means the transmitted information $\mathbf{w}$ is free of (one-bit) errors, if and only if $\mathbf{w} \in \mathrm{Ker}(H)$, a condition which can be easily tested.

Suppose $\mathbf{w} \notin \mathrm{Ker}(H)$ so that the information is faulty. How can the error be corrected? If bit number i has changed in $\mathbf{w}$ the correct original vector is $\mathbf{v} = \mathbf{w} - \mathbf{e}_i$. Since $\mathbf{v} \in \mathrm{Ker}(H)$ it follows that $H\mathbf{w} = H\mathbf{e}_i = \mathbf{H}^i$. Consequently, if $H\mathbf{w}$ equals column i of H then we should flip bit number i in $\mathbf{w}$ to correct for the error.

Let us carry all this out for an explicit example. Suppose that the transmitted message is $\mathbf{w} = (1,1,0,0,0,1,1)^T$ and that it contains at most one error. Then we work out

$$H\mathbf{w} = \begin{pmatrix} 0 \\ 1 \\ 0 \end{pmatrix} = \mathbf{H}^2 \ .$$

First, $\mathbf{w}$ is not in the kernel of H so an error has indeed occurred. Secondly, the vector $H\mathbf{w}$ corresponds to the second column vector of H so we should flip the second bit to correct for the error. This means, $\mathbf{v} = (1,0,0,0,0,1,1)^T$ and the original information (which is contained in the first four entries of $\mathbf{v}$) is $\boldsymbol{\beta} = (1,0,0,0)^T$.

By paying the price of enhancing the transmitted information from four bits (in $\boldsymbol{\beta}$) to seven bits (in $\mathbf{v}$) both a detection and correction of one-bit errors can be carried out with this method. Compare this with the naive method of transmitting the information in $\boldsymbol{\beta}$ twice which corresponds to an enhancement from four to eight bits. In this case, a one-bit error has occurred if the two transmissions differ. However, without further information it is unclear which transmission is the correct one, so there is no method for error correction.

Exercises

(†=challenging)

14.1 *Image, kernel, and rank of matrices*
Find the images, kernels, and ranks of the following matrices:

$$A = \begin{pmatrix} 1 & 2 \\ -1 & 1 \end{pmatrix}, \quad B = \begin{pmatrix} 1 & -2 & -4 \\ 0 & 1 & 3 \\ -5 & 2 & -4 \end{pmatrix}$$

$$C = \begin{pmatrix} 0 & \frac{1}{2} \\ -\frac{4}{5} & 1 \\ -\frac{5}{2} & 0 \end{pmatrix} \quad D = \begin{pmatrix} \frac{1}{3} & -1 & 0 \\ 0 & 2 & 1 \end{pmatrix},$$

seen as linear maps $\mathbb{R}^n \to \mathbb{R}^m$. Verify the rank theorem in each case.

14.2 *Rank of 3×3 matrices*
Consider matrices $A \in \mathcal{M}_{3,3}(\mathbb{R})$.
(a) Show that $\mathrm{rk}(A) < 3$ iff $\det(A) = 0$.
(b) Work out for which values of $a, b \in \mathbb{R}$ the matrix

$$A = \begin{pmatrix} 1 & -4 & 2 \\ -2 & 0 & -1 \\ 3 & -4 & a \end{pmatrix}, \quad B = \begin{pmatrix} 0 & -3 & 1 \\ a & 2 & -4 \\ 1 & 1 & b \end{pmatrix}$$

have maximal rank and are, hence, invertible.

14.3 *Image, kernel, and rank for differential operators*
Consider the space $\mathcal{P}_k(\mathbb{R})$ of polynomials with degree less equal k and $D = d/dx$. Find the image, kernel, and rank of the following maps:
(a) $D : V_3 \to V_2$
(b) $D : V_k \to V_{k-1}$ for $k \in \mathbb{Z}^{>0}$
(c) $D^p : V_k \to V_k$ for $p, k \in \mathbb{Z}^{>0}$.
Are the maps injective or surjective? Verify the rank theorem in each case.

14.4 *Rank of matrices*
Determine the rank of the matrices

$$A = \begin{pmatrix} 1 & 2 \\ a & 1 \end{pmatrix}, \quad B = \begin{pmatrix} 1 & 0 & b \\ a & a & -1 \end{pmatrix}$$

for all $a, b \in \mathbb{R}$.

14.5 *Some properties of matrix ranks*
For matrices A, B of suitable size, show that
(a) $\mathrm{rk}(AB) \leq \min(\mathrm{rk}(A), \mathrm{rk}(B))$
(b) $\mathrm{rk}(AB) \geq \mathrm{rk}(A) + \mathrm{rk}(B) - m$ where m is the number of columns of A
(c) $\mathrm{rk}(A^T A) = \mathrm{rk}(A)$.

14.6 *Direct sums and maps*
For $V = U_1 \oplus U_2$ and linear maps $f_1 : U_1 \to W$, $f_2 : U_2 \to W$ show the following.
(a) There exists a unique linear map $f : V \to W$ with $f|_{U_1} = f_1$ and $f|_{U_2} = f_2$.
(b) $f(V) = f_1(U_1) + f_2(U_2)$.
(c) $\mathrm{rk}(f) \leq \mathrm{rk}(f_1) + \mathrm{rk}(f_2)$ with equality iff $f_1(U_1) \cap f_2(U_2) = \{\mathbf{0}\}$.

14.7 *Coding theory*
Use the Hamming code described in Application 14.1 to decide whether the information contained in the vectors

$$\mathbf{w}_1 = (1, 1, 1, 1, 1, 1, 1)^T$$
$$\mathbf{w}_2 = (1, 1, 1, 0, 1, 1, 1)^T$$
$$\mathbf{w}_3 = (1, 1, 1, 0, 0, 1, 1)^T$$

has one-bit errors and find the correct original information in each case.

14.8 *Sequences†*
For vector spaces V_0, V_1, V_2 we have linear maps

$$\{\mathbf{0}\} \xrightarrow{f_0} V_0 \xrightarrow{f_1} V_1 \xrightarrow{f_2} V_2 \xrightarrow{f_3} \{\mathbf{0}\}$$

which satisfy $f_{i+1} \circ f_i = 0$ for $i = 0, \dots, 2$.
(a) Show that $\mathrm{Im}(f_i) \subset \mathrm{Ker}(f_{i+1})$ for $i = 0, \dots, 2$.
(b) If $\mathrm{Im}(f_i) = \mathrm{Ker}(f_{i+1})$ for $i = 0, \dots, 2$ show that f_1 is injective and f_2 is surjective.
(c) Under the same assumptions as in part (b), show that $\dim_{\mathbb{F}}(V_1) = \dim_{\mathbb{F}}(V_0) + \dim_{\mathbb{F}}(V_2)$.

15
Linear maps in terms of matrices

At this point, we have a full understanding of linear maps between coordinate vectors which, as we have seen, can be identified with matrices. This is possible thanks to the existence of a canonical basis, the standard unit vector basis. On the other hand, linear maps between general vector spaces might still seem somewhat abstract. Can we find a more 'hands-on' description of linear maps, in general? For general vector spaces no preferred choice of basis is available. However, we can still, arbitrarily, choose a basis on the domain and co-domain vector space. It turns out, relative to such a choice of bases, a linear map between abstract vector spaces can be described by a matrix. We will now discuss how this works.

15.1 Matrices representing linear maps

Summary 15.1 *A linear map $f : V \to W$ can be represented by a matrix, relative to a choice of bases on V and W. This matrix maps coordinate vectors in the same way as f maps the associated vectors. It can be computed by working out the images of the basis vectors for V under f.*

15.1.1 Basis choice

Start with a linear map $f : V \to W$ between two (finite-dimensional) vector spaces V and W over $\mathbb{F}$. On each vector space we introduce a basis and a coordinate map, as defined in Section 12.2.1. The resulting set-up is summarized in Table 15.1. The

Table 15.1 Set-up to represent a linear map $f : V \to W$ by a matrix.

Vector space	Dimension	Basis	Coordinate map	Coordinate vector
V	n	$(\mathbf{v}_1, \ldots, \mathbf{v}_n)$	$\varphi : \mathbb{F}^n \to V$	$\boldsymbol{\alpha} \in \mathbb{F}^n$
W	m	$(\mathbf{w}_1, \ldots, \mathbf{w}_m)$	$\psi : \mathbb{F}^m \to W$	$\boldsymbol{\beta} \in \mathbb{F}^m$

diagram below indicates the idea for how to assign a matrix A to the linear map f. The $m \times n$ matrix A should act on coordinate vector 'in the same way' as f acts on the associated vectors.

$$\begin{array}{ccc}
\varphi(\boldsymbol{\alpha}) \in V & \xrightarrow{\quad f \quad} & W \ni \psi(\boldsymbol{\beta}) \\
\Big\uparrow \ \uparrow{\scriptstyle\varphi} & & \uparrow{\scriptstyle\psi} \ \Big\uparrow \\
\boldsymbol{\alpha} \ \in \mathbb{F}^n & \xrightarrow[\ A = \psi^{-1} \circ f \circ \varphi\ ]{} & \mathbb{F}^m \ni \boldsymbol{\beta}
\end{array} \tag{15.1}$$

More precisely, if the coordinate vectors are related by $\boldsymbol{\beta} = A\boldsymbol{\alpha}$, then the corresponding vectors $\psi(\boldsymbol{\beta})$ and $\varphi(\boldsymbol{\alpha})$ should be related by $\psi(\boldsymbol{\beta}) = f(\varphi(\boldsymbol{\alpha}))$. In mathematical parlance, this is expressed by saying that the above diagram *commutes*: the result does not depend on the path chosen in the diagram. This means the desired matrix A can be written as

$$A = \psi^{-1} \circ f \circ \varphi \,. \tag{15.2}$$

This can be seen by going from F^n to F^m in the above diagram using the 'upper path', via V and W.

The matrix A in Eq. (15.2) is said to be the matrix which represents the linear map f relative to the bases $(\mathbf{v}_1, \ldots, \mathbf{v}_n)$ of V and $(\mathbf{w}_1, \ldots, \mathbf{w}_m)$ of W. An important special case is the one of a linear map $f : V \to V$ with the same domain and co-domain vector spaces and the 'in-basis' and 'out-basis' chosen to be the same, so $\mathbf{w}_i = \mathbf{v}_i$. In this case, A is simply referred to as the representing matrix for f relative to the basis $(\mathbf{v}_1, \ldots, \mathbf{v}_n)$ of V.

15.1.2 Computing the representing matrix

How do we find the representing matrix A explicitly? The images $f(\mathbf{v}_j)$ of the V basis vectors can always be written as a linear combination of the basis vectors for W so we have

$$f(\mathbf{v}_j) = \sum_{i=1}^{m} a_{ij}\mathbf{w}_i \tag{15.3}$$

for some coefficients $a_{ij} \in \mathbb{F}$. In fact, we know from Theorem 12.1 that the linear map is uniquely characterized by these images. We denote the standard unit vectors on $\mathbb{F}^n$ by $\mathbf{e}_i$ and the ones on $\mathbb{F}^m$ by $\tilde{\mathbf{e}}_i$. Their images under the coordinate maps

$$\varphi(\mathbf{e}_i) = \mathbf{v}_i \,, \qquad \psi(\tilde{\mathbf{e}}_i) = \mathbf{w}_i \tag{15.4}$$

are precisely the basis vectors for V and W (see Eq. (12.8)). Following Eq. (13.1), we find the entries of the matrix A by acting on the standard unit vectors.

$$A\mathbf{e}_j \overset{(15.2)}{=} \psi^{-1} \circ f \circ \varphi(\mathbf{e}_j) \overset{(15.4)}{=} \psi^{-1} \circ f(\mathbf{v}_j)$$

$$\overset{(15.3)}{=} \psi^{-1}\left(\sum_{i=1}^{m} a_{ij}\mathbf{w}_i\right) = \sum_{i=1}^{m} a_{ij}\psi^{-1}(\mathbf{w}_i) \overset{(15.4)}{=} \sum_{i=1}^{m} a_{ij}\tilde{\mathbf{e}}_i$$

Hence, the entries of A are the coefficients a_{ij} which appear in the expansion (15.3) of the images of the basis vectors. We summarize this result in the following theorem.

Theorem 15.1 *(Matrix describing a linear map) Let V and W be vector spaces over $\mathbb{F}$ with bases $(\mathbf{v}_1, \dots, \mathbf{v}_n)$ and $(\mathbf{w}_1, \dots, \mathbf{w}_m)$, respectively, and $f : V \to W$ be a linear map. The entries A_{ij} of the $m \times n$ matrix $A : \mathbb{F}^n \to \mathbb{F}^m$ representing f relative to this choice of bases can be read off from the images of the basis vectors via*

$$f(\mathbf{v}_j) = \sum_{i=1}^{m} A_{ij} \mathbf{w}_i \ . \tag{15.5}$$

The ranks of f and its representing matrix A are equal, $\mathrm{rk}(f) = \mathrm{rk}(A)$.

Proof It remains to show the final statement on the equality of ranks. From Eq. (15.2) and $\mathrm{Im}(f) = f(V)$ we have

$$\mathrm{Im}(A) = \mathrm{Im}(\psi^{-1} \circ f \circ \varphi) = \psi^{-1}(f(\varphi(\mathbb{F}^n))) = \psi^{-1}(f(V)) = \psi^{-1}(\mathrm{Im}(f)) \ .$$

Cor. 14.1 applied to ψ^{-1} and $U = \mathrm{Im}(f)$ then implies that $\dim_{\mathbb{F}}(\psi^{-1}(\mathrm{Im}(f))) = \dim_{\mathbb{F}}(\mathrm{Im}(f)) = \mathrm{rk}(f)$, which completes the proof. $\square$

While the above discussion might seem somewhat abstract it has led to a practical method to extract the matrix A which describes a linear map f relative to a choice of bases. Simply work out the images of the domain basis vectors and express them as linear combinations of the co-domain basis, as in Eq. (15.5). The coefficients which appear in this way are the entries of the desired matrix. More precisely, by careful inspection of the indices in Eq. (15.5), it follows that the coefficients which appear in the image of the jth basis vector form the jth *column* of the matrix A. This simple rule is one of the most useful ones in linear algebra. (See Exercises 15.6 and 15.7 for more on the relationship between linear maps and matrices.)

15.1.3 Examples for matrices describing linear maps

Note that the above theorem is even relevant for linear maps between coordinate vector spaces. In this case, we have a canonical identification between linear maps and matrices by choosing the standard unit vector bases. However, other bases choices on coordinate vector spaces are possible and we might like to know about the associated representing matrix. This is illustrated in our first exercise.

Problem 15.1 *(Matrix describing a linear map between coordinate vectors)*

Consider the linear map $B : \mathbb{R}^2 \to \mathbb{R}^2$ defined by the matrix

$$B = \begin{pmatrix} 1 & 0 \\ 0 & -2 \end{pmatrix} \ .$$

For simplicity, choose the same basis for the domain and the co-domain, namely $\mathbf{v}_1 = \mathbf{w}_1 = (1, 2)^T$ and $\mathbf{v}_2 = \mathbf{w}_2 = (-1, 1)^T$. Compute the matrix B' which represents the linear map B relative to this choice of basis. Discuss how the diagram (15.1) specializes for this example.

Solution: The first step is to work out the images of the domain basis vector and write them as linear combinations of the co-domain basis.

$$B\mathbf{v}_1 = \begin{pmatrix} 1 \\ -4 \end{pmatrix} = -1\mathbf{v}_1 - 2\mathbf{v}_2 \,, \qquad B\mathbf{v}_2 = \begin{pmatrix} -1 \\ -2 \end{pmatrix} = -1\mathbf{v}_1 + 0\mathbf{v}_2 \,.$$

Arranging the coefficients from $B\mathbf{v}_1$ into the first column of a matrix and the coefficients from $B\mathbf{v}_2$ into the second column we find

$$B' = \begin{pmatrix} -1 & -1 \\ -2 & 0 \end{pmatrix} \,.$$

This is the matrix representing the linear map B relative to the basis $(\mathbf{v}_1, \mathbf{v}_2)$.

   To see what exactly this means it is useful to think about the diagram (15.1). We start by working out the coordinate map $\varphi : \mathbb{R}^2 \to \mathbb{R}^2$ associated to the basis $(\mathbf{v}_1, \mathbf{v}_2)$. For a coordinate vector $\boldsymbol{\alpha} = (\alpha_1, \alpha_2)^T$ we have, using linearity,

$$\varphi(\boldsymbol{\alpha}) = \varphi(\alpha_1 \mathbf{e}_1 + \alpha_2 \mathbf{e}_2) = \alpha_1 \varphi(\mathbf{e}_1) + \alpha_2 \varphi(\mathbf{e}_2) \stackrel{(15.4)}{=} \alpha_1 \mathbf{v}_1 + \alpha_2 \mathbf{v}_2 = \begin{pmatrix} \alpha_1 - \alpha_2 \\ 2\alpha_1 + \alpha_2 \end{pmatrix} \,.$$

The diagram below is the specialized version of diagram (15.1) and it captures the meaning of the representing matrix B'. If B' maps between two coordinate vectors then the original linear map B maps between the corresponding vectors under the coordinate map φ.

$$
\begin{array}{ccc}
\begin{pmatrix} \alpha_1 - \alpha_2 \\ 2\alpha_1 + \alpha_2 \end{pmatrix} = \varphi(\boldsymbol{\alpha}) & \xrightarrow{\quad B \quad} & \varphi(\boldsymbol{\beta}) = \begin{pmatrix} \alpha_1 - \alpha_2 \\ -2(2\alpha_1 + \alpha_2) \end{pmatrix} \\[2mm]
\varphi \Big\uparrow & & \Big\uparrow \varphi \\[2mm]
\begin{pmatrix} \alpha_1 \\ \alpha_2 \end{pmatrix} = \boldsymbol{\alpha} & \xrightarrow[\quad B' \quad]{} & \boldsymbol{\beta} = B'\boldsymbol{\alpha} = \begin{pmatrix} -\alpha_1 - \alpha_2 \\ -2\alpha_1 \end{pmatrix}
\end{array}
$$

Said another way, starting with the vector $\boldsymbol{\alpha}$ in the lower left corner of the diagram, both the upper and the lower path lead to the same vector $\varphi(\boldsymbol{\beta})$ in the upper right corner.

In order to develop a better conceptual understanding it might be useful to discuss a linear map which is not defined by a matrix.

Problem 15.2 (*The matrix representing a differential operator*)

Consider the vector space $V = \mathcal{P}_2(\mathbb{R})$ of at most quadratic polynomials and the linear map

$$D = \frac{d}{dx} : V \to V \,,$$

obtained by taking the first derivative. Write down the coordinate map $\varphi : \mathbb{R}^3 \to V$ for the basis $(1, x, x^2)$ of monomials and find the matrix A, which represents D relative to this basis. Work out the first derivative of the polynomial $p \in V$, where $p(x) = 7 + 3x + 5x^2$, by using the matrix A.

Solution: The coordinate map reads explicitly

$$\varphi(\boldsymbol{\alpha}) = \alpha_0 + \alpha_1 x + \alpha_2 x^2 \,.$$

To find the matrix A which represents D we compute the first derivatives of the basis monomials and write them as linear combinations of the same monomials.

$$D(1) = 0 = 0 \cdot 1 + 0 \cdot x + 0 \cdot x^2$$
$$D(x) = 1 = 1 \cdot 1 + 0 \cdot x + 0 \cdot x^2$$
$$D(x^2) = 2x = 0 \cdot 1 + 2 \cdot x + 0 \cdot x^2$$

Arranging the coefficients in each row into the columns of a matrix we arrive at

$$A = \begin{pmatrix} 0 & 1 & 0 \\ 0 & 0 & 2 \\ 0 & 0 & 0 \end{pmatrix} .$$

This matrix generates the first derivative of quadratic polynomials relative to the standard monomial basis.

To understand the meaning of this matrix, we consider the specific polynomial $p(x) = \varphi((7,3,5)^T) = 7 + 3x + 5x^2$ with first derivative $p'(x) = \varphi((3,10,0)^T) = 3 + 10x$. Since the matrix A describes the first derivative map it must map the coordinate vector $(7,3,5)^T$ for p into the coordinate vector $(3,10,0)^T$ for p'. This is easily checked.

$$A \begin{pmatrix} 7 \\ 3 \\ 5 \end{pmatrix} = \begin{pmatrix} 3 \\ 10 \\ 0 \end{pmatrix} .$$

The correspondence between operators acting on functions and matrices acting on coordinate vectors illustrated in the previous example is at the heart of quantum mechanics. Historically, Schrödinger's formulation of quantum mechanics is in terms of (wave) functions and operators, while Heisenberg's formulation is in terms of vectors and matrices. The relation between those two formulations is precisely as in the above example.

The next two exercises, provide interesting applications of linear maps and their representing matrices to geometry.

Problem 15.3 *(The cross product as a linear map)*

Define a map $f : \mathbb{R}^3 \to \mathbb{R}^3$ by $f(\mathbf{v}) = \mathbf{n} \times \mathbf{v}$, where $\mathbf{n} = (n_1, n_2, n_3)^T \in \mathbb{R}^3$ is a fixed vector. Show that this map is linear and find its representing matrix A relative to the basis of standard unit vectors. Assume that $\mathbf{n}$ is a unit vector and consider an ortho-normal basis $(\mathbf{n}, \mathbf{u}_1, \mathbf{u}_2)$ of $\mathbb{R}^3$, where $\mathbf{u}_2 = \mathbf{n} \times \mathbf{u}_1$. Compute the matrix A' which represent f relative to this basis.

Solution: The map is linear since the cross product is linear in its second argument. To find the representing matrix A we work out the action of f on the standard unit vectors.

$$f(\mathbf{e}_1) = \mathbf{n} \times \mathbf{e}_1 = n_3 \mathbf{e}_2 - n_2 \mathbf{e}_3$$
$$f(\mathbf{e}_2) = \mathbf{n} \times \mathbf{e}_2 = -n_3 \mathbf{e}_1 + n_1 \mathbf{e}_3$$
$$f(\mathbf{e}_3) = \mathbf{n} \times \mathbf{e}_3 = n_2 \mathbf{e}_1 - n_1 \mathbf{e}_2 .$$

Arranging the coefficients which appear in $f(\mathbf{e}_j)$ into the jth column we get the matrix

$$A = \begin{pmatrix} 0 & -n_3 & n_2 \\ n_3 & 0 & -n_2 \\ -n_2 & n_1 & 0 \end{pmatrix} . \tag{15.6}$$

It follows that $f(\mathbf{v}) = \mathbf{n} \times \mathbf{v} = A\mathbf{v}$ for all vectors $\mathbf{v} \in \mathbb{R}^3$. The interesting conclusion is that vector products with a fixed vector $\mathbf{n}$ can also be represented by multiplication with an anti-symmetric matrix of the form (15.6). Everything is much more elegant in index notation where

$$A_{ij} = [f(\mathbf{e}_j)]_i = [\mathbf{n} \times \mathbf{e}_j]_i = \epsilon_{ikl} n_k [\mathbf{e}_j]_l = \epsilon_{ikl} n_k \delta_{jl} = \epsilon_{ikj} n_k , \tag{15.7}$$

so that $A_{ij} = \epsilon_{ikj} n_k$, in agreement with Eq. (15.6).

For the images of the ortho-normal basis $(\mathbf{n}, \mathbf{u}_1, \mathbf{u}_2)$ we find

$$\begin{aligned} f(\mathbf{n}) &= \mathbf{n} \times \mathbf{n} = \mathbf{0} = 0\mathbf{n} + 0\mathbf{u}_1 + 0\mathbf{u}_2 \\ f(\mathbf{u}_1) &= \mathbf{n} \times \mathbf{u}_1 = \mathbf{u}_2 = 0\mathbf{n} + 0\mathbf{u}_1 + 1\mathbf{u}_2 \\ f(\mathbf{u}_2) &= \mathbf{n} \times \mathbf{u}_2 = \mathbf{n} \times (\mathbf{n} \times \mathbf{u}_1) = \underbrace{(\mathbf{n} \cdot \mathbf{u}_1)}_{=0} \mathbf{n} - |\mathbf{n}|^2 \mathbf{u}_1 = 0\mathbf{n} + (-1)\mathbf{u}_1 + 0\mathbf{u}_2 \end{aligned}$$

Filling the coefficients from each equation into the columns gives the representing matrix

$$A' = \begin{pmatrix} 0 & 0 & 0 \\ 0 & 0 & -1 \\ 0 & 1 & 0 \end{pmatrix} .$$

Problem 15.4 *(Reflections in $\mathbb{R}^3$ as linear maps)*

Define the map $f : \mathbb{R}^3 \to \mathbb{R}^3$ by

$$f(\mathbf{v}) = \mathbf{v} - 2(\mathbf{n} \cdot \mathbf{v})\mathbf{n} , \tag{15.8}$$

where $\mathbf{n} \in \mathbb{R}^3$ is a unit vector. Show that f is linear and that it satisfies $f \circ f = \text{id}_{\mathbb{R}^3}$. Find the representing matrix A for f relative the standard unit vector basis $(\mathbf{e}_1, \mathbf{e}_2, \mathbf{e}_3)$. Next consider a basis $(\mathbf{u}_1, \mathbf{u}_2, \mathbf{n})$ of $\mathbb{R}^3$ with $\mathbf{n} \cdot \mathbf{u}_1 = \mathbf{n} \cdot \mathbf{u}_2 = 0$ and determine the representing matrix B for f relative to this basis. Interpret your results geometrically.

Solution: This map is linear due to linearity of the dot product. To show that it squares to the identity we can carry out an explicit computation.

$$f \circ f(\mathbf{v}) = f(\mathbf{v} - 2(\mathbf{n} \cdot \mathbf{v})\mathbf{n}) = \mathbf{v} - 2(\mathbf{n} \cdot \mathbf{v})\mathbf{n} - 2(\mathbf{n} \cdot (\mathbf{v} - 2(\mathbf{n} \cdot \mathbf{v})\mathbf{n}))\mathbf{n} = \mathbf{v}$$

Next, we determine the matrix A which represents f relative to the standard unit vector basis $(\mathbf{e}_1, \mathbf{e}_2, \mathbf{e}_2)$. Working out their images, $f(\mathbf{e}_j) = \mathbf{e}_j - 2(\mathbf{n} \cdot \mathbf{e}_j)\mathbf{n} = \mathbf{e}_j - 2n_j\mathbf{n} = (\delta_{ij} - 2n_i n_j)\mathbf{e}_i$, we find

$$A_{ij} = \delta_{ij} - 2n_i n_j , \qquad A = \begin{pmatrix} 1 - 2n_1^2 & -2n_1 n_2 & -2n_1 n_3 \\ -2n_1 n_2 & 1 - 2n_2^2 & -2n_2 n_3 \\ -2n_1 n_3 & -2n_2 n_3 & 1 - 2n_3^2 \end{pmatrix} .$$

Now consider the basis $(\mathbf{u}_1, \mathbf{u}_2, \mathbf{n})$. What is the matrix B representing f relative to this basis? Inserting these basis vectors into the definition, Eq. (15.8), of f and using $\mathbf{n} \cdot \mathbf{u}_1 = \mathbf{n} \cdot \mathbf{u}_2 = 0$ as well as $|\mathbf{n}| = 1$ we find

$$f(\mathbf{u}_1) = \mathbf{u}_1 , \quad f(\mathbf{u}_2) = \mathbf{u}_2 , \quad f(\mathbf{n}) = -\mathbf{n} ,$$

so the representing matrix is

$$B = \begin{pmatrix} 1 & 0 & 0 \\ 0 & 1 & 0 \\ 0 & 0 & -1 \end{pmatrix} .$$

This matrix makes the geometrical interpretation of f obvious. While the coordinates relative to the vectors $\mathbf{u}_1, \mathbf{u}_2$ remain unchanged, the coordinate relative to $\mathbf{n}$ is inverted. This means that f describes a reflection on the plane perpendicular to $\mathbf{n}$. Of course this explains why f squares to the identity — performing two reflections successively leaves a vector unchanged. The more complicated matrix A above describes the same reflection but relative to the basis of standard unit vectors.

15.2 Change of basis

Summary 15.2 *Two matrices A and A' which describe a linear map $f : V \to W$ relative to different basis choices on V and W are related by $A' = QAP^{-1}$. The matrices P and Q are invertible and describe the change of bases. For a linear map $f : V \to V$ this specializes to $A' = PAP^{-1}$.*

As we have seen, a linear map can be described by a matrix which depends on a choice of bases. We would like to understand how the representing matrix for a given linear map changes if the bases are changed.

15.2.1 General case

As usual, we start with a linear map $f : V \to W$ between two (finite-dimensional) vector spaces V and W over the field $\mathbb{F}$. The set-up for the choice of bases, coordinate vectors and maps is summarized in Table 15.2.

Relative to the unprimed basis, f is represented by the $m \times n$ matrix A and relative to the primed basis by the $m \times n$ matrix A'. Our goal is the work out the relationship between those two matrices. From Eq. (15.2) they can be written as $A = \psi^{-1} \circ f \circ \varphi$

Table 15.2 Set-up to represent a linear map $f : V \to W$ by a matrix.

Vector space	Dimension	Basis	Coordinate map	Coordinate vector
V	n	$(\mathbf{v}_1, \ldots, \mathbf{v}_n)$	$\varphi : \mathbb{F}^n \to V$	$\boldsymbol{\alpha} \in \mathbb{F}^n$
		$(\mathbf{v}'_1, \ldots, \mathbf{v}'_n)$	$\varphi' : \mathbb{F}^n \to V$	$\boldsymbol{\alpha}' \in \mathbb{F}^n$
W	m	$(\mathbf{w}_1, \ldots, \mathbf{w}_m)$	$\psi : \mathbb{F}^m \to W$	$\boldsymbol{\beta} \in \mathbb{F}^m$
		$(\mathbf{w}'_1, \ldots, \mathbf{w}'_m)$	$\psi' : \mathbb{F}^m \to W$	$\boldsymbol{\beta}' \in \mathbb{F}^m$

and $A' = \psi'^{-1} \circ f \circ \varphi'$. A short calculation gives

$$A' = \psi'^{-1} \circ f \circ \varphi' = \psi'^{-1} \circ \underbrace{\psi \circ \psi^{-1}}_{=\text{id}_W} \circ f \circ \underbrace{\varphi \circ \varphi^{-1}}_{=\text{id}_V} \circ \varphi'$$

$$= \underbrace{\psi'^{-1} \circ \psi}_{=:Q} \circ \underbrace{\psi^{-1} \circ f \circ \varphi}_{=A} \circ \underbrace{\varphi^{-1} \circ \varphi'}_{=:P^{-1}} = QAP^{-1} .$$

All we have done is to insert two identity maps in the second step and then combine maps differently in the third step. The result

$$A' = QAP^{-1} \tag{15.9}$$

describes the basis transformation of a matrix and is one of the key equations of linear algebra.

What is the interpretation of the matrices $Q = \psi'^{-1} \circ \psi : \mathbb{F}^m \to \mathbb{F}^m$ and $P = \varphi'^{-1} \circ \varphi : \mathbb{F}^n \to \mathbb{F}^n$ in this equation? Focusing on P for now, consider a vector $\mathbf{v} \in V$ with coordinate vectors $\boldsymbol{\alpha} = \varphi^{-1}(\mathbf{v})$ and $\boldsymbol{\alpha}' = \varphi'^{-1}(\mathbf{v})$ relative to the two choices of bases. Then, $\boldsymbol{\alpha}' = \varphi'^{-1}(\mathbf{v}) = \varphi'^{-1} \circ \varphi(\boldsymbol{\alpha}) = P\boldsymbol{\alpha}$ and a similar argument for Q gives $\boldsymbol{\beta}' = Q\boldsymbol{\beta}$. In summary, the matrices P and Q can be viewed as coordinate transformations, relating the coordinate vectors relative to the primed and unprimed bases, that is

$$\boldsymbol{\alpha}' = P\boldsymbol{\alpha} , \qquad \boldsymbol{\beta}' = Q\boldsymbol{\beta} , \tag{15.10}$$

provided that $\varphi(\boldsymbol{\alpha}) = \varphi'(\boldsymbol{\alpha}')$ and $\psi(\boldsymbol{\beta}) = \psi'(\boldsymbol{\beta}')$.

Given this interpretation of P and Q Eq. (15.9) can be understood intuitively. When acting on a coordinate vector $\boldsymbol{\alpha}'$, the matrix P^{-1} on the right-hand side of Eq. (15.9) first converts this coordinate vector into its unprimed counterpart $\boldsymbol{\alpha}$, on which the matrix A can sensibly act. Finally, Q converts the result back into a coordinate vector relative to the primed bases. Altogether, this reproduces the action of A'.

To compute the entries of P we can start with the equation $P\mathbf{e}_j = \sum_i P_{ij}\mathbf{e}_i$, apply φ' to both sides and use Eq. (15.4). This results in $\mathbf{v}_j = \sum_i P_{ij}\mathbf{v}'_i$ and, from a similar argument, we have $\mathbf{w}_j = \sum_i Q_{ij}\mathbf{w}'_i$. In conclusion, the entries of P and Q are obtained by expanding the unprimed basis vector in terms of the primed ones.

$$\mathbf{v}_j = \sum_i P_{ij}\mathbf{v}'_i , \qquad \mathbf{w}_j = \sum_i Q_{ij}\mathbf{w}'_i . \tag{15.11}$$

The above results are summarized in the following theorem.

Theorem 15.2 *(General basis change) Let V and W be vector spaces over $\mathbb{F}$ and $f : V \to W$ be a linear map. Suppose relative to bases $(\mathbf{v}_1, \ldots, \mathbf{v}_n)$ of V and $(\mathbf{w}_1 \ldots, \mathbf{w}_m)$ of W the map f is represented by a matrix A and relative to bases $(\mathbf{v}'_1, \ldots, \mathbf{v}'_n)$ of V and $(\mathbf{w}'_1 \ldots, \mathbf{w}'_m)$ of W by a matrix A'. Then the two matrices are related by*

$$A' = QAP^{-1} \quad \text{where} \quad \mathbf{v}_j = \sum_i P_{ij}\mathbf{v}'_i \text{ and } \mathbf{w}_j = \sum_i Q_{ij}\mathbf{w}'_i . \tag{15.12}$$

Proof This follows from the above arguments. □

15.2.2 Identical domain and co-domain

An important special case of our general discussion above arises when the vector spaces V and W are equal and we choose the same bases on domain and co-domain, so that $\mathbf{v}_i = \mathbf{w}_i$ and $\mathbf{v}'_i = \mathbf{w}'_i$. Then, Theorem 15.2 specializes to the following statement:

Corollary 15.1 *(Basis change for identical domain and co-domain) Let V be a vector space and $f : V \to V$ be a linear map. Suppose relative to a basis $(\mathbf{v}_1, \dots, \mathbf{v}_n)$ of V the map f is represented by a matrix A and relative to a basis $(\mathbf{v}'_1, \dots, \mathbf{v}'_n)$ of V it is represented by A'. Then the two matrices are related by*

$$A' = PAP^{-1} \quad where \quad \mathbf{v}_j = \sum_i P_{ij} \mathbf{v}'_i . \tag{15.13}$$

Proof This follows immediately from Theorem 15.2 by setting $W = V$, $\mathbf{w}_i = \mathbf{v}_i$, $\mathbf{w}'_i = \mathbf{v}'_i$ and $Q = P$. □

Eq. (15.13) is another key equation of linear algebra which describes the basis transformation of a matrix for same bases changes on domain and co-domain.

For linear maps $f : \mathbb{F}^n \to \mathbb{F}^n$ between coordinate vector spaces we can consider an interesting special case. In this case, we can choose the unprimed basis to be the standard unit vector basis, $(\mathbf{v}_1, \dots, \mathbf{v}_n) = (\mathbf{e}_1, \dots, \mathbf{e}_n)$, so that the representing matrix A is the one canonically identified with f. Then, the second Eq. (15.13) turns into $\mathbf{v}'_j = \sum_i (P^{-1})_{ij} \mathbf{e}_i$. Hence, the matrix A identified with f is transformed to a matrix A' relative to a new basis $(\mathbf{v}'_1, \dots, \mathbf{v}'_n)$ by

$$A' = PAP^{-1} \quad where \quad P^{-1} = (\mathbf{v}'_1, \dots, \mathbf{v}'_n) . \tag{15.14}$$

The point is that, in this case, the basis transformation P^{-1} is easily written down since its columns are the new basis vectors $\mathbf{v}'_i$.

Problem 15.5 *(Basis transformation of a matrix)*

Relative to the unprimed basis $(\mathbf{v}_1 = \mathbf{e}_1, \mathbf{v}_2 = \mathbf{e}_2)$ of standard unit vectors, a linear map $A : \mathbb{R}^2 \to \mathbb{R}^2$ is described by the matrix $A = \mathrm{diag}(1, -1)$. Find the representing matrix A' for this linear map relative to the basis $\mathbf{v}'_1 = (1, -1)/\sqrt{2}$, $\mathbf{v}'_2 = (1, 1)/\sqrt{2}$.

Solution: One way to proceed is as before, by applying Theorem 15.1, and compute the images of the basis vectors in order to read off A'. This leads to

$$A\mathbf{v}'_1 = 0\mathbf{v}'_1 + 1\mathbf{v}'_2 , \qquad A\mathbf{v}'_2 = 1\mathbf{v}'_1 + 0\mathbf{v}'_2 ,$$

and arranging the coefficients on the right-hand sides into the column of a matrix gives

$$A' = \begin{pmatrix} 0 & 1 \\ 1 & 0 \end{pmatrix} .$$

Alternatively, we can determine A' from Eq. (15.14). This leads to

$$P^{-1} = (\mathbf{v}'_1, \mathbf{v}'_2) = \frac{1}{\sqrt{2}} \begin{pmatrix} 1 & 1 \\ -1 & 1 \end{pmatrix} \quad \Rightarrow \quad P = \frac{1}{\sqrt{2}} \begin{pmatrix} 1 & -1 \\ 1 & 1 \end{pmatrix} ,$$

and applying this basis transformation gives

$$A' = PAP^{-1} = \frac{1}{2}\begin{pmatrix} 1 & -1 \\ 1 & 1 \end{pmatrix}\begin{pmatrix} 1 & 0 \\ 0 & -1 \end{pmatrix}\begin{pmatrix} 1 & 1 \\ -1 & 1 \end{pmatrix} = \begin{pmatrix} 0 & 1 \\ 1 & 0 \end{pmatrix},$$

in accordance with the earlier result.

Problem 15.6 *(Basis change for a differential operators)*

In Exercise 15.2 we have considered the vector space $V = \mathcal{P}_2(\mathbb{R})$ of at most quadratic polynomials and the derivative map $D = \frac{d}{dx} : V \to V$. We have shown that, relative to the monomial basis $(\mathbf{v}_1 = 1, \mathbf{v}_2 = x, \mathbf{v}_3 = x^2)$ of V, the derivative D is represented by the matrix A in Eq. (15.2). What is its representing matrix A' relative to the basis $(\mathbf{v}'_1 = 1 + x, \mathbf{v}'_2 = 1 - x, \mathbf{v}'_3 = 1 + x + x^2)$?

Solution: A direct computation from Theorem 15.1 leads to

$$\left.\begin{array}{rl} D(\mathbf{v}'_1) = & 1 = \frac{1}{2}\mathbf{v}'_1 + \frac{1}{2}\mathbf{v}'_2 + 0\mathbf{v}'_3 \\ D(\mathbf{v}'_2) = & -1 = -\frac{1}{2}\mathbf{v}'_1 - \frac{1}{2}\mathbf{v}'_2 + 0\mathbf{v}'_3 \\ D(\mathbf{v}'_3) = 1 + 2x = \frac{3}{2}\mathbf{v}'_1 - \frac{1}{2}\mathbf{v}'_2 + 0\mathbf{v}'_3 \end{array}\right\} \quad \Rightarrow \quad A' = \frac{1}{2}\begin{pmatrix} 1 & -1 & 3 \\ 1 & -1 & -1 \\ 0 & 0 & 0 \end{pmatrix}.$$

On the other hand, by comparison with the second Eq. (15.13), expanding the primed basis vectors in terms of the unprimed ones gives

$$\left.\begin{array}{l} \mathbf{v}'_1 = \mathbf{v}_1 + \mathbf{v}_2 \\ \mathbf{v}'_2 = \mathbf{v}_1 - \mathbf{v}_2 \\ \mathbf{v}'_3 = \mathbf{v}_1 + \mathbf{v}_2 + \mathbf{v}_3 \end{array}\right\} \quad \Rightarrow \quad P^{-1} = \begin{pmatrix} 1 & 1 & 1 \\ 1 & -1 & 1 \\ 0 & 0 & 1 \end{pmatrix} \quad \Rightarrow \quad P = \frac{1}{2}\begin{pmatrix} 1 & 1 & -2 \\ 1 & 1 & -1 \\ 0 & 0 & 2 \end{pmatrix}.$$

Performing the basis transformation from Eq. (15.13) gives

$$A' = PAP^{-1} = \frac{1}{2}\begin{pmatrix} 1 & 1 & -2 \\ 1 & 1 & -1 \\ 0 & 0 & 2 \end{pmatrix}\begin{pmatrix} 0 & 1 & 0 \\ 0 & 0 & 2 \\ 0 & 0 & 0 \end{pmatrix}\begin{pmatrix} 1 & 1 & 1 \\ 1 & -1 & 1 \\ 0 & 0 & 1 \end{pmatrix} = \frac{1}{2}\begin{pmatrix} 1 & -1 & 3 \\ 1 & -1 & -1 \\ 0 & 0 & 0 \end{pmatrix},$$

in agreement with the earlier result.

15.2.3 Conjugate matrices

We say that two matrices $A, A' \in \mathcal{M}_{n,n}(\mathbb{F})$ are *conjugate* if they are related by a basis transformation (15.13), so if there exists an invertible matrix $P \in \mathrm{GL}(\mathbb{F}^n)$ such that $A' = PAP^{-1}$.

Proposition 15.1 *Conjugacy of matrices is an equivalence relation.*

Proof We need to show reflexivity, symmetry, and transitivity.

'Reflexivity': To see that every matrix is conjugate to itself simply choose $P = \mathbb{1}_n$.

'Symmetry': If A is conjugate to A', so $A' = PAP^{-1}$, then $A = P^{-1}A'P$ so A is conjugate to A'.

'Transitivity': If A is conjugate to B and B is conjugate to C, so that $B = PAP^{-1}$ and $C = QBQ^{-1}$, then $C = (PQ)A(PQ)^{-1}$. This means A is conjugate to C. $\square$

This means that $\mathcal{M}_{n,n}(\mathbb{F})$ partitions into disjoint equivalence classes, also called *conjugacy classes*. Each conjugacy class contains all the matrices related by a basis transformation, so all the matrices which describe the same linear map.

Many problems in linear algebra are motivated by understanding these conjugacy classes better. For example, we would like to be able to decide whether two given matrices belong to the same class or not. A useful tool for this are *class functions* — functions of matrices which only depend on the conjugacy class but not on the particular element within each class. As we will see in Chapter 18, the determinant is an example of such a class function. Another important problem is to find the 'simplest' matrix in each conjugacy class — this leads to normal forms and diagonalization of matrices (see Part VI).

Exercises

($\dagger$=challenging)

15.1 A linear map in $\mathbb{R}^2$ is given by the action of the matrix

$$A = \begin{pmatrix} 1 & 2 \\ 0 & 1 \end{pmatrix} .$$

(a) Work out the matrix A' which represents this linear map relative to the basis $\mathbf{v}_1 = (1,1)^T$ and $\mathbf{v}_2 = (0,1)^T$.
(b) Find a 2×2 matrix P such that $A' = PAP^{-1}$.
(c) What is the interpretation of the matrix P?

15.2 A linear map $f : \mathbb{R}^3 \to \mathbb{R}^3$ is defined by $f(\mathbf{v}) = (\mathbf{n} \cdot \mathbf{v})\mathbf{n} + \mathbf{v}$, where $\mathbf{n} \in \mathbb{R}^3$ is a unit vector.
(a) Find the representing matrix A of f relative to the standard unit vector basis.
(b) Find the representing matrix A' of f relative to an ortho-normal basis $(\mathbf{n}, \mathbf{u}_1, \mathbf{u}_2)$ of $\mathbb{R}^3$.
(c) What is the geometrical interpretation of f?

15.3 Consider the vector space $V = \mathcal{P}_2(\mathbb{R})$ of at most quadratic polynomials and define the map $L : V \to V$ by

$$L(p) = x\frac{d^2p}{dx^2} + (1 - x)\frac{dp}{dx} + 2p .$$

(a) Why is this map linear?
(b) Work out the matrix, A, which

represents L relative to the standard monomial basis $(1, x, x^2)$ of V.
(c) Find the kernel of the matrix A, that is, the vectors $\mathbf{v}$ satisfying $A\mathbf{v} = 0$. Which polynomials p correspond to these vectors?
(d) Show by explicitly applying L that the polynomials p found in part (c) satisfy $L(p) = 0$.

15.4 A linear map $f : \mathbb{R}^4 \to \mathbb{R}^4$ is represented by a matrix

$$A = \begin{pmatrix} -\frac{1}{2} & \frac{1}{2} & -\frac{1}{2} & -\frac{1}{2} \\ 1 & 0 & 1 & 1 \\ 2 & -1 & 2 & 0 \\ -\frac{3}{2} & \frac{3}{2} & -\frac{3}{2} & \frac{1}{2} \end{pmatrix}$$

relative to the basis of standard unit vectors.
(a) Find the matrix A' which represents f relative to the basis $(\mathbf{v}'_1, \mathbf{v}'_2, \mathbf{v}'_3, \mathbf{v}'_4)$, where

$$\mathbf{v}'_1 = (1,1,1,-1)^T$$
$$\mathbf{v}'_2 = (1,-1,1,1)^T$$
$$\mathbf{v}'_3 = (1,1,-1,-1)^T$$
$$\mathbf{v}'_4 = (1,1,-1,1)^T$$

Also find the matrix P with $A' = PAP^{-1}$.
(b) Use the results from part (a) to find the matrix A_k which represents f^k for $k \in \mathbb{N}$ relative to the basis of standard unit vectors.

15.5 *Cross product as a linear map*[†]

For $\mathbf{n} \in \mathbb{R}^3$ define the linear map $f_{\mathbf{n}} : \mathbb{R}^3 \to \mathbb{R}^3$ by $f_{\mathbf{n}}(\mathbf{v}) = \mathbf{n} \times \mathbf{v}$, as in Exercise 15.3.

(a) Write down the matrix $A_{\mathbf{n}}$ which represents $f_{\mathbf{n}}$ relative to the basis of standard unit vectors. Show that the map $\mathbf{n} \mapsto A_{\mathbf{n}}$ defines an isomorphism $\mathbb{R}^3 \to \mathcal{A}_3(\mathbb{R})$ with $A_{\mathbf{n}}(\mathbf{v}) = \mathbf{n} \times \mathbf{v}$.

(b) Show that the matrices B_k which represents odd powers, $f_{\mathbf{n}}^{2k+1}$ are antisymmetric. Find the vectors $\mathbf{n}_k$ which correspond to B_k under the isomorphism in (b).

15.6 *Isomorphism between linear maps and matrices*[†]

Let V, W be vector spaces over $\mathbb{F}$, each with a fixed basis, relative to which we consider all representing matrices.

(a) Linear maps $f, g \in \mathrm{Hom}(V, W)$ are described by matrices A and B. Show that the linear map $\alpha f + \beta g$, where $\alpha, \beta \in \mathbb{F}$, is described by the matrix $\alpha A + \beta B$.

(b) Denote by A_f the matrix which describes $f \in \mathrm{Hom}(V, W)$. Show that the map $\imath : \mathrm{Hom}(V, W) \to \mathrm{Hom}(\mathbb{F}^n, \mathbb{F}^m)$ defined by $f \mapsto A_f$ is a vector space isomorphism.

(c) Use the result from part (b) to proof the dimension formula (12.7).

15.7 *Relation between* $\mathrm{GL}(V)$ *and* $\mathrm{GL}(\mathbb{F}^n)$[†]

Let V, W and U be three vector spaces over $\mathbb{F}$, each with a fixed basis, relative to which we consider all representing matrices.

(a) Two linear maps $f : V \to W$ and $g : W \to U$ are described by matrices A and B. Show that the linear map $g \circ f$ is described by the matrix BA.

(ii) If $f \in \mathrm{GL}(V)$ is described by a matrix A, show that f^{-1} is described by A^{-1}.

(iii) Denote by A_f the matrix which describes $f \in \mathrm{End}(V)$. Show that the map $\imath : \mathrm{GL}(V) \to \mathrm{GL}(n)$ defined by $f \mapsto A_f$ is a group isomorphism.

Part V

Linear systems and algorithms

We have now covered the theoretical foundations of vector spaces and linear maps. They have been illustrated with many examples, mostly in a small number of dimensions. However, we still have to develop efficient and systematic methods for calculation which will also work for higher-dimensional cases. The key question is how to compute with linear maps. Linear maps can always be described by matrices, relative to a choice of bases, so what we require is methods to calculate with matrices.

As we will see, the basic ingredient of these methods are row operations on matrices. In Chapter 16, we set up algorithms based on row operations in order to calculate the rank and the inverse of a matrix. Linear systems and systems of linear equation are studied in Chapter 17. As we explain, the algorithm for solving systems of linear equations is also based on row operations.

Determinants are another method for calculating with matrices. In Section 10.2 we have already introduced the determinant for 3×3 matrices and Chapter 18 generalizes this discussion to matrices with arbitrary size. Determinants can be used to decide whether a matrix is invertible, to calculate the matrix inverse and to solve (certain) systems of linear equations. But they are also of theoretical importance and play a key role in the theory of eigenvalues and eigenvector which will be developed in Part VI.

Part V

Linear systems and algorithms

16
Computing with matrices

How do we compute the rank of a matrix and the matrix inverse? The rank has been determined by counting the number of linearly independent column vectors while the inverse has been computed by inserting an Ansatz into the defining equation (13.24). Either method is cumbersome for larger matrices. The key ingredients of systematic, algorithmic methods are row operations on matrices. These will first be introduced and then applied to the calculation of the matrix rank and inverse.

16.1 Row operations

Summary 16.1 *There are three elementary row operations for matrices: exchange of rows, adding a multiple of one row to another and multiplying a row with a non-zero scalar. Elementary row operations leave the span of the matrix row vectors unchanged. They form the basic steps of an algorithm, called Gaussian elimination, which can be used to bring any matrix into upper echelon form.*

16.1.1 Definition of row operations

We are working with elements in $\mathcal{M}_{n.m}(\mathbb{F})$, so matrices of size $n \times m$ with entries in a field $\mathbb{F}$. At the heart of algorithmic methods for computing with such matrices are *elementary row operations* which are defined as follows.

Definition 16.1 *The following manipulations of a matrix are called elementary row operations:*

(R1) Exchange two rows.
(R2) Add a scalar multiple of one row to another.
(R3) Multiply a row with a non-zero scalar.
Elementary column operations are analogous but carried out on the columns.

In the following, we will focus on elementary row operations but most of our statements have analogues for elementary column operations. At any rate, we can think of elementary column operations as row operations carried out on the transpose of the matrix.

The key property of the elementary row operations is that they leave the span, $\mathrm{Span}(\mathbf{A}_1, \ldots, \mathbf{A}_n)$, of the row vectors of an $n \times m$ matrix A unchanged. This is immediately clear for row operations (R1) and (R3). To check this for (R2) consider changing the first row $\mathbf{A}_1$ to $\mathbf{A}_1 + \beta \mathbf{A}_2$, so by adding a multiple of the second row.

$$\mathrm{Span}(\mathbf{A}_1, \mathbf{A}_2, \ldots, \mathbf{A}_n) = \{\alpha_1 \mathbf{A}_1 + \alpha_2 \mathbf{A}_2 + \cdots \mid \alpha_i \in \mathbb{F}\}$$
$$= \{\alpha_1(\mathbf{A}_1 + \beta \mathbf{A}_2) + (\alpha_2 - \beta \alpha_1)\mathbf{A}_2 + \cdots \mid \alpha_i \in \mathbb{F}\}$$
$$= \mathrm{Span}(\mathbf{A}_1 + \beta \mathbf{A}_2, \mathbf{A}_2, \ldots, \mathbf{A}_n)$$

By a similar argument column operations leave the span, $\mathrm{Span}(\mathbf{A}^1, \ldots, \mathbf{A}^n)$, of the column vectors unchanged.

16.1.2 Upper echelon form

The main purpose of row operations is to bring a matrix to a simpler, more convenient form where certain properties can be read off easily. One important such form is the *upper echelon form* which is useful to compute the rank and the matrix inverse. A matrix A is in upper echelon form if it has the following structure.

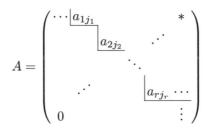

Here the entries a_{ij_i} are non-zero for all $i = 1, \ldots, r$, all other entries above the solid line are arbitrary (indicated by the $*$) and all entries below the solid line are zero. To be clear, we formulate this in a definition.

Definition 16.2 *A matrix is said to be in upper echelon form if the following conditions are satisfied:*

(E1) All entirely zero rows appear at the bottom of the matrix.
(E2) For each non-zero row i, its first non-zero entry (a_{ij_i} in the above matrix) appears strictly to the right of the first non-zero entry in row $i-1$.

The left-most non-zero entry in a row (a_{ij_i} in the above matrix) is called a pivot.

An important feature of the upper echelon form is that the non-zero rows are, in fact, linearly independent. To see this, consider the equation $\sum_{i=1}^{r} \alpha_i \mathbf{A}_i = \mathbf{0}$ where $\mathbf{A}_i$ are the non-zero row vectors of the above matrix in upper echelon form. The pivot in the first row vector $\mathbf{A}_1$ is in component j_1 but no other row vector has a non-zero j_1 entry. It follows immediately that $\alpha_1 = 0$. We can then repeat this argument with the second row and continue until we have concluded that all $\alpha_i = 0$.

16.1.3 Algorithm to bring a matrix into upper echelon form

Our next step is to show that every $n \times m$ matrix can be brought into upper echelon form by a sequence of row operations, using an algorithm called *Gaussian elimination* or *row reduction*.

Algorithm *(Gaussian elimination) The algorithm proceeds row by row. Let us assume that we have already dealt with the first $i-1$ rows of the matrix. Then, for the ith row we should carry out three steps.*

(1) *Find the left-most column j which has at least one non-zero entry in rows $i, \ldots, n$.*
(2) *If the (i, j) entry is zero exchange row i with one of the rows $i+1, \ldots, n$ (the one which contains the non-zero entry identified in step 1) so that the new (i, j) entry is non-zero.*
(3) *Subtract suitable multiples of row i from all rows $i+1, \ldots, n$ such that all entries $(i+1, j), \ldots, (n, j)$ in column j and below row i vanish.*

Continue with the next row until no more non-zero entries can be found in step (1).

The procedure is probably best explained with an example.

Problem 16.1 *(Bringing a matrix into upper echelon form)*

Bring the 3×3 matrix

$$A = \begin{pmatrix} 0 & 1 & -1 \\ 2 & 3 & -2 \\ 2 & 1 & 0 \end{pmatrix} ,$$

into upper echelon form by Gaussian elimination.

Solution:

$$\begin{pmatrix} 0 & 1 & -1 \\ 2 & 3 & -2 \\ 2 & 1 & 0 \end{pmatrix} \xrightarrow{R_1 \leftrightarrow R_3} \begin{pmatrix} 2 & 1 & 0 \\ 2 & 3 & -2 \\ 0 & 1 & -1 \end{pmatrix} \xrightarrow{R_2 \rightarrow R_2 - R_1} \begin{pmatrix} 2 & 1 & 0 \\ 0 & 2 & -2 \\ 0 & 1 & -1 \end{pmatrix} \xrightarrow{R_3 \rightarrow R_3 - R_2/2} \begin{pmatrix} 2 & 1 & 0 \\ 0 & 2 & -2 \\ 0 & 0 & 0 \end{pmatrix}$$

We have indicated the row operation from one step to the next above the arrow, referring to the ith row by R_i.

16.2 Rank of a matrix

Summary 16.2 *The column rank and the row rank of a matrix are equal. The rank can be determined by converting the matrix into upper echelon form, using Gaussian elimination. In upper echelon form, the rank equals the number of non-zero rows.*

16.2.1 Row and column rank

Our first application of row operations is to compute the rank of a matrix. For an $n \times m$ matrix $A \in \mathcal{M}_{n,m}(\mathbb{F})$ we can define the row and column rank

$$\text{rowrk}(A) = \dim_{\mathbb{F}}(\text{Span}(\mathbf{A}_1, \ldots, \mathbf{A}_n)) \tag{16.1}$$

$$\text{colrk}(A) = \dim_{\mathbb{F}}(\text{Span}(\mathbf{A}^1, \ldots, \mathbf{A}^m)) . \tag{16.2}$$

The column rank is of course the rank of the linear map defined by A while the row rank is the rank of the linear map associated to A^T. Having two types of ranks available for a matrix seems awkward but fortunately these two ranks are the same.

Theorem 16.1 *Row and column rank are equal, so $\text{rk}(A) = \text{colrk}(A) = \text{rowrk}(A)$, for any matrix A.*

Proof   Suppose one row, say $\mathbf{A}_1$, of a matrix A can be written as a linear combination of the others. Then, by dropping $\mathbf{A}_1$ from A we arrive at a matrix with one less row, but its row rank unchanged from that of A. The key observation is that the column rank also remains unchanged under this operation. This can be seen as follows. Write

$$\mathbf{A}_1 = \sum_{j=2}^{n} \alpha_j \mathbf{A}_j \,, \qquad \boldsymbol{\alpha} = \begin{pmatrix} \alpha_2 \\ \vdots \\ \alpha_n \end{pmatrix}$$

with some coefficients $\alpha_2, \ldots, \alpha_n$ which we have arranged into the vector $\boldsymbol{\alpha}$. Further, let us write the column vectors of A as

$$\mathbf{A}^i = \begin{pmatrix} a_i \\ \mathbf{b}_i \end{pmatrix} \,,$$

that is, we split off the entries in the first row, denoted by a_i, from the entries in the remaining $n-1$ rows which are contained in the vectors $\mathbf{b}_i$. It follows that $a_i = A_{1i} = (\mathbf{A}_1)_i = \sum_{j=2}^{n} \alpha_j A_{ji} = \sum_{j=2}^{n} \alpha_j (\mathbf{A}^i)_j = \boldsymbol{\alpha} \cdot \mathbf{b}_i$, so that the column vectors can also be written as

$$\mathbf{A}^i = \begin{pmatrix} \boldsymbol{\alpha} \cdot \mathbf{b}_i \\ \mathbf{b}_i \end{pmatrix} \,.$$

Hence, the entries in the first row are not relevant for the linear independence of the column vectors $\mathbf{A}^i$ — merely using the vectors $\mathbf{b}_i$ will lead to the same conclusions for linear independence. As a result we can drop a linearly dependent row without changing the row and the column rank of the matrix. Clearly, an argument similar to the above can be made if we drop a linearly dependent column vectors — again, both the row and column rank remain unchanged.

In this way, we can continue dropping linearly dependent row and column vectors from A until we arrive at a (generally smaller) matrix A' which has linearly independent row and column vectors and the same row and column ranks as A. On purely dimensional grounds, a matrix with all row vectors and all column vectors linearly independent must be a square matrix (for example, consider a 3×2 matrix. Its three 2-dimensional row vectors cannot be linearly independent). Therefore, row and column rank are the same for A' and, hence, for A.   □

This theorem implies that the rank of an $n \times m$ matrix A cannot exceed the number of its rows and columns, so that

$$\mathrm{rk}(A) \leq \min(n, m) \,. \tag{16.3}$$

We have already obtained this bound in Eq. (14.10) from arguments based on the structure of linear maps, but we have now re-derived it purely from matrix properties.

It is sometimes useful to talk about the 'generic' rank of matrices. We will not attempt to define this precisely, but loosely we mean the rank that 'most' matrices in $\mathcal{M}_{n,m}(\mathbb{F})$ have. It should be intuitively clear that this generic rank is, in fact, the maximal rank, $\mathrm{rk}(A) = \min(n, m)$. Indeed, a smaller value requires linear dependencies between the rows or columns of the matrix which amount to specific choices of entries.

Application 16.1 *More on internet search*

In Application 1.1 we have described a network with n pages, labelled by an index $k = 1, \ldots, n$, with the kth page containing n_k links to some of the other pages and being linked to by the pages $L_k \subset \{1, \ldots, n\}$. It was proposed that the page ranks x_k satisfy the equations

$$x_k = \sum_{j \in L_k} \frac{x_j}{n_j} \,, \tag{16.4}$$

where $k = 1, \ldots, n$. These equations form a homogeneous linear system which can also be written in matrix-vector form as $A\mathbf{x} = \mathbf{0}$, where $\mathbf{x} = (x_1, \ldots, x_n)^T$ is a vector containing the page ranks and the $n \times n$ matrix A has entries

$$A_{ij} = \delta_{ij} - \begin{cases} \frac{1}{n_j} & \text{if } j \in L_i \\ 0 & \text{if } j \neq L_i \end{cases} .$$

The solutions to this linear system are given by the kernel, $\text{Ker}(A)$. Of course, this always contains the trivial solution, $\mathbf{x} = \mathbf{0}$, but this is of no use for the purpose of ranking sites. For the set-up to make sense, it is crucial that $\text{Ker}(A) \neq \{\mathbf{0}\}$. To see that this is, in fact, always the case we note that the sum of the rows of A,

$$\sum_{i=1}^{n} A_{ij} = \sum_{i=1}^{n} \delta_{ij} - \sum_{i:j \in L_i} \frac{1}{n_j} = 1 - \frac{1}{n_j} \sum_{i:j \in L_i} 1 = 0 \,,$$

vanishes. This means the rows of A are linearly dependent so $\text{rk}(A) < n$. From the dimension formula (14.13) we conclude that $\dim_{\mathbb{R}} \text{Ker}(A) = n - \text{rk}(A) > 0$ and, hence, that the kernel of A is indeed non-trivial. This property is part of the magic of the page rank formula (16.4) — it guarantees the existence of a non-trivial solution which is crucial for ranking the pages of the network.

16.2.2   Computing the rank

Another important conclusion from Theorem 16.1 is that we can focus on the rows or the columns of a matrix to compute its rank. Since we want to work with elementary row operations we opt for rows. We have seen that elementary row operations leave the span of the row vectors and, therefore, the rank unchanged. Moreover, the non-zero rows of a matrix in upper echelon form are linearly independent, so

$$A \text{ in upper echelon form} \quad \Rightarrow \quad \text{rk}(A) = \text{number of non-zero rows} .$$

Altogether, this implies the following algorithm to compute the rank of a matrix.

Algorithm *(Computing the rank of a matrix)*

To compute the rank of a matrix A, carry out the following steps:

(1) *Bring A to upper echelon form using row reduction.*

(2) *Read off the number of non-zero rows. This number equals the rank.*

For example, the matrix A in Exercise 16.1 has rank 2.

Problem 16.2 *(Computing the matrix rank)*

Compute the rank of the following matrix:

$$A = \begin{pmatrix} 0 & 2 & 1 & -1 \\ 2 & 1 & 3 & 0 \\ 4 & 0 & -2 & 1 \\ 0 & -4 & 3 & 2 \end{pmatrix}.$$

Solution: We first bring A to upper echelon form.

$$\begin{pmatrix} 0 & 2 & 1 & -1 \\ 2 & 1 & 3 & 0 \\ 4 & 0 & -2 & 1 \\ 0 & -4 & 3 & 2 \end{pmatrix} \xrightarrow{R_1 \leftrightarrow R_2} \begin{pmatrix} 2 & 1 & 3 & 0 \\ 0 & 2 & 1 & -1 \\ 4 & 0 & -2 & 1 \\ 0 & -4 & 3 & 2 \end{pmatrix} \xrightarrow{R_3 \to R_3 - 2R_1} \begin{pmatrix} 2 & 1 & 3 & 0 \\ 0 & 2 & 1 & -1 \\ 0 & -2 & -8 & 1 \\ 0 & -4 & 3 & 2 \end{pmatrix} \xrightarrow[R_4 \to R_4 + 2R_2]{R_3 \to R_3 + R_2}$$

$$\begin{pmatrix} 2 & 1 & 3 & 0 \\ 0 & 2 & 1 & -1 \\ 0 & 0 & -7 & 0 \\ 0 & 0 & 5 & 0 \end{pmatrix} \xrightarrow{R_4 \to R_4 + 5R_3/7} \begin{pmatrix} 2 & 1 & 3 & 0 \\ 0 & 2 & 1 & -1 \\ 0 & 0 & -7 & 0 \\ 0 & 0 & 0 & 0 \end{pmatrix}$$

The last matrix is in upper echelon form and it has three non-zero rows. Hence, $\mathrm{rk}(A) = 3$.

Application 16.2 *Back to magic squares*

We now return to our discussion of magic squares. We have seen in Application 7.1 that all 3×3 magic squares form a vector space, and we have shown that the three specific magic squares M_1, M_2, M_3 in Eq. (7.10) are linearly independent. It remains to be shown that these matrices form a basis of the magic square vector space as asserted earlier. To do this it suffices to show that the dimension of the magic square vector space is three (see Theorem 7.2 (iii)).

We begin with an arbitrary 3×3 matrix

$$S = \begin{pmatrix} a & b & c \\ d & e & f \\ g & h & i \end{pmatrix}.$$

Recall that, for S to be a magic square, its rows, columns and both diagonals have to sum up to the same total. These conditions can be cast into the seven linear equations,

$$\begin{cases} d + e + f = a + b + c \\ g + h + i = a + b + c \\ a + d + g = a + b + c \\ b + e + h = a + b + c \\ c + f + i = a + b + c \\ a + e + i = a + b + c \\ c + e + g = a + b + c \end{cases} \Leftrightarrow \begin{cases} -a - b - c + d + e + g = 0 \\ -a - b - c + g + h + i = 0 \\ -b - c + d + g = 0 \\ -a - c + e + h = 0 \\ -a - b + f + i = 0 \\ -b - c + e + i = 0 \\ -a - b + e + g = 0 \end{cases}.$$

In matrix form, this system of equations can be written as follows:

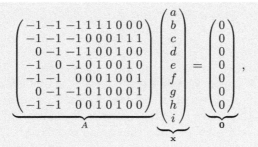

or, in short, $A\mathbf{x} = \mathbf{0}$. The magic squares are precisely the solutions to this homogeneous linear system which shows that the magic square vector space is the kernel, $\mathrm{Ker}(A)$, of the matrix A. By Gaussian elimination and with a bit of calculation, the matrix A can be brought into upper echelon form and the rank can be read off as $\mathrm{rk}(A) = 6$. Then, the dimension formula (14.13) leads to $\dim_{\mathbb{Q}}(\mathrm{Ker}(A)) = 9 - \mathrm{rk}(A) = 3$ and, hence, the dimension of the magic square vector space is indeed three.

In summary, the three matrices M_1, M_2, M_3 in Eq. (7.10) form a basis of the magic square vector space and every magic square is given as a (unique) linear combination of these three matrices.

16.3 Matrix inverse

Summary 16.3 *Elementary row operations can be generated by multiplying with certain elementary matrices from the left. This leads to an algorithm for computing the matrix inverse. Using elementary row operations, the matrix is first converted into upper echelon form and then into the unit matrix. Carrying out the same row operations on the unit matrix produces the matrix inverse.*

Row operations can be used to calculate the matrix inverse. To see how this works it is useful to re-formulate row operations in terms of matrix multiplication.

16.3.1 The elementary matrices

At first sight, the elementary row operations look somewhat artificial. But it turns out they are well-adapted to the structure of matrices, in that they can be generated by multiplying with certain, specific matrices from the left. In other words, to perform a certain row operation on a matrix A, we can find a suitable matrix P such that the row operation is generated by $A \mapsto PA$. As a simple example, consider 2×2 matrices.

$$A = \begin{pmatrix} a & b \\ c & d \end{pmatrix}, \ P = \begin{pmatrix} 1 & \lambda \\ 0 & 1 \end{pmatrix} \quad \Rightarrow \quad A \mapsto PA = \begin{pmatrix} 1 & \lambda \\ 0 & 1 \end{pmatrix} \begin{pmatrix} a & b \\ c & d \end{pmatrix} = \begin{pmatrix} a + \lambda c & b + \lambda d \\ c & d \end{pmatrix}$$

Evidently, multiplication with the matrix P from the left has generated the elementary row operation $R_1 \to R_1 + \lambda R_2$ on the arbitrary 2×2 matrix A. This works in general and the appropriate matrices, generating the three types of elementary row operations in Def. 16.1, are given by

$$P_{R_i \leftrightarrow R_j} = \mathbb{1} - E_{(ii)} - E_{(jj)} + E_{(ij)} + E_{(ji)}$$
$$P_{R_i \rightarrow R_i + \lambda R_j} = \mathbb{1} + \lambda E_{(ij)} \qquad (16.5)$$
$$P_{R_i \rightarrow \lambda R_i} = \mathbb{1} + (\lambda - 1)E_{(ii)} \ .$$

To see that these matrices do indeed produce the desired result, we first note that $E_{(ij)}A$ is a matrix with all rows zero except the ith one which contains the jth row of A. Using this rule and the distributive laws (13.22) it is easy to reason out what happens when one of the above matrices multiplies a matrix A from the left. For example, $P_{R_i \rightarrow R_i + \lambda R_j}A = A + \lambda E_{(ij)}A$ and the second matrix, $\lambda E_{(ij)}A$, is zero except for its ith row which contains $\lambda \mathbf{A}_j$. Adding this matrix to A produces the desired effect, $R_i \rightarrow R_i + \lambda R_j$.

It is intuitively clear that the elementary matrices are invertible since the row operations they generate can be undone by another row operation. Again, focusing on $P_{R_i \rightarrow R_i + \lambda R_j}$, we would expect $P_{R_i \rightarrow R_i - \lambda R_j}$ to be its inverse. Noting that $E_{(ij)}E_{(ij)} = 0$ for $i \neq j$, this can be easily verified:

$$P_{R_i \rightarrow R_i + \lambda R_j} P_{R_i \rightarrow R_i - \lambda R_j} = (\mathbb{1} + \lambda E_{(ij)})(\mathbb{1} - \lambda E_{(ij)}) = \mathbb{1} \ .$$

16.3.2   Algorithm to calculate the matrix inverse

Our next task is to devise an algorithm to compute the inverse of an $n \times n$ matrix A, using elementary row operations. (Recall that only square matrices can have an inverse.) To do this we attempt to convert the matrix into the unit matrix using row operations. Schematically, this works as follows:

$$A \xrightarrow{\text{row red.}} \begin{pmatrix} a'_{11} & & * \\ & a'_{22} & \\ & & \ddots & \\ 0 & & & a'_{nn} \end{pmatrix} \xrightarrow{\text{(R1), (R2)}} \begin{pmatrix} a'_{11} & & 0 \\ & a'_{22} & \\ & & \ddots & \\ 0 & & & a'_{nn} \end{pmatrix} \xrightarrow{\text{(R3)}} \begin{pmatrix} 1 & & 0 \\ & \ddots & \\ 0 & & 1 \end{pmatrix} = \mathbb{1}_n \ .$$

In the first step, we bring A into upper echelon form, by the algorithm already discussed, and we read off its rank. In $\text{rk}(A) < n$ then, from Cor. 14.2, the matrix is not invertible and we can stop. Otherwise, if $\text{rk}(A) = n$, all pivots must be along the diagonal so that $a'_{ii} \neq 0$ for all $i = 1, \ldots, n$. This means we can apply further row operations to set the entries above the diagonal to zero. We start with the last column and subtract suitable multiples of the last row from the others until all entries in the last column except a'_{nn} are zero. We proceed in a similar way, column by column from the right to the left, using row operations of type (R1) and (R2). In this way we arrive at a diagonal matrix, with diagonal entries $a'_{ii} \neq 0$ which, in the final step, can be converted into the unit matrix by row operations of type (R3).

This discussion implies we can find elementary matrices $P_1, \ldots, P_k$, as defined in Eq. (16.5), generating elementary row operations, such that

$$\mathbb{1}_n = \underbrace{P_1 \cdots P_k}_{A^{-1}} A \quad \Rightarrow \quad A^{-1} = P_1 \cdots P_k \mathbb{1}_n \ . \qquad (16.6)$$

These equations imply an explicit algorithm to compute the inverse of a square matrix. We convert A into the unit matrix $\mathbb{1}_n$ using elementary row operations as described above, and then simply carry out the same operations on $\mathbb{1}_n$ in parallel. When we are done the unit matrix will have been converted into A^{-1}.

Algorithm *(Computing the matrix inverse)*

To find the inverse of an $n \times n$ matrix A, carry out the following steps:

(1) *Bring A to upper echelon form and read off the rank. If $\mathrm{rk}(A) < n$ then A is not invertible so there is nothing more to do.*

(2) *Otherwise, if $\mathrm{rk}(A) = n$, use row operations (R2) to set the entries above the diagonal to zero, starting with the last column and proceeding from right to left. The result is a diagonal matrix.*

(3) *Use row operations (R3) to convert the diagonal matrix into a unit matrix.*

(4) *Carry out, in the same order, all above row operations on the unit matrix $\mathbb{1}_n$. This converts $\mathbb{1}_n$ into A^{-1}.*

Problem 16.3 *(Computing the matrix inverse with row operations)*

Using row operations, compute the inverse of the matrix below.

$$A = \begin{pmatrix} 1 & 0 & -2 \\ 0 & 3 & -2 \\ 1 & -4 & 0 \end{pmatrix}$$

Solutions: We follow the above algorithm and carry all row operation out on A and $\mathbb{1}_3$ in parallel.

$$A = \begin{pmatrix} 1 & 0 & -2 \\ 0 & 3 & -2 \\ 1 & -4 & 0 \end{pmatrix} \qquad \mathbb{1}_3 = \begin{pmatrix} 1 & 0 & 0 \\ 0 & 1 & 0 \\ 0 & 0 & 1 \end{pmatrix}$$

$$R_3 \to R_3 - R_1 : \quad \begin{pmatrix} 1 & 0 & -2 \\ 0 & 3 & -2 \\ 0 & -4 & 2 \end{pmatrix} \qquad \begin{pmatrix} 1 & 0 & 0 \\ 0 & 1 & 0 \\ -1 & 0 & 1 \end{pmatrix}$$

$$R_3 \to R_3 + \frac{4}{3} R_2 : \quad \begin{pmatrix} 1 & 0 & -2 \\ 0 & 3 & -2 \\ 0 & 0 & -\frac{2}{3} \end{pmatrix} \leftarrow \mathrm{rk}(A) = 3 \qquad \begin{pmatrix} 1 & 0 & 0 \\ 0 & 1 & 0 \\ -1 & \frac{4}{3} & 1 \end{pmatrix}$$

$$R_2 \to R_2 - 3R_3 : \quad \begin{pmatrix} 1 & 0 & -2 \\ 0 & 3 & 0 \\ 0 & 0 & -\frac{2}{3} \end{pmatrix} \qquad \begin{pmatrix} 1 & 0 & 0 \\ 3 & -3 & -3 \\ -1 & \frac{4}{3} & 1 \end{pmatrix}$$

$$R_1 \to R_1 - 3R_3 : \quad \begin{pmatrix} 1 & 0 & 0 \\ 0 & 3 & 0 \\ 0 & 0 & -\frac{2}{3} \end{pmatrix} \qquad \begin{pmatrix} 4 & -4 & -3 \\ 3 & -3 & -3 \\ -1 & \frac{4}{3} & 1 \end{pmatrix}$$

$$R_2 \to \frac{R_2}{3} : \quad \begin{pmatrix} 1 & 0 & 0 \\ 0 & 1 & 0 \\ 0 & 0 & -\frac{2}{3} \end{pmatrix} \qquad \begin{pmatrix} 4 & -4 & -3 \\ 1 & -1 & -1 \\ -1 & \frac{4}{3} & 1 \end{pmatrix}$$

$$R_3 \to -\frac{3}{2} R_3 : \quad \begin{pmatrix} 1 & 0 & 0 \\ 0 & 1 & 0 \\ 0 & 0 & 1 \end{pmatrix} = \mathbb{1}_3 \qquad \begin{pmatrix} 4 & -4 & -3 \\ 1 & -1 & -1 \\ \frac{3}{2} & -2 & -\frac{3}{2} \end{pmatrix} = A^{-1}$$

As a final check we show that

$$AA^{-1} = \begin{pmatrix} 1 & 0 & -2 \\ 0 & 3 & -2 \\ 1 & -4 & 0 \end{pmatrix} \begin{pmatrix} 4 & -4 & -3 \\ 1 & -1 & -1 \\ \frac{3}{2} & -2 & -\frac{3}{2} \end{pmatrix} = \begin{pmatrix} 1 & 0 & 0 \\ 0 & 1 & 0 \\ 0 & 0 & 1 \end{pmatrix} = \mathbb{1}_3 \ \checkmark$$

and thus confirm that we have correctly computed the inverse of A.

Exercises

(†=challenging, ††=difficult, wide-ranging)

16.1 *Computing the rank of a 3×3 matrix*
Use row operations to compute the rank of the matrix

$$A = \begin{pmatrix} 2 & 0 & -3 \\ 4 & 1 & 2 \\ 2 & 1 & a \end{pmatrix}$$

for all $a \in \mathbb{R}$. (Hint: Keep in mind that the rank of A may depend on the value of a so a case distinction may be required.)

16.2 *Computing the rank of a 4×4 matrix*
Use row operation to work out the rank of the matrix

$$A = \begin{pmatrix} 1 & 1 & 0 & -1 \\ 0 & 2 & -2 & 1 \\ 3 & 2 & 0 & -4 \\ 1 & -2 & a & 0 \end{pmatrix},$$

where $a \in \mathbb{R}$.

16.3 *Matrices for column operations*
(a) Explicitly verify for 2×2 matrices that the matrices in Eq. (16.5) generate row operations by multiplication from the left.
(b) Suppose a row operation is generated by multiplying from the left with one of the matrices P in Eq. (16.5). Show that the corresponding column operation is generated by multiplying with P^T from the right.
(c) Verify the statement from part (b) explicitly for 2×2 matrices.

16.4 *Computing the matrix inverse*
Use row operations to find the inverse of the matrix

$$A = \begin{pmatrix} 1 & 0 & -1 \\ 2 & 1 & -2 \\ 1 & -3 & 0 \end{pmatrix}.$$

Check your result!

16.5 *Semi-magic squares — again*[†]
A 3×3 semi-magic square is a 3×3 matrix of such that all rows and columns sum up to the same total.
(a) Verify that

$$A = \begin{pmatrix} 3 & 2 & 1 \\ 2 & 2 & 2 \\ 1 & 2 & 3 \end{pmatrix},$$

is a semi-magic square.
(b) Why do the semi-magic squares form a vector space?
(c) Show that the matrices

$$M_1 = \begin{pmatrix} 1 & 1 & 1 \\ 1 & 1 & 1 \\ 1 & 1 & 1 \end{pmatrix}$$

$$M_2 = \begin{pmatrix} 1 & -1 & 0 \\ -1 & 1 & 0 \\ 0 & 0 & 0 \end{pmatrix}$$

$$M_3 = \begin{pmatrix} 0 & 1 & -1 \\ 0 & -1 & 1 \\ 0 & 0 & 0 \end{pmatrix}$$

$$M_4 = \begin{pmatrix} 0 & 0 & 0 \\ 1 & -1 & 0 \\ -1 & 1 & 0 \end{pmatrix}$$

$$M_5 = \begin{pmatrix} 0 & 0 & 0 \\ 0 & 1 & -1 \\ 0 & -1 & 1 \end{pmatrix}$$

form a basis of the semi-magic squares. Write the semi-magic square from part (a) as a linear combination of this basis.

16.6 *Code for matrix rank*[††]
Write a programme in your favourite programming language which computes the rank of a matrix. To avoid numerical problems work with matrices in $\mathcal{M}_{n,m}(\mathbb{F}_p)$.

17
Linear systems

17.1 Abstract linear systems

Summary 17.1 *For a linear map $f : V \to W$ and $\mathbf{b} \in W$ an inhomogeneous linear system is defined by the equation $f(\mathbf{x}) = \mathbf{b}$. The solutions set of the associated homogeneous system, $f(\mathbf{x}) = \mathbf{0}$, is given by $\mathrm{Ker}(f)$, with dimension $k = \dim_{\mathbb{F}}(V) - \mathrm{rk}(f)$. The inhomogeneous system has a solution $\mathbf{x}_0$ if and only if $\mathbf{b} \in \mathrm{Im}(f)$. In this case, the solution set to the inhomogeneous system is the affine k-plane $\mathbf{x}_0 + \mathrm{Ker}(f)$.*

17.1.1 Definition of linear systems

Linear systems are the fundamental equations which arise and need to be solved in the context of vector spaces. They are determined by a linear map $f : V \to W$ between two vector spaces V and W and a vector $\mathbf{b} \in W$. Given this data, a linear system and its solution space are defined by

$$f(\mathbf{x}) = \mathbf{b} , \qquad \mathrm{Sol}(f, \mathbf{b}) = \{ \mathbf{x} \in V \mid f(\mathbf{x}) = \mathbf{b} \} . \tag{17.1}$$

If $\mathbf{b} = \mathbf{0}$ the linear system is called *homogeneous* otherwise it is called *inhomogeneous* with *imhomogeneity* $\mathbf{b}$. For an inhomogeneous linear system it is instructive to consider the associated homogeneous linear system

$$f(\mathbf{x}) = \mathbf{0} , \qquad \mathrm{Sol}(f, \mathbf{0}) = \{ \mathbf{x} \in V \mid f(\mathbf{x}) = \mathbf{0} \} , \tag{17.2}$$

obtained by setting $\mathbf{b} = \mathbf{0}$ in Eq. (17.1).

The solution space of a homogeneous linear system is non-empty. It always contains the zero vector, $\mathbf{0} \in \mathrm{Sol}(f, \mathbf{0})$, since $f(\mathbf{0}) = \mathbf{0}$ for any linear map. The zero vector is often referred to as the 'trivial solution' of the homogeneous linear system and all other solutions as 'non-trivial'.

17.1.2 Structure of solution space

The following theorem provides useful information about the existence of non-trivial solutions for a homogeneous linear system.

Theorem 17.1 *Let $f : V \to W$ be a linear map. The solution space to the homogeneous linear system $f(\mathbf{x}) = \mathbf{0}$ is the vector subspace $\mathrm{Sol}(\mathbf{f}, \mathbf{0}) = \mathrm{Ker}(f)$ of V. If V and W are finite-dimensional then*

$$\dim_{\mathbb{F}}(\mathrm{Sol}(f, \mathbf{0})) = \dim_{\mathbb{F}}(V) - \mathrm{rk}(f) . \qquad (17.3)$$

Proof The equality $\mathrm{Sol}(\mathbf{f}, \mathbf{0}) = \mathrm{Ker}(f)$ follows trivially from the definition of the kernel of a linear map and we know from Lemma 14.1 that the kernel is a vector subspace. The formula (17.3) for the dimension of the solution space is a direct consequence of the rank formula (14.13) for linear maps. □

In other words, the solution space of a homogeneous linear system is a vector space, the kernel of the linear map. Its dimension is an important piece of information which can be computed from the rank using Eq. (17.3). In particular, if $\mathrm{rk}(f) = \dim_{\mathbb{F}}(V)$ then the solution space is zero-dimensional, and the only solution is the trivial one (the zero vector). Otherwise, if $\mathrm{rk}(f) < \dim_{\mathbb{F}}(V)$, there are non-trivial solutions.

The solution space of an inhomogeneous linear system is closely related to the one for the associated homogeneous system.

Theorem 17.2 *Let $f : V \to W$ be a linear map and $\mathbf{b} \in W$. The linear system $f(\mathbf{x}) = \mathbf{b}$ has a solution, $\mathbf{x}_0$, if and only if $\mathbf{b} \in \mathrm{Im}(f)$. In this case, the solution sets of the inhomogeneous and the associated homogeneous linear systems are related by*

$$\mathrm{Sol}(f, \mathbf{b}) = \mathbf{x}_0 + \mathrm{Sol}(f, \mathbf{0}) . \qquad (17.4)$$

If V and W are finite-dimensional, this represents an affine k-plane in V with $k = \dim_{\mathbb{F}}(V) - \mathrm{rk}(f)$.

Proof If $\mathbf{x}_0$ is a solution of the inhomogeneous equation, then $f(\mathbf{x}_0) = \mathbf{b}$ which shows that $\mathbf{b} \in \mathrm{Im}(f)$. On the other hand, if $\mathbf{b} \in \mathrm{Im}(f)$, then, by definition of the image, there exists an $\mathbf{x}_0 \in V$ with $f(\mathbf{x}_0) = \mathbf{b}$.

To show the equality (17.4), assume the existence of a solution $\mathbf{x}_0 \in V$ with $f(\mathbf{x}_0) = \mathbf{b}$.

$$\mathbf{x} \in \mathrm{Sol}(A, \mathbf{b}) \quad \Leftrightarrow \quad f(\mathbf{x}) = \mathbf{b} = f(\mathbf{x}_0) \quad \Leftrightarrow \quad f(\mathbf{x} - \mathbf{x}_0) = \mathbf{0}$$
$$\Leftrightarrow \quad \mathbf{x} - \mathbf{x}_0 \in \mathrm{Ker}(f) \quad \Leftrightarrow \quad \mathbf{x} \in \mathbf{x}_0 + \mathrm{Sol}(f, \mathbf{0})$$

Finally, provided we are in the finite-dimensional case, it is clear that Eq. (17.4) is an affine k-plane, since $\mathrm{Sol}(f, \mathbf{0})$ is a vector subspace. The dimension k of this affine plane is defined to be the dimension of $\mathrm{Sol}(f, \mathbf{0})$ which, from Eq. (17.3), is indeed given by $\dim_{\mathbb{F}}(V) - \mathrm{rk}(f)$. □

The content of the previous theorem is often paraphrased by saying that the 'general solution to the inhomogeneous system' is obtained by adding to a 'special solution of the inhomogeneous system' all solution of the homogeneous system. We will see this more concretely in the next section when we discuss systems of linear equations. But the structure of linear systems outlined above is also instructive in other contexts, as the next example shows.

Example 17.1 *(Solution to inhomogenous linear differential equation)*

The previous theorems have a prominent application to inhomogeneous, linear (second order, say) differential equations. Introduce the linear second-order differential operator $L : \mathcal{C}^\infty([a, b], \mathbb{R}) \to \mathcal{C}^\infty([a, b], \mathbb{R})$ by

$$L := p\frac{d^2}{dx^2} + q\frac{d}{dx} + r \,,$$

where where $p, q, r \in \mathcal{C}^\infty([a, b], \mathbb{R})$ are fixed functions. For a function $b \in \mathcal{C}^\infty([a, b], \mathbb{R})$ an inhomogeneous linear system and its homogeneous counterpart are defined by

$$L(y) = b \,, \qquad L(y) = 0 \,.$$

The solution space of the homogeneous system is given by $\mathrm{Sol}(L, 0) = \mathrm{Ker}(L)$. Provided we can find a solution y_0 to the inhomogeneous system, so $L(y_0) = b$, then the set of solutions to the inhomogeneous system is given by $\mathrm{Sol}(L, b) = y_0 + \mathrm{Ker}(L)$, from Eq. (17.4).

To make this more concrete, consider the specific differential operator $L = \frac{d^2}{dx^2} + 1$ and a function $b(x) - x$ which lead to an inhomogeneous differential equation and associated homogeneous equation

$$\frac{d^2y}{dx^2} + y = x \,, \qquad \frac{d^2y}{dx^2} + y = 0 \,.$$

It is easy to see that the inhomogeneous equation is solved by $y_0(x) = x$ while the solution to the homogeneous equation is given by $\mathrm{Sol}(L, 0) = \mathrm{Ker}(L) = \mathrm{Span}(\sin, \cos)$. Hence, the general solution to the inhomogeneous equation is

$$y(x) = x + a\sin(x) + b\cos(x) \,,$$

for $a, b \in \mathbb{R}$ arbitrary. □

17.2 Systems of linear equations

Summary 17.2 *A system $A\mathbf{x} = \mathbf{b}$ of m linear equations in n variables has a solution if and only if $\mathbf{b} \in \mathrm{Im}(A)$. If a solution $\mathbf{x}_0$ exists the solution set is the affine k-plane $\mathrm{Sol}(A, \mathbf{b}) = \mathbf{x}_0 + \mathrm{Ker}(A)$, where $k = n - \mathrm{rk}(A)$. The solution of a system of linear equations can be computed by an algorithm based on elementary row operations.*

17.2.1 Definition

Every linear system $f(\xi) = \beta$, where $f : V \to W$ and $\beta \in W$ (for finite-dimensional vector spaces V and W over $\mathbb{F}$) can be converted into a system of linear equations, relative to a choice of bases $(\mathbf{v}_1, \ldots, \mathbf{v}_n)$ and $(\mathbf{w}_1, \ldots, \mathbf{w}_m)$ for V and W. To see this, write $\xi = \sum_{j=1}^n x_j \mathbf{v}_j$, $\beta = \sum_{i=1}^m b_i \mathbf{w}_i$ and recall that the $m \times n$ matrix A with entries

a_{ij} which represents f relative to this basis choice is obtained by $f(\mathbf{v}_j) = \sum_{i=1}^{m} a_{ij}\mathbf{w}_j$. Then, the left- and right-hand sides of the linear system can be written as

$$f(\boldsymbol{\xi}) = f\left(\sum_{j=1}^{n} x_j\mathbf{v}_j\right) = \sum_{j=1}^{n} x_j f(\mathbf{v}_j) = \sum_{i,j} a_{ij}x_j\mathbf{w}_i , \qquad \boldsymbol{\beta} = \sum_i b_i\mathbf{w}_i ,$$

and matching the coordinates of $\mathbf{w}_i$ shows that the equation $f(\boldsymbol{\xi}) = \beta$ is equivalent to $A\mathbf{x} = \mathbf{b}$. In other words, every linear system (for finite-dimensional vector spaces) can be converted into a system of linear equations of the form

$$\begin{array}{llll} A\mathbf{x} = \mathbf{b} & \mathrm{Sol}(A, \mathbf{b}) = \{\mathbf{x} \in \mathbb{F}^n \mid A\mathbf{x} = \mathbf{b}\} & \text{(inhomogeneous)} \\ A\mathbf{x} = \mathbf{0} & \mathrm{Sol}(A, \mathbf{0}) = \{\mathbf{x} \in \mathbb{F}^n \mid A\mathbf{x} = \mathbf{0}\} & \text{(homogeneous) .} \end{array} \qquad (17.5)$$

Writing out the entries of the matrix A and the components of $\mathbf{x}$ and $\mathbf{b}$ explicitly

$$A = \begin{pmatrix} a_{11} & \cdots & a_{1n} \\ \vdots & & \vdots \\ a_{m1} & \cdots & a_{mn} \end{pmatrix} , \qquad \mathbf{x} = \begin{pmatrix} x_1 \\ \vdots \\ x_n \end{pmatrix} , \qquad \mathbf{b} = \begin{pmatrix} b_1 \\ \vdots \\ b_m \end{pmatrix} ,$$

the Eqs. (17.5) can be cast into the form

$$\begin{array}{cc} \text{(inhomogeneous)} & \text{(homogeneous)} \\ \begin{array}{ccc} a_{11}x_1 + \cdots + a_{1n}x_n &=& b_1 \\ \vdots \quad \vdots \quad \vdots & & \vdots \\ a_{m1}x_1 + \cdots + a_{mn}x_n &=& b_m \end{array} & \begin{array}{ccc} a_{11}x_1 + \cdots + a_{1n}x_n &=& 0 \\ \vdots \quad \vdots \quad \vdots & & \vdots \\ a_{m1}x_1 + \cdots + a_{mn}x_n &=& 0 \end{array} \end{array} \qquad (17.6)$$

of m linear equations in n variables $x_1, \ldots, x_n$.

17.2.2 Solutions of homogeneous system

Everything we have said in Theorems 17.1 about the solution structure of homogeneous linear systems directly applies to systems of linear equations. To recap, for the solution space of the homogeneous system $A\mathbf{x} = \mathbf{0}$ of m equations in n variables we have

$$\mathrm{Sol}(A, \mathbf{0}) = \mathrm{Ker}(A) , \qquad \dim_{\mathbb{F}}(\mathrm{Sol}(A, \mathbf{0})) = n - \mathrm{rk}(A) . \qquad (17.7)$$

To determine the dimension of the solution space — the number of free parameters required to describe the solution — all we need is the rank of A.

Problem 17.1 (*Homogeneous linear system in $\mathbb{R}^2$*)

Find the solution of the simple homogeneous linear system

$$\begin{array}{lll} (E1): & x_1 - x_2 = 0 \\ (E2): & ax_1 + 3x_3 = 0 \end{array}$$

of two equations in two variables for all values of the parameter $a \in \mathbb{R}$.

Solution: We can write this system in matrix vector form, $A\mathbf{x} = \mathbf{0}$, with

$$A = \begin{pmatrix} 1 & -1 \\ a & 3 \end{pmatrix}, \qquad \mathbf{x} = \begin{pmatrix} x_1 \\ x_2 \end{pmatrix}.$$

The rank of A depends on the value of the parameter a. For the generic choice $a \neq -3$ the two columns of A are linearly independent so that $\mathrm{rk}(A) = 2$. From Eq. (17.7) this means the solution space is zero-dimensional so the zero vector is the only solution. On the other hand, if $a = -3$ the rank decreases to $\mathrm{rk}(A) = 1$, so that the solution space is one-dimensional.

To see this more explicitly combine the two equations in order to eliminate x_2.

$$3(E1) + (E2): \qquad (3 + a)x_1 = 0. \tag{17.8}$$

For $a \neq -3$ we can divide by $3 + a$ so that $x_1 = 0$ and, by inserting this into (E1), we find that $x_2 = 0$. This is precisely the case when $\mathrm{rk}(A) = 2$ and the system only has the trivial solution, $\mathrm{Sol}(A, \mathbf{0}) = \{\mathbf{0}\}$.

If $a = 3$ then Eq. (17.8) becomes trivial or, equivalently, (E1) and (E2) are multiples of each other. In this case, the solution consists of all $\mathbf{x} = (x_1, x_2)^T$ with $x_1 = x_2$, so that the solution space $\mathrm{Sol}(A, \mathbf{0}) = \mathrm{Span}((1, 1)^T)$ is one-dimensional, as expected.

The geometry of the solution spaces for these two cases is illustrated in the figure below.

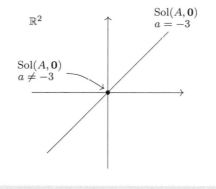

17.2.3 Solution of inhomogeneous system

Let us discuss the solution structure of an inhomogeneous linear system $A\mathbf{x} = \mathbf{b}$ of m equations in n variables, so $A : \mathbb{F}^n \to \mathbb{F}^m$, $\mathbf{x} \in \mathbb{F}^n$ and $\mathbf{b} \in \mathbb{F}^m$, following Theorem 17.2. Much can be said about the qualitative solution structure based on the three dimensions n, m and $\mathrm{rk}(A)$.

It is worth stressing that linear systems do not need to have a solution at all. A solution exists if and only if $\mathbf{b} \in \mathrm{Im}(A)$. If $\mathrm{rk}(A) = m$ then, from Prop. (14.2), A is surjective and, hence, a solution exists whatever the inhomogeneity $\mathbf{b}$. On the other hand, if $\mathrm{rk}(A) < m$ then the image, $\mathrm{Im}(A)$, is a proper vector subspace of the co-domain $\mathbb{F}^m$. Typical choices for $\mathbf{b}$ will not be in this subspace and in this case a solution does not exist. For example, if $m = 3$ and $\mathrm{rk}(A) = 2$ then the image of A is a plane in a three-dimensional space and we need to choose $\mathbf{b}$ to lie in this plane for a solution to exist. Clearly this corresponds to a very special choice of $\mathbf{b}$.

If a solution $\mathbf{x}_0$ to the inhomogeneous system of linear equations exists then the solution set is the affine k-plane

$$\mathrm{Sol}(A, \mathbf{b}) = \mathbf{x}_0 + \mathrm{Ker}(A) \qquad (17.9)$$

with dimension $k = n - \mathrm{rk}(A)$. This is just the solution $\mathrm{Sol}(A, \mathbf{0}) = \mathrm{Ker}(A)$ of the associated homogeneous system 'shifted' by $\mathbf{x}_0$ (see Fig. 17.1). The solution is unique (that is, the solution set is zero-dimensional) iff the associated homogeneous system only has the trivial solution or, equivalently, iff $\mathrm{rk}(A) = n$. In summary, we can classify

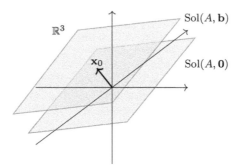

Fig. 17.1 The relation $\mathrm{Sol}(A, \mathbf{b}) = \mathbf{x}_0 + \mathrm{Sol}(A, \mathbf{0})$ between the solution set of an inhomogeneous and associated homogeneous system of linear equations.

the solution structure as follows:

(1) $\mathrm{rk}(A) = m$
   In this case there exists a solution, $\mathbf{x}_0$, for any choice of $\mathbf{b}$ and the general solution is given by the affine k-plane $\mathbf{x}_0 + \mathrm{Ker}(A)$. Its dimension (the number of free parameters in this solution) equals $k = n - m$.

(2) $\mathrm{rk}(A) < m$
   (a) $\mathbf{b} \in \mathrm{Im}(A)$ (this requires special choices for $\mathbf{b}$)
      There exists a solution $\mathbf{x}_0$. The solution set is the affine k-plane $\mathbf{x}_0 + \mathrm{Ker}(A)$ with dimension $k = n - \mathrm{rk}(A)$.
   (b) $\mathbf{b} \notin \mathrm{Im}(A)$ (true for generic choices of $\mathbf{b}$)
      There is no solution.

A common special situation is that of a system of n linear equations in n variables, so that $n = m$. In this case, the above classification becomes slightly more specific.

(1) $\mathrm{rk}(A) = n$
   A solution exists for any choice of $\mathbf{b}$ and there are no free parameters since $\dim_{\mathbb{F}}(\mathrm{Ker}(A)) = n - n = 0$. Hence, the solution is unique. Indeed, in this case, the matrix A is invertible (see Cor. 14.2) and the unique solution is given by $\mathbf{x} = A^{-1}\mathbf{b}$. Hence, the solution can be found by computing the inverse matrix A^{-1}.

(2) $\mathrm{rk}(A) < n$

(a) $\mathbf{b} \in \text{Im}(A)$ (this requires special choices for $\mathbf{b}$)
There exists a solution $\mathbf{x}_0$. The solution set is the affine k-plane $\mathbf{x}_0 + \text{Ker}(A)$ with dimension $k = n - \text{rk}(A)$.
(b) $\mathbf{b} \notin \text{Im}(A)$ (true for generic choices of $\mathbf{b}$)
There is no solution.

17.2.4   Examples with explicit calculation

We should illustrate the above structure with a few examples. For now, we follow a 'pedestrian' approach, solving the systems of linear equations by adding multiples of the equations in order to eliminate variables. This is what we have done so far and it is often the most efficient method in a small number of dimensions. A systematic method, based on row operations, will be introduced afterwards.

Problem 17.2 (*Inhomogeneous linear system in* $\mathbb{R}^2$)

Find the (real) solutions of the inhomogeneous linear system of equations

$$\begin{aligned}(E1): &\quad x_1 - x_2 = 3 \\ (E2): &\quad ax_1 + 3x_3 = b\end{aligned}$$

for all values of the paramters $a, b \in \mathbb{R}$.

Solution: The system can be written in matrix vector form, $A\mathbf{x} - \mathbf{b}$ with

$$A = \begin{pmatrix} 1 & -1 \\ a & 3 \end{pmatrix}, \qquad \mathbf{x} = \begin{pmatrix} x_1 \\ x_2 \end{pmatrix}, \qquad \mathbf{b} = \begin{pmatrix} 3 \\ b \end{pmatrix}.$$

Note that the associated homogeneous system has already been solved in Exercise 17.1. Without much calculation, we can already make qualitative statements about the solution following the above classification.

(1) $a \neq 3$: In this case, $\text{rk}(A) = 2$, so there is a unique solution.
(2a) $a = -3$, b special so that $\mathbf{b} \in \text{Im}(A)$: Now $\text{rk}(A) = 1$, so the solution is an affine line.
(2b) $a = -3$, b generic so that $\mathbf{b} \notin \text{Im}(A)$: There is no solution.

To verify these expectations by explicit calculation we first eliminate x_2.

$$3(E1) + (E2): \qquad (3 + a)x_1 = 9 + b, \tag{17.10}$$

This equation already reveals the expected case distinction.

(1) $a \neq -3$: We can divide Eq. (17.10) by $3 + a$ to find x_1 and then insert into (E2) to find x_2.

$$\text{Sol}(A, \mathbf{b}) = \{(x_1, x_2)^T\} \quad \text{where} \quad x_1 = \frac{b + 9}{a + 3}, \quad x_2 = \frac{b + 9}{a + 3} - 3 = \frac{b - 3a}{a + 3}.$$

Hence, the solution is unique and exists for all choices of b, as predicted.

(2a) $a = -3$, $b = -9$: For this choice of parameters, Eq. (17.10) becomes trivial or, equivalently, (E1) and (E2) are multiples of each other. Every $\mathbf{x}$ with $x_1 = x_2 + 3$ is a solution in this case, so we can write

$$\text{Sol}(A, \mathbf{b}) = \left\{ \begin{pmatrix} t + 3 \\ t \end{pmatrix} \mid t \in \mathbb{R} \right\} = \mathbf{x}_0 + \text{Sol}(A, \mathbf{0}), \qquad \mathbf{x}_0 = \begin{pmatrix} 3 \\ 0 \end{pmatrix},$$

where $\mathbf{x}_0 = (3, 0)^T$ is a solution of the inhomogeneous system and $\text{Sol}(A, \mathbf{0}) = \text{Span}((1, 1)^T)$ is the solution set of the associated homogeneous system from Exercise 17.1. As predicted, the solution is an affine line.

(2b) $a = -3$, $b \neq -9$: Inserting these parameter choices into Eq. (17.10) leads to a contradiction, so the linear system has no solution, $\mathrm{Sol}(A, \mathbf{b}) = \{\}$.

The solution structure for these case is illustrated in the figure below.

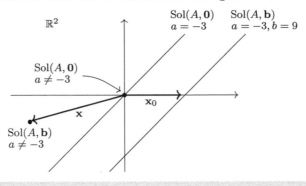

Problem 17.3 *(An inhomogeneous linear system in $\mathbb{R}^3$)*

Find the real solutions of the linear system

$$
\begin{aligned}
(E1): &\quad 2x_1 + 3x_2 - x_3 = -1 \\
(E2): &\quad -x_1 - 2x_2 + x_3 = 3 \\
(E3): &\quad ax_1 + x_2 - 2x_3 = b
\end{aligned}
$$

for all values of the parameters $a, b \in \mathbb{R}$.

Solution: We can write this system in vector matrix form, $A\mathbf{x} = \mathbf{b}$, with

$$
A = \begin{pmatrix} 2 & 3 & -1 \\ -1 & -2 & 1 \\ a & 1 & -2 \end{pmatrix}, \qquad \mathbf{x} = \begin{pmatrix} x_1 \\ x_2 \\ x_3 \end{pmatrix}, \qquad \mathbf{b} = \begin{pmatrix} -1 \\ 3 \\ b \end{pmatrix}.
$$

As usual, we begin by predicting the qualitative solution structure. The columns $\mathbf{A}^2$, $\mathbf{A}^3$ of A are clearly linearly independent so the rank of A is at least two. For a specific value of $a = a_0$ (to be determined shortly) we expect $\mathbf{A}^1$ to be in the plane spanned by $\mathbf{A}^2$, $\mathbf{A}^3$, and in this case $\mathrm{rk}(A) = 2$. Generically, we have $a \neq a_0$ so that $\mathrm{rk}(A) = 3$.

(1) $a \neq a_0$: We have $\mathrm{rk}(A) = 3$ so there is a unique solution for every value of b.
(2a) $a = a_0$, b special so that $\mathbf{b} \in \mathrm{Im}(A)$: The solution is an affine line.
(2b) $a = a_0$, b generic so that $\mathbf{b} \notin \mathrm{Im}(A)$: There is no solution.

To confirm these expectations we start by eliminating x_3 from two combinations of the three equations.

$$
\begin{aligned}
(E1') = (E1) + (E2): &\qquad x_1 + x_2 = 2 \\
(E3') = (E3) + 2(E2): &\qquad (a - 2)x_1 - 3x_2 = b + 6
\end{aligned}
$$

(While these are simple calculations in principle mistakes can easily slip in. It is, therefore, important to keep the calculation organized and keep track of the steps performed.) Finally, we combine the above two equations to eliminate x_2.

$$
3(E1') + (E3'): \qquad (a + 1)x_1 = b + 12. \tag{17.11}
$$

Case distinctions come into such calculations when divisions by (parameter-dependent) quantities which may be zero have to be carried out. It is helpful to avoid such divisions for as long as possible. This is the reason why we have eliminated x_2 and x_3 first. The case distinction can now be read off from Eq. (17.11).

(1) $a \neq -1$: We can divide Eq. (17.11) by $a + 1$ to obtain x_1, insert this result into $(E1')$ to obtain x_2 and, finally, use (E1) to get x_3. This leads to

$$x_1 = \frac{b + 12}{a + 1}, \quad x_2 = \frac{2a - b - 10}{a + 1}, \quad x_3 = \frac{7a - b - 5}{a + 1},$$

and, hence, we have a unique solution for every $a \neq -1$ and every $b \in \mathbb{R}$. We know that the solution to the associated homogeneous system in this case must be trivial.

(2a) $a = -1, b = -12$: For this choice of parameters, Eq. (17.11) becomes trivial which indicates that we have only two independent equations. We can solve the equations (E1') and (E1) for x_1 and x_3 in terms of x_2 which leads to $x_1 = 2 - x_2$ and $x_3 = 5 + x_2$. Setting $t = x_2$, the solutions space is the affine line

$$\text{Sol}(A, \mathbf{b}) = \left\{ \begin{pmatrix} 2 - t \\ t \\ 5 + t \end{pmatrix} \mid t \in \mathbb{R} \right\} = \begin{pmatrix} 2 \\ 0 \\ 5 \end{pmatrix} + \text{Span}\left(\begin{pmatrix} -1 \\ 1 \\ 1 \end{pmatrix} \right). \tag{17.12}$$

The above span is the solution space of the associated homogeneous system.

(2b) $a = -1, b \neq -12$: Inserting these values into Eq. (17.11) leads to a contradiction so there is no solution, $\text{Sol}(A, \mathbf{b}) = \{\}$.

17.2.5 Row operations for linear equations

A good strategy for solving systems of linear equations is the successive elimination of variables by adding multiples of equations. So far we have carried this out in a somewhat ad-hoc fashion which is efficient and works well for low dimensions. But for larger systems it is worth developing an algorithm and this is where row reduction comes into play.

As before, we start with a linear system $A\mathbf{x} = \mathbf{b}$ with n variables and m equations, so $\mathbf{x} \in \mathbb{F}^n$, $\mathbf{b} \in \mathbb{F}^m$ and $A \in \mathcal{M}_{m,n}(\mathbb{F})$. Suppose we multiply both sides of the linear system with an invertible $m \times m$ matrix $P \in \text{GL}(\mathbb{F}^m)$, so we arrive at the new linear system $PA\mathbf{x} = P\mathbf{b}$. This operation can be undone by multiplying with the inverse, P^{-1}, so the solution spaces of the two linear systems must be equal,

$$\text{Sol}(PA, P\mathbf{b}) = \text{Sol}(A, \mathbf{b}). \tag{17.13}$$

Now recall that elementary row operations can be generated by multiplying with the elementary matrices in Eq. (16.5) and that these matrices are invertible. If we take P to be one of these matrices then the operation

$$A\mathbf{x} = \mathbf{b} \mapsto PA\mathbf{x} = P\mathbf{b} \tag{17.14}$$

corresponds to a row operation simultaneously carried out on A and $\mathbf{b}$ and, from Eq. (17.13), this does not change the solutions space of the linear system. Note that such row operation on a linear system are really just a formal restatement of the steps involved in adding up multiples of equations. Row operations of type (R1) in Def. 16.1

exchange two columns of A (as well as the two corresponding entries of $\mathbf{b}$) and simply amount to exchanging two equations. Row operations of type (R2) correspond to adding a multiple of one equation to another and row operations of type (R3) multiple an equation with a non-zero number. The idea is now to simplify the linear system by a suitable sequence of such row operations until the solution set can be easily read off.

This is facilitated by introducing the *augmented matrix*

$$A' = (A|\mathbf{b}) \,, \tag{17.15}$$

an $m \times (n+1)$ matrix which consists of the matrix A and one additional, final column given by the vector $\mathbf{b}$. The augmented matrix is really just an efficient way to collect the data which determines the linear system into a single matrix. Row operations (17.14) on the linear system then translate into row operations $A' \mapsto PA'$ on the augmented matrix.

Before we formulate an explicitly algorithm, we note a useful criterion in terms of the augmented matrix which helps us to decide whether or not the linear system has solutions.

Proposition 17.1 $\mathbf{b} \in \mathrm{Im}(A) \iff \mathrm{rk}(A) = \mathrm{rk}(A')$

Proof '$\Rightarrow$': If $\mathbf{b} \in \mathrm{Im}(A)$ it is a linear combination of the column vectors of A and adding it to the matrix does not increase the rank.
'$\Leftarrow$': If $\mathrm{rk}(A) = \mathrm{rk}(A')$ the rank does not increase when $\mathbf{b}$ is added to the matrix. Therefore, $\mathbf{b} \in \mathrm{Span}(\mathbf{A}^1, \dots, \mathbf{A}^n) = \mathrm{Im}(A)$. $\square$

This means a solution exists iff the matrix A and the augmented matrix A' have the same rank!

17.2.6   Algorithm for solving linear equations

We are now ready to describe the solution algorithm for linear systems.

Algorithm *(Solving systems of linear equation by row reduction)*

(1) First we perform row operations on the augmented matrix A' in order to bring A into upper echelon form. This works exactly as described in Section 16.1.3 and converts the augmented matrix into the following form

$$A' \mapsto \begin{pmatrix} \cdots & a_{1j_1} & & & & * & \bigg| & b'_1 \\ & & a_{2j_2} & & \cdot\cdot & & \bigg| & \vdots \\ & & & \ddots & & & \bigg| & \vdots \\ & & & & a_{rj_r} & \cdots & \bigg| & b'_r \\ & 0 & & & & & \bigg| & b'_{r+1} \\ & & & & & & \bigg| & \vdots \\ & & & & & & \bigg| & b'_m \end{pmatrix} \,,$$

where the pivots a_{ij_i} are non-zero for $i = 1, \dots r$ and the star indicates arbitrary entries. Recall that the number of non-zero rows $r = \mathrm{rk}(A)$ equals the rank of A. If any of the entries b'_i with $i > r$ is non-zero, then $\mathrm{rk}(A') > \mathrm{rk}(A)$ and, from Prop. 17.1, we conclude that there is no solution. This can also be seen much more directly by converting the above matrix back into a system of linear equations. Any of the rows $i > r$ correspond to an equation $0 = b'_i$ which is a contradiction unless $b'_i = 0$. We only need to carry on if $b'_i = 0$ for all $i > r$, so that $\mathrm{rk}(A') = \mathrm{rk}(A)$ and the linear system has a solution.

(2) To proceed, we assume that $b'_i = 0$ for all $i > r$. For ease of notation we also permute the columns of A (this corresponds to a permutation of the variables that we will have to keep track of) so that the columns with pivots become the first r of the matrix. The resulting matrix has the following structure.

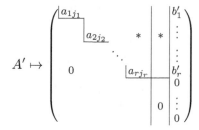

(3) By further row operations we can convert the $r \times r$ matrix in the upper left corner of the previous matrix into a unit matrix $\mathbb{1}_r$ using the same steps we have used in the algorithm to calculate the inverse of a matrix.. Schematically, the result is

$$A'_{\mathrm{fin}} = \left(\begin{array}{c|c|c} \mathbb{1}_r & B & \mathbf{c} \\ \hline 0 & 0 & 0 \end{array} \right) \tag{17.16}$$

where B is an $r \times (n-r)$ matrix and $\mathbf{c}$ is an r-dimensional vector.

(4) We are now ready to convert the augmented matrix A'_{fin} back into a system of linear equations. To do this it is useful to split the vector $\mathbf{x}$ which contains the n variables x_i up into an r dimensional vector $\boldsymbol{\xi}$ and an $(n-r)$-dimensional vector $\mathbf{t}$, in accordance with the structure of the matrix A'_{fin}. Hence, writing

$$\mathbf{x} = \begin{pmatrix} \boldsymbol{\xi} \\ \mathbf{t} \end{pmatrix} \tag{17.17}$$

the linear system associated to A'_{fin} takes the simple form

$$\boldsymbol{\xi} + B\mathbf{t} = \mathbf{c} . \tag{17.18}$$

Note that this is a system of r linear equations in n variables. It follows that every linear system of m equations in n variables can be reduced to an equivalent system with $r = \mathrm{rk}(A) \leq m$ equations in n variables. More importantly, the system (17.18) can be easily solved for $\boldsymbol{\xi}$ in terms of $\mathbf{t}$, giving $\boldsymbol{\xi} = \mathbf{c} - B\mathbf{t}$, and this was really the point of the exercise. Hence, the solution space is

$$\mathrm{Sol}(A, \mathbf{b}) = \left\{ \begin{pmatrix} \mathbf{c} - B\mathbf{t} \\ \mathbf{t} \end{pmatrix} \mid \mathbf{t} \in \mathbb{F}^{n-r} \right\} = \begin{pmatrix} \mathbf{c} \\ \mathbf{0} \end{pmatrix} + \mathrm{Ker}(A) \tag{17.19}$$

$$\mathrm{Ker}(A) = \mathrm{Span}\left(\begin{pmatrix} -B^1 \\ \mathbf{e}_1 \end{pmatrix}, \ldots, \begin{pmatrix} -B^{n-r} \\ \mathbf{e}_{n-r} \end{pmatrix} \right) \tag{17.20}$$

where $\mathbf{B}^i$ are the columns of the matrix B and $\mathbf{e}_i$ are the standard unit vectors in $\mathbb{F}^{n-r}$. Note that the solution (17.19) does indeed represent an affine $k = n - r$ plane of the general form (17.9).

Problem 17.4 *Solving a linear system with row reduction*

Use row reduction to solve the following linear system

$$
\begin{aligned}
x_1 + x_2 - 2x_3 &= 1 \\
2x_1 - x_2 + 3x_3 &= 0 \\
-x_1 - 4x_2 + 9x_3 &= b
\end{aligned}
$$

in $\mathbb{R}^3$, for all values of the parameter $b \in \mathbb{R}$.

Solution: The augmented matrix for the above system reads

$$
A' = \left(\begin{array}{ccc|c}
1 & 1 & -2 & 1 \\
2 & -1 & 3 & 0 \\
-1 & -4 & 9 & b
\end{array} \right) ,
$$

We proceed in the four steps outlined above.

(1) First we perform row operations on A' to bring A into upper echelon form.

$$
A' \mapsto \left(\begin{array}{ccc|c}
1 & 1 & -2 & 1 \\
0 & -3 & 7 & -2 \\
0 & 0 & 0 & b+3
\end{array} \right)
$$

We conclude that the rank of A is $r = \mathrm{rk}(A) = 2$. If $b \neq -3$ we have $\mathrm{rk}(A') = 3 > 2 = \mathrm{rk}(A)$ so in this case there is no solutions.

(2) Setting $b = -3$ we have the matrix

$$
A' \mapsto \left(\begin{array}{ccc|c}
1 & 1 & -2 & 1 \\
0 & -3 & 7 & -2 \\
0 & 0 & 0 & 0
\end{array} \right) .
$$

As it happens, we do not have to permute columns since the pivots are contained in the first two columns.

(3) By further elementary row operations we convert the 2×2 matrix in the upper left corner into a unit matrix.

$$
A'_{\mathrm{fin}} = \left(\begin{array}{ccc|c}
1 & 0 & \frac{1}{3} & \frac{1}{3} \\
0 & 1 & -\frac{7}{3} & \frac{2}{3} \\
0 & 0 & 0 & 0
\end{array} \right)
$$

(4) We split $\mathbf{x}$ up into an $r = 2$ dimensional vector $\boldsymbol{\xi}$ with components x_1, x_2 and an $n - r = 3 - 2 = 1$ dimensional vector $\mathbf{t}$ with a single component t as

$$
\mathbf{x} = \left(\begin{array}{c}
x_1 \\
x_2 \\
t
\end{array} \right) .
$$

Converting A'_{fin} into a linear system in those variables results in

$$x_1 + \frac{1}{3}t = \frac{1}{3}, \qquad x_2 - \frac{7}{3}t = \frac{2}{3}.$$

This can be easily solved for x_1, x_2 in terms of t which was the point of the exercise. This leads to the solution space

$$\mathrm{Sol}(A, \mathbf{b}) = \left\{ \begin{pmatrix} (1-t)/3 \\ (2+7t)/3 \\ t \end{pmatrix} \mid t \in \mathbb{R} \right\} = \begin{pmatrix} 1/3 \\ 2/3 \\ 1 \end{pmatrix} + \mathrm{Span}\left(\begin{pmatrix} -1/3 \\ 7/3 \\ 1 \end{pmatrix} \right).$$

Hence, the solution is an affine line.

Application 17.1 *Linear algebra and circuits*

Electrical circuits with batteries and resistors, such as the circuit below, can be described using methods from linear algebra.

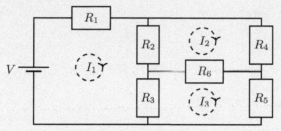

To do this, first assume that the circuit contains n loops and assign ('mesh') currents I_i, where $i = 1, \ldots, n$, to each loop. Then, applying Ohm's law and Kirchhoff's voltage low ('The voltages along a closed loop must sum to zero') to each loop leads to the linear system

$$
\begin{aligned}
R_{11}I_1 + \cdots + R_{1n}I_n &= V_1 \\
\vdots \qquad\qquad \vdots \ \ \vdots \\
R_{n1}I_i + \cdots + R_{nn}I_n &= V_n,
\end{aligned}
\tag{17.21}
$$

where R_{ij} describe the various resistors and V_i correspond to the voltages of the batteries. If we introduce the $n \times n$ matrix R with entries R_{ij}, the current vector $\mathbf{I} = (I_1, \ldots, I_n)^T$ and the voltage vector $\mathbf{V} = (V_1, \ldots, V_n)^T$ this system can also be written in the form of a generalized Ohm's law as

$$R\mathbf{I} = \mathbf{V}. \tag{17.22}$$

This is an $n \times n$ linear system, where we think of the resistors and battery voltages as given, while the currents $I_1, \ldots, I_n$ are a priori unknown and can be determined by solving the system. Of course any of the methods previously discussed can be used to solve this linear system and determine the currents I_i.

For example, consider the above circuit. To its three loops we assign the currents I_1, I_2, I_3 as indicated in the figure. Kirchhoff's voltage law applied to the three loops then leads to

$$
\begin{aligned}
R_1I_1 + R_2(I_1 - I_2) + R_3(I_1 - I_3) &= V \\
R_2(I_2 - I_1) + R_4I_2 + R_6(I_2 - I_3) &= 0 \\
R_3(I_3 - I_1) + R_6(I_3 - I_2) + R_5I_3 &= 0
\end{aligned}
\iff
\begin{aligned}
(R_1 + R_2 + R_3)I_1 - R_2I_2 - R_3I_3 &= V \\
-R_2I_1 + (R_2 + R_4 + R_6)I_2 - R_6I_3 &= 0 \\
-R_3I_1 - R_6I_2 + (R_3 + R_5 + R_6)I_3 &= 0.
\end{aligned}
$$

With the current and voltage vectors $\mathbf{I} = (I_1, I_2, I_3)^T$ and $\mathbf{V} = (V, 0, 0)^T$ the matrix R in Eq. (17.22) is then given by

$$R = \begin{pmatrix} R_1 + R_2 + R_3 & -R_2 & -R_3 \\ -R_2 & R_2 + R_4 + R_6 & -R_6 \\ -R_3 & -R_6 & R_3 + R_5 + R_6 \end{pmatrix}.$$

For example, for resistors $(R_1, \dots, R_6) = (3, 10, 4, 2, 5, 1)$ (in units of Ohm) we have the resistance matrix

$$R = \begin{pmatrix} 17 & -10 & -4 \\ -10 & 13 & -1 \\ -4 & -1 & 10 \end{pmatrix}.$$

For a battery voltage $V = 12$ (in units of volt) we can write down the augmented matrix

$$R' = \left(\begin{array}{ccc|c} 17 & -10 & -4 & 12 \\ -10 & 13 & -1 & 0 \\ -4 & -1 & 10 & 0 \end{array} \right),$$

and solve the linear system by row reduction. This leads to the solution

$$\mathbf{I} = \frac{1}{905} \begin{pmatrix} 1548 \\ 1248 \\ 744 \end{pmatrix}$$

for the currents (in units of Ampere).

17.3   Applications to geometry

Summary 17.3 *Solution sets to systems of linear equations are affine k-planes and, conversely, every affine k-plane can be described as a solution to a system of linear equations. An affine k_1-plane and an affine k_2-plane in $\mathbb{F}^n$ either have zero intersection or they intersect in an affine k-plane. For generic affine planes the dimension of their intersection is $k = k_1 + k_2 - n$.*

In Chapter 11, we have looked at lines and planes in $\mathbb{R}^2$ and $\mathbb{R}^3$, their parametric and Cartesian forms and some of their geometry properties, such as intersections. Perhaps somewhat frivolously, we have earlier used methods such as the dot and cross product to do this. We will now study the generalization to arbitrary dimensions, so to affine k-planes in $\mathbb{F}^n$, using only the results for systems of linear equations.

17.3.1   Parametric and Cartesian form

We have seen that the non-empty solution sets to systems of linear equations are, in fact, affine k-planes. An obvious question is if, conversely, every affine k-plane can be obtained as the solution set of a system of linear equations. The following theorem shows that the answer is 'yes' and that this amounts to what we have called the Cartesian form.

Theorem 17.3 *An affine k-plane $P = \mathbf{p} + W \subset \mathbb{F}^n$, where $\mathbf{p} \in \mathbb{F}^n$ and W is a k-dimensional vector subspace of $\mathbb{F}^n$, can be described as follows.*

(i) $P = \left\{ \mathbf{p} + \sum_{i=1}^{k} t_i \mathbf{w}_i \mid t_i \in \mathbb{F} \right\}$ *(parametric form)*
      where $(\mathbf{w}_1, \ldots, \mathbf{w}_k)$ *is a basis of* W
(ii) $P = \{ \mathbf{x} \in \mathbb{F}^n \mid N\mathbf{x} = N\mathbf{p} \} = \text{Sol}(N, N\mathbf{p})$ *(Cartesian form)*
      where N *is a* $(n-k) \times n$ *matrix with* $\text{Ker}(N) = W$.

Proof The existence of the parametric form is immediate — all we need to do is choose a basis $(\mathbf{w}_1, \ldots, \mathbf{w}_k)$ of W.

The Cartesian form is a little less obvious. Let us first assume that a matrix N with the stated properties exists. Then the system of $n-k$ linear equations in n variables defined by $N\mathbf{x} = N\mathbf{p}$ is solved by $\mathbf{x} = \mathbf{p}$ and the solution of the associated homogeneous system, $N\mathbf{x} = \mathbf{0}$, is $\text{Ker}(N) = W$. Hence, $\text{Sol}(N, N\mathbf{p}) = \mathbf{p} + W = P$, as required.

We still need to show that a matrix N with the stated properties exists. A basis $(\mathbf{w}_1, \ldots \mathbf{w}_k)$ of W can be completed to a basis $(\mathbf{w}_1, \ldots \mathbf{w}_k, \mathbf{w}_{k+1}, \ldots, \mathbf{w}_n)$ of $\mathbb{F}^n$. Define a linear map $N : \mathbb{F}^n \to \mathbb{F}^{n-k}$ by

$$N\mathbf{w}_i = \begin{cases} \mathbf{0} & \text{for } i = 1, \ldots, k \\ \mathbf{e}_{i-k} & \text{for } i = k+1, \ldots, n \end{cases}$$

We know from Theorem (12.1) that such a map does indeed exist. We clearly have $W \subset \text{Ker}(N)$ and since $\text{Im}(N) = \mathbb{F}^{n-k}$ it follows that $\text{rk}(N) = n - k$. Hence, the rank formula (14.13) implies that $\dim_{\mathbb{F}}(\text{Ker}(N)) = n - (n-k) = k$. As a result $\text{Ker}(N)$ has the same dimension, k, as W and since the latter is included in the former, they must be equal.   □

The parametric form of an affine k-plane requires k parameters. For this reason it is usually more practical for small k. For example, the parametric form of an affine line is $L = \{ \mathbf{p} + t\mathbf{w} \mid t \in \mathbb{F} \} \subset \mathbb{F}^n$. On the other hand, the Cartesian form of an affine k-plane amounts to a linear system with $n - k$ equations (since N is an $(n-k) \times n$ matrix). Therefore, affine k-planes with a large dimensions k or, equivalently, a small co-dimensions $n - k$, are easier described in Cartesian form. For example, an affine hyperplane ($k = n - 1$) is described by a single equation, $N\mathbf{x} = N\mathbf{p}$, where N is a $1 \times n$ matrix, that is, a row vector.

17.3.2   Intersection of affine k-planes

The intersection of two affine k-planes can be easily described in Cartesian form. All we need to do is combine the linear equations of each Cartesian form into a single linear system. Specifically, consider the two affine k_i-planes

$$P_i = \mathbf{p}_i + W_i = \{ \mathbf{x} \in \mathbb{F}^n \mid N_i \mathbf{x} = N_i \mathbf{p}_i \} \quad \text{where} \quad i = 1, 2 \,,$$

and N_i are matrices of size $(n - k_i) \times n$ with $\text{Ker}(N_i) = W_i$. Then the intersection

$$P_1 \cap P_2 = \{ \mathbf{x} \in \mathbb{F}^n \mid N\mathbf{x} = \mathbf{p} \} \quad \text{where} \quad N = \begin{pmatrix} N_1 \\ N_2 \end{pmatrix}, \quad \mathbf{p} = \begin{pmatrix} N_1 \mathbf{p}_1 \\ N_2 \mathbf{p}_2 \end{pmatrix},$$

equals the solution set of the linear system $N\mathbf{x} = \mathbf{p}$ of $2n - k_1 - k_2$ equations in n variables, obtained by combining the linear systems for P_1 and P_2. Of course, if

$\mathbf{p} \notin \text{Im}(N)$ this intersection is empty. Otherwise, Theorem 17.2 tells us that it is an affine k-plane, where $k = n - \text{rk}(N)$. To summarize, we have found that P_1 and P_2 either do not intersect or intersect in another affine k-plane.

Can we say more about the dimension, k, of the intersection? We know that $\text{rk}(N_1) = n - k_1$ and $\text{rk}(N_2) = n - k_2$ so it is certainly clear that

$$\text{rk}(N) \geq \max(n - k_1, n - k_2) \,. \tag{17.23}$$

(The number of linearly independent row vectors cannot decrease when we combine N_1 and N_2.) On the other hand, given that N is a matrix of size $(2n - k_1 - k_2, n)$, Eq. (16.3) implies that

$$\text{rk}(N) \leq \min(2n - k_1 - k_2, n) \,. \tag{17.24}$$

Combining Eqs. (17.23) and (17.24) with $\text{rk}(N) = n - k$ gives

$$\max(k_1 + k_2 - n, 0) \leq k \leq \min(k_1, k_2) \,, \tag{17.25}$$

and this is the desired constraint on the dimension of the intersection. We summarize these results in the following theorem.

Theorem 17.4 *Let* $P_i = \{\mathbf{x} \in \mathbb{F}^n \,|\, N_i \mathbf{x} = N_i \mathbf{p}_i\}$ *be affine k_i-planes, where $i = 1, 2$. Then the intersection $P_1 \cap P_2$ is either empty or it is the affine k-plane which is the solution of the combined system of linear equations, $N\mathbf{x} = \mathbf{p}$, where*

$$N = \begin{pmatrix} N_1 \\ N_2 \end{pmatrix} \,, \quad \mathbf{p} = \begin{pmatrix} N_1 \mathbf{p}_1 \\ N_2 \mathbf{p}_2 \end{pmatrix} \,.$$

The dimension of the intersection is $k = n - \text{rk}(N)$ and this dimension is constrained by $\max(k_1 + k_2 - n, 0) \leq k \leq \min(k_1, k_2)$.

The dimension of the intersection $P_1 \cap P_2$ can vary, depending on the specific affine planes, as indicated in the theorem, but there is a 'generic' value which holds for 'most" affine planes of given dimensions k_1 and k_2. This arises when the matrix N has its generic (that is maximal) rank which, from Eq. (17.24), is $\text{rk}(N) = \min(2n - k_1 - k_2, n)$. Hence there are two possibilities for the value of $\text{rk}(N)$. If $n < 2n - k_1 - k_2$ or, equivalently, $k_1 + k_2 - n < 0$ then N is not surjective, so the linear system $N\mathbf{x} = \mathbf{p}$ has no solution for generic $\mathbf{p}$. Hence, this case cannot correspond to the generic situation. In the opposite case, when $\text{rk}(N) = 2n - k_1 - k_2$, the map N is surjective so that a solution, an affine k-plane with $k = \dim_{\mathbb{F}}(\text{Ker}(N)) = n - (2n - k_1 - k_2) = k_1 + k_2 - n$, always exists. In summary, the generic dimension of the intersection is

$$k_{\text{gen}} = k_1 + k_2 - n \,, \tag{17.26}$$

and if this number is negative the intersection is generically empty.

Problem 17.5 *(Intersection of affine k-planes)*

Find the possible and generic dimensions of (i) the intersection of two affine planes in $\mathbb{R}^3$ and (ii) the intersection of an affine plane and an affine hyperplane in $\mathbb{R}^4$.

Solution: (i) The answer in this case is clear from geometrical intuition but let us follow Theorem 17.4 and Eq. (17.26) to convince ourselves that these statements make sense. For two planes in $\mathbb{R}^3$ we have $k_1 = k_2 = 2$ and $n = 3$. From Theorem (17.26) the intersection is either empty or an affine k-plane with $1 \le k \le 2$. The generic intersection dimension is $k_{\text{gen}} = 2 + 2 - 3 = 1$. Of course all this fits with geometrical intuition. Two 'typical' planes in $\mathbb{R}^3$ intersect in a line ($k = 1$) but, if they are identical, they can intersect in a plane ($k = 2$) or not intersect at all, if they are parallel.

(ii) Geometrical intuition is not particularly helpful in this four-dimensional case. We have $k_1 = 2$ for a plane, $k_2 = 3$ for a hyperplane and $n = 4$. If the intersection in not empty then, from Theorem 17.4, it is an affine k-plane with $1 \le k \le 3$ and, generically, $k_{\text{gen}} = 2 + 3 - 4 = 1$ from Eq. (17.26). So a plane and a hyperplane in $\mathbb{R}^4$ generically intersect in a line.

17.3.3 Intersections and linear systems

The geometry of affine k-planes leads to a geometrical interpretation of systems of linear equations. Consider such a system, $A\mathbf{x} = \mathbf{b}$ of m equation in n variables. We can break this up into m equations as

$$A\mathbf{x} = \mathbf{b} \quad \leftrightarrow \quad \begin{cases} \mathbf{A}_1 \cdot \mathbf{x} = b_1 \\ \vdots \quad \vdots \quad \vdots \\ \mathbf{A}_m \cdot \mathbf{x} = b_m \end{cases}.$$

If any of the row vectors $\mathbf{A}_i$ of A is zero and $b_i = 0$ then this equation is trivial and can be dropped. On the other hand, if $\mathbf{A}_i = \mathbf{0}$ and $b_i \ne 0$ for any i then the system of linear equations does not have a solution. If we discard these two cases and assume that $\mathbf{A}_i \ne \mathbf{0}$ for all $i = 1, \ldots, m$ then each equation on the right defines an affine hyperplane in $\mathbb{F}^n$. This means the solution set of a system of m linear equations in n variables is the intersection of m affine hyperplanes in $\mathbb{F}^n$. This provides us with a geometrical way to think about the solutions to a system of linear equations.

Problem 17.6 *Solutions to linear systems as hyperplane intersection*

Using geometrical arguments, discuss the various qualitative possibilities for the solutions of a system of linear equations with (i) two equations and two variables, (ii) two equations and three variables, (iii) three equations and three variables.

Solution: (i) A system of two equations in two variables corresponds to two hyperplanes, that is lines, in $\mathbb{F}^2$. Such two lines can intersection in a point (generic case), a line or have an empty intersection. This corresponds to the system of linear equations having a unique solution, a solution line or no solution at all (see Fig. 11.2).

(ii) A system of two equations in three variables corresponds to two hyperplanes, that is two planes in $\mathbb{F}^3$. Two such planes can intersect in a line (generic case), in a plane or have an empty intersection. Correspondingly, the system of linear equations can have a solution line, a solution plane or have no solution at all (see Fig. 11.5).

(iii) A system of three equations in three variables corresponds to three planes in $\mathbb{F}^3$. From

Fig. 11.7 they can intersect in a point (generic case), a line, a plane or have an empty inter-section. This corresponds to the possible solution structure of the system of linear equations.

Exercises

17.1 For functions $y \in C^2(\mathbb{R})$, consider the linear inhomogeneous differential equation

$$\frac{d^2 y}{dx^2} + 4y = 8x^2 \ .$$

(a) Find a solution to this differential equation. (Hint: Try a quadratic polynomial.)
(b) Find the general solution to the associated homogeneous differential equation.
(c) Write down the general solution to the inhomogeneous equation.

17.2 Consider the linear system of equations

$$(2 - \lambda)x + y + 2z = 0$$
$$x + (4 - \lambda)y - z = 0$$
$$2x - y + (2 - \lambda)z = 0 \ ,$$

where $x, y, z \in \mathbb{R}$ are the variables and $\lambda \in \mathbb{R}$ is a parameter.
(a) What is the rank of the associated coefficient matrix, A, depending on the value of λ?
(b) Based on the result in (a) what is the expectation for the qualitative structure of the solution.
(c) Confirm this expectation by an explicit calculation.

17.3 The linear system

$$\begin{aligned} x + y + z &= 1 \quad (E_1) \\ x + 2y + 4z &= \eta \quad (E_2) \\ x + 4y + 10z &= \eta^2 \quad (E_3) \end{aligned}$$

with variables $x, y, z \in \mathbb{R}$ depends on the real parameter η.
(a) Show that the rank of the coefficient matrix is two. What does this imply for the qualitative structure of the solution?
(b) Explicitly solve the system for the cases where a solution exists.

17.4 Solve the linear system

$$\begin{aligned} x + 2y + 3z &= 2 \\ \alpha y + z &= \beta \\ 2x + 2y &= 1 \ , \end{aligned}$$

with variables $x, y, z \in \mathbb{R}$ for all values of the parameters $\alpha, \beta \in \mathbb{R}$.

17.5 Solve the linear system

$$\begin{aligned} 3x + 2y - z &= 10 \\ 5x - y - 4z &= 17 \\ x + 5y + \alpha z &= \beta \ , \end{aligned}$$

with variables $x, y, z \in \mathbb{R}$ for all values of the parameters $\alpha, \beta \in \mathbb{R}$, using row reduction.

17.6 Solve the linear system

$$\begin{aligned} x_1 + \tfrac{1}{2}x_2 - 3x_3 &= 1 \\ 2x_2 - 4x_3 + \tfrac{1}{3}x_4 &= 2 \\ x_1 + \tfrac{1}{4}x_2 - x_3 &= 0 \\ x_1 - \tfrac{5}{4}x_2 - x_3 + ax_4 &= b \end{aligned}$$

with variables $x_1, x_2, x_3, x_4 \in \mathbb{R}$ for all values of the parameters $a, b \in \mathbb{R}$, using row reduction.

17.7 Following Application 17.1, write down the linear system for the circuit below.

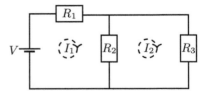

Solve this system for the currents I_1, I_2 for general values of the resistors R_1, R_2, R_3, and the voltage V.

17.8 What are the possible and generic dimensions for the intersections of the following affine k-planes.

(a) Two hyperplanes in four dimensions.
(b) Three hyperplanes in four dimensions.
(c) A plane and a hyperplane in five dimensions.
(d) Two hyperplanes in five dimensions.

17.9 (a) What are the possible and generic intersection dimensions of k hyperplanes in n dimensions.
(b) Use the result from (a) to predict the possible dimensions of the solution space for a linear system with k equations and n variables.

18
Determinants

In Section 10.2 we have introduced the three-dimensional determinant as the triple product of three vectors. We have observed that it is multi-linear and anti-symmetric and we will now use these properties to define the determinant axiomatically and in all dimensions. The determinant is quite a complicated object, compared to linear maps which are linear in only a single argument. From this point of view, it is appropriate for the determinant to appear quite late in a build-up of linear algebra. On the other hand, it is a very useful tool and it plays a crucial role in the computation of eigenvalues, as we will see in Part VI.

In the next section we show that multi-linearity and anti-symmetry, together with a normalization condition, fixes the determinant uniquely. More advanced properties of the determinant, such as its behaviour under matrix transposition and matrix multiplication, will be covered in Section 18.2. A crucial conclusion from these results is that the determinant is invariant under basis transformations (15.13) and this allows introducing the determinant of a linear map.

In Section 18.3 we study various computational aspect of determinants. To facilitate calculating determinants we introduce Laplace's rule as well as a method based on elementary row operations. The determinant can also be used to compute the inverse of a matrix and to solve certain systems of linear equations.

18.1 Existence and uniqueness

Summary 18.1 *The determinant in n dimensions maps n vectors in $\mathbb{F}^n$ to a number. It is multi-linear, anti-symmetric, and evaluates to one on the standard unit vectors. These properties define the determinant uniquely. The explicit formula for the n-dimensional determinant involves a sum over the $n!$ permutations in the permutation group S_n. Alternatively, this formula can also be written in terms of the generalized Levi-Civita symbol in n dimensions.*

18.1.1 Definition of determinant

In Prop. 10.3 we have shown that the three-dimensional determinant is linear in each argument, that it changes sign under the exchange of any two arguments and that it evaluates to one on the basis of standard unit vectors. Now we revert the logic and use these properties to define the determinant axiomatically.

Definition 18.1 *A map* $\det : \mathbb{F}^n \times \cdots \times \mathbb{F}^n \to \mathbb{F}$ *with* $(\mathbf{a}_1, \ldots, \mathbf{a}_n) \mapsto \det(\mathbf{a}_1, \ldots, \mathbf{a}_n)$ *is called a determinant if it satisfies the following properties for all* $\mathbf{a}, \mathbf{b} \in \mathbb{F}^n$ *and all* $\alpha, \beta \in \mathbb{F}$.

(D1)	$\det(\cdots, \alpha \mathbf{a} + \beta \mathbf{b}, \cdots) = \alpha \det(\cdots, \mathbf{a}, \cdots)$	*(multi-linearity)*
	$\qquad\qquad\qquad\qquad + \beta \det(\cdots, \mathbf{b}, \cdots)$	
(D2)	$\det(\cdots, \mathbf{a}, \cdots, \mathbf{b}, \cdots) = -\det(\cdots, \mathbf{b}, \cdots, \mathbf{a}, \cdots)$	*(anti-symmetry)*
(D3)	$\det(\mathbf{e}_1, \ldots, \mathbf{e}_n) = 1$	*(normalization)*

The dots in (D1) and (D2) stand for arguments which remain unchanged. The determinant of an $n \times n$ *matrix* $A \in \mathcal{M}_{n,n}(\mathbb{F})$ *is defined as the determinant of its column vectors, so* $\det(A) := \det(\mathbf{A}^1, \ldots, \mathbf{A}^n)$.

An easy but important conclusion from (D2) is that a determinant with two same arguments must vanish, so

$$\det(\cdots, \mathbf{a}, \cdots, \mathbf{a}, \cdots) = 0 \, . \tag{18.1}$$

We know that an object with the above properties exists for $n = 3$ but not yet in other dimensions.

To address this problem we need to use some facts about permutations from Section 3.2. Recall that the group S_n of permutations of n objects consists of bijective maps $\{1, \ldots, n\} \to \{1, \ldots, n\}$. A permutation is called a transposition if it only swaps two integers and leaves all others unchanged. We have seen that any permutation $\sigma \in S_n$ can be written as a composition $\sigma = \tau_i \circ \cdots \circ \tau_k$ of transpositions τ_i and that the sign of the permutation is given by $\mathrm{sgn}(\sigma) = (-1)^k$. Permutations with $\mathrm{sgn}(\sigma) = 1$ ($\mathrm{sgn}(\sigma) = -1$) are called even (odd).

The reason permutations are relevant for the discussion of determinants is the anti-symmetry property (D2) in Def. 18.1. We can ask how the value of the determinant $\det(\mathbf{a}_1, \mathbf{a}_2, \ldots, \mathbf{a}_n)$ changes when we permute the arguments by a permutation σ, so if we consider $\det(\mathbf{a}_{\sigma(1)}, \mathbf{a}_{\sigma(2)}, \ldots, \mathbf{a}_{\sigma(n)})$. If σ is a transposition, then, from (D2), the two determinants are related by a factor of -1. Hence, for an arbitrary permutation $\sigma \in S_n$, written as a product of transpositions, $\sigma = \tau_1 \circ \cdots \circ \tau_k$, the determinant changes by $(-1)^k = \mathrm{sgn}(\sigma)$, so that

$$\det(\mathbf{a}_{\sigma(1)}, \mathbf{a}_{\sigma(2)}, \ldots, \mathbf{a}_{\sigma(n)}) = \mathrm{sgn}(\sigma) \det(\mathbf{a}_1, \mathbf{a}_2, \ldots, \mathbf{a}_n) \, , \tag{18.2}$$

for all $\sigma \in S_n$.

18.1.2 The general formula for the determinant

To show uniqueness, we start with an $n \times n$ matrix $A \in \mathcal{M}_{n,n}(\mathbb{F})$ whose column vectors we expand as

$$\mathbf{A}^i = \sum_j A_{ji} \mathbf{e}_j \, , \tag{18.3}$$

in terms of standard unit vectors. The remainder of the argument really just involves inserting this expansion into the determinant and applying the axioms (D1), (D2), and (D3). But the proliferation of indices in multi-linear objects takes getting used to.

$$\det(A) \overset{(18.3)}{=} \det\left(\sum_{j_1=1}^{n} A_{j_1 1}\mathbf{e}_{j_1}, \cdots, \sum_{j_n=1}^{n} A_{j_n n}\mathbf{e}_{j_n}\right)$$

$$\overset{(D1)}{=} \sum_{j_1,\cdots,j_n} A_{j_1 1}\cdots A_{j_n n} \det(\mathbf{e}_{j_1}, \cdots, \mathbf{e}_{j_n})$$

$$\overset{(18.1)}{=} \sum_{\sigma \in S_n} A_{\sigma(1)1}\cdots A_{\sigma(n)n} \det(\mathbf{e}_{\sigma(1)}, \cdots, \mathbf{e}_{\sigma(n)})$$

$$\overset{(18.2)}{=} \sum_{\sigma \in S_n} \mathrm{sgn}(\sigma) A_{\sigma(1)1}\cdots A_{\sigma(n)n} \det(\mathbf{e}_1, \cdots, \mathbf{e}_n)$$

$$\overset{(D3)}{=} \sum_{\sigma \in S_n} \mathrm{sgn}(\sigma) A_{\sigma(1)1}\cdots A_{\sigma(n)n}$$

In the second line, only terms with all indices j_i different contribute to the sum due to Eq. (18.1). This means that the sum over $j_1,\ldots,j_n$ effectively runs over all permutations of $\{1,\ldots,n\}$ and can be replaced by a sum over all $\sigma \in S_n$. This can be done by setting $j_i = \sigma(i)$, as has been done in the third line.

Theorem 18.1 *The determinant defined in Def. 18.1 exists, it is unique and given by the expression*

$$\det(A) = \det(\mathbf{A}^1, \cdots, \mathbf{A}^n) = \sum_{\sigma \in S_n} \mathrm{sgn}(\sigma) A_{\sigma(1)1}\cdots A_{\sigma(n)n} . \qquad (18.4)$$

Proof Uniqueness is shown by the above calculation. To verify existence we have to show that the formula in Eq. (18.4) does indeed satisfy the three conditions in Def. 18.1.

(D1) This is apparent since the right-hand side of Eq. (18.4) depends linearly on the entries A_{ij} of each column j.

(D2) We verify the equivalent statement (18.2):

$$\det(\mathbf{A}^{\sigma(1)}, \ldots, \mathbf{A}^{\sigma(n)}) = \sum_{\rho \in S_n} \mathrm{sgn}(\rho) A_{\sigma(1)\rho(1)} \cdots A_{\sigma(n)\rho(n)}$$

$$= \sum_{\rho \in S_n} \mathrm{sgn}(\rho) A_{\sigma(\rho^{-1}(1))1} \cdots A_{\sigma(\rho^{-1}(n))n}$$

$$\overset{\tau=\sigma\circ\rho^{-1}}{=} \sum_{\tau \in S_n} \mathrm{sgn}(\tau^{-1}\circ\sigma) A_{\tau(1)1} \cdots A_{\tau(n)n}$$

$$\overset{\text{Thm. 3.1}}{=} \mathrm{sgn}(\sigma) \det(\mathbf{A}^1, \ldots, \mathbf{A}^n) .$$

(D3) Insert $\mathbf{A}^i = \mathbf{e}_i$ in Eq. (18.4). Since the components of the standard unit vectors satisfy $e_{i\sigma(i)} = 0$ unless $\sigma(i) = i$, we see that only the trivial permutation, $\sigma = \mathrm{id}$, can contribute to the sum. Hence

$$\det(\mathbf{e}_1, \ldots, \mathbf{e}_n) = \mathrm{sgn}(\mathrm{id})\, e_{11}\cdots e_{nn} = 1 .$$

$\square$

Note that the sum on the right-hand side of Eq. (18.4) runs over all permutations in S_n and, therefore, has $n!$ terms. A useful way to think about this sum is as follows. From the $n \times n$ matrix A, choose n entries such that no two of these appear in the same column or in the same row. A term in Eq. (18.4) consists of the product of these n entries (times the sign of the permutation) and the sum amounts to all possible ways of choosing them.

18.1.3 The Levi-Civita symbol

In Eq. (10.7) we have introduced the three-dimensional Levi-Civita symbol ϵ_{ijk} and we have seen that both the vector product and the three-dimensional determinant can be written concisely in terms of this symbol. In order to write the general determinant in a similar way we introduce the n-dimensional generalization of the Levi-Civita symbol by

$$
\epsilon_{i_1 \cdots i_n} = \begin{cases} +1 \text{ if } i_1 = \sigma(1), \dots, i_n = \sigma(n) \text{ with an even permutation } \sigma \\ -1 \text{ if } i_1 = \sigma(1), \dots, i_n = \sigma(n) \text{ with an odd permutation } \sigma \\ 0 \text{ otherwise} \end{cases} . \quad (18.5)
$$

Essentially, the Levi-Civita tensor plays the same role as the sign of the permutation but, in addition, it vanishes if it has an index appearing twice, in which case $(i_1, \dots, i_n)$ is not actually a permutation of $(1, \dots, n)$. Then, the formula (18.4) for the determinant can alternatively be written as

$$
\det(A) = \epsilon_{i_1 \cdots i_n} A_{i_1 1} \cdots A_{i_n n} , \quad (18.6)
$$

with sums over the n indices $i_1, \dots, i_n$ implied. This formulate for the determinant is frequently used in a physics context.

18.1.4 The determinant in low dimensions

The determinant of a 1×1 matrix $A = (a)$ is obviously given by its single entry, so $\det(A) = a$.

For the two-dimensional case we find from Eq. (18.6) that

$$
\det \begin{pmatrix} a_1 & b_1 \\ a_2 & b_2 \end{pmatrix} = \epsilon_{ij} a_i b_j = \epsilon_{12} a_1 b_2 + \epsilon_{21} a_2 b_1 = a_1 b_2 - a_2 b_1 . \quad (18.7)
$$

The two terms on the right-hand side correspond to the two permutations of $\{1, 2\}$.

In three dimensions we find

$$
\det \begin{pmatrix} a_1 & b_1 & c_1 \\ a_2 & b_2 & c_2 \\ a_3 & b_3 & c_3 \end{pmatrix} = \epsilon_{ijk} a_i b_j c_k = a_1 b_2 c_3 + a_2 b_3 c_1 + a_3 b_1 c_2 - a_2 b_1 c_3 - a_3 b_2 c_1 - a_1 b_3 c_2
$$

$$
= \mathbf{a} \cdot (\mathbf{b} \times \mathbf{c}) \quad (18.8)
$$

The six terms which arise correspond to the six permutations of $\{1, 2, 3\}$. In particular, we see explicitly that the three-dimensional version of the determinant coincides with our earlier definition in Section 10.2. (This was already clear from uniqueness.)

The expression for the n-dimensional determinant consists of $n!$ terms. For $n = 4$ this leads to $4! = 24$ terms which is already rather impractical to write down. Clearly, to work with higher-dimensional determinants we require more sophisticated methods.

18.1.5 Determinants for triangular matrices

An interesting special class of matrices for which the determinant is simple consists of *upper triangular matrices*, that is, matrices with all entries below the diagonal vanishing. In this case

$$\det \begin{pmatrix} a_1 & & * \\ & \ddots & \\ 0 & & a_n \end{pmatrix} = a_1 \cdots a_n \, , \tag{18.9}$$

so the determinant is simply the product of the diagonal elements. (An analogous statement holds for lower triangular matrices, of course.) This can be seen from Eq. (18.4). We should consider all ways of choosing one entry per column such that no two entries appear in the same row. For an upper triangular matrix, the only non-zero choice in the first column is the first entry, so that the first row is 'occupied'. In the second column the only available non-trivial choice is, therefore, the entry in the second row etc. In conclusion, from the $n!$ terms in Eq. (18.4) only the term for the identity permutation, $\sigma = \mathrm{id}$, which corresponds to the product of the diagonal elements is non-zero.

Problem 18.1 *(Computing determinants)*

Compute the determinants of the matrices

$$A = \begin{pmatrix} 3 & -2 \\ 4 & -5 \end{pmatrix} , \qquad B = \begin{pmatrix} 1 & -2 & 0 \\ 3 & 2 & -1 \\ 4 & 2 & 5 \end{pmatrix} , \qquad C = \begin{pmatrix} 1 & -2 & 0 & 5 \\ 0 & 2 & 1 & -3 \\ 0 & 0 & -4 & 1 \\ 0 & 0 & 0 & -1 \end{pmatrix} .$$

Solution: From Eq. (18.7), the determinant of A is

$$\det(A) = \det \begin{pmatrix} 3 & -2 \\ 4 & -5 \end{pmatrix} = 3 \cdot (-5) - (-2) \cdot 4 = -7 \, .$$

For the determinant of B, Eq. (18.8) gives

$$\det(B) = \det \begin{pmatrix} 1 & -2 & 0 \\ 3 & 2 & -1 \\ 4 & 2 & 5 \end{pmatrix} = \begin{aligned} & +1 \cdot 2 \cdot 5 + (-2) \cdot (-1) \cdot 4 + 0 \cdot 3 \cdot 2 \\ & - 0 \cdot 2 \cdot 4 - (-2) \cdot 3 \cdot 5 - 1 \cdot (-1) \cdot 2 \end{aligned}$$
$$= 10 + 8 + 30 + 2 = 50 \, .$$

Finally, C is an upper diagonal matrix so from Eq. (18.9) its determinant is the product of the diagonal entries.

$$\det(C) = \det \begin{pmatrix} 1 & -2 & 0 & 5 \\ 0 & 2 & 1 & -3 \\ 0 & 0 & -4 & 1 \\ 0 & 0 & 0 & -1 \end{pmatrix} = 1 \cdot 2 \cdot (-4) \cdot (-1) = 8 \, .$$

18.2   Properties of the determinant

Summary 18.2 *The determinant is unchanged under transposition of matrices and the determinant of a matrix product equals the product of the two determinants. Since the determinant is invariant under basis transformation it is a class function. This fact facilitates defining the determinant of a linear map. A linear map is invertible iff its determinant is non-zero.*

The explicit expression for the determinant becomes complicated quickly as the dimension increases. To be able to work with determinants we need to explore some of their more sophisticated properties.

18.2.1   Determinant and transposition

We begin by showing that a matrix and its transpose have the same determinant.

Proposition 18.1 *For any $n \times n$ matrix $A \in \mathcal{M}_{n,n}(\mathbb{F})$ we have $\det(A) = \det(A^T)$.*

Proof By setting $j_a = \sigma(a)$, for a permutation $\sigma \in S_n$ we can re-write a term in the sum (18.4) for the determinant as $A_{\sigma(1)1} \cdots A_{\sigma(n)n} = A_{j_1 \sigma^{-1}(j_1)} \cdots A_{j_n \sigma^{-1}(j_n)} = A_{1\sigma^{-1}(1)} \cdots A_{n\sigma^{-1}(n)}$, where the last equality follows simply be re-ordering the factors, given that $j_1, \ldots, j_n$ is a permutation of $1, \ldots, n$. From this observation the determinant (18.4) can be written as

$$\det(A) = \sum_{\sigma \in S_n} \mathrm{sgn}(\sigma) A_{1\sigma^{-1}(1)} \cdots A_{n\sigma^{-1}(n)} \overset{(3.8)}{=} \sum_{\sigma^{-1} \in S_n} \mathrm{sgn}(\sigma^{-1}) A_{1\sigma^{-1}(1)} \cdots A_{n\sigma^{-1}(n)}$$

$$\overset{\rho=\sigma^{-1}}{=} \sum_{\rho \in S_n} \mathrm{sgn}(\rho)(A^T)_{\rho(1)1} \cdots (A^T)_{\rho(n)n} = \det(A^T) \,.$$

$\square$

Recall that the determinant sums all products of n entries of the matrix, chosen such that no two of these n entries appear in the same row or column. This statement treats rows and columns on equal footing, so it should not come as a surprise that transposition does not change the determinant.

18.2.2   Determinant and matrix multiplication

Another obvious question is about the relation between the determinant and matrix multiplication. Fortunately, there is a simply and beautiful answer.

Theorem 18.2 *For two $n \times n$ matrices $A, B \in \mathcal{M}_{n,n}(\mathbb{F})$ we have*

$$\det(AB) = \det(A)\det(B) \,. \tag{18.10}$$

Proof We begin by writing the jth column $(AB)^j$ of the matrix product AB in a suitable way.

$$(AB)_{ij} = \sum_k A_{ik}B_{kj} \quad \Rightarrow \quad (AB)^j = \sum_k B_{kj}\mathbf{A}^k \ .$$

Inserting these expressions into the determinant and using multi-linearity gives

$$\det(AB) = \det((AB)^1, \cdots, (AB)^n) = \det\left(\sum_{k_1} B_{k_1 1}\mathbf{A}^{k_1}, \cdots, \sum_{k_n} B_{k_n n}\mathbf{A}^{k_n}\right)$$

$$\overset{(D1)}{=} \sum_{k_1,\cdots,k_n} B_{k_1 1}\cdots B_{k_n n}\det(\mathbf{A}^{k_1},\cdots,\mathbf{A}^{k_n})$$

$$\overset{k_a=\sigma(a)}{=} \sum_{\sigma\in S_n} B_{\sigma(1)1}\cdots B_{\sigma(n)n}\det(\mathbf{A}^{\sigma(1)},\cdots,\mathbf{A}^{\sigma(n)})$$

$$\overset{(18.2)}{=} \underbrace{\sum_{\sigma\in S_n} \operatorname{sgn}(\sigma)B_{\sigma(1)1}\cdots B_{\sigma(n)n}}_{\det(B)}\underbrace{\det(\mathbf{A}^1,\cdots,\mathbf{A}^n)}_{\det(A)} = \det(A)\det(B) \ .$$

$\square$

This simple multiplication rule for determinants of matrix products has a number of profound consequences. First, we can prove a criterion for invertibility of a matrix, based on the determinant, a generalization of Theorem 10.1.

Theorem 18.3 *An $n \times n$ matrix $A \in \mathcal{M}_{n,n}(\mathbb{F})$ has the following properties.*
(i) A is invertible if and only if $\det(A) \neq 0$.
(ii) If A is invertible then $\det(A^{-1}) = (\det(A))^{-1}$.

Proof (i) '$\Rightarrow$': If A is bijective it has an inverse A^{-1} and $1 = \det(\mathbb{1}_n) = \det(AA^{-1}) = \det(A)\det(A^{-1})$. This implies that $\det(A) \neq 0$ and that $\det(A^{-1}) = (\det(A))^{-1}$ which also proves (ii).
(i) '$\Leftarrow$': We prove this indirectly, so we start by assuming that A is not bijective. From Corollary 14.2 this means that $\operatorname{rk}(A) < n$, so the rank of A is less than maximal. Hence, at least one of the column vectors of A, say $\mathbf{A}^1$ for definiteness, can be expressed as a linear combination of the others, so that $\mathbf{A}^1 = \sum_{i=2}^n \alpha_i \mathbf{A}^i$ for some coefficients α_i. For the determinant of A this means

$$\det(A) = \det(\mathbf{A}^1, \mathbf{A}^2, \ldots, \mathbf{A}^n) = \det\left(\sum_{i=2}^n \alpha_i \mathbf{A}^i, \mathbf{A}^2, \ldots, \mathbf{A}^n\right)$$

$$\overset{(D1)}{=} \sum_{i=2}^n \alpha_i \det(\mathbf{A}^i, \mathbf{A}^2, \ldots, \mathbf{A}^n) \overset{(18.1)}{=} 0$$

$\square$

Problem 18.2 *(Using the determinant to check if a matrix is invertible)*

Find the values of the parameter $a \in \mathbb{R}$ for which the following matrix is invertible.

$$A = \begin{pmatrix} 1 & -1 & a \\ 0 & a & -3 \\ -2 & 0 & 1 \end{pmatrix}$$

Solution: Computing the determinant is straightforward and leads to $\det(A) = 2a^2 + a - 6$. This vanishes precisely when $a = -2$ or $a = 3/2$. Hence, for these values of a the matrix A is singular and $\mathrm{rk}(A) < 3$. In fact, the first two column of A are always linearly independent so $\mathrm{rk}(A) = 2$. For all other values, $a \neq -2$ and $a \neq 3/2$, it is invertible and $\mathrm{rk}(A) = 3$. Note that the rank reduces only for specific values of a, so maximal rank is the generic case.

18.2.3   Determinant and basis transformation

In Section 15.2.2 we have derived the formula, Eq. (15.13), for the transformation of a matrix under a change of basis. What happens to the determinant under such a basis change?

Corollary 18.1 *The determinant is invariant under basis change, so*

$$\det(PAP^{-1}) = \det(A), \tag{18.11}$$

for all matrices $A \in \mathcal{M}_{n,n}(\mathbb{F})$ and $P \in \mathrm{GL}(\mathbb{F}^n)$.

Proof   This follows easily from Eq. (18.2) and Theorem 18.3 (ii).

$$\det(PAP^{-1}) = \det(P)\det(A)\det(P^{-1}) = \det(P)\det(A)\,(\det(P))^{-1} = \det(A).$$

$\square$

This statement is of immense significance and is one of the 'magical' properties of the determinant. Recall that matrices related by a basis transformation are called conjugate and that conjugation is an equivalence relation. The associated equivalence classes, called conjugacy classes, contain all matrices related by basis transformation. The invariance (18.11) means that the determinant is a class function: it only depends on the conjugacy class but not the specific matrix within each class. This property allows us to define the determinant of linear maps.

Definition 18.2 *For a linear maps $f : V \to V$, the determinant, $\det(f)$, is defined to be the determinant of any of the matrices describing f relative to a choice of basis.*

This definition makes sense since the determinant is a class function, so that the value of $\det(f)$ is the same for whichever representing matrix is chosen. In this way, many of the properties of the matrix determinant straightforwardly transfer to the determinant for linear maps.

Proposition 18.2 *For $f, g \in \mathrm{End}(V)$, the determinant has the following properties.*

(i) $\det(f \circ g) = \det(f) \det(g)$.
(ii) f *is invertible iff* $\det(f) \neq 0$.
(iii) *If f is invertible then* $\det(f^{-1}) = \det(f)^{-1}$.

Proof   Let A and B be matrices which described f and g relative to a basis of V
(i) Since $f \circ g$ is represented by AB we have

$$\det(f \circ g) = \det(AB) \overset{\mathrm{Eq.\ (18.10)}}{=} \det(A)\det(B) = \det(f)\det(g) \, .$$

(ii)

$$f \text{ invertible} \overset{Thm\ 15.1}{\Longleftrightarrow} A \text{ invertible} \overset{Thm\ 18.3}{\Longleftrightarrow} \det(A) = \det(f) \neq 0 \, .$$

(iii) This follows directly from (i), by setting $g = f^{-1}$ and using that $\det(\mathrm{id}_V) = \det(\mathbb{1}_n) = 1$. $\square$

Problem 18.3 *(Determinant and basis transformations)*

Show that the two matrices

$$A = \begin{pmatrix} 1 & -3 \\ 2 & 1 \end{pmatrix}, \qquad B = \begin{pmatrix} -1 & 4 \\ -2 & 2 \end{pmatrix}$$

cannot be related by a basis transformation.

Solution: One approach is to show that no invertible matrix P with $B = PAP^{-1}$ exists but this would entail an awkward calculation. Instead, work out the two determinants $\det(A) = 1 \cdot 1 - (-3) \cdot 2 = 7$ and $\det(B) = (-1) \cdot 2 - 4 \cdot (-2) = 6$. Since they are different (18.11) implies that A and B cannot be related by a basis transformation.

18.2.4   Orientation

Orientation or *handedness* is an important property of a coordinate system which plays a role in many applications. As we will now see, this notion follows naturally from the properties of linear maps, bases and determinants.

The first observation is that Prop. 18.2 provides us with a way of characterizing the general linear group $\mathrm{GL}(V)$ of a vector space V. It consists of all endomorphisms with non-zero determinant, so

$$\mathrm{GL}(V) = \{f \in \mathrm{End}(V) \,|\, \det(f) \neq 0\} \, . \tag{18.12}$$

For the remainder of this subsection we assume V is a vector space over an ordered field $\mathbb{F}$, for example $\mathbb{F} = \mathbb{R}$. Then the general linear group splits into two disjoint subsets of positive and negative determinant.

$$\mathrm{GL}(V) = \mathrm{GL}_+(V) \cup \mathrm{GL}_-(V) \, , \quad \mathrm{GL}_\pm(V) = \{f \in \mathrm{GL}(V) \,|\, \pm \det(f) > 0\} \, . \tag{18.13}$$

It is easy to check (using Prop. 18.2 and the criteria in Section 3.1.3) that $\mathrm{GL}_+(V)$ is a sub-group of $\mathrm{GL}(V)$ (see Exercise 18.7). We call a linear map $f \in \mathrm{GL}(V)$ *orientation*

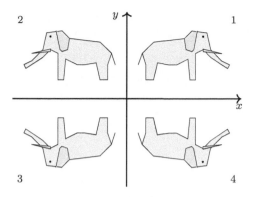

Fig. 18.1 The elephant in quadrant 1 has been mapped to quadrants 1 and 4 by the orientation reversing maps $\mathbf{x} \mapsto A\mathbf{x}$, with $A = \mathrm{diag}(\pm 1, \mp 1)$, and to quadrant 3 by the orientation preserving map $\mathbf{x} \mapsto -\mathbf{x}$, where $\mathbf{x} = (x, y)^T$.

preserving if $f \in \mathrm{GL}_+(V)$ and *orientation reversing* otherwise (see Fig. 18.1).

Suppose we have two bases $(\mathbf{v}_1, \ldots, \mathbf{v}_n)$ and $(\mathbf{v}'_1, \ldots, \mathbf{v}'_n)$ of V. Then, from Theorem 12.1 there exists a unique linear map $f \in \mathrm{GL}(V)$ with $f(\mathbf{v}_i) = \mathbf{v}'_i$ for $i = 1, \ldots, n$. We say the two bases $(\mathbf{v}_1, \ldots, \mathbf{v}_n)$ and $(\mathbf{v}'_1, \ldots, \mathbf{v}'_n)$ have the same orientation if the so-defined linear map f is orientation preserving, that is, if $\det(f) > 0$.

Theorem 18.4 *Having the same orientation defines an equivalence relation on the set of all bases on a vector space V over an ordered field $\mathbb{F}$. There are precisely two equivalence classes.*

Proof   Any basis of V is related to itself since $\det(\mathrm{id}_V) = 1 > 0$, so the relation is reflexive. If a map $f \in \mathrm{GL}_+(V)$ relates two bases via $\mathbf{v}'_i = f(\mathbf{v}_i)$, then $\mathbf{v}_i = f^{-1}(\mathbf{v}'_i)$ and $\det(f^{-1}) = \det(f)^{-1} > 0$. This shows the relation is symmetric. Finally, transitivity follows since $\det(f) > 0$ and $\det(g) > 0$ imply that $\det(f \circ g) = \det(f)\det(g) > 0$.

If we fix a basis $(\mathbf{v}_1, \ldots, \mathbf{v}_n)$ of V then the two equivalence classes are

$$\mathcal{B}_{\pm}(V) = \{(\mathbf{w}_1, \ldots, \mathbf{w}_n) \text{ basis of } V \mid \mathbf{w}_i = f(\mathbf{v}_i) \text{ and } \pm\det(f) > 0\} \qquad (18.14)$$

It is clear from Theorem (12.1) that $\mathcal{B}_+(V) \cup \mathcal{B}_-(V)$ contains all bases on V. It is also easy to show that any two bases in $\mathcal{B}_+(V)$ ($\mathcal{B}_-(V)$) are related, while any basis in $\mathcal{B}_+(V)$ is unrelated to any basis in $\mathcal{B}_-(V)$. □

The two equivalence classes in Eq. (18.14) are referred to as *orientations* since they each contain all the bases with the same orientation. At this point neither of the two orientations is preferred.

However, on coordinate vector spaces $\mathbb{F}^n$ (over an ordered field $\mathbb{F}$), we have a preferred basis of standard unit vectors $(\mathbf{e}_1, \ldots, \mathbf{e}_n)$ and we can be more explicit. Any other basis can be obtained from the standard unit vectors by the action of a linear map $A \in \mathrm{GL}(\mathbb{F}^n)$ via $(A\mathbf{e}_1, \ldots, A\mathbf{e}_n) = (\mathbf{A}^1, \ldots, \mathbf{A}^n)$. Such a basis has the same orientation as

$(e_1, \ldots, e_n)$ if $\det(A) = \det(\mathbf{A}^1, \ldots, \mathbf{A}^n) > 0$ and the opposite orientation if $\det(A) = \det(\mathbf{A}^1, \ldots, \mathbf{A}^n) < 0$. This means the two orientations are

$$\mathcal{B}_{\pm}(\mathbb{F}^n) = \{(\mathbf{v}_1, \ldots, \mathbf{v}_n) \text{ basis of } \mathbb{F}^n \mid \pm \det(\mathbf{v}_1, \ldots, \mathbf{v}_n) > 0\}, \qquad (18.15)$$

and we can refer to the bases in $\mathcal{B}_+(\mathbb{F}^n)$ which have the same orientation as the standard unit vector basis as positively oriented or right-handed and to the others in $\mathcal{B}_-(\mathbb{F}^n)$ as negatively oriented or left-handed.

Problem 18.4 *Orientation of a basis*

What is the orientation of the $\mathbb{R}^3$ bases (e_1, e_2, e_3), $(-e_1, e_2, e_3)$, (e_2, e_1, e_3) and $(\mathbf{v}_1, \mathbf{v}_1, \mathbf{v}_3)$, where $\mathbf{v}_1 = (-2, 0, 1)^T$, $\mathbf{v}_2 = (1, 1, 0)^T$ and $\mathbf{v}_3 = (0, 0, -1)^T$? Do the bases $(1, x, x^2)$ and $(1 - x, 1 + x, x - x^2)$ on $\mathcal{P}_2(\mathbb{R})$ have the same or different orientation?

Solution: Since $\det(e_1, e_2, e_3) = 1$ the basis (e_1, e_2, e_3) is right-handed. (The three vectors correspond to the first three fingers of your right hand, hence the term right-handed.) However, $\det(-e_1, e_2, e_3) = -1$, so $(-e_1, e_2, e_3)$ is left-handed. (These vectors correspond to the first three fingers of your left hand.) The basis (e_2, e_1, e_3) is left-handed as well, since $\det(e_2, e_1, e_3) = -1$. Finally, from

$$\det(\mathbf{v}_1, \mathbf{v}_2, \mathbf{v}_3) = \det \begin{pmatrix} -2 & 1 & 0 \\ 0 & 1 & 0 \\ 1 & 0 & -1 \end{pmatrix} = 2$$

it follows that $(\mathbf{v}_1, \mathbf{v}_2, \mathbf{v}_3)$ is right-handed.

To compare the two bases of $\mathcal{P}_2(\mathbb{R})$ we consider the map $f : \mathcal{P}_2(\mathbb{R}) \to \mathcal{P}_2(\mathbb{R})$ defined by $f(1) = 1 - x$, $f(x) = 1 + x$ and $f(x^2) = x - x^2$. The representing matrix B of f, relative to the monomial basis $(1, x, x^2)$, is

$$B = \begin{pmatrix} 1 & 1 & 0 \\ -1 & 1 & 1 \\ 0 & 0 & -1 \end{pmatrix} \quad \Rightarrow \quad \det(f) = \det(B) = -2.$$

Since $\det(f) < 0$, the map f is an orientation-reversing map and it follows that $(1, x, x^2)$ and $(1 - x, 1 + x, x - x^2)$ have opposite orientations.

18.3 Computing with determinants

Summary 18.3 *The Laplace expansion can be used to express the determinant in terms of determinants of sub-matrices. Alternatively, determinants can be calculated by an algorithm based on row operations. The determinant also provides a formula for the matrix inverse in terms of the co-factor matrix and a formula for the solution of certain systems of linear equations.*

18.3.1 The co-factor matrix

Our next goal is to find a recursive method to calculate the determinant, essentially by writing the determinant of a matrix in terms of determinants of sub-matrices. To this end, we define for an $n \times n$ matrix A, the associated $n \times n$ matrices

$$
\tilde{A}_{(ij)} =
\begin{pmatrix}
& & 0 & & \\
\text{'}A\text{'} & \vdots & \text{'}A\text{'} \\
& & 0 & & \\
0 \cdots 0 & 1 & 0 \cdots 0 \\
& & 0 & & \\
\text{'}A\text{'} & \vdots & \text{'}A\text{'} \\
& & 0 & &
\end{pmatrix}
\leftarrow i^{\text{th}} \text{ row} .
$$

$$\uparrow$$
$$j^{\text{th}} \text{ column}$$

$$(18.16)$$

They are obtained from A by setting the (i,j) entry to 1, the other entries in row i and column j to zero and keeping the rest of the matrix unchanged. Note that the subscripts (i,j) indicate the row and column which have been changed rather than specific entries of the matrix (hence the bracket notation). We can now define the co-factor matrix.

Definition 18.3 *(Co-factor matrix) For an $n \times n$ matrix $A \in \mathcal{M}_{n,n}(\mathbb{F})$, the $n \times n$ matrix C with entires*

$$C_{ij} := \det(\tilde{A}_{(ij)}) , \qquad (18.17)$$

and the matrices $\tilde{A}_{(ij)}$ defined in Eq. (18.16), is called the co-factor matrix of A.

To find a more elegant expression for the co-factor matrix, we also introduce the $(n-1) \times (n-1)$ matrices $A_{(ij)}$ which are obtained from A by simply removing the ith row and the jth column. It takes $i-1$ swaps of neighbouring rows in (18.16) to move row i to the first row (without changing the order of any other rows) and a further $j-1$ swaps to move column j to the first column. After these swaps the matrix $\tilde{A}_{(ij)}$ becomes

$$
B_{(ij)} =
\begin{pmatrix}
1 & 0 & \cdots & 0 \\
0 & & & \\
\vdots & & A_{(ij)} & \\
0 & & &
\end{pmatrix} .
\qquad (18.18)
$$

From Def. 18.1 (D2) and Lemma 18.1 it is clear that $\det(\tilde{A}_{(ij)}) = (-1)^{i+j} \det(B_{(ij)})$, since we need a total of $i+j-2$ swaps of rows and columns to convert one matrix into the other. Further, the explicit form of the determinant (18.4) implies that $\det(B_{(ij)}) = \det(A_{(ij)})$, as the only non-trivial choice of entry in the first column of $B_{(ij)}$ is the 1 in the first row (see also Exercise 18.9). Combining these observations we get the following formula for the co-factor matrix.

Lemma 18.1 *(Formula for co-factor matrix) For an $n \times n$ matrix $A \in \mathcal{M}_{n,n}(\mathbb{F})$, the entries of the co-factor matrix C are given by*

$$C_{ij} = (-1)^{i+j} \det(A_{(ij)}) , \qquad (18.19)$$

where the $(n-1) \times (n-1)$ matrices $A_{(i,j)}$ are obtained from A by removing the ith row and the jth column.

Proof This follows from the arguments above. $\qquad \square$

Hence, the co-factor matrix contains, up to signs, the determinants of the $(n-1) \times (n-1)$ sub-matrices of A, obtained by deleting one row and one column from A. As we will see, for explicit calculations, it is useful to note that the signs in Eq. (18.19) follow a 'chess board pattern', that is, the matrix with entries $(-1)^{i+j}$ has the form

$$
\begin{pmatrix}
+ & - & + & \cdots \\
- & + & - & \cdots \\
+ & - & + & \cdots \\
\vdots & \vdots & \vdots & \ddots
\end{pmatrix} .
\tag{18.20}
$$

The crucial statement which makes the co-factor matrix a useful object is the following lemma.

Lemma 18.2 *For an $n \times n$ matrix A with associated co-factor matrix C, as given in Eq. (18.19), we have*

$$
C^T A = \det(A) \mathbb{1}_n .
\tag{18.21}
$$

Proof This follows from the definition of the co-factor matrix, more or less by direct calculation.

$$
(C^T A)_{ij} = \sum_k (C^T)_{ik} A_{kj} = \sum_k A_{kj} C_{ki} \overset{(18.17)}{=} \sum_k A_{kj} \det(\tilde{A}_{(ki)})
$$

$$
\overset{(18.16)}{=} \sum_k A_{kj} \det(\mathbf{A}^1, \cdots, \mathbf{A}^{i-1}, \mathbf{e}_k, \mathbf{A}^{i+1}, \cdots, \mathbf{A}^n)
$$

$$
\overset{(D1)}{=} \det\left(\mathbf{A}^1, \cdots, \mathbf{A}^{i-1}, \sum_k A_{kj} \mathbf{e}_k, \mathbf{A}^{i+1}, \cdots, \mathbf{A}^n \right)
$$

$$
= \det(\mathbf{A}^1, \cdots, \mathbf{A}^{i-1}, \mathbf{A}^j, \mathbf{A}^{i+1}, \cdots, \mathbf{A}^n) \overset{(18.1)}{=} \delta_{ij} \det(A)
$$

$\square$

18.3.2 Laplace expansion of determinant

The first main conclusion from Eq. (18.21) is the desired recursive formula for the determinant.

Theorem 18.5 *(Laplace expansion of determinant) For an $n \times n$ matrix $A \in \mathcal{M}_{n,n}(\mathbb{F})$ we have*

$$
\det(A) = \sum_{i=1}^n (-1)^{i+j} A_{ij} \det(A_{(ij)}) = \sum_{l=1}^n (-1)^{k+l} A_{kl} \det(A_{(kl)})
\tag{18.22}
$$

for any $j, k = 1, \ldots, n$. The sub-matrices $A_{(ij)}$ are obtained by deleting row i and column j from A.

Proof To prove the first equality in (18.22), we focus on the diagonal (jj) entry of Eq. (18.21).

$$\det(A) = (C^T A)_{jj} = \sum_i (C^T)_{ji} A_{ij} = \sum_i C_{ij} A_{ij} = \sum_i (-1)^{i+j} A_{ij} \det(A_{(ij)}).$$

The second equality (18.21) follows from the first by using the invariance of the determinant under transposition. $\square$

Eq. (18.22) is referred to as *Laplace expansion* of the determinant. It realizes our goal of expressing the determinant of A in terms of determinants of the sub-matrices $A_{(i,j)}$. More specifically, in the first part of Eq. (18.22) we can choose any column j and compute the determinant of A by summing over the entries i in this column times the determinants of the corresponding sub-matrices $A_{(ij)}$ (taking into account the sign). This is also referred to as expanding the determinant 'along the jth column'. The second part of Eq. (18.22) says that we can carry out an analogous process by expanding 'along the kth row'. To see how this works in practice we consider the following exercise.

Problem 18.5 *(Laplace expansion of determinant)*

Compute the determinant of the matrix

$$A = \begin{pmatrix} 2 & -1 & 0 \\ 1 & 2 & -2 \\ 0 & 3 & 4 \end{pmatrix}$$

by a Laplace expansion along its 1st column.

Solution: From Eq. (18.22), taking into account the signs as indicated in (18.20), we find

$$\det(A) = A_{11} \det(A_{(1,1)}) - A_{21} \det(A_{(2,1)}) + A_{31} \det(A_{(3,1)})$$
$$= 2 \cdot \det \begin{pmatrix} 2 & -2 \\ 3 & 4 \end{pmatrix} - 1 \cdot \det \begin{pmatrix} -1 & 0 \\ 3 & 4 \end{pmatrix} + 0 \cdot \det \begin{pmatrix} -1 & 0 \\ 2 & -2 \end{pmatrix}$$
$$= 2 \cdot 14 - 1 \cdot (-4) + 0 \cdot 2 = 32$$

Note that the efficiency of the calculation can be improved by expanding along the row or column with the most zeros.

18.3.3 Matrix inverse from co-factors

The second important result from Eq. (18.21) is a new method to compute the matrix inverse.

Theorem 18.6 *For an invertible $n \times n$ matrix $A \in \mathcal{M}_{n,n}(\mathbb{F})$ the inverse is given by*

$$A^{-1} = \frac{1}{\det(A)} C^T , \tag{18.23}$$

where C is the co-factor matrix of A.

Proof Since A is invertible we know that $\det(A) \neq 0$ from Theorem 18.3. We can, hence, divide Eq. (18.21) by $\det(A)$ and obtain

$$\frac{1}{\det(A)} C^T A = \mathbb{1}_n \qquad \Rightarrow \qquad A^{-1} = \frac{1}{\det(A)} C^T .$$

$\square$

Problem 18.6 (*Inverse of a 2 × 2 matrix using the co-factor method*)

Using the co-factor methods, find the inverse of a general 2×2 matrix

$$A = \begin{pmatrix} a & b \\ c & d \end{pmatrix} .$$

Solution: The co-factor matrix of A is easily obtained by switching around the diagonal and non-diagonal entries and inverting the signs of the latter:

$$C = \begin{pmatrix} d & -c \\ -b & a \end{pmatrix} . \tag{18.24}$$

With $\det(A) = ad - cb$ (which should be different from zero for the inverse to exist) we have for the inverse

$$A^{-1} = \frac{1}{\det(A)} C^T = \frac{1}{ad - cb} \begin{pmatrix} d & -b \\ -c & a \end{pmatrix} . \tag{18.25}$$

Note that this provides a rule for inverting 2×2 matrices which is relatively easy to remember: Exchange the diagonal elements, invert the signs of the off-diagonal elements and divide by the determinant.

Problem 18.7 (*Inverse of a 3 × 3 matrix using the co-factor method*)

Using the co-factor method, find the inverse of the matrix A from Exercise 18.5.

Solution: Applying Eq. (18.19) gives the associated co-factor matrix

$$C = \begin{pmatrix} 14 & -4 & 3 \\ 4 & 8 & -6 \\ 2 & 4 & 5 \end{pmatrix} .$$

For example, the $(1, 1)$ entry of C is computed from $A_{(11)}$, the 2×2 matrix obtained from A by dropping the first row and first column.

$$A_{(11)} = \begin{pmatrix} 2 & -2 \\ 3 & 4 \end{pmatrix} , \qquad C_{11} = \det(A_{(11)}) = 14 .$$

The other entries of C are computed analogously. With $\det(A) = 32$ and Eq. (18.23) the inverse is given by

$$A^{-1} = \frac{1}{\det(A)} C^T = \frac{1}{32} \begin{pmatrix} 14 & 4 & 2 \\ -4 & 8 & 4 \\ 3 & -6 & 5 \end{pmatrix} .$$

18.3.4 Determinant and row operations

Despite our improved methods, the calculation of determinants for large matrices remains a problem, essentially because of the aforementioned $n!$ growth of the number of terms in Eq. (18.4). Using a Laplace expansion will improve matters only if the matrix in question has many zeros. How can we compute the determinants of large matrices? The key is to understand the relationship between determinants and row operations.

Proposition 18.3 *Under row operations, Def. 16.1, on an $n \times n$ matrix A the determinant of A behaves as follows:*

(R1) Adding a multiple of one row to another does not change the determinant.
(R2) Exchanging two rows changes the determinant by a factor -1.
(R3) Multiplying a row with a number $\alpha \neq 0$ changes the determinant by a factor α.

Proof Since the determinant does not change under transposition we can proof the analogues of the above statements for columns.
(R1) Linearity of the determinant and Eq. (18.1) implies that

$$\det(\cdots, \mathbf{a}_i, \cdots, \mathbf{a}_j + \alpha \mathbf{a}_i, \cdots) \overset{(D1)}{=} \det(\cdots, \mathbf{a}_i, \cdots, \mathbf{a}_j, \cdots) + \alpha \det(\cdots, \mathbf{a}_i, \cdots, \mathbf{a}_i, \cdots)$$
$$\overset{(18.1)}{=} \det(\cdots, \mathbf{a}_i, \cdots, \mathbf{a}_j, \cdots) .$$

(R2) This is the anti-symmetry property of the determinant, Def. 18.1 (D2).

(R3) $\det(\cdots, \alpha\, \mathbf{a}, \cdots) \overset{(D1)}{=} \alpha \det(\cdots, \mathbf{a}, \cdots)$ □

Suppose we use row operations of type (R1) and (R2) to bring an $n \times n$ matrix A into upper echelon form, by applying the algorithm described in Section 16.1.3. From the previous lemma this process changes the value of the determinant only by a factor $(-1)^k$, where k is the number of row swaps (R2) used. A square matrix in upper echelon form is also an upper triangular matrix and, from Eq. (18.9), its determinant is simply the product of its diagonal entries. In summary, the value of the determinant is given by

$$\det(A) = (-1)^k a_1 \cdots a_n , \tag{18.26}$$

where $a_1, \ldots, a_n$ are the diagonal entries in the upper echelon form of A and k is the number of rows swaps used to reach the upper echelon form. Computing the upper echelon form for an $n \times n$ matrix needs $\sim n^3$ algebraic operations so for large matrices this method is much more efficient than using the formula (18.4) which involves $n!$ terms. (Even on a modern computers, working out $n!$ terms is impossible even for moderately large n while n^3 operations are still feasible.)

Problem 18.8 *(Computing determinants via row operations)*

Compute the determinant of the matrix

$$A = \begin{pmatrix} 0 & 1 & 2 \\ -1 & 3 & -2 \\ 2 & 0 & 5 \end{pmatrix}$$

by bringing it to upper triangular form, using elementary row operations.

Solution: Bringing A to upper echelon form with row operations gives

$$A = \begin{pmatrix} 0 & 1 & 2 \\ -1 & 3 & -2 \\ 2 & 0 & 5 \end{pmatrix} \xrightarrow{R_1 \leftrightarrow R_2} \begin{pmatrix} -1 & 3 & -2 \\ 0 & 1 & 2 \\ 2 & 0 & 5 \end{pmatrix} \xrightarrow{R_3 \to R_3 + 2R_1} \begin{pmatrix} -1 & 3 & -2 \\ 0 & 1 & 2 \\ 0 & 6 & 1 \end{pmatrix} \xrightarrow{R_3 \to R_3 - 6R_2} \begin{pmatrix} -1 & 3 & -2 \\ 0 & 1 & 2 \\ 0 & 0 & -11 \end{pmatrix}$$

The matrix on the right is in upper echelon and upper triangular form with diagonal entries $(a_1, a_2, a_3) = (-1, 1, -11)$ and we can now compute the determinant from Eq. (18.26). We note that we have used one row exchange, so $k = 1$. Then, $\det(A) = (-1)^k a_1 a_2 a_3 = (-1) \cdot (-1) \cdot 1 \cdot (-11) = -11$. Using the explicit formula (18.8) for the determinant of course leads to the same answer.

For practical reasons we have illustrated this method with a small matrix. However, its main relevance is for computer calculations of large determinants where the basic formula (18.4) or a Laplace expansion is inefficient.

Example 18.1 (Vandermonde determinant)

The Vandermonde determinant (of order n) is the determinant of the $n \times n$ matrix

$$A_n = \begin{pmatrix} 1 & 1 & \cdots & 1 \\ a_1 & a_2 & \cdots & a_n \\ a_1^2 & a_2^2 & \cdots & a_n^2 \\ \vdots & \vdots & \vdots & \vdots \\ a_1^{n-1} & a_2^{n-1} & \cdots & a_n^{n-1} \end{pmatrix}, \tag{18.27}$$

where $a_1, \ldots, a_n \in \mathbb{F}$. The claim is that this determinant is explicitly given by

$$\det(A_n) = \prod_{1 \le i < j \le n} (a_j - a_i) \tag{18.28}$$

This can be shown in a number of ways but we opt for a proof based on induction in n. The basis of the induction, for $n = 2$, is certainly true since $\det(A_2) = a_2 - a_1$. The induction assumption is that the formula (18.28) is true for $n - 1$. To make the induction step, we add multiplies of the first $n - 1$ rows of A_n to the last row to get

$$\tilde{A}_n = \begin{pmatrix} 1 & 1 & \cdots & 1 \\ a_1 & a_2 & \cdots & a_n \\ a_1^2 & a_2^2 & \cdots & a_n^2 \\ \vdots & \vdots & \vdots & \vdots \\ p(a_1) & p(a_2) & \cdots & p(a_n) \end{pmatrix} \quad \text{with} \quad p(x) = \prod_{i=1}^{n-1} (x - a_i),$$

where the specified polynomial p can be achieved by adding suitable row multiples. Of course, these row operations have not changed the determinant. In addition we know that $p(a_i) = 0$ for all $i < n$, so the only non-zero entry in the bottom row of $\tilde{A}_n$ is the last one. This suggest we should try a Laplace expansion along the last row of $\tilde{A}_n$ which gives

$$\det(A_n) = p(a_n)\det(A_{n-1}) = \prod_{i=1}^{n-1}(a_n - a_i) \prod_{1 \le i < j \le n-1} (a_j - a_i) = \prod_{1 \le i < j \le n} (a_j - a_i),$$

where the induction assumption has been used for the second equality. This completes the argument.

Note that the Vandermonde determinant is non-zero iff the numbers $a_1, \ldots, a_n$ are pairwise different. $\qquad\square$

18.3.5 Minors

We have seen how to use row operations to compute the rank of a matrix but can determinants be used for the same purpose? The determinant does provide some limited information about the rank of an $n \times n$ matrix A. From Theorem 18.3 $\det(A) \neq 0$ implies that $\mathrm{rk}(A) = n$. On the other hand, for $\det(A) = 0$ we know that the rank of A is less than maximal, $\mathrm{rk}(A) < n$, but we have no further information about its value. Of course the determinant is only defined for square matrices so even these limited statements cannot be directly applied to non-square matrices.

To extract more information we need to consider the square sub-matrices which can be extracted from a given matrix $A \in \mathcal{M}_{n,m}(\mathbb{F})$ as well as their determinants. Suppose by a sequence of row and column swaps A can be brought into the form

$$
A \xrightarrow{\text{row,col. swaps}} \tilde{A} = \begin{pmatrix} \overset{k}{A'} & \overset{m-k}{B} \\ C & D \end{pmatrix} \begin{matrix} k \\ n-k \end{matrix}, \qquad (18.29)
$$

where the size of the blocks has been indicated on top and to the right. The determinant $\det(A')$ of the $k \times k$ block in the upper left is called a *minor of order k* of A.

Theorem 18.7 *The maximal order of non-zero minors of a matrix A equals its rank.*

Proof We begin by showing that any non-zero minor of order k satisfies $k \leq \mathrm{rk}(A)$. To this end, assume that $\det(A') \neq 0$ for the $k \times k$ matrix A' in Eq. (18.29). Then $\mathrm{rowrk}(A') = k$ and, hence,

$$
\mathrm{rk}(A) = \mathrm{rk}(\tilde{A}) \geq \mathrm{rk}\begin{pmatrix} A' \\ C \end{pmatrix} = \mathrm{rowrk}\begin{pmatrix} A' \\ C \end{pmatrix} \geq \mathrm{rowrk}(A') = k \,.
$$

It remains to be shows that A has a non-zero minor of rank $k = \mathrm{rk}(A)$. Clearly, A has $\mathrm{rk}(A)$ linearly independent column vectors which can be swapped into the first k columns of $\tilde{A}$. Suppose this has been done in Eq. (18.29) so that

$$
\mathrm{rk}(A) = \mathrm{rk}(\tilde{A}) = \mathrm{colrk}\begin{pmatrix} A' \\ C \end{pmatrix} = \mathrm{rowrk}\begin{pmatrix} A' \\ C \end{pmatrix} \,.
$$

The $\mathrm{rk}(A)$ linearly independent rows in the matrix on the right can be brought to the top by suitable row swaps and in this way we obtain a $\mathrm{rk}(A) \times \mathrm{rk}(A)$ matrix A'' which leads to a non-zero minor, $\det(A'') \neq 0$, of order $\mathrm{rk}(A)$. $\qquad\square$

This theorem can be used to determine the rank of any matrix by computing its minors.

Problem 18.9 *Matrix rank from minors*

Using minors, determine the rank of the matrix $A \in \mathcal{M}_{3,3}(\mathbb{R})$ given by

$$A = \begin{pmatrix} 1 & a & b \\ -2 & b & 1 \\ 2 & -3 & b \end{pmatrix}$$

for all $a, b \in \mathbb{R}$.

Solution: We have $\det(A) = 2ab + 2a - b^2 + 6b + 3$ and for all pairs (a, b) for which this is non-zero we have $\mathrm{rk}(A) = 3$. On the other hand, the determinant vanishes for all $b \neq -1$ and

$$a = \frac{b^2 - 6b - 3}{2(b+1)} .$$

In this case, $\mathrm{rk}(A) < 3$. However, one minor of order two is obtained from the second and third row and column,

$$A' = \begin{pmatrix} b & 1 \\ -3 & b \end{pmatrix} \quad \Rightarrow \quad \det(A') = b^2 + 3 ,$$

and since this never vanishes we have $\mathrm{rk}(A) = 2$ whenever $\mathrm{rk}(A) < 3$.

18.3.6   Cramer's rule

For a system of n linear equations in n variables with a unique solution a formula for the solution can be written down in terms of determinants. This formula is known as *Cramer's rule*.

Theorem 18.8 *(Cramer's rule) Let $A \in \mathcal{M}_{n,n}(\mathbb{F})$ be invertible and $\mathbf{b} \in \mathbb{F}^n$. Then, the unique solution of the linear system $A\mathbf{x} = \mathbf{b}$ is given by*

$$x_i = \frac{\det(B_{(i)})}{\det(A)} \quad where \quad B_{(i)} := (\mathbf{A}^1, \cdots, \mathbf{A}^{i-1}, \mathbf{b}, \mathbf{A}^{i+1}, \cdots, \mathbf{A}^n) . \tag{18.30}$$

Proof The linear system $A\mathbf{x} = \mathbf{b}$ can also be written as

$$\sum_j x_j \mathbf{A}^j = \mathbf{b} , \tag{18.31}$$

where $\mathbf{A}^j$ are the columns of A. A short calculation shows that

$$\det(B_{(i)}) = \det(\mathbf{A}^1, \cdots, \mathbf{A}^{i-1}, \mathbf{b}, \mathbf{A}^{i+1}, \cdots, \mathbf{A}^n)$$

$$\stackrel{(18.31)}{=} \det(\mathbf{A}^1, \cdots, \mathbf{A}^{i-1}, \sum_j x_j \mathbf{A}^j, \mathbf{A}^{i+1}, \cdots, \mathbf{A}^n)$$

$$\stackrel{(D1)}{=} \sum_j x_j \det(\mathbf{A}^1, \cdots, \mathbf{A}^{i-1}, \mathbf{A}^j, \mathbf{A}^{i+1}, \cdots, \mathbf{A}^n)$$

$$\stackrel{(18.1)}{=} x_i \det(\mathbf{A}^1, \cdots, \mathbf{A}^{i-1}, \mathbf{A}^i, \mathbf{A}^{i+1}, \cdots, \mathbf{A}^n) = x_i \det(A) .$$

Since A is invertible Theorem 18.3 implies that $\det(A) \neq 0$. Dividing by $\det(A)$ then gives the desired result. □

Problem 18.10 *(Cramer's rule)*

Using Cramer's rule find the solution of the linear system $A\mathbf{x} = \mathbf{b}$ with

$$A = \begin{pmatrix} 2 & -1 & 0 \\ 1 & 2 & -2 \\ 0 & 3 & 4 \end{pmatrix}, \qquad \mathbf{b} = \begin{pmatrix} 1 \\ 2 \\ 0 \end{pmatrix}.$$

Solution: The three matrices $B_{(i)}$ in Cramer's rule (18.30) are obtained by replacing the ith column of A with the vector $\mathbf{b}$.

$$B_{(1)} = \begin{pmatrix} 1 & -1 & 0 \\ 2 & 2 & -2 \\ 0 & 3 & 4 \end{pmatrix}, \quad B_{(2)} = \begin{pmatrix} 2 & 1 & 0 \\ 1 & 2 & -2 \\ 0 & 0 & 4 \end{pmatrix}, \quad B_{(3)} = \begin{pmatrix} 2 & -1 & 1 \\ 1 & 2 & 2 \\ 0 & 3 & 0 \end{pmatrix}.$$

By straightforward computation, for example using a Laplace expansion, it follows that $\det(A) = 32$, $\det(B_{(1)}) = 22$, $\det(B_{(2)}) = 12$ and $\det(B_{(3)}) = -9$. From Eq. (18.30) this leads to the solution

$$\mathbf{x} = \frac{1}{32} \begin{pmatrix} 22 \\ 12 \\ -9 \end{pmatrix}.$$

Exercises

(†=challenging)

18.1 Calculate the determinants of the matrices

$$A = \begin{pmatrix} 0 & -i & i \\ i & 0 & -i \\ -i & i & 0 \end{pmatrix}$$

$$B = \frac{1}{\sqrt{8}} \begin{pmatrix} \sqrt{3} & -\sqrt{2} & -\sqrt{3} \\ 1 & \sqrt{6} & -1 \\ 2 & 0 & 2 \end{pmatrix}.$$

18.2 For which values of the parameters $a, b \in \mathbb{R}$ is the matrix

$$A = \begin{pmatrix} a & 1 & a \\ 1 & b & -1 \\ 0 & -1 & a \end{pmatrix}$$

not invertible? Determine the rank of A for all $a, b \in \mathbb{R}$.

18.3 Invert the matrix

$$A = \begin{pmatrix} 1 & -2 & 0 & 3 \\ 0 & 4 & -1 & 1 \\ 2 & -1 & 0 & 3 \\ 5 & 4 & 1 & -2 \end{pmatrix}$$

using the co-factor method.

18.4 Solve the system of linear equations

$$\begin{aligned} x + 2y + 3z &= 2 \\ 3x + 4y + 5z &= 4 \\ x + 3y + 4z &= 6 \end{aligned}$$

by (a) the matrix inverse, (b) Cramer's method and (c) row reduction.

(d) If you had to write a computer program solving systems of linear equations (of arbitrary and possibly large size) which of the above methods would you base it on?

18.5 *Rank of non-square matrices*
(a) For a 2×3 matrix A show that A has non-maximal rank iff

$\det(\mathbf{A}^1, \mathbf{A}^2) = \det(\mathbf{A}^2, \mathbf{A}^3) = \det(\mathbf{A}^1, \mathbf{A}^3) = 0$.

(b) Formulate and proof the analogous statement for $2 \times n$ matrices, where $n \geq 3$.

(c) Determine the rank of the matrix

$$A = \begin{pmatrix} 1 & a & -2 \\ b & 3 & -a \end{pmatrix}$$

for all values of $a, b \in \mathbb{R}$, using the criterion from part (a).

18.6 *Determinant of linear map*

On the vector space $V = \mathcal{P}_3(\mathbb{R})$ of polynomials with degree less equal three consider the linear map $L : V \to V$ defined by

$$L = \frac{d}{dx} + 1 \, .$$

(a) Compute $\det(L)$. Why is L invertible even though $L(y) = 0$ must have a solution?

(b) Find L^{-1}.

(c) Use the result from part (b) to find a solution y to $L(y)(x) = x^3$.

18.7 *Subgroups of* $\mathrm{GL}(V)$

(a) For a vector space V over $\mathbb{R}$, show that the set $\mathrm{GL}_+(V)$ of all linear maps $f \in \mathrm{GL}(V)$ with $\det(f) > 0$ forms a subgroup of $\mathrm{GL}(V)$.

(b) Let V be a vector space over $\mathbb{F}$ and $\mathbb{F}^* = \mathbb{F} \setminus \{0\}$ is the multiplicative group of the field $\mathbb{F}$. Show that $\det : \mathrm{GL}(V) \to \mathbb{F}^*$ is a group homomorphism.

(c) Show that the set $\mathrm{SL}(V)$ of all linear maps $f \in \mathrm{GL}(V)$ with $\det(f) = 1$ forms a subgroup of $\mathrm{GL}(V)$. (This group is called the *special linear group*.)

18.8 *Orientation*

(a) For two linearly independent vec-

tors $\mathbf{v}_1, \mathbf{v}_2 \in \mathbb{R}^3$ show that $(\mathbf{v}_1 \times \mathbf{v}_2, \mathbf{v}_1, \mathbf{v}_2)$ is a positively oriented basis.

(b) For a permutation $\sigma \in S_n$, show that the bases $(\mathbf{v}_1, \ldots, \mathbf{v}_n)$ and $(\mathbf{v}_{\sigma(1)}, \ldots, \mathbf{v}_{\sigma(n)})$ of V have the same orientation iff $\mathrm{sgn}(\sigma) = 1$.

18.9 *Determinant of block-diagonal matrices*

Matrixes $A \in \mathcal{M}_{n,n}(\mathbb{F})$ and $B \in \mathcal{M}_{m,m}(\mathbb{F})$ are arranged into the block-diagonal matrix

$$C = \begin{pmatrix} A & 0 \\ 0 & B \end{pmatrix} \, ,$$

with size $(n+m) \times (n+m)$. Show that $\det(C) = \det(A)\det(B)$.

18.10 *Generalization of cross product*

For linearly independent vectors $\mathbf{v}_1, \ldots, \mathbf{v}_{n-1} \in \mathbb{R}^n$, define the vector $\mathbf{w} \in \mathbb{R}^n$ with components $w_i = \det(\mathbf{v}_1, \ldots, \mathbf{v}_{n-1}, \mathbf{e}_i)$. Show that

(a) $\mathbf{w}$ is orthogonal with respect to the dot product to all vectors $\mathbf{v}_a$.

(b) $(\mathbf{v}_1, \ldots, \mathbf{v}_{n-1}, \mathbf{w})$ is a basis of $\mathbb{R}^n$.

(c) $|\mathbf{w}|^2 = \det(\mathbf{v}_1, \ldots, \mathbf{v}_{n-1}, \mathbf{w})$.

(d) for $\mathbf{v}_a = \mathbf{e}_a$ we have $\mathbf{w} = \mathbf{e}_n$.

(e) for $n = 3$ the vector $\mathbf{w}$ can be written in terms of a cross product.

18.11 *Determinant formulae*[†]

For two matrices $A, B \in \mathcal{M}_{n,m}(\mathbb{R})$ show the following:

(a) $\det(AB^T) = 0$ if $n > m$.

(b) $\det(AB^T) = \sum \det(\mathbf{A}^{a_1}, \ldots, \mathbf{A}^{a_n}) \cdot \det(\mathbf{B}^{a_1}, \ldots, \mathbf{B}^{a_n})$ if $n \leq m$, where the sum runs over all $a_1, \ldots, a_n$ with $1 \leq a_1 < \cdots < a_n \leq m$. (Hint: Write the determinants in terms of Levi-Civita tensors and go wild with indices.)

(c) $\det(AA^T) = \sum \det(\mathbf{A}^{a_1}, \ldots, \mathbf{A}^{a_n})^2$ if $n \leq m$ and with the sum over the a_k as in part (b).

Part VI

Eigenvalues and eigenvectors

In Chapter 15, we have seen that an endomorphism $f \in \text{End}(V)$ can be represented by a square matrix A, relative to a choice of basis on V. For a different basis, the same linear map is represented by another matrix A', related to A by the basis transformation $A' = PAP^{-1}$, as in Eq. (15.13). Two matrices related by such a basis transformation are called conjugate and we have seen that conjugation is an equivalence relation. Its equivalence classes, the conjugacy classes, consist of all matrices which represent the same linear map, relative to different basis choices. This structure suggests a set of questions. How can we find a basis for which the representing matrix of a linear map is particularly simply, for example diagonal? Does such a diagonal matrix exist in each conjugacy class? If not, what is the simplest possible choice? As we will see, eigenvectors and eigenvalues are the key to answering these questions.

Bringing matrices into a simple form by a basis transformation can be immensely helpful for solving or simplifying a wide range of problems. For this reason, eigenvectors and eigenvalues are of great practical importance and have numerous applications. For example, eigenvectors and eigenvalues are at the heart of quantum mechanics. We will encounter a number of these applications as we go along.

As a simple motivational example for why diagonalizing or otherwise simplifying matrices can be useful consider a sequence $\mathbf{x}_0, \mathbf{x}_1, \ldots \in \mathbb{F}^n$ of vectors which are determined recursively, by an equation $\mathbf{x}_{k+1} = M\mathbf{x}_k$, where $M \in \mathcal{M}_{n,n}(\mathbb{F})$ is a fixed matrix. This might describe a discrete process, such as the evolution of a population, where the index k labels the time step, $\mathbf{x}_k$ describes a distribution of certain characteristics within the population at time k and the evolution to the next time step is accomplished by multiplying with M. Once we fix the initial vector $\mathbf{x}_0$, the entire sequence is determined by the equation

$$\mathbf{x}_k = M^k \mathbf{x}_0 . \tag{18.32}$$

Therefore, if we want to understand how the system evolves, we have to work out powers, M^k, of the matrix M. Matrix multiplication is generally a complicated operation and repeating it many times may well be very difficult, even for small matrices M. However, for a diagonal $M = \text{diag}(\lambda_1, \ldots, \lambda_n)$ the calculation is easy and leads to $M^k = \text{diag}(\lambda_1^k, \ldots, \lambda_n^k)$. If we can somehow bring Eq. (18.32) into an equivalent form where M becomes diagonal the problem can be solved. Eigenvector and eigenvalues will help to do this, as we will see.

In the next chapter, we begin by developing the basics of eigenvalues and eigenvectors, their definition, structure and computation. In particular, the *characteristic polynomial*, a central object in the theory of eigenvalues, is introduced. We finish with the theorem of Cayley–Hamilton which states that every endomorphism inserted into its own characteristic polynomial gives the zero map.

In Chapter 20, we derive criteria for when endomorphisms can be diagonalized and show how eigenvalues and eigenvectors can be used to compute the diagonal form. While this can be achieved frequently we will also see that there are some conjugacy classes which do not contain a diagonal matrix.

For such cases, the strategy is to find a basis in which the representing matrix is as close to diagonal as possible. The resulting structure is called the *Jordan normal form* which leads to matrices with zeros everywhere except possibly along the diagonal and the entries just above the diagonal. The Jordan normal form will be derived in Chapter 21.

19
Basics of eigenvalues

Eigenvectors and eigenvalues play a role in many areas of pure and applied mathematics. In this chapter, we develop the basics, define eigenvalues and eigenvectors and explain how to compute them. Every endomorphism $f : V \to V$ has an associated polynomial χ_f, called the *characteristic polynomial*. The eigenvalues of f are precisely the zeros of χ_f. As a by-product, we obtain new class functions for matrices which follow from the invariance of the characteristic polynomial under basis transformations. We discuss in detail the most important one of those, the *trace* of a matrix which is given by the sum of its diagonal entries. We end the chapter by proving the Cayley—Hamilton theorem.

19.1 Eigenvalues and eigenspaces

Summary 19.1 *For a linear map $f : V \to V$, an eigenvector $\mathbf{v} \in V$ is a non-zero vector which scales under f, so that $f(\mathbf{v}) = \lambda \mathbf{v}$. The scalar λ is the associated eigenvalue. The eigenvectors for each eigenvalue λ are collected in the eigenspace $\mathrm{Eig}_f(\lambda)$ which consists of all solutions to the homogeneous linear system $(f - \lambda \, \mathrm{id}_V)\mathbf{v} = \mathbf{0}$. The dimension of the eigenspace is called the degeneracy of the eigenvalue.*

19.1.1 Definition of eigenvalues and eigenvectors

Recall from Theorem 15.1 that the matrix A which represents a linear map $f : V \to V$ relative to a basis $(\mathbf{v}_1, \ldots, \mathbf{v}_n)$ of V is determined by

$$f(\mathbf{v}_j) = \sum_{i=1}^{n} A_{ij} \mathbf{v}_i \, . \tag{19.1}$$

Suppose we had somehow succeeded in choosing the basis $(\mathbf{v}_1, \ldots, \mathbf{v}_n)$ such that $A = \mathrm{diag}(a_1, \ldots, a_n)$ is diagonal. In this case, Eq. (19.1) turns into

$$f(\mathbf{v}_i) = a_i \mathbf{v}_i \, . \tag{19.2}$$

Hence, a diagonal representing matrix requires basis vectors which are multiplied by a scalar under the action of f. Vectors with such a scaling behaviour are called *eigenvectors* and the scalars a_i multiplying them are called *eigenvalues*. Formally, they are defined as follows.

Definition 19.1 *Let $f : V \to V$ be a linear map on a vector space V over $\mathbb{F}$. A scalar $\lambda \in \mathbb{F}$ is called an eigenvalue of f if there exists a non-zero vector $\mathbf{v} \in V$ such that*

$$f(\mathbf{v}) = \lambda \mathbf{v} . \tag{19.3}$$

Such a vector $\mathbf{v}$ is called an eigenvector of f with eigenvalue λ.

Note that, while $\lambda = 0$ is a perfectly acceptable eigenvalue, eigenvectors are always non-zero. This requirement is, in fact, crucial. If the zero vector was allowed as an eigenvector, then the eigenvalue equation (19.3) would be satisfied for every scalar λ, since $f(\mathbf{0}) = \mathbf{0} = \lambda\mathbf{0}$.

19.1.2   Degeneracy and eigenspaces

For a given eigenvalue λ, the eigenvector is not unique. For example, for an eigenvector $\mathbf{v}$, every multiple $\alpha\mathbf{v}$, where $\alpha \neq 0$, is also an eigenvector, since $f(\mathbf{v}) = \lambda\mathbf{v}$ implies that $f(\alpha\mathbf{v}) = \alpha f(\mathbf{v}) = \lambda(\alpha\mathbf{v})$. For this reason it makes sense to collect all eigenvectors for a given eigenvalue into a set, called the *eigenspace*. Since Eq. (19.3) can be re-written as $(f - \lambda \operatorname{id}_V)\mathbf{v} = \mathbf{0}$ the eigenspace for a scalar $\lambda \in \mathbb{F}$ is defined as

$$\operatorname{Eig}_f(\lambda) := \operatorname{Ker}(f - \lambda \operatorname{id}_V) . \tag{19.4}$$

As a kernel of a linear map, the eigenspace is a vector subspace of V. Note that the eigenspace for eigenvalue 0 is the kernel of the linear map, so $\operatorname{Eig}_f(0) = \operatorname{Ker}(f)$. From Def. 19.1, the scalar λ is an eigenvalue if and only if the eigenspace $\operatorname{Eig}_f(\lambda)$ is non-trivial, so we have

$$\lambda \text{ eigenvalue of } f \quad \Leftrightarrow \quad \operatorname{Eig}_f(\lambda) \neq \{\mathbf{0}\} \quad \Leftrightarrow \quad \dim_{\mathbb{F}}(\operatorname{Eig}_f(\lambda)) > 0 . \tag{19.5}$$

The dimension of the eigenspace is an important property of the eigenvalue for which we introduce the following terminology.

Definition 19.2 *An eigenvalue $\lambda \in \mathbb{F}$ for a linear map $f \in \operatorname{End}(V)$ is called non-degenerate if $\dim_{\mathbb{F}}(\operatorname{Eig}_f(\lambda)) = 1$ and, otherwise, if $\dim_{\mathbb{F}}(\operatorname{Eig}_f(\lambda)) > 1$, it is called degenerate. The dimension $\dim_{\mathbb{F}}(\operatorname{Eig}_f(\lambda))$ is called the degeneracy of the eigenvalue λ.*

What is the intersection of the eigenspaces for two different eigenvalues λ and λ'? Suppose we have a vector $\mathbf{v} \in \operatorname{Eig}_f(\lambda) \cap \operatorname{Eig}_f(\lambda')$. Then the eigenvalue equation (19.3) implies that $f(\mathbf{v}) = \lambda\mathbf{v} = \lambda'\mathbf{v}$ and, hence, since $\lambda \neq \lambda'$, that $\mathbf{v} = \mathbf{0}$. In conclusions, eigenspaces for different eigenvalues intersect trivially so their sum is direct (see Section 8.1.4).

19.2   The characteristic polynomial

Summary 19.2 *The characteristic polynomial χ_f of a linear map $f : V \to V$ is a polynomial of degree $\dim_{\mathbb{F}}(V)$ whose zeros are the eigenvalues of f. The characteristic polynomial of a matrix is invariant under basis transformations, so that all* *coefficients in χ_f are class functions. In particular this shows that the trace of a matrix is a class function. The degeneracy of an eigenvalue λ is bounded from above by the multiplicity of λ in the characteristic polynomial.*

19.2.1 Definition of characteristic polynomial

How do we compute eigenvalues and eigenvectors for a given linear map? The key observation comes from the equivalence (19.5). It says that λ is an eigenvalue iff $\text{Ker}(f - \lambda \, \text{id}_V)$ is non-trivial. This is the case iff $f - \lambda \, \text{id}_V$ is not invertible which is equivalent to $\det(f - \lambda \, \text{id}_V) = 0$. This last condition is crucial since it gives an equation for the eigenvalues and it motivates the following definition.

Definition 19.3 *For a linear map $f : V \to V$ the map $\chi_f : \mathbb{F} \to \mathbb{F}$ defined by*

$$\chi_f(\lambda) := \det(f - \lambda \, \text{id}_V) \tag{19.6}$$

is called the characteristic polynomial of f.

As suggested by the above discussion, the eigenvalues are the zeros of the characteristic polynomial.

Theorem 19.1 *Let $f \in \text{End}(V)$ be a linear map on a vector space V over $\mathbb{F}$ with characteristic polynomial $\chi_f : \mathbb{F} \to \mathbb{F}$. The scalar $\lambda \in \mathbb{F}$ is an eigenvalue of f if and only if $\chi_f(\lambda) = 0$.*

Proof

λ eigenvalue of f $\overset{(19.5)}{\Longleftrightarrow}$ $\dim_\mathbb{F}(\text{Ker}(f - \lambda \, \text{id}_V)) > 0$ $\overset{Cor.\ 14.2}{\Longleftrightarrow}$ $f - \lambda \, \text{id}_V$ not invertible

$\overset{Thm.\ 18.3}{\Longleftrightarrow}$ $\det(f - \lambda \, \text{id}_V) = 0$

$\square$

19.2.2 Properties of the characteristic polynomial

To get a handle on eigenvalues we should understand the characteristic polynomial better, including why it actually is a polynomial.

Proposition 19.1 *For a linear map $f : V \to V$, where V is an n-dimensional vector space over $\mathbb{F}$, and any representing matrix $A \in \mathcal{M}_{n,n}(\mathbb{F})$ of f the characteristic polynomial χ_f has the following properties.*

(i) $\chi_f = \chi_A$
(ii) χ_f *is a polynomial of degree n*
(iii) $\chi_{PAP^{-1}} = \chi_A$ *for any matrix $P \in \text{Gl}(\mathbb{F}^n)$*
(iv)   *If we write the characteristic polynomial as*

$$\chi_A(\lambda) = c_n \lambda^n + c_{n-1} \lambda^{n-1} + \cdots + c_1 \lambda + c_0 \tag{19.7}$$

then the coefficients c_i are invariant under basis transformations.
(v)   *For the coefficients we have*

$$c_n = (-1)^n , \qquad c_{n-1} = (-1)^{n-1} \sum_{i=1}^{n} A_{ii} , \qquad c_0 = \det(A) . \qquad (19.8)$$

Proof (i) This follows from Def. 18.2 which defined the determinant of a linear map as the determinant of any of its representing matrices.

(ii) To find the characteristic polynomial, we need to compute the determinant

$$\chi_f(\lambda) = \det \begin{pmatrix} A_{11} - \lambda & A_{12} & \cdots & A_{1,n-1} & A_{1n} \\ A_{21} & A_{22} - \lambda & \cdots & A_{2,n-1} & A_{2n} \\ \vdots & \vdots & \ddots & \vdots & \vdots \\ A_{n-1,1} & A_{n-1,2} & \cdots & A_{n-1,n-1} - \lambda & A_{n-1,n} \\ A_{n1} & A_{n2} & \cdots & A_{n,n-1} & A_{nn} - \lambda \end{pmatrix} . \qquad (19.9)$$

The general expression (18.4) for the determinant of an $n \times n$ matrix is a degree n polynomial in the entries. Since the entries of the matrix in Eq. (19.9) are at most linear in λ is follows that $\chi_f(\lambda)$ is a polynomial in λ of degree n or less. Result (v) shows that the coefficient of λ^n is, in fact, always non-zero so the degree equals n.

(iii) This follows from (i) since the characteristic polynomial is the same for any representing matrix of f but it can also be checked explicitly.

$$\chi_{PAP^{-1}}(\lambda) \overset{(19.6)}{=} \det(PAP^{-1} - \lambda \mathbb{1}_n) = \det(PAP^{-1} - \lambda P \mathbb{1}_n P^{-1})$$
$$= \det(P(A - \lambda \mathbb{1}_n)P^{-1}) \overset{(18.11)}{=} \det(A - \lambda \mathbb{1}_n) \overset{(19.6)}{=} \chi_A(\lambda)$$

(iv) This follows from (iii). If the entire characteristic polynomial is invariant under basis transformations then so are its coefficients.

(v) The formula for c_0 follows easily from $c_0 = \chi_f(0) = \det(f)$. Powers λ^n and λ^{n-1} in the determinant (19.9) can only arise from the product of the diagonal elements, so

$$\chi_A(\lambda) = \prod_{i=1}^{n} (A_{ii} - \lambda) + \mathcal{O}(\lambda^{n-2}) = (-1)^n \lambda^n + (-1)^{n-1} \left(\sum_{i=1}^{n} A_{ii} \right) \lambda^{n-1} + \mathcal{O}(\lambda^{n-2})$$

Reading off the factors in front of λ^n and λ^{n-1} gives the desired results. $\qquad \square$

19.2.3   Examples

Combining the above results leads to an algorithm for computing eigenvalues and eigenvectors which can be summarized as follows.

Algorithm *(Computing eigenvalues and eigenvectors)*
(1) Compute the characteristic polynomial $\chi_f(\lambda) = \det(f - \lambda \operatorname{id}_V)$ of f.
(2) Find the zeros, λ, of χ_f. They are the eigenvalues of f.
(3) For each eigenvalue λ compute the eigenspace $\operatorname{Eig}_f(\lambda) = \operatorname{Ker}(f - \lambda \operatorname{id}_V)$ by finding all vectors $\mathbf{v}$ which solve the homogeneous linear system

$$(f - \lambda \operatorname{id}_V)(\mathbf{v}) = \mathbf{0} . \qquad (19.10)$$

Things are simple for diagonal matrices $A = \text{diag}(a_1, \ldots, a_n)$ since their characteristic polynomial equals $\chi_A(\lambda) = (a_1 - \lambda) \cdots (a_n - \lambda)$. Hence, the eigenvalues are precisely the diagonal entries a_i and, provided they are pairwise different, they are non-degenerate with eigenspaces $\text{Eig}_A(a_i) = \text{Span}(e_i)$. (If some of the a_i are the same then there is degeneracy and the eigenspaces enhance to the span of the unit vectors in those directions.) In general, finding eigenvalues and eigenvectors is not as simple and requires a calculation.

Problem 19.1 *Computing eigenvalues and eigenvectors*

Compute the eigenvalues, eigenvectors (and eigenspaces) of the linear map $A : \mathbb{R}^3 \to \mathbb{R}^3$ defined by

$$A = \begin{pmatrix} 1 & -1 & 0 \\ -1 & 2 & -1 \\ 0 & -1 & 1 \end{pmatrix} . \tag{19.11}$$

Solution: The characteristic polynomial is

$$\chi_A(\lambda) = \det \begin{pmatrix} 1 - \lambda & -1 & 0 \\ -1 & 2 - \lambda & -1 \\ 0 & -1 & 1 - \lambda \end{pmatrix} = \lambda(\lambda - 1)(\lambda - 3) ,$$

so we have three eigenvalues, $\lambda_1 = 0$, $\lambda_2 = 1$ and $\lambda_3 = 3$. Writing $\mathbf{v} = (x, y, z)^T$, we compute the eigenvectors for each of these eigenvalues in turn.

$$\lambda_1 = 0 : (A - 0\mathbb{1})\mathbf{v} = \begin{pmatrix} 1 & -1 & 0 \\ -1 & 2 & -1 \\ 0 & -1 & 1 \end{pmatrix} \begin{pmatrix} x \\ y \\ z \end{pmatrix} = \begin{pmatrix} x - y \\ -x + 2y - z \\ -y + z \end{pmatrix} \overset{!}{=} 0 \quad \Leftrightarrow \quad x = y = z$$

$$\lambda_2 = 1 : (A - 1\mathbb{1})\mathbf{v} = \begin{pmatrix} 0 & -1 & 0 \\ -1 & 1 & -1 \\ 0 & -1 & 0 \end{pmatrix} \begin{pmatrix} x \\ y \\ z \end{pmatrix} = \begin{pmatrix} -y \\ -x + y - z \\ -y \end{pmatrix} \overset{!}{=} 0 \quad \Leftrightarrow \quad y = 0, \, x = -z$$

$$\lambda_3 = 3 : A - 3\mathbb{1})\mathbf{v} = \begin{pmatrix} -2 & -1 & 0 \\ -1 & -1 & -1 \\ 0 & -1 & -2 \end{pmatrix} \begin{pmatrix} x \\ y \\ z \end{pmatrix} = \begin{pmatrix} -2x - y \\ -x - y - z \\ -y - 2z \end{pmatrix} \overset{!}{=} 0 \quad \Leftrightarrow \quad y = -2x, \, z = x$$

Hence, the eigenspaces are given by $\text{Eig}_A(\lambda_i) = \text{Span}(\mathbf{v}_i)$ with

$$\mathbf{v}_1 = \begin{pmatrix} 1 \\ 1 \\ 1 \end{pmatrix} , \quad \mathbf{v}_2 = \begin{pmatrix} -1 \\ 0 \\ 1 \end{pmatrix} , \quad \mathbf{v}_3 = \begin{pmatrix} 1 \\ -2 \\ 1 \end{pmatrix} .$$

They are one-dimensional so all eigenvalues are non-degenerate.

Counting in the previous exercise is rather suggestive: we have three dimensions and three non-degenerate eigenvalues. Unfortunately, things are not always so straightforward as the following exercise shows.

Problem 19.2 *(Eigenvalues and eigenvectors — more examples)*

Find the eigenvalues and eigenvectors (eigenspaces) for the linear maps $\mathbb{R}^2 \to \mathbb{R}^2$ defined by

$$\mathbb{1}_2 \,, \qquad A = \begin{pmatrix} 1 & 1 \\ 0 & 1 \end{pmatrix} \,, \qquad B = \begin{pmatrix} 0 & 1 \\ -1 & 0 \end{pmatrix} \,, \qquad C = \begin{pmatrix} 2 & 1 \\ 1 & 2 \end{pmatrix} \,. \tag{19.12}$$

Solution: For the characteristic polynomials we find

$$\chi_{\mathbb{1}_2}(\lambda) = \det \begin{pmatrix} 1-\lambda & 0 \\ 0 & 1-\lambda \end{pmatrix} = (1-\lambda)^2 \qquad \chi_A(\lambda) = \det \begin{pmatrix} 1-\lambda & 1 \\ 0 & 1-\lambda \end{pmatrix} = (1-\lambda)^2$$

$$\chi_B(\lambda) = \det \begin{pmatrix} -\lambda & 1 \\ -1 & -\lambda \end{pmatrix} = \lambda^2 + 1 \qquad \chi_C(\lambda) = \det \begin{pmatrix} 2-\lambda & 1 \\ 1 & 2-\lambda \end{pmatrix} = (2-\lambda)^2 - 1 \,.$$

The map $\mathbb{1}_2$ has only one eigenvalue, $\lambda = 1$, with a two-dimensional eigenspace $\mathrm{Eig}_{\mathbb{1}_2}(1) = \mathbb{R}^2$. The map A has the same characteristic polynomial as $\mathbb{1}_2$, so also has only one eigenvalue, $\lambda = 1$. To find the eigenvectors we work out

$$(A - \mathbb{1}_2)\mathbf{v} = \begin{pmatrix} 0 & 1 \\ 0 & 0 \end{pmatrix} \begin{pmatrix} x \\ y \end{pmatrix} \overset{!}{=} \mathbf{0} \qquad \Leftrightarrow \qquad y = 0 \,.$$

Hence, unlike for the previous case, the eigenspace $\mathrm{Eig}_A(1) = \mathrm{Span}(\mathbf{e}_1)$ is one-dimensional and the eigenvalue is non-degenerate.

The characteristic polynomial for B has no zeros over $\mathbb{R}$, so there are no eigenvalues. However, if we view B as a map $\mathbb{C}^2 \to \mathbb{C}^2$ then we have the two eigenvalues $\lambda_\pm = \pm i$ and the associated eigenvectors are determined by

$$\lambda_\pm = \pm i : \ (B \mp i\mathbb{1})\mathbf{v} = \begin{pmatrix} \mp i & 1 \\ -1 & \mp i \end{pmatrix} \begin{pmatrix} x \\ y \end{pmatrix} = \begin{pmatrix} \mp ix + y \\ -x \mp iy \end{pmatrix} \overset{!}{=} \mathbf{0} \qquad \Leftrightarrow \qquad y = \pm ix$$

This shows both eigenvalues are non-degenerated with eigenspaces $\mathrm{Eig}_B(\pm i) = \mathrm{Span}((1, \pm i)^T)$. Finally, for C we have two eigenvalues $\lambda_1 = 1$ and $\lambda_2 = 3$ with eigenvalues determined by

$$\lambda_1 = 1 : \ (C - \mathbb{1})\mathbf{v} = \begin{pmatrix} 1 & 1 \\ 1 & 1 \end{pmatrix} \begin{pmatrix} x \\ y \end{pmatrix} = \begin{pmatrix} x+y \\ x+y \end{pmatrix} \overset{!}{=} \mathbf{0} \qquad \Leftrightarrow \qquad y = -x$$

$$\lambda_2 = 3 : \ (C - 3\mathbb{1})\mathbf{v} = \begin{pmatrix} -1 & 1 \\ 1 & -1 \end{pmatrix} \begin{pmatrix} x \\ y \end{pmatrix} = \begin{pmatrix} x-y \\ x-y \end{pmatrix} \overset{!}{=} \mathbf{0} \qquad \Leftrightarrow \qquad y = x$$

Both eigenvalues are non-degenerate with eigenspaces $\mathrm{Eig}_C(1) = \mathrm{Span}((1, -1)^T)$ and $\mathrm{Eig}_C(3) = \mathrm{Span}((1, 1)^T)$.

19.2.4   Degeneracy and multiplicity

In Section 4.4 we have introduced various basic features of polynomials. It will help our discussion of eigenvalues and eigenvectors to consider these for the characteristic polynomial. First, recall that the existence of zeros and factorization of polynomials depends on the choice of the underlying field $\mathbb{F}$. That is why we have to be careful about the field $\mathbb{F}$ when we discuss eigenvalues and eigenvectors, as Problem 19.2 illustrates.

We have also seen that different values for the degeneracy of eigenvalues are possible. However, the degeneracy can never exceed the multiplicity of the eigenvalue in the characteristic polynomial, as we now prove.

Proposition 19.2 *Let $f : V \to V$ be a linear map on a (finite-dimensional) vector space V over $\mathbb{F}$. If λ_0 is a degeneracy d eigenvalue of f whose multiplicity in χ_f is m then $1 \leq d \leq m$.*

Proof   The eigenvalue λ_0 has an associated eigenvector so it is clear that $1 \leq d$.

To proof the upper bound choose a basis $(\mathbf{v}_1, \ldots, \mathbf{v}_d)$ of $\mathrm{Eig}_f(\lambda_0)$ and complete to a basis $(\mathbf{v}_1, \ldots, \mathbf{v}_d, \mathbf{v}_{d+1}, \ldots, \mathbf{v}_n)$ of V. Then $f(\mathbf{v}_j) = \lambda_0 \mathbf{v}_j$ for $j = 1 \ldots d$ and

$$f(\mathbf{v}_j) = \sum_{i=1}^{d} B_{ij}\mathbf{v}_i + \sum_{i=d+1}^{n} C_{ij}\mathbf{v}_i \, ,$$

for $j = d+1, \ldots, n$ and suitable matrices B and C. This means the representing matrix A for f relative to the chosen basis has the form

$$A = \begin{pmatrix} \lambda_0 \mathbb{1}_d & B \\ 0 & C \end{pmatrix} \quad \Rightarrow \quad \chi_f(\lambda) = \det(A - \lambda\mathbb{1}_n) = (\lambda - \lambda_0)^d \chi_C(\lambda) \, .$$

The above result for χ_f means the degeneracy of λ_0 is at least d, so $d > m$ leads to a contradiction (see Def. 4.3). It follows that $d \leq m$. $\square$

19.2.5   Class functions

In Section 18.2.3 we have seen that the determinant is a class function — it takes the same value on any two matrices in the same conjugacy class. It is, therefore, a property of the underlying linear map and we have used this fact to define the determinant for linear maps. Class functions are very important since they can tell us about properties of matrices which are independent of the choice of basis. We should, therefore, take note that Prop. 19.1 tells use about the existence of a whole range of class functions: the coefficients c_i in the characteristic polynomial (19.7) for f. In fact, the coefficient $c_0 = \det(A)$ is the determinant and $c_n = (-1)^n$ is trivial but all other c_i are new.

Of particular interest is the coefficient c_{n-1} in Eq. (19.7) which is proportional to the sum of all diagonal entries. This sum is called the *trace* of a matrix and for $A \in \mathcal{M}_{n,n}(\mathbb{F})$ it is defined by

$$\mathrm{tr}(A) := \sum_{i=1}^{n} A_{ii} \, . \tag{19.13}$$

Proposition 19.3 *(Properties of the trace) For matrices $A, B \in \mathcal{M}_{n,n}(\mathbb{F})$ and $P \in \mathrm{GL}(\mathbb{F}^n)$ and scalars $\alpha, \beta \in \mathbb{F}$, the trace has the following properties.*

   (i) $\mathrm{tr}(\alpha A + \beta B) = \alpha \, \mathrm{tr}(A) + \beta \, \mathrm{tr}(B)$ *(linearity)*
   (ii) $\mathrm{tr}(AB) = \mathrm{tr}(BA)$ *(commutativity)*
   (iii) $\mathrm{tr}((PAP^{-1})^k) = \mathrm{tr}(A^k)$ *for* $k = 1, 2, \ldots$ *(class function)*

Proof   (i) $\operatorname{tr}(\alpha A + \beta B) = \sum_i (\alpha A + \beta B)_{ii} = \alpha \sum_i A_{ii} + \beta \sum_i B_{ii} = \alpha \operatorname{tr}(A) + \beta \operatorname{tr}(B)$

(ii) $\operatorname{tr}(AB) = \sum_{i,j} A_{ij} B_{ji} = \sum_{i,j} B_{ji} A_{ij} = \operatorname{tr}(BA)$

(iii) For $k = 1$ this follows from Prop. 19.1 (iv). The general case follows from direct calculation.

$$\operatorname{tr}((PAP^{-1})^k) = \operatorname{tr}(PA\overbrace{P^{-1}P}^{\mathbb{1}_n}AP^{-1}\cdots PAP^{-1}) = \operatorname{tr}(PA^kP^{-1})$$

$$\overset{(ii)}{=} \operatorname{tr}(P^{-1}PA^k) = \operatorname{tr}(A^k)$$

$\square$

Since the trace is a class function, we can define the trace of a linear map as the trace of any of its representing matrices, in analogy with what we did for the determinant. Of course the linearity and commutativity properties of the matrix trace from Prop. 19.3 directly transfer to the trace for linear maps. The trace is of particular importance since it is a class function which is linear and it plays a role in many other areas of mathematics. Also note that we have obtained an entire sequence $\operatorname{tr}(A^k)$, where $k = 1, 2, \ldots$, of class functions.

Problem 19.3 *(Basis independence of determinant and trace)*

Are the matrices

$$A = \begin{pmatrix} 1 & 2 \\ -1 & -3 \end{pmatrix}, \qquad B = \begin{pmatrix} -1 & 1 \\ 3 & -2 \end{pmatrix}.$$

related by a basis transformation?

Solution: The two matrices have the same determinant, $\det(A) = \det(B) = -1$, which is inconclusive. However, their traces $\operatorname{tr}(A) = 1 - 3 = -2$ and $\operatorname{tr}(B) = -1 - 2 = -3$ are different so the matrices are not related by a basis transformation.

19.3   The theorem of Cayley–Hamilton*

Summary 19.3 *Endomorphisms $f \in \operatorname{End}(V)$ can be inserted into polynomials p which results in endomorphisms $p(f) \in \operatorname{End}(V)$. The Cayley–Hamilton theorem states that any endomorphism f inserted into its own characteristic polynomial χ_f gives the zero map, so $\chi_f(f) = 0$.*

The Cayley–Hamilton theorem establishes a profound relationship between an endomorphism and its characteristic polynomial. It is, admittedly, of limited use in applications of linear algebra but needs to be covered nevertheless because of its mathematical importance. We require a bit of preparation to formulate the statement.

19.3.1   Polyomials of endomorphisms

We start with an endomorphism $f \in \operatorname{End}(V)$ on a vector space V over $\mathbb{F}$ and a polynomial p with coefficients in $\mathbb{F}$. Our goal is to make sense of the expressions $p(f)$, obtained by 'inserting' f into the polynomial. From Section 12.1.4 we know that powers $f^k = f \circ \cdots \circ f \in \mathrm{End}(V)$ are again endomorphisms. We also recall that $\mathrm{End}(V)$ is a vector space (see Section 12.1.3) so endomorphisms can be added and scalar multiplied. This is really all that is required to see that the formal replacement

$$p(x) = \sum_{i=0}^{k} c_i x^i \qquad \xrightarrow{\;x \mapsto f\;} \qquad p(f) = \sum_{i=0}^{k} c_i f^i \tag{19.14}$$

of the polynomial's argument x by an endomorphism f makes sense, and leads to a new endomorphisms $p(f) \in \mathrm{End}(V)$.

In particular, we can insert square matrices $A \in \mathrm{End}(\mathbb{F}^n)$ into a polynomial. A useful observation is that this process commutes with basis transformations, that is,

$$p(P^{-1}AP) = P^{-1}p(A)P \,. \tag{19.15}$$

This follows from

$$p(PAP^{-1}) = \sum_{i=0}^{k} c_i \, PA\underbrace{P^{-1}P}_{=\,\mathbb{1}_n}\overbrace{AP^{-1}\cdots PAP^{-1}}^{k\ \mathrm{times}}$$

$$= \sum_{i=0}^{k} c_i PA^i P^{-1} = P\left(\sum_{i=0}^{k} c_i A^i\right) P^{-1} = P\, p(A)P^{-1} \,.$$

19.3.2 The minimal polynomial

For a given endomorphism $f \in \mathrm{End}(V)$ it is natural to consider the set

$$I_f := \{p \in \mathcal{P}(\mathbb{F}) \,|\, p(f) = 0\} \tag{19.16}$$

of polynomials for which $p(f)$ is the zero map. It is easy to see that this set contains non-zero polynomials. Consider the $n^2 + 1$ linear maps $f^0, f^1, f^2, \ldots, f^{n^2}$, where $n = \dim_{\mathbb{F}}(V)$. Since $\dim_{\mathbb{F}}(\mathrm{End}(V)) = n^2$ we know that these must be linearly dependent, so there are $\alpha_i \in \mathbb{F}$ with at least one $\alpha_i \neq 0$ such that

$$\sum_{i=0}^{n^2} \alpha_i f^i = 0 \,. \tag{19.17}$$

This means the polynomial $\sum_i \alpha_i x^i$ is contained in I_f. Since I_f is closed under addition and scalar multiplication of polynomials it is, in fact, a vector subspace of $\mathcal{P}(\mathbb{F})$. What is more, if $p(x) \in I_f$, then $q(x)p(x) \in I_f$ for any polynomial $q(x)$ and this property makes I_f into what is called an *ideal* (see, for example, Lang 2000). At first sight, I_f appears to be a complicated object but it is, in fact, easily described.

Proposition 19.4 *There is a unique monic polynomial $\mu_f \in I_f$ such that every $p \in I_f$ can be written as $p = q\,\mu_f$, where $q \in \mathcal{P}(\mathbb{F})$.*

Proof There is a minimal non-zero degree m which arises in I_f. Choose a monic polynomial μ with this degree. By polynomial division, every other polynomial $p \in I_f$ can be written as $p = q\,\mu + r$, where $\det(r) < m$ (see Theorem 4.1). But $r = p - q\,\mu \in I_f$ which contradicts degree minimality of μ, unless $r = 0$. Hence, every $p \in I_f$ can be written as $p = q\,\mu$.

For uniqueness consider another monic polynomial $\tilde{\mu} \in I_f$ with degree m. From the earlier statement it must be a multiple $\tilde{\mu} = a\,\mu$, where $a \in \mathbb{F}$, but since both $\tilde{\mu}$ and μ are monic it follows that $a = 1$. $\qquad\square$

The polynomial μ_f from this proposition is called the *minimal polynomial* of the endomorphism f. The set I_f is now easily described as all the polynomial multiples of the minimal polynomial μ_f, so

$$I_f = \{q\,\mu_f \,|\, q \in \mathcal{P}(\mathbb{F})\}\,.$$

19.3.3 The theorem

We have expressed the set I_f for $f \in \text{End}(V)$ in terms of a minimal polynomial but we do not yet know what its degree is. Eq. (19.17) shows it is certainly less equal than n^2, where $n = \dim_{\mathbb{F}}(V)$. The Cayley–Hamilton theorem states that the characteristic polynomial of f is in I_f, so this tightens the upper bound on the degree of the minimal polynomial to n.

Theorem 19.2 *(Cayley–Hamilton) For $f \in \text{End}(V)$, we have $\chi_f(f) = 0$.*

Proof We represent f by a matrix $A \in \mathcal{M}_{n,n}(\mathbb{F})$ relative to some basis of V and we want to proof that $\chi_A(A) = 0$. To do this define the matrix $M(\lambda) = (A - \lambda \mathbb{1}_n)^T$, so that $\det(M(\lambda)) = \chi_A(\lambda)$. Its entries are polynomials in λ, linear on the diagonal, constant otherwise. If we evaluate each of these polynomials on the matrix A, that is replace λ by A (and constants by $\mathbb{1}_n$), we get the $n^2 \times n^2$ matrix:

$$M(A) = \begin{pmatrix} A_{11}\mathbb{1}_n - A & A_{21}\mathbb{1}_n & \cdots & A_{n1}\mathbb{1}_n \\ \vdots & \vdots & \vdots & \vdots \\ A_{1n}\mathbb{1}_n & A_{2n}\mathbb{1}_n & \cdots & A_{nn}\mathbb{1}_n - A \end{pmatrix}.$$

This matrix has the remarkable property

$$\mathbf{v} := \begin{pmatrix} \mathbf{e}_1 \\ \vdots \\ \mathbf{e}_n \end{pmatrix}, \qquad M(A)\mathbf{v} = \begin{pmatrix} A_{11}\mathbf{e}_1 - A\mathbf{e}_1 + A_{21}\mathbf{e}_2 + \cdots A_{n1}\mathbf{e}_n \\ \vdots \\ A_{1n}\mathbf{e}_1 + A_{2n}\mathbf{e}_2 + \cdots A_{nn}\mathbf{e}_n - A\mathbf{e}_n \end{pmatrix} = \begin{pmatrix} \mathbf{0} \\ \vdots \\ \mathbf{0} \end{pmatrix},$$

of vanishing on the above vector $\mathbf{v}$. Denote by $C(\lambda)$ the co-factor matrix of $M(\lambda)$ so that, from Eq. (18.21)

$$C(\lambda)^T M(\lambda) = \det(M(\lambda))\mathbb{1}_n = \chi_A(\lambda)\mathbb{1}_n\,.$$

If we replace λ by A in this equation and let it act on $\mathbf{v}$, using $M(A)\mathbf{v} = \mathbf{0}$, we get

$$\begin{pmatrix} \mathbf{0} \\ \vdots \\ \mathbf{0} \end{pmatrix} = C(\lambda)M(\lambda)\mathbf{v} = \begin{pmatrix} \chi_A(A) & \cdots & 0 \\ \vdots & \vdots & \vdots \\ 0 & \cdots & \chi_A(A) \end{pmatrix}\mathbf{v} = \begin{pmatrix} \chi_A(A)\mathbf{e}_1 \\ \vdots \\ \chi_A(A)\mathbf{e}_n \end{pmatrix}.$$

This means $\chi_A(A)\mathbf{e}_i = \mathbf{0}$ for all $i = 1, \ldots, n$, so $\chi_A(A)$ vanishes on a basis. This is only possible if $\chi_A(A) = 0$. $\qquad\square$

An immediate conclusion form the Cayley–Hamilton theorem and Prop. 19.4 is that the characteristic polynomial must be a multiple of the minimal polynomial, so

$$\chi_f = q\,\mu_f \qquad\qquad (19.18)$$

for $q \in \mathcal{P}(\mathbb{F})$. This means zeros of the minimal polynomial are zeros of the characteristic polynomial and, hence, eigenvalues of f. More concretely, if these polynomials fully factorize (as is always the case if $\mathbb{F} = \mathbb{C}$) then

$$\chi_f(x) = (\lambda_1 - x)^{m_1} \cdots (\lambda_k - x)^{m_k} , \qquad \mu_f(x) = (x - \lambda_1)^{s_1} \cdots (x - \lambda_k)^{s_k} , \quad (19.19)$$

where $\lambda_1, \ldots, \lambda_k$ are the pairwise different eigenvalues of f with multiplicities m_i and $s_i \leq m_i$. In many cases, the minimal and characteristic polynomials are equal (up to a possible factor -1), but, as the following problem shows, this is not always the case.

Problem 19.4 *(Minimal and characteristic polynomials)*

Check the Cayley–Hamilton theorem for the matrices $A = \mathrm{diag}(0, 1, -1)$ and $B = \mathrm{diag}(-1, 1, 1)$ and compare their characteristic and minimal polynomials.

Solution: We have $\chi_A(x) = x(1 - x)(1 + x)$, so clearly $\chi_A(A) = 0$. There is no factor which can be dropped from χ_A while preserving the vanishing on A so in this case $\mu_A(x) = -\chi_A(x)$.

For the matrix B we have $\chi_B(x) = -(x+1)(x-1)^2$ and $\chi_B(B) = 0$ is immediate. But in this case one factor of $x - 1$ can be dropped, so $\mu_B(x) = (x+1)(x-1)$ is the minimal polynomial since $\mu_B(B) = 0$. This shows the minimal polynomial can indeed have a lower degree than the characteristic polynomial.

We will be able to be more precise about the multiplicities s_i in the minimal polynomial — and under which circumstances it differs from the characteristic polynomial — when we discuss the Jordan normal form in Chapter 21.

Exercises

19.1 *Eigenvalues for 2×2 matrices*
Find the eigenvalues, eigenspaces and degeneracies for the matrices

$$A = \begin{pmatrix} 1 & 2 \\ 3 & 1 \end{pmatrix}, \ B = \begin{pmatrix} -5 & 0 \\ 1 & 1 \end{pmatrix}$$

$$C = \begin{pmatrix} 1 & -2 \\ 1 & 1 \end{pmatrix}, \ D = \begin{pmatrix} 0 & i \\ 0 & 0 \end{pmatrix}$$

where $A, B \in \mathcal{M}_{2,2}(\mathbb{R})$ and $C, D \in \mathcal{M}_{2,2}(\mathbb{C})$.

19.2 *Eigenvalues for real 3×3 matrices*
Find the eigenvalues, eigenspaces and degeneracies for the linear map $A : \mathbb{R}^3 \to \mathbb{R}^3$ given by

$$A = \begin{pmatrix} 4 & \frac{3}{2} & \frac{3}{2} \\ -9 & -\frac{7}{2} & -\frac{9}{2} \\ -3 & -\frac{3}{2} & -\frac{1}{2} \end{pmatrix} .$$

19.3 *Eigenvalues for complex 3×3 matrices*
Find the eigenvalues, eigenspaces, and degeneracies for the linear map $A : \mathbb{C}^3 \to \mathbb{C}^3$ given by

$$B = \begin{pmatrix} 1+i & \frac{1}{2}+\frac{i}{2} & 1+i \\ -1+i & -1 & -1+i \\ -1 & -\frac{1}{2}-\frac{i}{2} & -1 \end{pmatrix}$$

19.4 *Eigenvalues of a differential operator*
Consider the space $V_k = \mathcal{P}_k(\mathbb{R})$ of polynomials with degree less equal k and the linear maps $V_k \to V_k$ given by

$$D = \frac{d}{dx} , \quad L = xD + 1 .$$

(a) For $k = 2$, find the eigenvalues and eigenvectors of D by solving the eigenvalue equation.
(b) For $k = 2$, find the representing matrix for D relative to the monomial

basis $(1, x, x^2)$ and use this to work out the eigenvalues and eigenvectors.
(c) Carry out the analogous tasks for the map L.
(d) Work out eigenvalues and eigenvectors for D and L for arbitrary k.

19.5 *Determinant in terms of traces*
(a) Find a formula for the determinant of matrices $A \in \mathcal{M}_{2,2}(\mathbb{F})$ in terms of $\operatorname{tr}(A)$ and $\operatorname{tr}(A^2)$.
(b) Find the analogous formula for matrices $A \in \mathcal{M}_{3,3}(\mathbb{F})$.

19.6 (a) Show that the characteristic polynomial of a matrix $A \in \mathcal{M}_{2,2}(\mathbb{F})$ can be written as

$$\chi_A(\lambda) = \lambda^2 - \operatorname{tr}(A)\lambda + \det(A) .$$

(b) Do the same for a matrix $A \in \mathcal{M}_{3,3}(\mathbb{F})$ and show that

$$\begin{aligned} \chi_A(\lambda) = &-\lambda^3 + \operatorname{tr}(A)\lambda^2 \\ &-\tfrac{1}{2}(\operatorname{tr}(A)^2 - \operatorname{tr}(A^2))\lambda \\ &+\det(A) . \end{aligned}$$

19.7 Show that the matrices in Exercise 19.1 are pairwise non-conjugate.

19.8 For the matrices B and D in Exercise 19.1 explicitly verify the Cayley–Hamilton theorem.

19.9 For a linear map $f : V \to V$ show that, provided f is invertible,
(a) all eigenvalues of f are non-zero.
(b) if λ is an eigenvalue for f then λ^{-1} is an eigenvalue for f^{-1}.
(c) $\operatorname{Eig}_f(\lambda) = \operatorname{Eig}_{f^{-1}}(\lambda^{-1})$.

19.10 For a linear map $f : V \to V$ prove the following statements.
(a) If λ is an eigenvalue of f then λ^k is an eigenvalue of f^k.
(b) $\operatorname{Eig}_f(\lambda) \subset \operatorname{Eig}_{f^k}(\lambda^k)$.
(c) Give an example to show that $\operatorname{Eig}_{f^k}(\lambda^k)$ can be larger than $\operatorname{Eig}_f(\lambda)$.

20
Diagonalizing linear maps

Probably the most important application of eigenvalues and eigenvectors is to the diagonalization of linear maps. Unfortunately, not all linear maps can be diagonalized so the first task is to spell out criteria for when this is possible. When it is we can formulate an algorithm for how to carry this out and in Section 20.2 this algorithm will be applied to a number of examples. Projectors are a class of endomorphisms with a range of interesting applications and, as we will see in Section 20.3, they can always be diagonalized. Sometimes, problems involve two or more linear maps and it may be desirable to diagonalize them simultaneously. This point will be addressed in the final section of this chapter.

20.1 Diagonalization

Summary 20.1 *We say a linear map $f : V \to V$ can be diagonalized if there is a basis of V relative to which f is described by a diagonal matrix. Not all linear maps $f : V \to V$ can be diagonalized. This is the case if and only if there is a basis $(\mathbf{v}_1, \dots, \mathbf{v}_n)$ of V which consists of eigenvectors of f, so that $f(\mathbf{v}_i) = \hat{\lambda}_i \mathbf{v}_i$. Then, the diagonal matrix which describes f relative to this basis is $\hat{A} = \mathrm{diag}(\hat{\lambda}_1, \dots, \hat{\lambda}_n)$.*

To be precise we start by defining what exactly we mean by saying that a linear map or a matrix can be diagonalized.

Definition 20.1 *(i) We say a linear map $f : V \to V$ can be diagonalized if there exist a basis of V relative to which f is described by a diagonal matrix.*

(ii) We say a matrix $A \in \mathcal{M}_{n,n}(\mathbb{F})$ can be diagonalized if there exists a matrix $P \in \mathrm{GL}(\mathbb{F}^n)$ such that the basis-transformed matrix $P^{-1}AP$ is diagonal.

20.1.1 Basic criteria

Here is a set of criteria to decide whether a linear map can be diagonalized.

Theorem 20.1 *Let $f : V \to V$ be a linear map on an n-dimensional vector space V. The degeneracies of the pairwise different eigenvalues λ_i of f are denoted by d_i and their multiplicity in χ_f by m_i, where $i = 1, \dots, k$. Then the following statements are equivalent.*

   (i)   f can be diagonalized.
   (ii)   There exists a basis of V which consists of eigenvectors of f.
   (iii) The degeneracies sum to the total dimensions, $\sum_{i=1}^{k} d_i = n$.
   (iv) $\sum_{i=1}^{k} m_i = n$ (χ_f fully factorizes) and $d_i = m_i$ for $i = 1, \ldots, k$.
   (v) $\bigoplus_{i=1}^{k} \mathrm{Eig}_f(\lambda_i) = V$

Proof (i) '$\Rightarrow$ (ii)': Assume that f can be diagonalized. From Def. 20.1 there exists a basis $(\mathbf{v}_1, \ldots, \mathbf{v}_n)$ relative to which f is described by a diagonal matrix $\hat{A} = \mathrm{diag}(\hat{\lambda}_1, \ldots, \hat{\lambda}_n)$. Given the general relationship (19.1) between a linear map and a representing matrix this implies $f(\mathbf{v}_j) = \sum_{i=1}^{n} \hat{A}_{ij}\mathbf{v}_i = \hat{\lambda}_j\mathbf{v}_j$. Hence, the $\mathbf{v}_i$ are eigenvectors of f with eigenvalues $\hat{\lambda}_i$ and $(\mathbf{v}_1, \ldots, \mathbf{v}_n)$ is a basis of eigenvectors.

(ii) '$\Rightarrow$ (iii)': Assume $(\mathbf{v}_1, \ldots, \mathbf{v}_n)$ is a basis of V consisting of eigenvectors of f, so that $f(\mathbf{v}_i) = \hat{\lambda}_i\mathbf{v}_i$. Group the eigenvalues $\hat{\lambda}_i$ into pairwise distinct ones, $\lambda_1, \ldots, \lambda_k$, each of which arises δ_i times in the list $(\hat{\lambda}_1, \ldots, \hat{\lambda}_n)$ and, hence, come with δ_i linearly independent eigenvectors. The span of these eigenvectors forms a δ_i-dimensional vector subspace of $\mathrm{Eig}_f(\lambda_i)$ so it follows that $\delta_i \leq d_i = \dim_{\mathbb{F}}(\mathrm{Eig}_f(\lambda_i))$ and $\sum_{i=1}^{k} d_i \geq \sum_{i=1}^{k} \delta_i = n$. But since the sum of eigenspaces for different eigenvalues is direct we also have $\sum_{i=1}^{k} d_i \leq n$. Combining these inequalities we get the desired statement $\sum_{i=1}^{k} d_i = n$ (and indeed $\delta_i = d_i$ for all $i = 1, \ldots, k$).

(iii) '$\Rightarrow$ (iv)': Assume that $\sum_{i=1}^{k} d_i = n$. Certainly, $\sum_{i=1}^{k} m_i \leq n$, but since $m_i \geq d_i$ from Prop. 19.2, it also follows that $\sum_{i=1}^{k} m_i \geq \sum_{i=1}^{k} d_i = n$. Hence, $\sum_{i=1}^{k} m_i = n$, so the characteristic polynomial fully factorizes. Given that $d_i \leq m_i$, if any $d_i < m_i$ then $\sum_{i=1}^{k} d_i < n$, so we must have $d_i = m_i$ for all $i = 1, \ldots, k$.

(iv) '$\Rightarrow$ (v)': Since the sum of eigenspaces for different eigenvalues is direct the condition $\sum_{i=1}^{k} d_i = n$ implies $\bigoplus_{i=1}^{k} \mathrm{Eig}_f(\lambda_i) = V$ (see Cor. 7.1 and Eq. (8.7)).

(v) '$\Rightarrow$ (i)': Given that $\bigoplus_{i=1}^{k} \mathrm{Eig}_f(\lambda_i) = V$, choose a basis for each eigenspace and combine these into a basis $(\mathbf{v}_1, \ldots, \mathbf{v}_n)$ of V with $f(\mathbf{v}_i) = \hat{\lambda}_i\mathbf{v}_i$ for $i = 1, \ldots, n$. (As above, the $\hat{\lambda}_i$ are the same as the eigenvalues λ_i but with repetitions to account for the degeneracies.) From Eq. (19.1) the representing matrix for f relative to this basis is the diagonal matrix $\hat{A} = \mathrm{diag}(\hat{\lambda}_1, \ldots, \hat{\lambda}_n)$ so f can indeed be diagonalized. $\square$

In particular, the theorem tells us that a linear map $f : V \to V$ on an n-dimensional vector space V can be diagonalized if it has n pairwise distinct eigenvalues. Indeed, in this case we have $m_i = d_i = 1$ for $i = 1, \ldots, n$ so that criterion (iv) is satisfied. But also note that finding less than n distinct eigenvalues does not imply that the linear map cannot be diagonalized — degeneracies might save the day. If multiplicities sum to less than the dimension, so $\sum_{i=1}^{k} m_i < n$, we can definitely say from (iv) that the map cannot be diagonalized. On the other hand, if $\sum_{i=1}^{k} m_i = n$, the map may still not be diagonalizable since it may happen that $d_i < m_i$ for some i. In essence, Theorem 20.1 says that, special cases apart, the number of eigenvalues is not the right quantity to decide whether a map can be diagonalized but what matters is the number of linearly independent eigenvectors.

20.1.2 The diagonal matrix

If a map can be diagonalized we should spell out exactly what this implies.

Corollary 20.1 *Relative to a basis* $(\mathbf{v}_1, \ldots, \mathbf{v}_n)$ *of eigenvectors, a map* $f \in \text{End}(V)$ *with eigenvalues* $\hat{\lambda}_1, \ldots, \hat{\lambda}_n$ *is described by the matrix* $\hat{A} = \text{diag}(\hat{\lambda}_1, \ldots, \hat{\lambda}_n)$. *If* f *is identified with a matrix* $A \in \text{End}(\mathbb{F}^n)$ *it is diagonalized by*

$$\hat{A} = \text{diag}(\hat{\lambda}_1, \ldots, \hat{\lambda}_n) = P^{-1}AP \quad \text{where} \quad P = (\mathbf{v}_1, \ldots, \mathbf{v}_n) . \tag{20.1}$$

Proof Since $f(\mathbf{v}_i) = \hat{\lambda}_i \mathbf{v}_i$, Eq. (19.1) implies that $\hat{A} = \text{diag}(\hat{\lambda}_1, \ldots, \hat{\lambda}_n)$ represents f relative to the basis $(\mathbf{v}_1, \ldots, \mathbf{v}_n)$. For a matrix A, the formula (20.1) follows from the result for basis transformations, Eq. (15.14). $\qquad\square$

Note that the diagonal matrix which describes f has the eigenvalues as its diagonal entries. Also, Eq. (20.1) is quite convenient for diagonalizing matrices. It tells us that the diagonalizing basis transformation P is the matrix whose columns are the eigenvectors.

20.1.3 Diagonalizing and class functions

Suppose an endomorphism $f \in \text{End}(V)$ can be diagonalized with diagonal matrix $\hat{A} = \text{diag}(\hat{\lambda}_1, \ldots, \hat{\lambda}_n)$ and eigenvalues $\hat{\lambda}_i$. Any class function from Section 19.2.5 evaluates to the same value in any basis, including the diagonalizing one. This implies

$$\det(f) = \prod_{i=1}^{n} \hat{\lambda}_i , \qquad \text{tr}(f) = \sum_{i=1}^{n} \hat{\lambda}_i , \qquad \text{tr}(f^k) = \sum_{i=1}^{n} \hat{\lambda}_i^k , \tag{20.2}$$

so the determinant is the product and the trace is the sum of the eigenvalues (included with degeneracies). These formulae can be useful for explicit calculations (see Exercise 20.3).

20.2 Examples

Summary 20.2 *We practice diagonalizing linear maps, including* 2×2, 3×3 *and* 4×4 *matrices and a linear differential operator.*

Before we tackle explicit examples we summarize the results be setting up an algorithm for diagonalizing linear maps.

Algorithm (*Diagonalizing linear maps*) *To diagonalize a linear map* $f : V \to V$ *where* V *is an* n-*dimensional vector space, proceed as follows:*
(1) *Find the pairwise different eigenvalues* $\lambda_1, \ldots, \lambda_k$ *of* f *and compute the eigenspace* $\text{Eig}_f(\lambda_i)$ *for each.*

(2) Define the degeneracies $d_i = \dim_{\mathbb{F}}(\mathrm{Eig}_f(\lambda_i))$. If $\sum_{i=1}^{k} d_i < n$ the map f cannot be diagonalized and there is nothing more to do. Otherwise, if $\sum_{i=1}^{k} d_i = n$ holds, f can be diagonalized.

(3) Choose a basis for each eigenspace $\mathrm{Eig}_f(\lambda_i)$ and combine these bases into a basis $(\mathbf{v}_1, \ldots, \mathbf{v}_n)$ of V with $f(\mathbf{v}_i) = \hat{\lambda}_i \mathbf{v}_i$. Relative to this basis f is described by the matrix $\hat{A} = \mathrm{diag}(\hat{\lambda}_1, \ldots, \hat{\lambda}_n)$.

(4) If $V = \mathbb{F}^n$ and f is a matrix A, then $\hat{A} = P^{-1}AP$, where $P = (\mathbf{v}_1, \ldots, \mathbf{v}_n)$.

Problem 20.1 *(Diagonalizing 2 × 2 matrices)*

Check if the 2×2 matrices from Exericse 19.2, seen as linear maps on $\mathbb{R}^2$ or $\mathbb{C}^2$, can be diagonalized and, if so, find the diagonalizing basis transformation and the diagonal matrix.

Solution: The matrix $\mathbb{1}_2$ is already diagonal so, clearly, it can be diagonalized.

The matrix A in Eq. (19.12) has only one eigenvalue, $\lambda = 1$ with multiplicity 2, which is non-degenerate. Hence, from Theorem 20.1 (iii) it cannot be diagonalized.

The matrix B in Eq. (19.12) has no eigenvalues over $\mathbb{R}$ so cannot be diagonalized in this case. Over $\mathbb{C}$ it has two non-degenerate eigenvalues, $\lambda_{\pm} = \pm i$ with eigenvalues $\mathbf{v}_{\pm} = (1, \pm i)^T$, so in this case it can be diagonalized. We already know that the diagonalized matrix must have the eigenvalues along the diagonal, so $\hat{A} = \mathrm{diag}(i, -i)$. This can also be checked explicitly by carrying out the basis transformation (20.1):

$$P = (\mathbf{v}_+, \mathbf{v}_-) = \begin{pmatrix} 1 & 1 \\ i & -i \end{pmatrix}, \quad P^{-1}BP = \frac{1}{2} \begin{pmatrix} 1 & -i \\ 1 & i \end{pmatrix} \begin{pmatrix} 0 & 1 \\ -1 & 0 \end{pmatrix} \begin{pmatrix} 1 & 1 \\ i & -i \end{pmatrix} = \mathrm{diag}(i, -i) = \hat{A}.$$

Finally, the matrix C in Eq. (19.12) has two eigenvalues $\lambda_1 = 1$, $\lambda_2 = 3$ with eigenvalues $\mathbf{v}_1 = (1, -1)^T$ and $\mathbf{v}_2 = (1, 1)^T$. Hence, it can be diagonalized and the diagonal matrix is $\hat{A} = \mathrm{diag}(1, 3)$. This can be verified by

$$P = (\mathbf{v}_1, \mathbf{v}_2) = \begin{pmatrix} 1 & 1 \\ -1 & 1 \end{pmatrix}, \quad P^{-1}CP = \frac{1}{2} \begin{pmatrix} 1 & -1 \\ 1 & 1 \end{pmatrix} \begin{pmatrix} 2 & 1 \\ 1 & 2 \end{pmatrix} \begin{pmatrix} 1 & 1 \\ -1 & 1 \end{pmatrix} = \mathrm{diag}(1, 3) = \hat{A}.$$

Problem 20.2 *(Diagonalizing a 3 × 3 matrix)*

Show that the matrix $A \in \mathrm{End}(\mathbb{R}^3)$ from Exercise 19.1 can be diagonalized. Find the diagonalizing basis transformation and the diagonal matrix.

Solution: The matrix A in Eq. (19.11) has eigenvalues $\lambda_1 = 0$, $\lambda_2 = 1$ and $\lambda_3 = 3$ with eigenvectors $\mathbf{v}_1 = (1, 1, 1)^T$, $\mathbf{v}_2 = (-1, 0, 1)^T$ and $\mathbf{v}_3 = (1, -2, 1)^T$. Since there are three different eigenvalues in three dimensions the matrix can be diagonalized. The diagonal matrix is $\hat{A} = \mathrm{diag}(0, 1, 3)$ and this is confirmed by a basis transformation.

$$P = (\mathbf{v}_1, \mathbf{v}_2, \mathbf{v}_3) = \begin{pmatrix} 1 & -1 & 1 \\ 1 & 0 & -2 \\ 1 & 1 & 1 \end{pmatrix}, \quad P^{-1} = \frac{1}{6} \begin{pmatrix} 2 & 2 & 2 \\ -3 & 0 & 3 \\ 1 & -2 & 1 \end{pmatrix} \quad \Rightarrow \quad P^{-1}AP = \mathrm{diag}(0, 1, 3).$$

Problem 20.3 *(Diagonalizing a 4×4 matrix)*

For the 4×4 matrix $A \in \text{End}(\mathbb{R}^4)$ defined by

$$A = \begin{pmatrix} -9 & -28 & 7 & -2 \\ 5 & 24 & -7 & 2 \\ 5 & 28 & -11 & 2 \\ -18 & -24 & 6 & 8 \end{pmatrix}$$

find the eigenvalues and eigenspaces. Check if A can be diagonalized and if so, find the diagonalizing basis transformation and the diagonal matrix.

Solution: The characteristic polynomial for A is

$$\chi_A(\lambda) = \det(A - \lambda \mathbb{1}_4) = (\lambda + 4)^2 (\lambda - 8)(\lambda - 12)\,,$$

so we have three eigenvalues $\lambda_1 = -4$, $\lambda_2 = 8$ and $\lambda_3 = 12$. This is fewer eigenvalues than the dimension of the space so to check whether A can be diagonalized we have to compute the eigenspaces and degeneracies. Writing $\mathbf{v} = (x, y, z, u)^T$ we have

$$\lambda_1 = -4: \quad (A - \lambda_1 \mathbb{1}_4)\mathbf{v} = \begin{pmatrix} -8 & -28 & 7 & -2 \\ 5 & 25 & -7 & 2 \\ 5 & 28 & -10 & 2 \\ -18 & -24 & 6 & 9 \end{pmatrix} \begin{pmatrix} x \\ y \\ z \\ u \end{pmatrix} \stackrel{!}{=} 0 \quad \Leftrightarrow \quad x = u,\ z - u - 4y = 0$$

$$\lambda_2 = 8: \quad (A - \lambda_1 \mathbb{1}_4)\mathbf{v} = \begin{pmatrix} -11 & -28 & 7 & -2 \\ 5 & 22 & -7 & 2 \\ 5 & 28 & -13 & 2 \\ -18 & -24 & 6 & 6 \end{pmatrix} \begin{pmatrix} x \\ y \\ z \\ u \end{pmatrix} \stackrel{!}{=} 0 \quad \Leftrightarrow \quad 2x = -2y = -2z = u$$

$$\lambda_3 = 12: \quad (A - \lambda_1 \mathbb{1}_4)\mathbf{v} = \begin{pmatrix} -12 & -28 & 7 & -2 \\ 5 & 21 & -7 & 2 \\ 5 & 28 & -14 & 2 \\ -18 & -24 & 6 & 5 \end{pmatrix} \begin{pmatrix} x \\ y \\ z \\ u \end{pmatrix} \stackrel{!}{=} 0 \quad \Leftrightarrow \quad x = -y = -z,\ u = 0$$

For λ_1 we have only two conditions on the vector $\mathbf{v}$ so the eigenspace is two-dimensional while the other two eigenspaces are one-dimensional. If we write

$$\text{Eig}_A(-4) = \text{Span}(\mathbf{v}_1, \mathbf{v}_2)\,, \quad \text{Eig}_A(8) = \text{Span}(\mathbf{v}_3)\,, \quad \text{Eig}_A(12) = \text{Span}(\mathbf{v}_4)$$

the vectors $\mathbf{v}_i$ can be taken as

$$\mathbf{v}_1 = \begin{pmatrix} 4 \\ -1 \\ 0 \\ 4 \end{pmatrix}\,, \quad \mathbf{v}_2 = \begin{pmatrix} 0 \\ 1 \\ 4 \\ 0 \end{pmatrix}\,, \quad \mathbf{v}_3 = \begin{pmatrix} 1 \\ -1 \\ -1 \\ 2 \end{pmatrix}\,, \quad \mathbf{v}_4 = \begin{pmatrix} -1 \\ 1 \\ 1 \\ 0 \end{pmatrix}\,.$$

These vectors form a basis of $\mathbb{R}^4$ (or, equivalently, the degeneracies of the eigenvalues sum up to four) so the matrix A can be diagonalized. A diagonalizing matrix is

$$P = (\mathbf{v}_1, \mathbf{v}_2, \mathbf{v}_3, \mathbf{v}_4) = \begin{pmatrix} 4 & 0 & 1 & -1 \\ -1 & 1 & -1 & 1 \\ 0 & 4 & -1 & 1 \\ 4 & 0 & 2 & 0 \end{pmatrix}\,, \quad P^{-1} = \frac{1}{8} \begin{pmatrix} 3 & 4 & -1 & 0 \\ -1 & -4 & 3 & 0 \\ -6 & -8 & 2 & 4 \\ -2 & 8 & -2 & 4 \end{pmatrix}$$

and it can be checked that indeed $P^{-1}AP = \text{diag}(-4, -4, 8, 12) = \hat{A}$. Note that, instead of $(\mathbf{v}_1, \mathbf{v}_2)$, we could have chosen any other basis of the two-dimensional eigenspace $\text{Eig}_A(-4)$.

Problem 20.4 *(Diagonalizing a differential operator)*

It is worth discussing eigenvalues and eigenvectors for a linear map which is not defined by a matrix. To this end, consider the vector space $V = \mathcal{P}_2(\mathbb{R})$ of at most quadratic polynomials and the differential operator

$$L = (1+x)\frac{d}{dx} : V \to V .$$

Find the eigenvalues and eigenvectors of L, check if it can be diagonalized and if so find its diagonal representing matrix.

Solution: The first step is to work out the characteristic polynomial χ_L. To do this we have to remember that the determinant of a linear map is defined to be the determinant of any of its representing matrices. So we choose the simple monomial basis $(1, x, x^2)$ for V and work out the representing matrix A for L, relative to this basis:

$$\left.\begin{array}{l} L(1) = 0 \\ L(x) = 1+x \\ L(x^2) = 2x + 2x^2 \end{array}\right\} \quad \Rightarrow \quad A = \begin{pmatrix} 0 & 1 & 0 \\ 0 & 1 & 2 \\ 0 & 0 & 2 \end{pmatrix} \quad \Rightarrow \quad \chi_L(\lambda) = \det(A - \lambda\,\mathbb{1}_3) = -\lambda(\lambda-1)(\lambda-2) .$$

Hence, we have three eigenvalues, $\lambda = 0, 1, 2$, and it follows that L can be diagonalized. To work out the eigenspaces, we first write out a general quadratic polynomial $p(x) = a_0 + a_1 x + a_2 x^2$ and compute $L(p)(x) = a_1 + (a_1 + 2a_2)x + 2a_2 x^2$. Then, the eigenvalue equation $L(p) = \lambda p$ for the three eigenvalues reads

$$L(p) = a_1 + (a_1 + 2a_2)x + 2a_2 x^2 = \left\{\begin{array}{l} 0 \\ a_0 + a_1 x + a_2 x^2 \\ 2a_0 + 2a_1 x + 2a_2 x^2 \end{array}\right. \left.\begin{array}{l} \lambda = 0 \\ \lambda = 1 \\ \lambda = 2 \end{array}\right\} \quad \Leftrightarrow \quad \left\{\begin{array}{l} a_1 = a_2 = 0 \\ a_0 = a_1, a_2 = 0 \\ 2a_0 = a_1 = 2a_2 \end{array}\right.$$

This means, the three eigenspaces are

$$\mathrm{Eig}_L(0) = \mathrm{Span}(1) , \quad \mathrm{Eig}_L(1) = \mathrm{Span}(1+x) , \quad \mathrm{Eig}_L(2) = \mathrm{Span}(1 + 2x + x^2) ,$$

and, relative to the basis $(1, 1+x, 1+2x+x^2)$, the operator L is diagonalized and described by the matrix $\hat{A} = \mathrm{diag}(0, 1, 2)$.

Example 20.1 *(Discrete linear process)*

Our motivational example at the beginning of the chapter was about a discrete process, with the state $\mathbf{x}_k \in \mathbb{R}^n$ of a system at time $k = 0, 1, \ldots$ given recursively by $\mathbf{x}_{k+1} = M\mathbf{x}_k$, where $M \in \mathcal{M}_{n,n}(\mathbb{R})$ is a matrix. As we have mentioned, we can write down a formal solution

$$\mathbf{x}_k = M^k \mathbf{x}_0 \tag{20.3}$$

but this is of little practical use unless we can evaluate the matrix powers M^k. Eigenvalues and eigenvectors can help us to do this. First, we need to assume that the matrix M can be diagonalized, so that $P^{-1}MP = \hat{M} = \mathrm{diag}(\hat{\lambda}_1, \ldots, \hat{\lambda}_n)$ with eigenvalues $\hat{\lambda}_i$ and a suitable basis transformation P. The key point is that the same matrix P also diagonalizes all matrix powers M^k. To see this note that

$$P^{-1}M^k P = P^{-1}M \underbrace{PP^{-1}}_{=\mathbb{1}_n} M \cdots \underbrace{PP^{-1}}_{=\mathbb{1}_n} MP = \hat{M}^k = \mathrm{diag}(\hat{\lambda}_1^k, \ldots, \hat{\lambda}_n^k) , \tag{20.4}$$

where the unit matrix in the form PP^{-1} has been inserted between the factors of M in the second step. This result also implies that the eigenvalues of M^k are $\hat{\lambda}_i^k$ (see

Exercise 19.10). Now introduce the vectors $\mathbf{y}_k = P^{-1}\mathbf{x}_k$ and multiply Eq. (20.3) with P^{-1} from the left.

$$\mathbf{y}_k = P^{-1}\mathbf{x}_k = P^{-1}M^k\mathbf{x}_0 = P^{-1}M^kPP^{-1}\mathbf{x}_0 = \hat{M}^k\mathbf{y}_0 \,.$$

This manoeuvre has diagonalized the equation and the components $y_{k,i}$ of the vector $\mathbf{y}_k$ can now easily be computed from $y_{k,i} = \hat{\lambda}_i^k y_{0,i}$. If an eigenvalue $\hat{\lambda}_i$ satisfies $|\hat{\lambda}_i| < 1$ then the corresponding component $y_{k,i}$ will go to zero for large k. For $\hat{\lambda}_i = 1$, the component $y_{k,i}$ remains constant and for $|\hat{\lambda}_i| > 1$ it grows unbounded. With $\mathbf{y}_k$ determined, the original variable $\mathbf{x}_k$ can be re-covered from $\mathbf{x}_k = P\mathbf{y}_k$. □

Problem 20.5 *(A specific discrete linear process)*

The fractions with which two features occur in a populations are described by the components of $\mathbf{x}_k \in \mathbb{R}^2$, were $k = 0, 1, \ldots$ is discrete time. Evolution is according to $\mathbf{x}_{k+1} = M\mathbf{x}_k$, where

$$M = \frac{1}{6}\begin{pmatrix} 5 & 2 \\ 1 & 4 \end{pmatrix} \,,$$

and, initially, only the first feature exists, so $\mathbf{x}_0 = (1,0)^T$. Find $\mathbf{x}_k$ for large k.

Solution: The characteristic polynomial for M is $\chi_M(\lambda) = \det(M - \lambda\mathbb{1}_2) = (\lambda - 1/2)(\lambda - 1)$ so we have eigenvalues $\lambda_1 = 1/2$ and $\lambda_2 = 1$. Associated eigenvectors are easily found to be $\mathbf{v}_1 = (1, -1)^T$ and $\mathbf{v}_2 = (2, 1)^T$, so that

$$P = (\mathbf{v}_1, \mathbf{v}_2) = \begin{pmatrix} 1 & 2 \\ -1 & 1 \end{pmatrix}, \quad P^{-1}MP = \frac{1}{18}\begin{pmatrix} 1 & -2 \\ 1 & 1 \end{pmatrix}\begin{pmatrix} 5 & 2 \\ 1 & 4 \end{pmatrix}\begin{pmatrix} 1 & 2 \\ -1 & 1 \end{pmatrix} = \mathrm{diag}(1/2, 1) = \hat{M} \,.$$

The power $\hat{M}^k = \mathrm{diag}((1/2)^k, 1)$ becomes $\hat{M}_\infty = \mathrm{diag}(0, 1)$ in the large k limit. Hence,

$$M_\infty = P\hat{M}_\infty P^{-1} = \frac{1}{3}\begin{pmatrix} 2 & 2 \\ 1 & 1 \end{pmatrix} \quad \Rightarrow \quad M_\infty \mathbf{x}_0 = \begin{pmatrix} 2/3 \\ 1/3 \end{pmatrix} \,.$$

For large k the fraction of the population with the first (second) feature is $2/3$ $(1/3)$.

Application 20.1 *Newton's equation with linear forces*

A common problem in classical mechanics is the motion under linear forces. Suppose, the system is described by n coordinates $\mathbf{q}(t) = (q_1(t), \ldots, q_n(t))^T \in \mathbb{R}^n$ which are functions of time $t \in \mathbb{R}$ and evolve according to the differential equation

$$\ddot{\mathbf{q}} = -M\mathbf{q} \,,$$

where $M \in \mathcal{M}_{n,n}(\mathbb{R})$ is an $n \times n$ matrix, and the dot indicates a time-derivative d/dt. Solving this differential equation is complicated by the presence of the matrix M which may be non-diagonal and may 'couple' the n components of the equation.

We can make significant progress if we assume that M can be diagonalized, so that there is an invertible $n \times n$ matrix P such that $P^{-1}MP = \hat{M} := \mathrm{diag}(m_1, \ldots, m_n)$ and, for simplicy, that the eigenvalues $m_i \in \mathbb{R}$ are real. (These assumption have to be checked for a given example, but as we will see in Chapter 24 they are automatic for symmetric matrices.) If we introduce new coordinates $\mathbf{s} = P^{-1}\mathbf{q}$ and multiply the differential equation with P^{-1} we find

$$\ddot{\mathbf{s}} = -\underbrace{P^T M P}_{=\hat{M}}\mathbf{s} \qquad \Leftrightarrow \qquad \ddot{s}_i = -m_i s_i \quad \text{for} \quad i = 1, \ldots, n \,.$$

By diagonalizing M and re-writing the system in terms of the new coordinates $\mathbf{s}$ we have decoupled the n components of the equation. Each component can now be solved separately and this leads to

$$s_i(t) = \begin{cases} a_i \sin(w_i t) + b_i \cos(w_i t) & \text{for } m_i > 0 \\ a_i e^{w_i t} + b_i e^{-w_i t} & \text{for } m_i < 0 \\ a_i t + b_i & \text{for } m_i = 0 \end{cases} \quad \text{where} \quad w_i = \sqrt{|m_i|} \,,$$

and $a_i, b_i \in \mathbb{R}$ are arbitrary constants. To obtain the solution in terms of the original coordinates all we have to do is work out $\mathbf{q} = P\mathbf{s}$. (In a physical context, the constants a_i, b_i can be fixed by imposing initial conditions, that is, be demanding specific values for $\mathbf{q}(t_0)$ and $\dot{\mathbf{q}}(t_0)$ at a given time t_0.)

An interesting observation which we would like to focus on is that the nature of the solution depends on the signs of the eigenvalues m_i of the matrix M. For a positive eigenvalue, the solution is oscillatory, for a negative one exponential and for a vanishing one linear. Physically, a negative or vanishing eigenvalue m_i indicates an instability. In this case, the corresponding solution for $s_i(t)$ becomes large at late times (except for special choices of the constants a_i, b_i). The lesson is that stability of the system can be analysed simply by looking at the eigenvalues of M. If they are all positive, the system is fully oscillatory and stable, if there are vanishing or negative eigenvalues the system generically 'runs away' in some directions.

As an explicit example, consider the coordinates $\mathbf{q}(t) = (q_1(t), q_2(t), q_3(t))^T$ with differential equations

$$\left.\begin{array}{l} \ddot{q}_1 = -q_1 + q_2 \\ \ddot{q}_2 = q_1 - q_2 + q_3 \\ \ddot{q}_3 = q_2 - q_3 \end{array}\right\} \quad \Rightarrow \quad M = \begin{pmatrix} 1 & -1 & 0 \\ -1 & 2 & -1 \\ 0 & -1 & 1 \end{pmatrix} \,.$$

This matrix M has already been studied in Exercise 20.2 (a) and we have seen that is has three eigenvalues, $m_1 = 0$, $m_2 = 1$ and $m_3 = 3$, and that it can be diagonalized. Inserting into the above general formulae, the solution in term of the variable $\mathbf{s}$ is

$$\mathbf{s}(t) = \begin{pmatrix} a_1 t + b_1 \\ a_2 \sin(t) + b_2 \cos(t) \\ a_3 \sin(\sqrt{3}t) + b_3 \cos(\sqrt{3}t) \end{pmatrix}$$

and we can rotate this to the original coordinates by $\mathbf{q} = P\mathbf{s}$ with the matrix P from Exercise 20.2 (a). We note that the system is oscillatory in the directions s_2 and s_3 but, due to the vanishing eigenvalue $m_1 = 0$, linear and, hence, unstable in the direction s_1.

For another explicit case, consider two variables $\mathbf{q}(t) = (q_1(1), q_2(t))^T$ with differential equations

$$\left.\begin{array}{l} \ddot{q}_1 = -(aq_1 + bq_2) \\ \ddot{q}_2 = -(bq_1 + aq_2) \end{array}\right\} \quad \Rightarrow \quad M = \begin{pmatrix} a & b \\ b & a \end{pmatrix}$$

where $a, b \in \mathbb{R}$ are parameters. The characteristic polynomial

$$\chi_M(\lambda) = \det \begin{pmatrix} a - \lambda & b \\ b & a - \lambda \end{pmatrix} = (\lambda - a - b)(\lambda - a + b) \,,$$

shows we have two eigenvalues, $\lambda_\pm = a \pm b$, and, hence, M can be diagonalized. For this system to be stable both eigenvalues need to be strictly positive which is the case if and only if $a > |b|$. The stable and unstable regions in the (a, b) parameter space are indicated in the figure below.

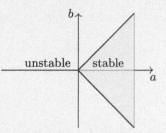

This result, simple as it may be, points to an important lesson about physical systems. Suppose that the parameters (a, b) of the system are slowly changing over time, as a result of changes in the environment into which our system is embedded. If we start out with values (a, b) in the stable region the system may remain stable for a long time until it reaches one of the 'critical lines', $a = |b|$. When this happens it becomes unstable and drastically changes its behaviour from oscillatory to linear and, after crossing the line, exponential. This indicates that physical systems can develop sudden 'catastrophic' behaviour when external conditions exceed certain limits.

20.3   Projectors

Summary 20.3 *Projectors are endomorphisms p characterized by the relation $p \circ p = p$. Their eigenvalues can only be zero or one and they can always be diagonalized. The rank of a projector equals its trace. The unique decomposition of vectors implied by a direct sum can be phrased in terms of projectors.*

20.3.1   Definition of projectors

An interesting class of endomorphism which can always be diagonalized are projectors. Special projectors in $\mathbb{R}^n$ have already been considered in Example 9.1 and Application 9.1. Projectors have many applications and are worth a closer look. We start with their abstract definition.

Definition 20.2 *An endomorphism $p \in \text{End}(V)$ with $p \circ p = p$ is called a projector.*

A projector is also called an *idempotent linear map*. The intuitive idea behind the relation $p \circ p = p$ is that a projection, once applied, should have no further effect.

Lemma 20.1 *A projector $p \in \text{End}(V)$ has the following properties.*

(i)   *The only possible eigenvalues of p are 0 and 1.*
(ii) *$p|_I = \text{id}_I$, where $I = \text{Im}(p)$.*
(iii) *$V = I \oplus K$, where $K = \text{Ker}(p)$.*

Proof   (i) For an eigenvalue λ of p with eigenvector $\mathbf{v}$ we have

$$\lambda\mathbf{v} = p(\mathbf{v}) = p \circ p(\mathbf{v}) = \lambda^2\mathbf{v}$$

and, since $\mathbf{v} \neq \mathbf{0}$, it follows that $\lambda^2 = \lambda$, so $\lambda \in \{0,1\}$.

(ii) If $\mathbf{w} \in I$ then there exists a $\mathbf{v} \in V$ with $\mathbf{w} = p(\mathbf{v})$. Then $p(\mathbf{w}) = p \circ p(\mathbf{v}) = p(\mathbf{v}) = \mathbf{w}$ which shows that $p|_I = \mathrm{id}_I$.

(iii) Suppose that $\mathbf{w} \in I \cap K$, so that $\mathbf{w} = p(\mathbf{v})$ for a $\mathbf{v} \in V$ and $p(\mathbf{w}) = \mathbf{0}$. Then

$$\mathbf{w} = p(\mathbf{v}) = p \circ p(\mathbf{v}) = p(\mathbf{w}) = \mathbf{0}$$

so it follows that $I \cap K = \{\mathbf{0}\}$. Hence, the sum $I + K$ is direct and since

$$\dim_{\mathbb{F}}(I \oplus K) \overset{\text{Eq. (8.7)}}{=} \dim_{\mathbb{F}}(K) + \dim_{\mathbb{F}}(I) \overset{\text{Eq. (14.8)}}{=} \dim_{\mathbb{F}}(V)$$

we conclude that $V = I \oplus K$, from Cor. 7.1. $\qquad\qquad\square$

20.3.2   Diagonalizing projectors

Lemma 20.1 already contains all the information we need to diagonalize a projector. In fact, since $p|_I = \mathrm{id}_I$ and $p|_K = 0$ it is clear that the image $I = \mathrm{Eig}_p(1)$ is the eigenspace for eigenvalue 1 and, as always, the kernel $K = \mathrm{Eig}_p(0)$ equals the eigenspace for eigenvalue 0. Therefore, $V = \mathrm{Eig}_p(1) \oplus \mathrm{Eig}_p(0)$ so Theorem 20.1 (v) implies that p can be diagonalized.

More explicitly, since $V = I \oplus K$, a basis $(\mathbf{w}_1, \dots \mathbf{w}_k)$ of the image I and a basis $(\mathbf{u}_1, \dots, \mathbf{u}_{n-k})$ of the kernel K combine to a basis $(\mathbf{w}_1, \dots \mathbf{w}_k, \mathbf{u}_1, \dots, \mathbf{u}_{n-k})$ of V. Since $p(\mathbf{w}_i) = \mathbf{w}_i$ for $i = 1, \dots, k$ and $p(\mathbf{u}_i) = \mathbf{0}$ for $i = 1, \dots, n-k$ the matrix which describes p relative to this basis is

$$\hat{A} = \mathrm{diag}(\underbrace{1, \dots, 1}_{k}, \underbrace{0, \dots, 0}_{n-k}) \ . \tag{20.5}$$

The 1's in this matrix correspond to the image I which is projected onto, while the 0's correspond to the kernel K. The rank of the projector which equals the dimension of the subspace projected onto can be computed in terms of the trace as

$$\mathrm{rk}(p) = \mathrm{rk}(\hat{A}) = k = \mathrm{tr}(\hat{A}) = \mathrm{tr}(p) \ . \tag{20.6}$$

We summarize these results in the following theorem.

Theorem 20.2 *A projector $p \in \mathrm{End}(V)$ can be diagonalized. If $(\mathbf{w}_1, \dots \mathbf{w}_k)$ is a basis of $\mathrm{Im}(p)$ and $(\mathbf{u}_1, \dots, \mathbf{u}_{n-k})$ is a basis of $\mathrm{Ker}(p)$ then $(\mathbf{w}_1, \dots \mathbf{w}_k, \mathbf{u}_1, \dots, \mathbf{u}_{n-k})$ is a basis of V relative to which p is described by the matrix (20.5). The rank is given by $\mathrm{rk}(p) = \mathrm{tr}(p)$.*

If p is a projector then so is $q := \mathrm{id}_V - p$ since

$$q \circ q = \mathrm{id}_V - 2p + \underbrace{p \circ p}_{=p} = \mathrm{id}_V - p = q \ .$$

It is clear that $\mathrm{Ker}(q) = \mathrm{Im}(p)$ and $\mathrm{Im}(q) = \mathrm{Ker}(p)$, so q is a projector with the kernel and image exchanged, relative to p. For this reason q is also called the *complementary projector* to p. The meaning of this terminology is apparent when we write down the matrix

$$\hat{B} = \operatorname{diag}(\underbrace{0,\ldots,0}_{k},\underbrace{1,\ldots,1}_{n-k}) \, ,$$

which represents q relative to the basis where p takes the form (20.5).

To a direct sum $V = W \oplus U$ we can always associate a projector which project onto W and its complement which projects onto U. Indeed, from Theorem 12.1, there exists a unique linear map p_W with $p|_W = \operatorname{id}_W$ and $p|_U = 0$. This is clearly a projector onto $W = \operatorname{Im}(p_W)$ with kernel $U = \operatorname{Ker}(p_W)$. The complementary projector $p_U = \operatorname{id}_V - p_W$ projects onto U. With these projectors, the unique decomposition for a vector $\mathbf{v} \in V$ implied by the direct sum (see Prop. 8.1) can be written as

$$\mathbf{v} = \operatorname{id}_V(\mathbf{v}) = (p_W + p_U)(\mathbf{v}) = p_W(\mathbf{v}) + p_U(\mathbf{v}) \, , \tag{20.7}$$

where $p_W(\mathbf{v}) \in W$ and $p_U(\mathbf{v}) \in U$.

Problem 20.6 (*Projectors in $\mathbb{R}^3$*)

Show that the map $A \in \operatorname{End}(\mathbb{R}^3)$ defined by

$$A = \begin{pmatrix} 2 & 1 & -1 \\ -1 & 0 & 1 \\ 1 & 1 & 0 \end{pmatrix}$$

is a projector, find the dimension of the subspace it projects onto, its image and its kernel. Write down the complementary projector B and decompose the vector $\mathbf{v} = (1,1,1)^T$ as in Eq. (20.7).

Solution: To show that A is a projector, verify that $A^2 = A$, which is easily done. Since $\operatorname{tr}(A) = 2$ it projects onto a plane in $\mathbb{R}^3$. This plane is the eigenspace $\operatorname{Eig}_A(1) = \operatorname{Ker}(A - \mathbb{1}_3) = \operatorname{Span}(\mathbf{w}_1, \mathbf{w}_2)$, where $\mathbf{w}_1 = (1,0,1)^T$ and $\mathbf{w}_2 = (1,-1,0)^T$. On the other hand, for the kernel one finds $\operatorname{Ker}(A) = \operatorname{Span}(\mathbf{u}_1)$ with $\mathbf{u}_1 = (1,-1,1)^T$. Everything can be checked explicitly by diagonalizing A.

$$P = (\mathbf{w}_1, \mathbf{w}_2, \mathbf{u}_1) = \begin{pmatrix} 1 & 1 & 1 \\ 0 & -1 & -1 \\ 1 & 0 & 1 \end{pmatrix} \, , \quad P^{-1} = \begin{pmatrix} 1 & 1 & 0 \\ 1 & 0 & -1 \\ -1 & -1 & 1 \end{pmatrix} \quad \Rightarrow \quad P^{-1}AP = \operatorname{diag}(1,1,0)$$

The complementary projector is given by

$$B = \mathbb{1}_3 - A = \begin{pmatrix} -1 & -1 & 1 \\ 1 & 1 & -1 \\ -1 & -1 & 1 \end{pmatrix} \, ,$$

and the desired decomposition of $\mathbf{v}$ is $\mathbf{v} = A\mathbf{v} + B\mathbf{v}$, where $A\mathbf{v} = (2,0,2)^T$ and $B\mathbf{v} = (-1,1,-1)^T$.

20.4 Simultaneous diagonalization*

Summary 20.4 *Restrictions of diagonalizable linear maps to vector subspaces are diagonalizable. Two or more linear maps are called simultaneously diagonalizable if there exists a common basis relative to which they are represented by diagonal matrices. Diagonalizable maps are simultaneously diagonalizable if and only if they commute.*

Being able to diagonalize a linear map or a matrix can lead to substantial simplifications. In a situation where two or more linear maps are involved, we might like to know whether all these maps can be brought into diagonal form simultaneously. We should be precise about what this means. We say that two endomorphisms $f, g \in \mathrm{End}(V)$ over a finite-dimensional vector space can be *diagonalized simultaneously* if there exists a basis of V relative to which both f and g are represented by a diagonal matrix. As we will see, this is by no means always possible, even if f and g can be diagonalized individually. However, there is a simple criterion whose derivation requires a bit of preparation.

20.4.1 Diagonalization of restricted maps

The following Lemma proves that restrictions of diagonalizable maps are diagonalizable.

Lemma 20.2 *Let V be a finite-dimensional vector space, $f \in \mathrm{End}(V)$ a diagonalizable endomorphisms and $W \subset V$ a vector subspace with $f(W) \subset W$. Then the restriction $f|_W$ is diagonalizable.*

Proof Let λ_i, where $i = 1, \ldots, k$, be the pairwise distinct eigenvalues of f and $W_i := \mathrm{Eig}_f(\lambda_i)$ the associated eigenspaces. Since f can be diagonalized we know from Theorem 20.1 (v) that $V = \bigoplus_{i=1}^k W_i$. This means any vector $\mathbf{w} \in W$ can be written as $\mathbf{w} = \sum_{j=1}^k \mathbf{w}_j$, where $\mathbf{w}_j \in W_j$. Since $f(\mathbf{w}_j) = \lambda_j \mathbf{w}_j$ it follows that

$$f^{i-1}(\mathbf{w}) = \sum_{j=1}^k \lambda_j^{i-1} \mathbf{w}_j = \sum_{j=1}^k A_{ij} \mathbf{w}_j \quad \text{where} \quad A_{ij} = \lambda_j^{i-1} \, ,$$

for $i = 1, \ldots, k$. The $k \times k$ matrix A is, in fact, of the Vandermonde type (see Example 18.1) and, since the eigenvalues are pairwise distinct, its determinant is non-zero. This means A is invertible and the vectors $\mathbf{w}_j$ can be written as linear combinations of the vectors $f^{i-1}(\mathbf{w}) \in W$. It follows that $\mathbf{w}_j \in W$ and, hence, that $W = \bigoplus_{i=1}^k (W_i \cap W)$. From Theorem 20.1 (v) this means that $f|_W$ can be diagonalized. $\square$

20.4.2 Criterion for simultaneous diagonalization

To formulate the main criterion we recall that the commutator of two linear maps $f, g \in \mathrm{End}(V)$ is defined as $[f, g] := f \circ g - g \circ f$. Clearly, the two maps commute iff the commutator vanishes.

Theorem 20.3 *Two diagonalizable linear maps* $f, g \in \text{End}(V)$ *on a finite-dimensional vector space can be simultaneously diagonalized if and only if they commute, that is, iff* $[f, g] = 0$.

Proof '⇒': This is the easy direction. Suppose f, g can be simultaneously diagonalized, so there is a basis $(\mathbf{v}_1, \ldots, \mathbf{v}_n)$ of V with $f(\mathbf{v}_i) = \hat{\lambda}_i \mathbf{v}_i$ and $g(\mathbf{v}_i) = \hat{\mu}_i \mathbf{v}_i$. A short calculation

$$(f \circ g)(\mathbf{v}_i) = f(g(\mathbf{v}_i)) = f(\hat{\mu}_i \mathbf{v}_i) = \hat{\mu}_i f(\mathbf{v}_i) = \hat{\mu}_i \hat{\lambda}_i \mathbf{v}_i = \hat{\mu}_i f(\mathbf{v}_i) = (g \circ f)(\mathbf{v}_i)$$

shows that f and g commute on a basis and, hence, they commute. Another way of saying the same things is that the matrices which represent f and g relative to the basis $(\mathbf{v}_1, \ldots, \mathbf{v}_n)$ are both diagonal and, therefore, commute.

'⇐': Now assume f and g can each be diagonalized and they commute. We can write $V = \bigoplus_{i=1}^{k} W_i$, where $W_i = \text{Eig}_f(\lambda_i)$ are the eigenspaces for the pairwise different eigenvalues λ_i of f. Consider an eigenvector $\mathbf{w} \in W_i$, so that $f(\mathbf{w}) = \lambda_i \mathbf{w}$. It follows from commutativity that

$$f(g(\mathbf{w})) = g(f(\mathbf{w})) = g(\lambda_i \mathbf{w}) = \lambda_i g(\mathbf{w})$$

and this shows that $g(W_i) \subset W_i$. In other words, the eigenspaces of f are invariant under g. We know from Lemma (20.2) that the restricted maps $g|_{W_i}$ can be diagonalized. Hence, choose, for each W_i, a basis $(\mathbf{w}_{i1}, \ldots, \mathbf{w}_{id_i})$ of eigenvectors of $g|_{W_i}$. Combining these bases into a single list $(\mathbf{w}_{11}, \ldots, \mathbf{w}_{1d_1}, \ldots, \mathbf{w}_{k1}, \ldots, \mathbf{w}_{kd_k})$ gives a basis for V which consists of common eigenvectors of f and g. So relative to this basis f and g are both described by diagonal matrices. □

Note that things are much easier if the eigenvalues λ_i of f are all non-degenerate. In this case, the eigenspaces $\text{Eig}_f(\lambda_i)$ are all one-dimensional. If we write $W_i = \text{Eig}_f(\lambda_i) = \text{Span}(\mathbf{v}_i)$, then invariance of W_i under g implies that $g(\mathbf{v}_i) = \hat{\mu}_i \mathbf{v}_i$, for some numbers $\hat{\mu}_i$ (which are, in fact, the eigenvalues of g). So, in this case, the eigenvectors of f are also eigenvectors of g, although for generally different eigenvalues. The situation for a degenerate eigenspace $\text{Eig}_f(\lambda)$ of f is more complicated. Not every basis on $\text{Eig}_f(\lambda)$ consists of eigenvector of g — we have to make the right choice.

Problem 20.7 *(Simultaneous diagonalization)*

Can the three matrices

$$A = \begin{pmatrix} 2 & -1 \\ -1 & 2 \end{pmatrix}, \qquad B = \begin{pmatrix} 3 & 2 \\ 2 & 3 \end{pmatrix}, \qquad C = \begin{pmatrix} 1 & 1 \\ 1 & 2 \end{pmatrix}$$

be diagonalized? Which of these matrices can be diagonalized simultaneously?

Solution: The characteristic polynomials for the three matrices are

$$\chi_A(\lambda) = (\lambda - 1)(\lambda - 3), \qquad \chi_B(\lambda) = (\lambda - 1)(\lambda - 5), \qquad \chi_C(\lambda) = (\lambda - \lambda_+)(\lambda - \lambda_-),$$

where $\lambda_\pm = (3 \pm \sqrt{5})/2$. This shows that all three matrices can be diagonalized. We could continue computing the eigenvectors and check if a set of common eigenvectors can be found for each two of the matrices. But it is quicker to use Theorem 20.3 and check the commutators.

$$[A, B] = 0 \,, \qquad [A, C] \neq 0 \,, \qquad [B, C] \neq 0 \,.$$

Hence, A, B can be diagonalized simultaneously but not A, C or B, C.

Eigenvectors for the eigenvalues $\lambda_1 = 1$ and $\lambda_2 = 3$ of A are given by $\mathbf{v}_1 = (1, 1)^T$ and $\mathbf{v}_2 = (1, -1)^T$. Since these eigenvalues are non-degenerate we know that they must also be eigenvectors of B and this is easy to check explicitly.

$$P = (\mathbf{v}_1, \mathbf{v}_2) = \begin{pmatrix} 1 & 1 \\ 1 & -1 \end{pmatrix} \quad \Rightarrow \quad P^{-1}AP = \mathrm{diag}(1, 3) \,, \quad P^{-1}BP = \mathrm{diag}(5, 1)$$

Problem 20.8 *(Simultaneous diagonalization of differential operators)*

On the space V_k of polynomials over $\mathbb{R}$ with degree less equal k consider the two linear differential operators

$$D = \frac{d}{dx} \,, \qquad L = x \frac{d}{dx} \,.$$

Can D and L be diagonalized simultaneously?

Solution: Check, on an arbitrary polynomial $p \in V$, if D and L commute.

$$[D, L](p)(x) = \frac{d}{dx}(xp'(x)) - x \frac{d}{dx}(p'(x)) = p'(x) + xp''(x) - xp''(x) = Dp(x) \quad \Rightarrow \quad [D, L] = D$$

In conclusion, D and L do not commute so they cannot be simultaneously diagonalized.

Exercises

20.1 Which of the matrices in Exercises 19.1, 19.2 and 19.3 can be diagonalized? For those which can be diagonalized, check your results by carrying out the diagonalizing basis transformations explicitly.

$$A = \begin{pmatrix} 1 & 1 & 0 \\ 0 & 0 & 1 \\ 0 & 0 & 0 \end{pmatrix}$$

cannot be diagonalized.

20.2 (a) Diagonalize the matrix A from Exercise 19.2 by performing a basis transformation.
(b) Do the same for the matrix B from Exercise 19.3.

20.3 A matrix $A \in \mathcal{M}_{2,2}(\mathbb{F})$ can be diagonalized. Find formulae for its eigenvalues λ_1 and λ_2 in terms of $\mathrm{tr}(A)$ and $\det(A)$.

20.4 Show that the linear map $A : \mathbb{C}^3 \to \mathbb{C}^3$ given by

20.5 The linear map $f : \mathbb{R}^3 \to \mathbb{R}^3$ is defined by

$$f(\mathbf{v}) = \alpha(\mathbf{n} \cdot \mathbf{v})\mathbf{n} + \beta(\mathbf{v} - (\mathbf{n} \cdot \mathbf{v})\mathbf{n}) \,,$$

where $\mathbf{n} \in \mathbb{R}^3$ is a unit vector and $\alpha, \beta \in \mathbb{R}$. Show that f can be diagonalized, find its eigenvalues, eigenspaces and associated diagonal matrix. Interpret the map geometrically.

20.6 Which of the matrices

$$A = \begin{pmatrix} 2 & -4 \\ 3 & -2 \end{pmatrix}, \quad B = \begin{pmatrix} -\frac{1}{2} & 1 \\ 1 & 1 \end{pmatrix}$$

$$C = \begin{pmatrix} \frac{7}{4} & -6 \\ \frac{9}{8} & -\frac{5}{2} \end{pmatrix}, \quad D = \begin{pmatrix} \frac{5}{2} & 2 \\ -\frac{3}{8} & \frac{9}{2} \end{pmatrix}$$

can be simultaneously diagonalized?

20.7 Assume $f \in \text{End}(V)$ is diagonalized relative to a basis $(\mathbf{v}_1, \ldots, \mathbf{v}_n)$ with diagonal matrix $\hat{A} = \text{diag}(\lambda_1, \ldots, \lambda_n)$. Prove the following statements.
(a) If f^{-1} exists it is diagonalized relative to the basis $(\mathbf{v}_1, \ldots, \mathbf{v}_n)$ with diagonal matrix $\hat{A}^{-1} = (\lambda_1^{-1}, \ldots, \lambda_n^{-1})$.
(b) If p is a polynomial then $p(f)$ is diagonalized relative to the basis $(\mathbf{v}_1, \ldots, \mathbf{v}_n)$ with diagonal matrix $p(\hat{A}) = \text{diag}(p(\lambda_1), \ldots, p(\lambda_n))$.
(c) Use part (b) for a simple proof of the Cayley–Hamilton theorem for linear maps which can be diagonalized.

20.8 Consider an endomorphism $f \in \text{End}(V)$ with $f \circ f = \text{id}_V$. Show that
(a) the only eigenvalues of f are ± 1.
(b) f can be diagonalized with diagonal matrix $\hat{A} = \text{diag}(1, \ldots, 1, -1 \ldots, -1)$.
(c) the degeneracy of the eigenvalues ± 1 equals $\frac{1}{2}(\dim_{\mathbb{F}}(V) \pm \text{tr}(f))$.

20.9 *Stability of linear Newton type equations*
Discuss the stability of the solutions $\mathbf{q}(t) = (q_1(t), q_2(t), q_3(t))^T$ to the system of differential equations

$$\begin{aligned} \ddot{q}_1 &= -(aq_1 + q_2) \\ \ddot{q}_2 &= -(q_1 + aq_2 + q_3) \\ \ddot{q}_3 &= -(q_2 + aq_3) \end{aligned}$$

as a function of the parameter $a \in \mathbb{R}$.

21
The Jordan normal form*

We have seen in the previous chapter that not all linear maps can be diagonalized. The Jordan normal form is the 'next best thing' if diagonalization is not possible — it leads to a representing matrix which is quite close to being diagonal.

Throughout this chapter, we will be working with endomorphisms $f \in \text{End}(V)$ on a n-dimensional vector space V over $\mathbb{F}$ and pairwise different eigenvalues $\lambda_1, \ldots, \lambda_k$ with multiplicities m_i. We also assume that the characteristic polynomial fully decomposes, so $\sum_{i=1}^{k} m_i = n$. As we will show, the Jordan normal form of f is an $n \times n$ block-diagonal matrix of the form

$$\begin{pmatrix} \tilde{J}_1 & 0 & \cdots & 0 \\ 0 & \tilde{J}_2 & \cdots & 0 \\ \vdots & \vdots & \ddots & \vdots \\ 0 & 0 & \cdots & \tilde{J}_k \end{pmatrix} \qquad \text{with} \qquad \tilde{J}_i = \begin{pmatrix} \lambda_i & * & 0 & \cdots & 0 \\ 0 & \lambda_i & * & \cdots & 0 \\ \vdots & \vdots & \ddots & \ddots & \vdots \\ 0 & 0 & 0 & \lambda_i & * \\ 0 & 0 & 0 & 0 & \lambda_i \end{pmatrix}. \qquad (21.1)$$

The $m_i \times m_i$ blocks $\tilde{J}_i$ contain the eigenvalues λ_i on the diagonal, the starred entries above the diagonal represent either 0 or 1 and all other entries are zero. Clearly, eigenvalues play a central role for the Jordan normal form as much as they do for diagonalization.

Given we assume that the characteristic polynomial fully decomposes, the potential obstruction to diagonalization is that some degeneracies $d_i = \dim_{\mathbb{F}}(\text{Eig}_f(\lambda_i))$ may be smaller than their maximal possible value m_i. When this happens the eigenspace $\text{Eig}_f(\lambda_i) = \text{Ker}(f - \lambda_i \, \text{id}_V)$ contains too few eigenvectors and, as a result, a basis of eigenvectors cannot be found. The Jordan normal form is based on the simple idea of replacing the eigenspaces by the *generalized eigenspaces*

$$W_i = \text{Ker}((f - \lambda_i \, \text{id}_V)^{m_i}). \qquad (21.2)$$

Taking a power of the map $f - \lambda_i \, \text{id}_V$ potentially increases the size of the kernel and, as it turns out, it does so precisely as required so that $\dim_{\mathbb{F}}(W_i) = m_i$. But there is a price to pay — the appearance of a nilpotent map — which manifests itself in the non-vanishing entries above the diagonal in Eq. (21.1).

While the basic idea is simple, deriving the Jordan normal form requires a bit of preparation, particularly on powers of endomorphisms and nilpotent maps, which makes the story somewhat more complicated than most of what we have seen so far.

21.1 Nilpotent endormorphisms*

Summary 21.1 *Endomorphisms* $\nu \in \text{End}(V)$ *are called nilpotent if* $\nu^q = 0$ *for some positive integer q. There exists a basis of V relative to which a nilpotent endomorphism is represented by a simple matrix with 0 or 1 in the entries just above the diagonal and zero everywhere else.*

21.1.1 Powers of endomorphisms

Our first step is to look at the structure which emerges from taking powers g^r of an endomorphism $g \in \text{End}(V)$ on an n-dimensional vector space V. Eventually, in view of Eq. (21.2), we will apply the results to endormorphisms of the form $g = f - \lambda_i \, \text{id}_V$ but for now we keep g general. The powers of g induce ascending and descending chains of kernels and images of the form

$$\{\mathbf{0}\} \subset \text{Ker}(g) \subset \text{Ker}(g^2) \subset \cdots \subset \text{Ker}(g^r) \subset \cdots$$
$$V \supset \text{Im}(g) \supset \text{Im}(g^2) \supset \cdots \supset \text{Im}(g^r) \supset \cdots \ . \tag{21.3}$$

Since we are working in a finite-dimensional vector spaces the kernels in the above chain cannot increase forever, so the lowest power

$$q = \min\{r \,|\, \text{Ker}(g^r) = \text{Ker}(g^{r+1})\} \tag{21.4}$$

for which the kernel does not increase is well-defined. More importantly, once the kernel has staid the same from one step to the next it remains constant thereafter, as the following lemma shows.

Lemma 21.1 *For an endomorphism* $g \in \text{End}(V)$ *on an* n-*dimensional vector space* V, *define* $K = \text{Ker}(g^q)$ *and* $I = \text{Im}(g^q)$, *with* q *as in Eq. (21.4). Then we have the following statements:*

(i) $\text{Ker}(g^r) = K$ *and* $\text{Im}(g^r) = I$ *for all* $r \geq q$.
(ii) $\dim_{\mathbb{F}}(K) \geq q$.
(iii) $V = K \oplus I$.

Proof (i) We show mutual inclusion of the two sets. For $r \geq q$ we have $K \subset \text{Ker}(g^r)$ from Eq. (21.3). On the other hand, from the definition of q in Eq. (21.4), we know that $g^{q+1}(\mathbf{v}) = \mathbf{0}$ for $\mathbf{v} \in V$ implies $g^q(\mathbf{v}) = \mathbf{0}$. Applying this statement $r - q$ times shows that $g^r(\mathbf{v}) = \mathbf{0}$ implies $g^q(\mathbf{v}) = \mathbf{0}$, so that $\text{Ker}(g^r) \subset K$.
    Since the dimensions of $\text{Ker}(g^r)$ and $\text{Im}(g^r)$ have to sum up to $\dim_{\mathbb{F}}(V)$ for every r it is clear that $\text{Im}(g^r)$ must remain constant when $\text{Ker}(g^r)$ does, so for $r \geq q$.

(ii) Given the definition of q, Eq. (21.4), every step in the kernel chain (21.3) must increase the dimension by at least one. This implies the claim.

(iii) Suppose $\mathbf{v} \in K \cap I$, so that $g^q \mathbf{v} = \mathbf{0}$ and $\mathbf{v} = g^q \mathbf{w}$ for some $\mathbf{w} \in V$. It follows that $g^{2q} \mathbf{w} = \mathbf{0}$ and, hence, from (i), that $\mathbf{w} \in \text{Ker}(g^{2q}) = K$. This means that

$\mathbf{0} = g^q(\mathbf{w}) = \mathbf{v}$ and we conclude that $K \cap I = \{\mathbf{0}\}$. Hence, the sum $K + I$ is direct. Since $K \oplus I \subset V$ and

$$\dim_{\mathbb{F}}(K \oplus I) \overset{(8.7)}{=} \dim_{\mathbb{F}}(K) + \dim_{\mathbb{F}}(I) \overset{(14.13)}{=} \dim_{\mathbb{F}}(V)$$

it follows from Cor. 7.1 that $K \oplus I = V$.   □

So, in essence, the kernel and image chains (21.3) become constant after q steps and, from thereon, kernel and image of g^r for $r \geq q$ directly sum to the total vector space.

21.1.2   Definition of nilpotentcy

How would you define an endomorphism which is close to the zero endomorphism but is not quite equal to it? One possible answer to this question leads to the following definition of nilpotent endomorphisms.

Definition 21.1 *(Nilpotent endomorphisms) An endomorphism $\nu \in \mathrm{End}(V)$ is called nilpotent if there exists a positive integer r such that $\nu^r = 0$. The smallest such r is called the order of the nilpotent endomorphism.*

Problem 21.1 *(Nilpotent endomorphisms)*
On a vector space V with basis $(\mathbf{v}_1, \ldots, \mathbf{v}_p)$ define an endomorphisms $f \in \mathrm{End}(V)$ by $f(\mathbf{v}_i) = \mathbf{v}_{i-1}$, for $i = 1, \ldots, p$, adopting the convention that $\mathbf{v}_i = \mathbf{0}$ for $i < 1$. Show that f is nilpotent of order p and find the matrix $\mathcal{N}_p$ which represents f relative to the basis $(\mathbf{v}_1, \ldots, \mathbf{v}_p)$. For the case $p = 4$, work out $\mathcal{N}_4$, $\mathcal{N}_4^2$, $\mathcal{N}_4^3$ and $\mathcal{N}_4^4$.
Solution: Since $f^r(\mathbf{v}_i) = \mathbf{v}_{i-r}$ (with $\mathbf{v}_i = \mathbf{0}$ for $i < 1$) it follows that $f^p = 0$ and this is the smallest integer for which this happens. Since $f(\mathbf{v}_i) = \mathbf{v}_{i-1}$ the representing matrix is

$$\mathcal{N}_p = (\mathbf{0}, \mathbf{e}_1, \mathbf{e}_2, \ldots, \mathbf{e}_{p-1}) = \begin{pmatrix} 0 & 1 & 0 & \cdots & 0 & 0 \\ 0 & 0 & 1 & \cdots & 0 & 0 \\ \vdots & \vdots & \ddots & \ddots & \vdots & \vdots \\ 0 & 0 & 0 & \cdots & 1 & 0 \\ 0 & 0 & 0 & \cdots & 0 & 1 \\ 0 & 0 & 0 & \cdots & 0 & 0 \end{pmatrix}. \tag{21.5}$$

In the four-dimensional case, $n = 4$, the matrix $\mathcal{N}_4$ and its powers are

$$\mathcal{N}_4 = \begin{pmatrix} 0 & 1 & 0 & 0 \\ 0 & 0 & 1 & 0 \\ 0 & 0 & 0 & 1 \\ 0 & 0 & 0 & 0 \end{pmatrix}, \quad \mathcal{N}_4^2 = \begin{pmatrix} 0 & 0 & 1 & 0 \\ 0 & 0 & 0 & 1 \\ 0 & 0 & 0 & 0 \\ 0 & 0 & 0 & 0 \end{pmatrix}, \quad \mathcal{N}_4^3 = \begin{pmatrix} 0 & 0 & 0 & 1 \\ 0 & 0 & 0 & 0 \\ 0 & 0 & 0 & 0 \\ 0 & 0 & 0 & 0 \end{pmatrix}, \quad \mathcal{N}_4^4 = \begin{pmatrix} 0 & 0 & 0 & 0 \\ 0 & 0 & 0 & 0 \\ 0 & 0 & 0 & 0 \\ 0 & 0 & 0 & 0 \end{pmatrix}.$$

Since the order of the nilpotent map is $n = 4$, the result $\mathcal{N}_4^4 = 0$ was, of course, expected. However, it is instructive to see how the non-zero diagonal 'propagates' towards the upper right of the matrix as successive powers are taken, until it disappears.

Comparison of the above matrices with Eq. (21.1) indicates why nilpotent maps are relevant for the Jordan normal form — they encode the degree to which the linear map cannot be diagonalized.

21.1.3 Structure of nilpotent endomorphisms

To get to the Jordan normal form we need to construct a basis which leads to a simply representing matrix for nilpotent endomorphisms. In fact, Example 21.1 is not too far away from the general case, as the following theorem shows.

Theorem 21.1 *(Nilpotent endomorphisms) Let $\nu \in \text{End}(V)$ be a nilpotent endomorphisms of order q on an n-dimensional vector space V. Then there exists a basis of V relative to which ν is represented by a block-matrix,*

$$\hat{N} = \text{diag}(\underbrace{\mathcal{N}_q, \ldots, \mathcal{N}_q}_{r_q}, \underbrace{\mathcal{N}_{q-1}, \ldots, \mathcal{N}_{q-1}}_{r_{q-1}}, \ldots, \underbrace{\mathcal{N}_1, \ldots, \mathcal{N}_1}_{r_1}) \tag{21.6}$$

with the matrices $\mathcal{N}_p$ from Eq. (21.5) appearing with multiplicity r_p along the diagonal and $\sum_{p=1}^{q} p\, r_p = n$.

Proof Our task is to construct the relevant basis. As before, we define the kernels $K_p = \text{Ker}(\nu^p)$ for $p = 0, \ldots, q$. From their definition, these kernels satisfy $\nu(K_p) \subset K_{p-1}$ and $K_{q-1} \subset K_q$, so that we have chains

$$V = K_q \xrightarrow{\nu} K_{q-1} \xrightarrow{\nu} \cdots \xrightarrow{\nu} K_1 \xrightarrow{\nu} K_0 = \{\mathbf{0}\}$$
$$V = K_q \supset K_{q-1} \supset \cdots \supset K_1 \supset K_0 = \{\mathbf{0}\} \, .$$

The right basis is obtained by focusing on the 'difference' between two successive kernels, so we write

$$K_q = K_{q-1} \oplus U_q \, , \tag{21.7}$$

for some suitable vector subspace U_q. An element $\mathbf{u} \in U_q$ must satisfy $\nu^{q-1}(\mathbf{u}) \neq \mathbf{0}$ (because otherwise it would be in K_{q-1}), so for its image $\mathbf{v} = \nu(\mathbf{u})$ we have $\nu^{q-2}(\mathbf{v}) = \nu^{q-1}(\mathbf{u}) \neq \mathbf{0}$. This means that $\mathbf{v} \notin K_{q-2}$ so that $\nu(U_q) \cap K_{q-2} = \{\mathbf{0}\}$. This means that we write $K_{q-1} = K_{q-2} \oplus U_{q-1}$ for a suitable subspace U_{q-1}, such that $\nu(U_q) \subset U_{q-1}$. Moreoever $\nu|_{U_q}$ is injective since U_q has a trivial intersection with $K_1 = \text{Ker}(\nu)$.

We can repeat this process of constructing the spaces U_p to end up with the structure

$$U_q \xrightarrow{\nu} U_{q-1} \xrightarrow{\nu} \cdots \xrightarrow{\nu} U_1 = \text{Ker}(\nu) \, , \qquad V = U_1 \oplus \cdots \oplus U_q \, , \tag{21.8}$$

and all the maps $\nu|_{U_p}$ are injective. If we define the dimensions $d_q = \dim_{\mathbb{F}}(U_q)$ and start by selecting a basis $(\mathbf{u}_1, \ldots, \mathbf{u}_{d_q})$ of U_q, the images $\nu(\mathbf{u}_1), \ldots, \nu(\mathbf{u}_{d_q})$ are linearly independent (since $\nu|_{U_p}$ is injective and with Cor. 14.1) and can be completed to a basis of U_{q-1}. We can continue in this way, recursively constructing a basis for all subspaces U_p, by completing the images of the U_{p+1} basis vectors to a basis for U_p. The resulting list of basis vectors looks as follows:

$$\begin{array}{llllll} U_q & \mathbf{u}_1 & , \ldots, & \mathbf{u}_{d_q} & & \\ U_{q-1} & \nu(\mathbf{u}_1) & , \ldots, & \nu(\mathbf{u}_{d_q}) & , & \mathbf{u}_{d_q+1} & , \ldots, & \mathbf{u}_{d_{q-1}} \\ \vdots & \vdots & \vdots & \vdots & \vdots & \vdots \\ U_1 & \nu^{q-1}(\mathbf{u}_1) & , \ldots, & \nu^{q-1}(\mathbf{u}_{d_q}) & , & \nu^{q-2}(\mathbf{u}_{d_q+1}) & , \ldots, & \nu^{q-2}(\mathbf{u}_{d_{q-1}}) & \cdots \end{array} \tag{21.9}$$

There are $r_q = d_q$ columns with length q each of which, with the basis vectors ordered from bottom to top, gives rise to a matrix $\mathcal{N}_q$. The next $r_{q-1} = d_{q-1} - d_q$ columns, with length $q-1$, each leads to a matrix $\mathcal{N}_{q-1}$ and so forth. The number r_p of $p \times p$ blocks can be determined by writing $K_p = K_{p-1} \oplus U_p = K_{p-1} \oplus \nu(U_{p+1}) \oplus R_p$, where R_p is the space spanned by the additional basis vector which have to be added to complete to a basis of U_p. It follows that

$$r_p = \dim_{\mathbb{F}}(R_p) = \dim_{\mathbb{F}}(K_p) - \dim_{\mathbb{F}}(K_{p-1}) - \dim(U_{p+1}) \,. \tag{21.10}$$

$\square$

The above proof contains an algorithm to work out the normal form of a nilpotent endomorphism which is worth summarizing.

Algorithm *(Normal form of a nilpotent endomorphism)*

To find the normal form of a nilpotent endomorphism $\nu \in \mathrm{End}(V)$ proceed as follows:

(1) Work out the powers ν^p and find the order, q, of ν (the smallest integer p such that $\nu^p = 0$).

(2) Compute the kernels $K_p = \mathrm{Ker}(\nu^p)$ for $p = 1, \ldots, q$.

(3) Determine the subspaces U_p with $K_p = K_{p-1} \oplus U_p$ and $\nu(U_p) \subset \nu(U_{p-1})$ for $p = 1, \ldots, q$.

(4) Choose a basis for U_q and bases for U_p with $p < q$ such that the images under ν of the U_p basis vectors are basis vectors of U_{p-1}, as indicated in Eq. (21.9).

(5) Combine the U_p bases from (4) into a single basis of V, with the ordering as indicated below Eq. (21.9). Relative to this basis ν is described by the normal form (21.6).

(6) In order to find the explicit normal form either work out ν relative to the basis from (5) or compute the multiplicities r_p of the various block sizes in Eq. (21.6) from Eq. (21.10).

21.1.4 Examples

Problem 21.2 *(Normal form of a nilpotent endomorphism)*

Show that the endomorphism $N \in \mathrm{End}(\mathbb{R}^4)$ given by

$$N = \begin{pmatrix} -2 & -3 & -5 & -6 \\ -1 & -1 & -2 & -2 \\ 0 & -1 & -1 & -2 \\ 1 & 2 & 3 & 4 \end{pmatrix}$$

is nilpotent of order 3. Find the normal form (21.6) of N and the associated basis (21.9).

Solution: (1) We begin by working out the powers of N.

$$N = \begin{pmatrix} -2 & -3 & -5 & -6 \\ -1 & -1 & -2 & -2 \\ 0 & -1 & -1 & -2 \\ 1 & 2 & 3 & 4 \end{pmatrix}, \quad N^2 = \begin{pmatrix} 1 & 2 & 3 & 4 \\ 1 & 2 & 3 & 4 \\ -1 & -2 & -3 & -4 \\ 0 & 0 & 0 & 0 \end{pmatrix}, \quad N^3 = \begin{pmatrix} 0 & 0 & 0 & 0 \\ 0 & 0 & 0 & 0 \\ 0 & 0 & 0 & 0 \\ 0 & 0 & 0 & 0 \end{pmatrix}$$

which shows that N is nilpotent of order 3.

(2) It is straightforward to work out the kernels $K_p = \text{Ker}(N^p)$ by solving the homogeneous linear systems $N^p \mathbf{v} = \mathbf{0}$. The result is

$$K_1 = \text{Span}(2\mathbf{e}_2 - \mathbf{e}_4, \mathbf{e}_1 + \mathbf{e}_2 - \mathbf{e}_3), \quad K_2 = \text{Span}(2\mathbf{e}_2 - \mathbf{e}_4, \mathbf{e}_1 + \mathbf{e}_2 - \mathbf{e}_3, 2\mathbf{e}_2 - \mathbf{e}_4), \quad K_3 = \mathbb{R}^4.$$

(3) Next we should work out the 'difference subspaces' U_p which satisfy $K_p = K_{p-1} \oplus U_p$. To satisfy $K_3 = K_2 \oplus U_2$ we can choose $U_2 = \text{Span}(\mathbf{e}_1)$. The space U_1 in $K_2 = K_1 \oplus U_1$ must contain the vector $N\mathbf{e}_1 = -2\mathbf{e}_1 - \mathbf{e}_2 + \mathbf{e}_4$ and is, in fact spanned by this vector and we have $U_1 = K_1$.

(4) So, writing down the basis vectors as in Eq. (21.9) gives

$$
\begin{array}{ll}
U_3 & \mathbf{e}_1 \\
U_2 & N\mathbf{e}_1 = -2\mathbf{e}_1 - \mathbf{e}_2 + \mathbf{e}_4 \\
U_1 & N^2\mathbf{e}_1 = \mathbf{e}_1 + \mathbf{e}_2 - \mathbf{e}_3, \quad 2\mathbf{e}_2 - \mathbf{e}_4.
\end{array}
$$

(5) Arranging these into a basis, proceeding from bottom to top and left to right, we have

$$
P = (N^2\mathbf{e}_1, N\mathbf{e}_1, \mathbf{e}_1, 2\mathbf{e}_2 - \mathbf{e}_4) = \begin{pmatrix} 1 & -2 & 1 & 0 \\ 1 & -1 & 0 & 2 \\ -1 & 0 & 0 & 0 \\ 0 & 1 & 0 & -1 \end{pmatrix} \quad \Rightarrow \quad P^{-1} = \begin{pmatrix} 0 & 0 & -1 & 0 \\ 0 & 1 & 1 & 2 \\ 1 & 2 & 3 & 4 \\ 0 & 1 & 1 & 1 \end{pmatrix}.
$$

(6) This matrix brings N into the normal form via a basis transformation, as can be checked explicitly.

$$
P^{-1}NP = \text{diag}(\mathcal{N}_3, \mathcal{N}_1) = \left(\begin{array}{ccc|c} 0 & 1 & 0 & 0 \\ 0 & 0 & 1 & 0 \\ 0 & 0 & 0 & 0 \\ \hline 0 & 0 & 0 & 0 \end{array} \right)
$$

For the multiplicities of the various block sizes this mean $(r_3, r_2, r_1) = (1, 0, 1)$, a result which can also be easily obtained by inserting the dimensions of K_p and U_p into Eq. (21.10).

Problem 21.3 (*A nilpotent differential operator*)

On the space $V = \mathcal{P}_2(\mathbb{R})$ of at most quadratic polynomials, consider the linear differential operator $D = d/dx$. Show that D is nilpotent with order 3 and find its normal form (21.6).

Solution: (1) For a general quadratic polynomial $p(x) = a_2 x^2 + a_1 x + a_0$ we have

$$Dp(x) = 2a_2 x + a_1, \quad D^2 p(x) = 2a_2, \quad D^3 p(x) = 0.$$

(2, 3) Hence, the kernels $K_p = \text{Ker}(D^p)$ and the spaces U_p in $K_p = K_{p-1} \oplus U_p$ are given by

$$
\begin{array}{lll}
K_1 = \text{Span}(1) & K_2 = \text{Span}(1, x) & K_3 = \text{Span}(1, x, x^2) \\
U_1 = \text{Span}(1) & U_2 = \text{Span}(x) & U_3 = \text{Span}(x^2)
\end{array}
$$

(4, 5, 6) The adapted basis (21.9) is then $(D^2(x^2) = 2, D(x^2) = 2x, x^2)$ and the matrix representing D relative to this basis

$$
N = \begin{pmatrix} 0 & 1 & 0 \\ 0 & 0 & 1 \\ 0 & 0 & 0 \end{pmatrix}
$$

is of the expected normal form with multiplicities $(r_3, r_2, r_1) = (1, 0, 0)$.

21.2   The Jordan form*

Summary 21.2 *The decomposition theorem states that, for any endomorphism $f \in$ End(V), the generalized eigenspaces W_i of f are invariant under f and directly sum to V. In each generalized eigenspace W_i with eigenvalue λ_i the map $\nu_i := (f - \lambda_i \, \mathrm{id})|_{W_i}$ is nilpotent. Combining these facts with the normal form for nilpotent endomorphisms leads to the Jordan normal form. The Jordan normal form is close to a diagonal matrix, with the eigenvalues λ_i in the diagonal, entries 0 or 1 in the entries above the diagonal and zeros everywhere else.*

21.2.1   The decomposition theorem

We are now ready to proof the crucial decomposition theorem which splits the vector space into the generalized eigenspaces (21.2).

Theorem 21.2 *Let $f \in$ End(V) be an endomorphism on an n-dimensional vector space, $\lambda_1, \ldots, \lambda_k$ its pairwise different eigenvalues with multiplicities m_i satisfying $\sum_{i=1}^{k} m_i = n$ and $W_i = \mathrm{Ker}((f - \lambda_i \, \mathrm{id}_V)^{m_i})$ the generalized eigenspaces. Then we have the following statements for all $i = 1, \ldots k$.*

(i) $f(W_i) \subset W_i$

(ii)   The maps $\nu_i := (f - \lambda_i \, \mathrm{id})|_{W_i}$ are nilpotent of order $q_i \leq m_i$.

(iii) $\dim_{\mathbb{F}}(W_i) = m_i$

(iv) $V = W_1 \oplus \cdots \oplus W_k$

(v) f can be written as $f = h + \nu$, where $h \in$ End(V) can be diagonalized and $\nu \in$ End(V) is nilpotent.

Proof   (i) A vector $\mathbf{v} \in W_i$ satisfies $(f - \lambda_i \, \mathrm{id}_V)^{m_i}(\mathbf{v}) = \mathbf{0}$ and, hence,

$$\mathbf{0} = f((f - \lambda_i \, \mathrm{id}_V)^{m_i}(\mathbf{v})) = (f - \lambda_i \, \mathrm{id}_V)^{m_i}(f(\mathbf{v}))$$

so that $f(\mathbf{v}) \in W_i$. It follows that $f(W_i) \subset W_i$.

(ii) The definition of W_i implies that $\nu_i^{m_i} = 0$, so ν_i is nilpotent of order $q_i \leq m_i$.

(iii) Set $d_i := \dim_{\mathbb{F}}(W_i)$ and, from (iii), write $f|_{W_i} = \lambda_i \, \mathrm{id}_{W_i} + \nu_i$. This means that there is a suitable basis of W_i where the matrix representing $f|_{W_i}$ has λ_i along the diagonal,, from the normal form (21.6) of nilpotent maps, 0 or 1 above the diagonal and zero everywhere else. Taking the determinant, this implies for the characteristic polynomial that $\chi_f(\lambda) = (\lambda - \lambda_i)^{d_i} \tilde{\chi}(\lambda)$, where $\tilde{\chi}$ is the remaining part of χ_f. Since χ_f cannot have more that m_i factors of $\lambda - \lambda_i$ is follow that $d_i \leq m_i$. If we apply Lemma 21.1 to $g = f - \lambda_i \, \mathrm{id}_V$ it follows that $V = W_i \oplus I$ and $\chi_f(\lambda) = (\lambda - \lambda_i)^{d_i} \chi_{f|_I}(\lambda)$. If $d_i < m_i$ at least one factor of $\lambda - \lambda_i$ must reside in $\chi_{f|_I}$, implying at least on eigenvector with eigenvalue λ_i in I. However, this is a contradiction since all such eigenvectors must be in W_i and we conclude that $d_i = m_i$.

(iv) We want to show that $W_i \cap W_j = \{\mathbf{0}\}$ for $i \neq j$. If $\mathbf{v} \in W_i \cap W_j$ is follows from (ii) that $f(\mathbf{v}) = \lambda_i \mathbf{v} + \nu_i(\mathbf{v})$ and $f(\mathbf{v}) = \lambda_j \mathbf{v} + \nu_j(\mathbf{v})$. Combining these two equations and taking the mth power gives $(\lambda_i - \lambda_j)^m(\mathbf{v}) = (\nu_j - \nu_i)^m(\mathbf{v})$ and for sufficiently large m

the right-hand side vanishes due to nilpotency of ν_i and ν_j. Since $\lambda_i \neq \lambda_j$ it follows that $\mathbf{v} = \mathbf{0}$. Hence, the sum $W_1 \oplus \cdots \oplus W_k$ is direct. Since

$$\dim_{\mathbb{F}}(W_1 \oplus \cdots \oplus W_k) = \sum_{i=1}^{k} \dim_{\mathbb{F}}(W_i) = \sum_{i=1}^{k} m_i = n \, ,$$

it follows that $W_1 \oplus \cdots \oplus W_k = V$.

(v) From (iv) every $\mathbf{v} \in V$ can be written as $\mathbf{v} = \sum_{i=1}^{k} \mathbf{w}_i$ with $\mathbf{w}_i \in W_i$ in a unique way. Then, the map $h \in \mathrm{End}(V)$ defined by $h(\mathbf{v}) = \sum_{i=1}^{k} \lambda_i \mathbf{w}_i$ can be diagonalized. On the other hand, $\nu \in \mathrm{End}(V)$ defined by $\nu(\mathbf{v}) = \sum_{i=1}^{k} \nu_i(\mathbf{w}_i)$ is nilpotent and, from (ii), we have $f = h + \nu$. $\qquad \square$

21.2.2 The theorem

This previous theorem tells us every endomorphism can be written as a sum of a diagonalizable and a nilpotent endomorphism. Combining these statements with the normal form for nilpotent maps from Theorem 21.1 leads to the Jordan normal form.

Theorem 21.3 *(Jordan normal form) Let $f \in \mathrm{End}(V)$ be an endomorphism on an n-dimensional vector space and $\lambda_1, \ldots, \lambda_k$ its pairwise different eigenvalues with multiplicities m_i satisfying $\sum_{i=1}^{k} m_i = n$. Then there exists a basis of V relative to which f is represented by a block matrix*

$$\hat{A} = \mathrm{diag}(\lambda_1 \mathbb{1}_{m_1} + \hat{N}_1, \ldots, \lambda_k \mathbb{1}_{m_k} + \hat{N}_k) \, , \tag{21.11}$$

where each $\hat{N}_i$ is a $m_i \times m_i$ nilpotent matrix of the normal form (21.6).

Proof From Theorem 21.2 we can write V as a direct sum $V = W_1 \oplus \cdots \oplus W_k$ of the generalized eigenspaces W_i and $f|_{W_i} = \lambda_i \, \mathrm{id}_{W_i} + \nu_i$ with ν_i nilpotent for $i = 1, \ldots, k$. On each generalized eigenspace W_i we choose the basis $(\mathbf{w}_1^{(i)}, \ldots, \mathbf{w}_{m_i}^{(i)})$, for which ν_i is represented by a matrix $\hat{N}_i$ in the standard form (21.6), as explained in Theorem 21.1. These bases are combined into a single basis $(\mathbf{w}_1^{(1)}, \ldots, \mathbf{w}_{m_1}^{(1)}, \ldots, \mathbf{w}_1^{(k)}, \ldots, \mathbf{w}_{m_k}^{(k)})$ of V and relative to this basis f is represented by the matrix (21.11). $\qquad \square$

Each block $\lambda_i \mathbb{1}_{m_i} + \hat{N}_i$ of the matrix (21.11) splits into smaller blocks of the form

$$\lambda_i \mathbb{1}_p + \mathcal{N}_p = \begin{pmatrix} \lambda_i & 1 & 0 & \cdots & 0 & 0 \\ 0 & \lambda_i & 1 & \cdots & 0 & 0 \\ \vdots & \vdots & \vdots & \ddots & \vdots & \vdots \\ 0 & 0 & 0 & \cdots & \lambda_i & 1 \\ 0 & 0 & 0 & \cdots & 0 & \lambda_i \end{pmatrix} , \tag{21.12}$$

where $\mathcal{N}_p$ are the matrices (21.5). Their sizes p are determined by the structure of the nilpotent map $\hat{N}_i$, as in Eq. (21.6) and they are in the range $p = 1, \ldots, q_i$, each with multiplicity $r_{i,p}$ and q_i the order of the nilpotent map $\hat{N}_i$. Of course the largest block with size, q_i, appears at least once, $r_{i,q_i} > 0$, since this block determines the order of $\hat{N}_i$. In addition, in order to match dimensions we must have

$$m_i = \sum_{p=1}^{q_i} r_{i,p}\, p \;. \tag{21.13}$$

The $p \times p$ matrices in Eq. (21.12) are also called *Jordan blocks*.

Evidently, the Jordan normal form (21.11) is an 'almost diagonal' form with the eigenvalues along the diagonal, the entries just above the diagonal 0 or 1 and all other entries vanishing.

21.2.3 Implications of Jordan normal form

The Jordan normal form allows us to generalize the formulae (20.2) for the determinant and trace in terms of the eigenvalues to all linear maps. For any endormorphism $f \in \text{End}(V)$ with eigenvalues $\hat{\lambda}_1, \ldots, \hat{\lambda}_n$ (listed with multiplicity) we have (see Exercise 21.5)

$$\text{tr}(f^k) = \sum_{i=1}^{n} \hat{\lambda}_i^k\;, \qquad \det(f) = \prod_{i=1}^{n} \hat{\lambda}_i\;. \tag{21.14}$$

Problem 21.4 *Eigenvalues in terms of trace and determinant*

Find a formula for the eigenvalues $\hat{\lambda}_\pm$ of a 2×2 matrix $A \in M_{2,2}(\mathbb{C})$ in terms of the trace and the determinant. Use the result to compute the eigenvalues of

$$B = \begin{pmatrix} 2 & -3 \\ 1 & 4 \end{pmatrix}\;.$$

Solution: Eq. (21.14) implies that $\hat{\lambda}_+ + \hat{\lambda}_- = \text{tr}(A)$ and $\hat{\lambda}_+\hat{\lambda}_- = \det(A)$. Solving these two equations for $\hat{\lambda}_\pm$ gives

$$\hat{\lambda}_\pm = \frac{1}{2}\left(\text{tr}(A) \pm \sqrt{\text{tr}(A)^2 - 4\det(A)}\right)\;.$$

For the explicit matrix B we have $\text{tr}(B) = 6$, $\det(B) = 11$ and inserting this into the above formula gives $\hat{\lambda}_\pm = 3 \pm i\sqrt{2}$.

The Jordan normal form also leads to a more specific statement about the minimal polynomial.

Corollary 21.1 *The minimal polynomial μ_f of an endomorphism $f \in \text{End}(V)$ is given by*

$$\mu_f(x) = (\lambda_1 - x)^{q_1} \cdots (\lambda_k - x)^{q_k}\;, \tag{21.15}$$

where λ_i are the pairwise different eigenvalues of f and q_i are the orders of the nilpotent matrices $\hat{N}_i$ which appear in the Jordan normal form 21.11.

Proof We know from Eq. (19.19) that the minimal polynomial must be of the form $p(x) = (x - \lambda_1)^{s_1} + \cdots (x - \lambda_k)^{s_k}$ with multiplicities $s_i \leq m_i$. Inserting the Jordan normal form (21.11) leads to

$$p(A) = (A - \lambda_1 \mathbb{1}_n)^{s_1} \cdots (A - \lambda_k \mathbb{1}_n)^{s_k}$$

$$= \mathrm{diag}\left(N_1^{s_1} \prod_{i \neq 1}((\lambda_1 - \lambda_i)\mathbb{1}_{m_1} - N_i)^{s_i}, \cdots, \prod_{i \neq k}((\lambda_k - \lambda_i)\mathbb{1}_{m_k} - N_i)^{s_i} N_k^{s_k} \right) .$$

This matrix vanishes iff every block vanishes and since the matrices $(\lambda_j - \lambda_i)\mathbb{1}_{m_j} - N_i$ for $i \neq j$ are invertible this is the case iff all $N_i^{s_i} = 0$. The smallest powers s_i for which this happens are the nilpotent orders q_i of the matrices N_i. □

An easy conclusion is this new criterion for an endomorphism to be diagonalizable.

Corollary 21.2 *The endomorphism $f \in \mathrm{End}(V)$ can be diagonalized iff all zeros of the minimal polynomial have multiplicity one.*

Proof If f can be diagonalized to a matrix $\hat{A} = \mathrm{diag}(\lambda_1 \mathbb{1}_{m_1}, \ldots, \lambda_k \mathbb{1}_{m_k})$, then inserting into Eq. (21.15) shows that $\mu_f(\hat{A}) = 0$ for all $q_i = 1$. Conversely, if all $q_i = 1$, then all $\hat{N}_i$ are nilpotent of order one, that is, $\hat{N}_i = 0$. From Eq. (21.11) this means the Jordan normal form is, in fact, diagonal and, hence, f can be diagonalized. □

21.3 Examples*

Algorithm *(Jordan normal form)*

Let $f \in \mathrm{End}(V)$ be an endomorphism on an n-dimensional vector space whose characteristic polynomial fully decomposes. To find the Jordan normal form of f proceed as follows.

(1) *Compute the characteristic polynomial χ_f and find the pairwise different eigenvalues λ_i and their multiplicities m_i, where $i = 1, \ldots, k$.*

(2) *Compute the generalized eigenspaces $W_i = \mathrm{Ker}((f - \lambda_i \, \mathrm{id}_V)^{m_i})$.*

(3) *For each nilpotent map $\nu_i := f|_{W_i} - \lambda_i \, \mathrm{id}_{W_i}$ find the basis $(\mathbf{w}_1^{(i)}, \ldots, \mathbf{w}_{m_i}^{(i)})$ relative to which ν_i is described by a matrix $\hat{N}_i$ in the normal form (21.6).*

(4) *Combine the bases from (3) into a single basis $(\mathbf{v}_1, \ldots, \mathbf{v}_n)$ of V. Relative to this basis f is described by the matrix $\hat{A} = \mathrm{diag}(\lambda_1 \mathbb{1}_{m_1} + \hat{N}_1, \ldots, \lambda_k \mathbb{1}_{m_k} + \hat{N}_k)$.*

(5) *If $f \in \mathrm{End}(\mathbb{F}^n)$ is given by matrix A then the Jordan normal form can be achieved by the basis transformation $\hat{A} = P^{-1}AP$, where $P = (\mathbf{v}_1, \ldots, \mathbf{v}_n)$.*

Problem 21.5 *(Jordan normal form for a 3 × 3 matrix)*

Compute the Jordan normal form for the endomorphism $A \in \mathrm{End}(\mathbb{R}^3)$ given by

$$A = \begin{pmatrix} 4 & -3 & -2 \\ 1 & 1 & -1 \\ 0 & -1 & 2 \end{pmatrix} .$$

Solution: (1) We have $\chi_A(\lambda) = -(\lambda - 3)(\lambda - 2)^2$ so that we have $k = 2$ eigenvalues, $\lambda_1 = 3$ and $\lambda_2 = 2$ with multiplicities $m_1 = 1$ and $m_2 = 2$.

(2) Solving $(A - 3\mathbb{1}_3)\mathbf{v} = \mathbf{0}$ gives $W_1 = \mathrm{Span}(\mathbf{v}_1)$ with $\mathbf{v}_1 = (1, 1, -1)^T$. Solving $(A - 2\mathbb{1}_3)\mathbf{v} = \mathbf{0}$ gives the one-dimensional eigenspace $\mathrm{Eig}_A(2) = \mathrm{Span}(\mathbf{v}_2)$ with $\mathbf{v}_2 = (-1, 0, -1)^T$. Hence, A cannot be diagonalized — there is no basis of eigenvectors. To find W_2 we have to solve $(A - 2\mathbb{1}_3)^2\mathbf{v} = \mathbf{0}$ and this leads to $W_2 = \mathrm{Span}(\mathbf{v}_2, \mathbf{v}_3)$, where $\mathbf{v}_3 = (1, 1, 0)^T$.

(3) Since W_1 is one-dimensional the nilpotent map on this subspace is trivial and we can use $(\mathbf{v}_1)$ as a basis. For W_2 we define the nilpotent map $N_2 = A - 2\mathbb{1}_2$ and compute

$$K_1 = \mathrm{Ker}(N_2) = \mathrm{Span}(\mathbf{v}_2) \qquad K_2 = \mathrm{Ker}(N_2^2) = W_2 = \mathrm{Span}(\mathbf{v}_2, \mathbf{v}_3)$$
$$U_1 = K_1 = \mathrm{Span}(\mathbf{v}_2) \qquad U_2 = \mathrm{Span}(\mathbf{v}_3)$$

Since $N_2\mathbf{v}_3 = \mathbf{v}_2$ we know that $(\mathbf{v}_2, \mathbf{v}_3)$ is the correct basis on W_2, which leads to the normal form for N_2.

(4) The correct basis is then

$$P = (\mathbf{v}_1, \mathbf{v}_2, \mathbf{v}_3) = \begin{pmatrix} 1 & -1 & 1 \\ 1 & 0 & 1 \\ -1 & -1 & 0 \end{pmatrix} \qquad \Rightarrow \qquad P^{-1} = \begin{pmatrix} 0 & 1 & 1 \\ 1 & -2 & -1 \\ 1 & -1 & -1 \end{pmatrix}$$

and the Jordan normal form for A is

$$P^{-1}AP = \mathrm{diag}(3, 2\mathbb{1}_2 + J_2) = \begin{pmatrix} 3 & 0 & 0 \\ 0 & 2 & 1 \\ 0 & 0 & 2 \end{pmatrix} .$$

Problem 21.6 *(Jordan normal form for a differential operator)*

On the space $V = \mathcal{P}_3(\mathbb{R})$ of at most cubic polynomials find the Jordan normal form for the linear differential operator

$$L = x^2 \frac{d^2}{dx^2} + \frac{d}{dx} .$$

Solution: (1) To compute the characteristic polynomial we represent L relative to the standard monomial basis $(1, x, x^2, x^3)$. With

$$L(1) = 0, \quad L(x) = 1, \quad L(x^2) = 2x^3 + 2x, \quad L(x^3) = 6x^3 + 3x^2$$

the representing matrix is

$$A = \begin{pmatrix} 0 & 1 & 0 & 0 \\ 0 & 0 & 2 & 0 \\ 0 & 0 & 2 & 3 \\ 0 & 0 & 0 & 6 \end{pmatrix} \qquad \Rightarrow \qquad \chi_L(\lambda) = \det(A - \lambda\mathbb{1}_4) = \lambda^2(\lambda - 2)(\lambda - 6) .$$

Hence, we have $k = 3$ eigenvalues, $\lambda_1 = 0$, $\lambda_2 = 2$ and $\lambda_3 = 6$, with multiplicities $m_1 = 2$, $m_2 = 1$, and $m_3 = 1$. The eigenspace for $\lambda_1 = 0$ is $\mathrm{Eig}_L(0) = \mathrm{Span}(\mathbf{e}_1)$, so the degeneracy $d_1 = 1$ of this eigenvalue is smaller than its multiplicity $m_1 = 2$. This shows that L cannot be diagonalized and we should proceed to computing the Jordan normal form.

(2) To find the generalized eigenspaces $W_i = \mathrm{Ker}((L - \lambda_i \, \mathrm{id}_V)^{m_i})$ we write down their defining equations for a general cubic $p(x) = a_3x^3 + a_2x^2 + a_1x + a_0$

$$L(p)(x) = 6a_3x^3 + (2a_2 + 3a^3)x^2 + 2a_2x + a_1 \overset{!}{=} \lambda(a_3x^3 + a_2x^2 + a_1x + a_0) , \quad \lambda = 2, 6$$
$$L^2(p)(x) = 36a_3x^3 + (4a_2 + 24a_3)x^2 + (4a_2 + 6a_3)x + 2a_2 \overset{!}{=} 0 \qquad\qquad , \quad \lambda = 0$$

Inserting the different eigenvalues and comparing monomial coefficients gives

$$W_1 = \text{Span}(1, x), \quad W_2 = \text{Span}(x^2 + x + 2), \quad W_3 = \text{Span}(24x^3 + 18x^2 + 6x + 1).$$

(3) There is nothing to do for the one-dimensional eigenspaces W_2 and W_3. For W_1 we should, in principle, apply the procedure to bring the nilpotent map $\nu_1 = L|_{W_1}$ into standard form, but the upper 2×2 block of above matrix A shows that the monomial basis $(1, x)$ is already the appropriate basis for W_1.

(4) The complete basis is then $(1, x, x^2 + x + 2, 24x^3 + 18x^2 + 6x + 1)$ and, relative to this basis, L is represented by

$$\hat{A} = \begin{pmatrix} 0 & 1 & 0 & 0 \\ 0 & 0 & 0 & 0 \\ 0 & 0 & 2 & 0 \\ 0 & 0 & 0 & 6 \end{pmatrix}.$$

Exercises

21.1 *Nilpotent matrix*
Show that the map $N : \mathbb{R}^4 \to \mathbb{R}^4$ with

$$N = \begin{pmatrix} -5 & 6 & 2 & -11 \\ 2 & -2 & -1 & 4 \\ -9 & 12 & 3 & -21 \\ 2 & -2 & -1 & 4 \end{pmatrix}$$

is nilpotent of order three and find its normal form.

21.2 *Jordan normal form for 3×3 matrix*
Find the Jordan normal form for the linear map $A : \mathbb{R}^3 \to \mathbb{R}^3$ with

$$A = \begin{pmatrix} \frac{4}{3} & -\frac{2}{3} & -1 \\ -\frac{5}{6} & \frac{8}{3} & -\frac{1}{2} \\ -\frac{2}{3} & \frac{4}{3} & 1 \end{pmatrix}.$$

21.3 *Jordan normal form for 3×3 matrix*
Find the Jordan normal form for the linear map $B : \mathbb{R}^3 \to \mathbb{R}^3$ with

$$B = \begin{pmatrix} 10 & -6 & 4 \\ 11 & -6 & 6 \\ -1 & 1 & 2 \end{pmatrix}.$$

21.4 *Jordan normal form for 4×4 matrix*
Find the Jordan normal form for the

linear map $A : \mathbb{R}^4 \to \mathbb{R}^4$ with

$$A = \begin{pmatrix} 5 & -\frac{8}{3} & \frac{2}{3} & \frac{1}{3} \\ \frac{7}{2} & -\frac{17}{6} & \frac{1}{6} & -\frac{7}{3} \\ \frac{5}{2} & -\frac{6}{5} & -\frac{1}{6} & \frac{8}{3} \\ -\frac{5}{2} & \frac{6}{5} & \frac{5}{2} & 4 \end{pmatrix}.$$

21.5 (a) Let A and B be upper triangular matrices with diagonal entries a_i and b_i. Show that AB is upper triangular with diagonal entries $a_i b_i$.
(b) Use the result from part (a) to prove Eqs. (21.14).

21.6 *Newton type equations and Jordan normal form*
The functions

$$\mathbf{q}(t) = (q_1(t), q_2(t), q_3(t))^T$$

satisfy the differential equation $\ddot{\mathbf{q}} = A\mathbf{q}$, where A is the matrix from Exercise 21.2. Discuss how the Jordan normal form can help to solve this equation and find its solutions.

21.7 *Cayley–Hamilton theorem*
Use the Jordan normal form to prove the Cayley–Hamilton theorem.

Part VII

Inner product vector spaces

A great deal has been said about maps which are linear in one vectorial argument. In this part we go one step further and talk about objects which are bi-linear, that is, they are linear in two vectorial argument (or sesqui-linear in the case of complex numbers). The most important such objects are scalar products $\langle \cdot, \cdot \rangle : V \times V \to \mathbb{F}$ which are generalizations of the dot product we have already encountered in Chapter 9. In fact, in much the same way that the general definition of vector spaces was motivated by the rules observed for coordinate vectors, the definition of scalar products takes its cues from the dot product and its properties.

Scalar products provide us with basic notions of geometry, such as length, angles, and orthogonality, on a vector space V, much as the dot product has done for $\mathbb{R}^n$. However, this is the first time we need to distinguish explicitly between different underlying fields $\mathbb{F}$. For $\mathbb{F} = \mathbb{R}$, we define the *real scalar product* which is the direct generalization of the dot product. In the complex case, for $\mathbb{F} = \mathbb{C}$, there is a small twist due to the existence of complex conjugation, which leads to the introduction of the *Hermitian scalar product*. In either case, a vector space equipped with such a scalar product is also called an *inner product vector space*.

An important property of inner product vector spaces is the existence of a special class of bases — the *ortho-normal bases*. A basis $(\mathbf{v}_1, \ldots, \mathbf{v}_n)$ on an inner product vector space V is called ortho-normal if all basis vectors have length one and if they are mutually orthogonal or, in short, if $\langle \mathbf{v}_i, \mathbf{v}_j \rangle = \delta_{ij}$ for all $i, j = 1, \ldots, n$. As we will see, such bases always exist (in the finite-dimensional case) and can be constructed using the *Gram–Schmidt procedure*. An ortho-normal basis leads to many simplifications compared to a general basis. For example, it is much easier to compute the coordinates of a vector or the representing matrix of a linear map relative to an ortho-normal basis. All this will be discussed in the next chapter.

The presence of a scalar product on V naturally singles out certain classes of endomorphisms. For one these are the *self-adjoint linear maps* (also called *Hermitian maps*) which can be moved from one argument of a scalar product into the other without changing the scalar product's value. Another important class are the *unitary maps* which leave a scalar product invariant. Both types of maps will be introduced in Chapter 23 and are of considerable importance, both in mathematics and in scientific applications. For example, Hermitian linear maps can always be diagonalized relative to an ortho-normal basis and their eigenvectors are real, as we will see in Chapter 24. In the context of quantum mechanics, these properties facilitate representing physical quantities by Hermitian maps. The set of all unitary endomorphisms on a given vector space with a scalar product forms a group, the *unitary group*. For the dot product on $\mathbb{R}^n$ this group is the *orthogonal group* $\mathrm{O}(n)$, which contains the *special orthogonal group* (or rotation group) $\mathrm{SO}(n)$ as a sub-group. The standard Hermitian scalar product on $\mathbb{C}^n$ leads to the unitary group $\mathrm{U}(n)$ and its sub-group, the *special unitary group* $\mathrm{SU}(n)$.

With new classes of endomorphisms singled out in this way it makes sense to re-visit the problem of diagonalization. In Part VI we have seen that not all endomorphisms can be diagonalized. Moreover, checking if diagonalization is possible usually requires computing the eigenvalues and eigenvectors which can be tedious. In Chapter 24 we will see that certain classes of endomorphisms, including self-adjoint and *normal* endomorphisms, can always be diagonalized. We will also discuss the problem of diagonalizing homomorphisms $V \to W$ between two generally different vector spaces. This leads to the *singular value decomposition*, an important method with a large range of scientific applications.

The final chapter of this part is devoted to *symmetric bi-linear forms* and *Hermitian sesqui-linear forms*, which generalize scalar products. We will see how these linear forms can be classified in terms of the signature (n_+, n_-). Perhaps the most prominent example is the *Minkowski product*, a symmetric bi-linear form with signature $(3, 1)$ which is one of the main ingredients in the theory of special relativity. We will also briefly consider a geometrical application of symmetric bi-linear forms and their associated quadratic forms. These can be used be used to define *quadratic hyper-surfaces* in $\mathbb{R}^n$ which turn out to be classified by the signature.

22
Scalar products

In this chapter, we cover the basics of scalar products, their definition, important classes of examples and the existence and construction of ortho-normal bases.

22.1 Real and Hermitian scalar products

Summary 22.1 *Real scalar products are defined on vector spaces V over $\mathbb{R}$. They are symmetric, bi-linear and positive maps $V \times V \to \mathbb{R}$ which are abstract generalizations of the dot product. Their counterparts for vector spaces over $\mathbb{C}$ are called Hermitian scalar products. Vector spaces with either type of scalar product are called inner product vector spaces. On such spaces there is a norm (or length) associated to the scalar product and the notion of orthogonality can be defined.*

22.1.1 Definition of norms

The most basic geometrical notion on a vector space is that of the *length* or *norm* of a vector. The following definition of norm is inspired by the properties of the Euklidean norm in Prop. 9.2.

Definition 22.1 *For a vector space V over $\mathbb{F} = \mathbb{R}$ or $\mathbb{F} = \mathbb{C}$, a length (norm) is a map $|\cdot| : V \to \mathbb{R}^{\geq 0}$ which has the following properties for all $\mathbf{v}, \mathbf{w} \in V$ and all $\alpha \in \mathbb{F}$.*

(N1)	$|\mathbf{v}| > 0$ *if* $\mathbf{v} \neq \mathbf{0}$	*(positivity)*
(N2)	$|\alpha \mathbf{v}| = |\alpha|\,|\mathbf{v}|$	*(scaling)*
(N3)	$|\mathbf{v} + \mathbf{w}| \leq |\mathbf{v}| + |\mathbf{w}|$	*(triangle inequality)*

A vector space with a norm is also called a normed vector space.

We will see more examples of norms shortly but for now we emphasize that the three properties above are what we would intuitively ask any notion of length to satisfy. Note that the modulus, $|\alpha|$, of the scalar in (N2) refers to the real (complex) modulus for $\mathbb{F} = \mathbb{R}$ ($\mathbb{F} = \mathbb{C}$) [1].

22.1.2 Definition of scalar products

The following definition of scalar products is motivated by the properties for the dot product in Prop. 9.1.

---

[1] We continue to be somewhat sloppy notationally, by using $|\cdot|$ to denote the modulus of scalars as well as the norm of vectors, letting the type of the argument indicate which one is referred to.

Definition 22.2 *A real (Hermitian) scalar product on a vector space V over $\mathbb{F} = \mathbb{R}$ (over $\mathbb{F} = \mathbb{C}$) is a map $\langle\,\cdot\,,\,\cdot\,\rangle : V \times V \to \mathbb{F}$ which satisfies the following rules for all $\mathbf{v}, \mathbf{u}, \mathbf{w} \in V$ and all $\alpha, \beta \in \mathbb{F}$,*

(S1)	$\langle \mathbf{v}, \mathbf{w} \rangle = \langle \mathbf{w}, \mathbf{v} \rangle$, *for a real scalar product,* $\mathbb{F} = \mathbb{R}$	*(symmetry)*
	$\langle \mathbf{v}, \mathbf{w} \rangle = \overline{\langle \mathbf{w}, \mathbf{v} \rangle}$, *for a Hermitian scalar product,* $\mathbb{F} = \mathbb{C}$	*(hermiticity)*
(S2)	$\langle \mathbf{v}, \alpha\mathbf{u} + \beta\mathbf{w} \rangle = \alpha\langle \mathbf{v}, \mathbf{u} \rangle + \beta\langle \mathbf{v}, \mathbf{w} \rangle$	*(linearity)*
(S3)	$\langle \mathbf{v}, \mathbf{v} \rangle > 0$ *if* $\mathbf{v} \neq \mathbf{0}$	*(positivity)*

A vector space with such a scalar product is also called an inner product vector space.

Let us discuss these properties, starting with the real scalar product, so $\mathbb{F} = \mathbb{R}$. In this case, combining symmetry (S1) with linearity in the second argument (S2), implies linearity

$$\langle \alpha\mathbf{v} + \beta\mathbf{u}, \mathbf{w} \rangle = \alpha\langle \mathbf{v}, \mathbf{w} \rangle + \beta\langle \mathbf{u}, \mathbf{w} \rangle \,, \tag{22.1}$$

in the first argument. In short, a real scalar product is *bi-linear*, just as the dot product on $\mathbb{R}^n$.

The situation is somewhat more complicated in the Hermitian case, so for $\mathbb{F} = \mathbb{C}$. First, hermiticity implies that $\langle \mathbf{v}, \mathbf{v} \rangle = \overline{\langle \mathbf{v}, \mathbf{v} \rangle}$. Hence, $\langle \mathbf{v}, \mathbf{v} \rangle$ is always real and it is for this reason the positivity condition (S3) makes sense in the complex case. This fact can also be seen as a motivation for including the complex conjugation in (S1). Combining linearity in the second argument with hermiticity leads to

$$\langle \alpha\mathbf{v} + \beta\mathbf{u}, \mathbf{w} \rangle = \bar{\alpha}\langle \mathbf{v}, \mathbf{w} \rangle + \bar{\beta}\langle \mathbf{u}, \mathbf{w} \rangle \,. \tag{22.2}$$

Evidently, sums in the first argument of a Hermitian scalar product can still be pulled apart as usual, but scalars are 'pulled out' with a complex conjugation. This property, together with the linearity in the second argument [2] is also called *sesqui-linearity*. This term means '1.5 linearity', which alludes to $\langle \cdot, \cdot \rangle$ being linear in the second argument and half — or *semi-linear* — in the first argument.

In the following we will often carry out proofs and calculations for the Hermitian case. The corresponding version for the real case is then obtained by omitting the complex conjugations.

22.1.3   The norm associated to a scalar product

For any inner product vector space, we can define a prospective length or norm by

$$|\mathbf{v}| := \sqrt{\langle \mathbf{v}, \mathbf{v} \rangle} \,. \tag{22.3}$$

The positivity condition (S3) on the scalar product ensures that the above square root is well-defined and that the positivity condition (N1) of a norm in Def. (22.1) is

---

[2] The convention adopted in some parts of the mathematics literature is to define a Hermitian scalar product as linear in the first argument, rather than the second one as we have done. All formulae can be easily converted between the two conventions.

satisfied. It is also easy to see that the scaling property, (N2), of a norm follows from the (sesqui-) linearity of the scalar product:

$$|\alpha\mathbf{v}|^2 = \langle \alpha\mathbf{v}, \alpha\mathbf{v} \rangle = \alpha\bar{\alpha}\langle \mathbf{v}, \mathbf{v} \rangle = |\alpha|^2|\mathbf{v}|^2 \ .$$

Proving that Eq. (22.3) satisfies the triangle inequality is a bit more difficult but proceeds exactly as in the case of the dot product and the Euklidean norm.

Proposition 22.1 *For an inner product vector space V with scalar product $\langle \cdot, \cdot \rangle$ and a map $|\cdot| : V \to \mathbb{R}^{\geq 0}$ defined by $|\mathbf{v}| := \sqrt{\langle \mathbf{v}, \mathbf{v} \rangle}$ the following relations hold.*

$$|\langle \mathbf{v}, \mathbf{w} \rangle| \leq |\mathbf{v}|\,|\mathbf{w}| \qquad \text{(Cauchy–Schwarz inequality)}$$
$$|\mathbf{v} + \mathbf{v}| \leq |\mathbf{v}| + |\mathbf{w}| \qquad \text{(triangle inequality)}$$

Proof The proof of the Cauchy–Schwarz inequality works exactly as in the dot product case (see Theorem 9.1) with the dot replaced by the general scalar product $\langle \cdot, \cdot \rangle$.

The proof of the triangle inequality in the real case is also exactly as for the Eulkidean length (see Eq. (9.9)). For the complex case, it is similar but with a slight twist due to hermiticity.

$$\begin{aligned}
|\mathbf{v} + \mathbf{w}|^2 &= \langle \mathbf{v} + \mathbf{w}, \mathbf{v} + \mathbf{w} \rangle = |\mathbf{v}|^2 + |\mathbf{w}|^2 + \langle \mathbf{v}, \mathbf{w} \rangle + \langle \mathbf{w}, \mathbf{v} \rangle \\
&= |\mathbf{v}|^2 + |\mathbf{w}|^2 + 2\,\Re(\langle \mathbf{v}, \mathbf{w} \rangle) \leq |\mathbf{v}|^2 + |\mathbf{w}|^2 + 2\,|\langle \mathbf{v}, \mathbf{w} \rangle| \\
&\leq |\mathbf{v}|^2 + |\mathbf{w}|^2 + 2\,|\mathbf{v}|\,|\mathbf{w}| = (|\mathbf{v}| + |\mathbf{w}|)^2
\end{aligned}$$

$\square$

We conclude that Eq. (22.3) does indeed satisfies all conditions of Def. 22.1 and, therefore, defines a norm. This norm is also called the *norm associated to the scalar product*.

Problem 22.1 *(Norm associated to a scalar product)*

Show that a scalar product is determined by its associated norm. Start with the case of a real scalar product and then tackle the Hermitian case.

Solution: For a real scalar product on V and $\mathbf{v}, \mathbf{w} \in V$ we have $|\mathbf{v}\pm\mathbf{w}|^2 = |\mathbf{v}|^2+|\mathbf{w}|^2\pm2\langle \mathbf{v}, \mathbf{w} \rangle$ and subtracting these two equations from each other gives the *polarization identity*:

$$\langle \mathbf{v}, \mathbf{w} \rangle = \frac{1}{4}\left(|\mathbf{v} + \mathbf{w}|^2 - |\mathbf{v} - \mathbf{w}|^2\right) \ . \tag{22.4}$$

This equation allows computation of the scalar product in terms of the associated norm.

For a Hermitian product, things are slightly more complicated since $\langle \mathbf{v}, \mathbf{w} \rangle = \overline{\langle \mathbf{w}, \mathbf{v} \rangle}$. However, taking an appropriate linear combination of the four equations

$$\begin{aligned}
|\mathbf{v} \pm \mathbf{w}|^2 &= \langle \mathbf{v} \pm \mathbf{w}, \mathbf{v} \pm \mathbf{w} \rangle = \langle \mathbf{v}, \mathbf{v} \rangle + \langle \mathbf{w}, \mathbf{w} \rangle \pm \langle \mathbf{v}, \mathbf{w} \rangle \pm \langle \mathbf{w}, \mathbf{v} \rangle \\
|\mathbf{v} \pm i\mathbf{w}|^2 &= \langle \mathbf{v} \pm i\mathbf{w}, \mathbf{v} \pm i\mathbf{w} \rangle = \langle \mathbf{v}, \mathbf{v} \rangle + \langle \mathbf{w}, \mathbf{w} \rangle \pm i\langle \mathbf{v}, \mathbf{w} \rangle \mp i\langle \mathbf{w}, \mathbf{v} \rangle
\end{aligned}$$

we find

$$\langle \mathbf{v}, \mathbf{w} \rangle = \frac{1}{4}\left(|\mathbf{v} + \mathbf{w}|^2 - |\mathbf{v} - \mathbf{w}|^2 - i|\mathbf{v} + i\mathbf{w}|^2 + i|\mathbf{v} - i\mathbf{w}|^2\right) \ . \tag{22.5}$$

22.1.4   Orthogonal vectors and angles

In analogy with the dot product, we can define two vectors $\mathbf{v}, \mathbf{w} \in V$ as *orthogonal*, denoted $\mathbf{v} \perp \mathbf{w}$, if their scalar product vanishes, so

$$\mathbf{v} \perp \mathbf{w} \quad :\Longleftrightarrow \quad \langle \mathbf{v}, \mathbf{w} \rangle = 0 \ . \tag{22.6}$$

In the real case, the Cauchy–Schwarz inequality facilitates introducing an angle between two non-zero vectors by

$$\cos(\sphericalangle(\mathbf{v}, \mathbf{w})) := \frac{\langle \mathbf{v}, \mathbf{w} \rangle}{|\mathbf{v}| \, |\mathbf{w}|} \ , \tag{22.7}$$

so that two non-zero vectors $\mathbf{v}$ and $\mathbf{w}$ are orthogonal iff $\sphericalangle(\mathbf{v}, \mathbf{w}) = \pi/2$. In the complex case, the scalar product $\langle \mathbf{v}, \mathbf{w} \rangle$ is, in general, complex, so that Eq. (22.7) is not a sensible definition of an angle.

No non-zero vector $\mathbf{v} \in V$ can be orthogonal to all vectors in V, for if $\langle \mathbf{u}, \mathbf{v} \rangle = 0$ for all $\mathbf{u} \in V$ then also $\langle \mathbf{v}, \mathbf{v} \rangle = 0$ which implies $\mathbf{v} = \mathbf{0}$ from positivity of the scalar product. This property is sometimes expressed by saying that a scalar product is *non-degenerate*.

22.2   Examples of scalar products

Summary 22.2 *The dot product is the standard scalar product on $\mathbb{R}^n$ and there is a corresponding standard scalar product on $\mathbb{C}^n$. Scalar products can also be introduced on matrix vector spaces, using the trace, and on function vector spaces, using integrals.*

Example 22.1 *(Standard scalar product on $\mathbb{R}^n$)*

Having motivated the general definition, the dot product on $\mathbb{R}^n$ is of course an example of a scalar product. It is also referred to as the standard scalar product on $\mathbb{R}^n$ and its associated norm is the Euklidean length:

$$\langle \mathbf{v}, \mathbf{w} \rangle = \mathbf{v} \cdot \mathbf{w} = \mathbf{v}^T \mathbf{w} = \sum_{i=1}^{n} v_i w_i \ , \qquad |\mathbf{v}| = \sqrt{\langle \mathbf{v}, \mathbf{v} \rangle} = \sqrt{\sum_{i=1}^{n} v_i^2} \ . \tag{22.8}$$

We know from Prop. 9.1 that the dot product satisfies the axioms for a real scalar product in Def. 22.2. $\qquad\qquad\square$

Example 22.2 *(Standard scalar product on $\mathbb{C}^n$)*

The standard scalar product on $\mathbb{C}^n$ (as a vector space over $\mathbb{C}$) is a Hermitian scalar product and it is the close cousin of the dot product. It is obtained by modifying the dot product (22.8) to include an additional complex conjugation, so

$$\langle \mathbf{v}, \mathbf{w} \rangle := \mathbf{v}^\dagger \mathbf{w} = \sum_{i=1}^{n} \bar{v}_i w_i \,, \qquad |\mathbf{v}| = \sqrt{\langle \mathbf{v}, \mathbf{v} \rangle} = \sqrt{\sum_{i=1}^{n} |v_i|^2} \,. \tag{22.9}$$

Note that including the complex conjugation in the first argument is crucial to ensure hermiticity (S2) as well as positivity of the expression for the associated norm. □

Example 22.3 *(Scalar product on $\mathcal{M}_{n,n}(\mathbb{R})$)*

We know that the $n \times n$ matrices $\mathcal{M}_{n,n}(\mathbb{R})$ with real entries form a vector space under matrix addition and multiplication of matrices with scalars. On this vector space, we can introduce a real scalar product by using the trace. Specifically, for two matrices $A, B \in \mathcal{M}_{n,n}(\mathbb{R})$ we can define

$$\langle A, B \rangle := \operatorname{tr}(A^T B) = \sum_{i,j=1}^{n} A_{ij} B_{ij} \,, \qquad |A| = \sqrt{\operatorname{tr}(A^T A)} = \sum_{i,j=1}^{n} A_{ij}^2 \,. \tag{22.10}$$

Bi-linearity and symmetry follows easily from the properties of the trace in Prop. 19.3 and positivity is explicit from the above expressions. In fact, the scalar product (22.10) is identical to the dot product (22.8), but with the sum running over two indices rather than one.

For complex matrices in $\mathcal{M}_{n,n}(\mathbb{C})$ we can also define a Hermitian scalar product by changing the transposition in Eq. (22.10) to a Hermitian conjugation, in complete analogy with the standard scalar product on $\mathbb{C}^n$. □

Example 22.4 *(Scalar product on function vector spaces)*

Our last example is a bit more surprising and it Illustrates why we have gone to the trouble of a general axiomatic definition. The vector space under consideration is the space $\mathcal{C}([a,b], \mathbb{F})$ of continuous functions on the interval $[a, b]$ which are either real-valued ($\mathbb{F} = \mathbb{R}$) or complex-valued ($\mathbb{F} = \mathbb{C}$). Given two functions $f, g \in \mathcal{C}([a, b], \mathbb{F})$ a scalar product can be defined by

$$\langle f, g \rangle = \int_a^b dx \, \overline{f(x)} g(x) \,, \qquad |f| = \sqrt{\langle f, f \rangle} = \sqrt{\int_a^b dx |f(x)|^2} \,. \tag{22.11}$$

Linearity in the second argument is obvious from the structure of the integrand and linearity of the integral and symmetry (hermiticity) follows immediately from the form of the integrand. Positivity is a bit harder to prove (and is really a question for analysis) but, intuitively, a continuous function f which is non-zero for some $x \in [a, b]$ must be non-zero in an entire neighbourhood of x and, hence, must leads to a a positive norm, $|f| > 0$. Intuitively, we can think of the scalar product (22.11) as a 'continuous' version of the standard scalar products on $\mathbb{R}^n$ or $\mathbb{C}^n$ where the sum has been replaced by an integral. Function scalar products of this kind are widely used in functional analysis (see, for example, Rynne and Youngson 2008) and they are of great importance in physics, particularly in the context of quantum mechanics (see, for example, Messiah 2014). □

22.3 Orthogonality and Gram–Schmidt procedure

Summary 22.3 *An ortho-normal basis of an inner product vector space is a basis of pairwise orthogonal unit vectors. Every finite-dimensional inner product vector space has ortho-normal bases which can be constructed via the Gram–Schmidt procedure. Ortho-normal bases simplify a number of task, such as computing coordinates of vectors and finding the matrix representing a linear map. For any vector subspace $W \subset V$ of an inner product vector space V, we have $V = W \oplus W^\perp$, where $W^\perp$ is the orthogonal complement of W.*

In the remainder of this part we will work with a general inner product vector space V with (real or Hermitian) scalar product $\langle \cdot, \cdot \rangle$ and associated norm $|\cdot|$, unless specified otherwise. In this way, all general results can be applied to any of the above examples and indeed many more.

22.3.1 Ortho-normal bases

A simple but important observation is that orthogonality implies linear independence.

Lemma 22.1 *Non-zero and pairwise orthogonal vectors $\mathbf{v}_1, \ldots, \mathbf{v}_k \in V$ are linearly independent.*

Proof We need to show that the equation $\sum_{i=1}^{k} \alpha_i \mathbf{v}_i = \mathbf{0}$ is only solved if all $\alpha_i = 0$. To do this, we take the scalar product of this equation with $\mathbf{v}_j$, for any $j = 1, \ldots, k$, using that $\langle \mathbf{v}_j, \mathbf{v}_i \rangle = 0$ for $i \neq j$. This results in $\alpha_j |\mathbf{v}_j|^2 = 0$ and since $\mathbf{v}_j \neq 0$ (so that $|\mathbf{v}_j| > 0$) it follows that $\alpha_j = 0$. □

This motivates the following definition of ortho-normal bases.

Definition 22.3 *A basis $(\boldsymbol{\epsilon}_1, \ldots, \boldsymbol{\epsilon}_n)$ of V is called ortho-normal if $\langle \boldsymbol{\epsilon}_i, \boldsymbol{\epsilon}_j \rangle = \delta_{ij}$ for all $i, j = 1, \ldots, n$.*

We have already used ortho-normal bases routinely without drawing attention to this property. For example, the standard unit vectors $(\mathbf{e}_1, \ldots, \mathbf{e}_n)$ form an ortho-normal basis on $\mathbb{R}^n$ and on $\mathbb{C}^n$, relative to their respective standard scalar products (22.8) and (22.9). However, the standard unit vectors are by no means the only ortho-normal bases, as the following problem shows.

Problem 22.2 *(Ortho-normal bases on $\mathbb{R}^2$ and $\mathbb{C}^2$)*

(i) On $\mathbb{R}^2$, show that $(\boldsymbol{\epsilon}_1, \boldsymbol{\epsilon}_2)$ with $\boldsymbol{\epsilon}_1 = (1, 1)^T / \sqrt{2}$ and $\boldsymbol{\epsilon}_2 = (1, -1)^T / \sqrt{2}$ form an ortho-normal basis relative to the dot product.

(ii) On $\mathbb{C}^2$, show that (ϵ_1, ϵ_2) with $\epsilon_1 = (2, i)^T/\sqrt{5}$ and $\epsilon_2 = (1, -2i)^T/\sqrt{5}$ form an ortho-normal basis, relative to the standard scalar product.

Solution: (i) Using the dot product (22.8) this is easily verified.

$$|\epsilon_1| = \frac{1}{\sqrt{2}}\sqrt{1^2 + 1^2} = 1 \,, \quad |\epsilon_2| = \frac{1}{\sqrt{2}}\sqrt{1^2 + (-1)^2} = 1 \,, \quad \epsilon_1 \cdot \epsilon_2 = \frac{1}{2}(1 \cdot 1 + 1 \cdot (-1)) = 0 \,.$$

(ii) Now we should use the standard scalar product (22.9), taking care to include the complex conjugation in the calculation.

$$|\epsilon_1| = \frac{1}{\sqrt{5}}\sqrt{|2|^2 + |i|^2} = 1, \; |\epsilon_2| = \frac{1}{\sqrt{5}}\sqrt{|1|^2 + |-2i|^2} = 1, \; \langle \epsilon_1, \epsilon_2 \rangle = \frac{1}{5}(\bar{2} \cdot 1 + \bar{i} \cdot (-2i)) = 0 \,.$$

It is important to appreciate that the concept of an ortho-normal basis is general and can be applied to more abstract vector spaces as well.

Problem 22.3 *Finite Fourier series*

On the interval $[-\pi, \pi]$, consider the vector space of finite Fourier series

$$V = \left\{ \frac{\alpha_0}{2} + \sum_{k=1}^{n}(\alpha_k \sin(kx) + \beta_k \cos(kx)) \,|\, \alpha_k, \beta_k \in \mathbb{R} \right\} \,,$$

with scalar product

$$\langle g, h \rangle = \int_{-\pi}^{\pi} dx\, g(x)h(x) \,.$$

Show that the functions $c_0(x) = \frac{1}{\sqrt{2\pi}}$, $s_k(x) = \frac{1}{\sqrt{\pi}}\sin(kx)$ and $c_k(x) = \frac{1}{\sqrt{\pi}}\cos(kx)$, where $k = 1, \ldots, n$, form an ortho-normal basis of V.

Solution: Explicit integration, using standard integrals for sine and cosine, gives $\langle s_k, s_l \rangle = \delta_{kl}$, $\langle c_k, c_l \rangle = \delta_{kl}$ and $\langle s_k, c_l \rangle = 0$. From Lemma 22.1 this means the functions s_k, c_k must be linearly independent and since they evidently span V they form an ortho-normal basis of V.

22.3.2 Existence of ortho-normal bases

Does every (finite-dimensional) inner product vector space space have an ortho-normal basis and, if so, how can such a basis be constructed? The *Gram–Schmidt* procedure answers both of these questions.

Theorem 22.1 *(Gram–Schmidt procedure) Let V be an inner product vector space with basis $(\mathbf{v}_1, \ldots, \mathbf{v}_n)$. Then V has an ortho-normal basis $(\epsilon_1, \ldots, \epsilon_n)$ with*

$$\text{Span}(\epsilon_1, \ldots, \epsilon_k) = \text{Span}(\mathbf{v}_1, \ldots, \mathbf{v}_k) \quad \text{for all} \quad k = 1, \ldots, n \,. \tag{22.12}$$

Proof The proof is constructive. The first vector of our prospective ortho-normal basis is obtained by simply normalizing $\mathbf{v}_1$, that is,

$$\epsilon_1 = \frac{\mathbf{v}_1}{|\mathbf{v}_1|} \,. \tag{22.13}$$

Clearly, $|\epsilon_1| = 1$ and $\text{Span}(\epsilon_1) = \text{Span}(\mathbf{v}_1)$. Suppose we have already constructed the first $k - 1$ vectors $\epsilon_1, \ldots, \epsilon_{k-1}$, mutually orthogonal, normalized, and such that

$\mathrm{Span}(\epsilon_1, \ldots, \epsilon_j) = \mathrm{Span}(\mathbf{v}_1, \ldots, \mathbf{v}_j)$ for all $j = 1, \ldots, k - 1$. The next vector, ϵ_k, is then constructed by first subtracting from $\mathbf{v}_k$ its projections onto $\epsilon_1, \ldots, \epsilon_{k-1}$ and then normalizing (see Fig. 22.1), so

$$\mathbf{v}'_k = \mathbf{v}_k - \sum_{i=1}^{k-1} \langle \epsilon_i, \mathbf{v}_k \rangle \epsilon_i, \qquad \epsilon_k = \frac{\mathbf{v}'_k}{|\mathbf{v}'_k|}. \qquad (22.14)$$

Note that $\mathbf{v}'_k$ can indeed be normalized to one for if $\mathbf{v}'_k = \mathbf{0}$ Eq. (22.14) implies that $\mathbf{v}_k$ is a linear combination of $\epsilon_1, \ldots, \epsilon_{k-1}$ and, hence, of $\mathbf{v}_1, \ldots, \mathbf{v}_{k-1}$ which contradicts the assumption that $(\mathbf{v}_1, \ldots, \mathbf{v}_n)$ is a basis. For any vector ϵ_j with $j < k$ we have

$$\langle \epsilon_j, \mathbf{v}'_k \rangle = \langle \epsilon_j, \mathbf{v}_k \rangle - \sum_{i=1}^{k-1} \langle \epsilon_i, \mathbf{v}_k \rangle \underbrace{\langle \epsilon_j, \epsilon_i \rangle}_{=\delta_{ij}} = \langle \epsilon_j, \mathbf{v}_k \rangle - \langle \epsilon_j, \mathbf{v}_k \rangle = 0.$$

Hence, ϵ_k is orthogonal to all vectors $\epsilon_1, \ldots, \epsilon_{k-1}$. Moreover, since $\mathrm{Span}(\epsilon_1, \ldots, \epsilon_{k-1}) = \mathrm{Span}(\mathbf{v}_1, \ldots, \mathbf{v}_{k-1})$ and $\mathbf{v}_k$ and ϵ_k only differ by a re-scaling and terms proportional to $\epsilon_1, \ldots, \epsilon_{k-1}$ is follows that $\mathrm{Span}(\epsilon_1, \ldots, \epsilon_k) = \mathrm{Span}(\mathbf{v}_1, \ldots, \mathbf{v}_k)$. □

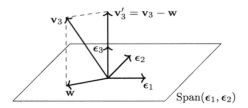

Fig. 22.1 The third step in a Gram–Schmidt procedure, with the first two ortho-normal vectors ϵ_1 and ϵ_2 already determined. The vector $\mathbf{w} = \langle \epsilon_1, \mathbf{v}_3 \rangle \epsilon_1 + \langle \epsilon_2, \mathbf{v}_3 \rangle \epsilon_2$ is the projection of $\mathbf{v}_3$ onto the plane $\mathrm{Span}(\epsilon_1, \epsilon_2)$.

22.3.3 Construction of ortho-normal bases

Since every finite-dimensional vector space has a basis the above theorem tells us that every finite-dimensional inner product vector space has an ortho-normal basis. What is more, it provides a method to construct this ortho-normal basis which is worth summarizing as an algorithm.

Algorithm *(Gram–Schmidt procedure)*

Given an inner product vector space V with a basis $(\mathbf{v}_1, \ldots, \mathbf{v}_n)$, an ortho-normal basis $(\epsilon_1, \ldots, \epsilon_n)$ of V with the property (22.12) can be obtained as follows.

(1) Compute ϵ_1 by normalizing $\mathbf{v}_1$, using Eq. (22.13).

(2) Assuming $(\epsilon_1, \ldots, \epsilon_{k-1})$ are known, work out ϵ_k from $\mathbf{v}_k$, using Eqs. (22.14).

(3) Repeat (2) until the basis is complete.

Problem 22.4 *(Gram–Schmidt procedure for $\mathbb{R}^3$)*

For $\mathbb{R}^3$ with the dot product and basis $(\mathbf{v}_1, \mathbf{v}_2, \mathbf{v}_3)$, where $\mathbf{v}_1 = (1,1,0)^T$, $\mathbf{v}_2 = (2,0,1)^T$ and $\mathbf{v}_3 = (1,-2,-2)^T$, find the associated Gram–Schmidt basis $(\epsilon_1, \epsilon_2, \epsilon_3)$.

Solution: (1) To find ϵ_1 use Eq. (22.13):

$$\epsilon_1 = \frac{\mathbf{v}_1}{|\mathbf{v}_1|} = \frac{1}{\sqrt{2}}\begin{pmatrix}1\\1\\0\end{pmatrix} .$$

(2) To find ϵ_2 use Eq. (22.14) for $k=2$:

$$\mathbf{v}_2' = \mathbf{v}_2 - \langle\epsilon_1, \mathbf{v}_2\rangle\epsilon_1 = \begin{pmatrix}2\\0\\1\end{pmatrix} - \begin{pmatrix}1\\1\\0\end{pmatrix} = \begin{pmatrix}1\\-1\\1\end{pmatrix} , \qquad \epsilon_2 = \frac{\mathbf{v}_2'}{|\mathbf{v}_2'|} = \frac{1}{\sqrt{3}}\begin{pmatrix}1\\-1\\1\end{pmatrix} .$$

(3) To find ϵ_3 use Eq. (22.14) for $k=3$:

$$\mathbf{v}_3' = \mathbf{v}_3 - \langle\epsilon_1, \mathbf{v}_3\rangle\epsilon_1 - \langle\epsilon_2, \mathbf{v}_3\rangle\epsilon_2 = \frac{7}{6}\begin{pmatrix}1\\-1\\-2\end{pmatrix} , \qquad \epsilon_3 = \frac{\mathbf{v}_3'}{|\mathbf{v}_3'|} = \frac{1}{\sqrt{6}}\begin{pmatrix}1\\-1\\-2\end{pmatrix} .$$

So, in summary, the ortho-normal basis is

$$\epsilon_1 = \frac{1}{\sqrt{2}}\begin{pmatrix}1\\1\\0\end{pmatrix} , \qquad \epsilon_2 = \frac{1}{\sqrt{3}}\begin{pmatrix}1\\-1\\1\end{pmatrix} , \qquad \epsilon_3 = \frac{1}{\sqrt{6}}\begin{pmatrix}1\\-1\\-2\end{pmatrix} .$$

It is easy (and always advisable) to check that indeed $\langle\epsilon_i, \epsilon_j\rangle = \delta_{ij}$.

Problem 22.5 *(Gram–Schmidt procedure for a function vector space)*

For a somewhat more adventurous application of the Gram–Schmidt procedure consider the vector space $V = \mathcal{P}_2(\mathbb{R})$ of at most quadratic polynomials on the interval $[-1,1]$ and a scalar product defined by

$$\langle p, q\rangle = \int_{-1}^{1} dx\, p(x)q(x) .$$

Find the ortho-normal basis (p_0, p_1, p_2) associated to the monomial basis (m_0, m_1, m_2), where $m_k = x^k$.

Solution: (1) To find p_0 we have to normalize the monomial m_0:

$$\langle m_0, m_0\rangle = \int_{-1}^{1} dx = 2 , \qquad p_0 = \frac{m_0}{|m_0|} = \frac{1}{\sqrt{2}} .$$

(2) To find p_1 first compute

$$\langle p_0, m_1\rangle = \int_{-1}^{1} dx\, \frac{x}{\sqrt{2}} = 0 , \qquad m_1' = m_1 - \langle p_0, m_1\rangle p_0 = x ,$$

and then normalize

$$\langle m_1', m_1' \rangle = \int_{-1}^{1} dx\, x^2 = \frac{2}{3} \qquad p_1 = \frac{m_1'}{|m_1'|} = \sqrt{\frac{3}{2}} x \;.$$

(3) Finally, to find p_2 first compute the integrals

$$\langle p_0, m_2 \rangle = \frac{1}{\sqrt{2}} \int_{-1}^{1} dx\, x^2 = \frac{\sqrt{2}}{3} \;, \qquad \langle p_1, m_2 \rangle = \sqrt{\frac{3}{2}} \int_{-1}^{1} dx\, x^3 = 0 \;,$$

and $m_2' = m_2 - \langle p_0, m_2 \rangle p_0 - \langle p_1, m_2 \rangle p_1 = x^2 - \frac{1}{3}$ and normalize

$$\langle m_2', m_2' \rangle = \int_{-1}^{1} dx \left(x^2 - \frac{1}{3} \right)^2 = \frac{8}{45} \;, \qquad p_2 = \frac{m_2'}{|m_2'|} = \sqrt{\frac{5}{8}} (3x^2 - 1) \;.$$

So, in summary, the ortho-normal polynomial basis is

$$p_0 = \frac{1}{\sqrt{2}} \;, \quad p_1 = \sqrt{\frac{3}{2}} x \;, \quad p_2 = \sqrt{\frac{5}{8}} (3x^2 - 1) \;.$$

These are the first three of an infinite family of ortho-normal polynomials, referred to as *Legendre polynomials*, which play an important role in mathematical physics (see, for example, Jackson 1962; Messiah 2014).

22.3.4 Properties of ortho-normal bases

An ortho-normal basis has many advantages compared to an arbitrary basis of a vector space. For example, consider the coordinates of a vector $\mathbf{v} \in V$ relative to an ortho-normal basis $(\epsilon_1, \ldots, \epsilon_n)$. Of course, we can write $\mathbf{v}$ as a linear combination $\mathbf{v} = \sum_{i=1}^{n} \alpha_i \epsilon_i$ with some coordinates α_i which can be determined by solving a linear system. However, for an ortho-normal basis there is a much simpler and faster method. We can just take the scalar product with ϵ_j, which leads to

$$\langle \epsilon_j, \mathbf{v} \rangle = \langle \epsilon_j \sum_{i=1}^{n} \alpha_i \epsilon_i \rangle = \sum_{i=1}^{n} \alpha_i \underbrace{\langle \epsilon_j, \epsilon_i \rangle}_{=\delta_{ij}} = \alpha_j \;.$$

In summary, the coordinates of a vector $\mathbf{v}$ relative to an ortho-normal basis $(\epsilon_1, \ldots, \epsilon_n)$ can be computed as

$$\mathbf{v} = \sum_{i=1}^{n} \alpha_i \epsilon_i \quad \Longleftrightarrow \quad \alpha_i = \langle \epsilon_i, \mathbf{v} \rangle \text{ for } i = 1, \ldots, n. \tag{22.15}$$

Problem 22.6 *(Coordinates relative to an ortho-normal basis)*

Consider an ortho-normal basis (ϵ_1, ϵ_2) and the vector $\mathbf{v} = (2, -3)^T = \alpha_1 \epsilon_1 + \alpha_2 \epsilon_2$. Compute the coordinates α_i

(i) in $\mathbb{R}^2$ with the dot product and $\epsilon_1 = (1,1)^T/\sqrt{2}$ and $\epsilon_2 = (1,-1)^T/\sqrt{2}$.

(ii) in $\mathbb{C}^2$ with the standard scalar product and $\epsilon_1 = (2,i)^T/\sqrt{5}$ and $\epsilon_2 = (1,-2i)^T/\sqrt{5}$.

Solution: (i) Use Eq. (22.15) and the dot product:

$$\alpha_1 = \epsilon_1^T \mathbf{v} = \frac{1}{\sqrt{2}} \begin{pmatrix} 1 \\ 1 \end{pmatrix}^T \begin{pmatrix} 2 \\ -3 \end{pmatrix} = -\frac{1}{\sqrt{2}} \,, \qquad \alpha_2 = \epsilon_2^T \mathbf{v} = \frac{1}{\sqrt{2}} \begin{pmatrix} 1 \\ -1 \end{pmatrix}^T \begin{pmatrix} 2 \\ -3 \end{pmatrix} = \frac{5}{\sqrt{2}} \,.$$

(ii) Use Eq. (22.15) and the standard Hermitian scalar product on $\mathbb{C}^2$:

$$\beta_1 = \epsilon_1^\dagger \mathbf{v} = \frac{1}{\sqrt{5}} \begin{pmatrix} 2 \\ i \end{pmatrix}^\dagger \begin{pmatrix} 2 \\ -3 \end{pmatrix} = \frac{4+3i}{\sqrt{5}} \,, \qquad \beta_2 = \epsilon_2^\dagger \mathbf{v} = \frac{1}{\sqrt{5}} \begin{pmatrix} 1 \\ -2i \end{pmatrix}^\dagger \begin{pmatrix} 2 \\ -3 \end{pmatrix} = \frac{2-6i}{\sqrt{5}} \,.$$

Note it is crucial to use the Hermitian conjugate, rather than the transpose in this calculation.

We would like to re-write the scalar product in terms of coordinates relative to an orth-normal basis $(\epsilon_1, \ldots, \epsilon_n)$. To do this we expand two vectors $\mathbf{v}, \mathbf{w} \in V$ as

$$\mathbf{v} = \sum_i \alpha_i \epsilon_i \,, \quad \alpha_i = \langle \epsilon_i, \mathbf{v} \rangle \,, \qquad\qquad \mathbf{w} = \sum_i \beta_i \epsilon_i \,, \quad \beta_i = \langle \epsilon_i, \mathbf{w} \rangle \,,$$

and compute their scalar product

$$\langle \mathbf{v}, \mathbf{w} \rangle = \sum_{i,j} \bar{\alpha}_i \beta_j \underbrace{\langle \epsilon_i, \epsilon_j \rangle}_{=\delta_{ij}} = \sum_i \bar{\alpha}_i \beta_i = \sum_i \langle \mathbf{v}, \epsilon_i \rangle \langle \epsilon_i, \mathbf{w} \rangle \,. \tag{22.16}$$

The result means that, in terms of the coordinates relative to an ortho-normal basis, every real scalar product looks like the dot product on $\mathbb{R}^n$ and every Hermitian scalar product like the standard scalar product on $\mathbb{C}^n$.

Finally, suppose we would like to compute the representing matrix A of a linear map $f : V \to W$ between two inner product spaces V and W with scalar products $\langle \cdot, \cdot \rangle_V$ and $\langle \cdot, \cdot \rangle_W$, relative to an ortho-normal bases $(\epsilon_1, \ldots, \epsilon_n)$ of V and $(\tilde{\epsilon}_1, \ldots, \tilde{\epsilon}_m)$ of W. In general, following Theorem 13.1, the entries A_{ij} of the matrix A can be obtained from

$$f(\epsilon_j) = \sum_k A_{kj} \tilde{\epsilon}_k \,. \tag{22.17}$$

Taking the scalar product of this equation with $\tilde{\epsilon}_i$ results in the simple formula

$$A_{ij} = \langle \tilde{\epsilon}_i, f(\epsilon_j) \rangle_W \,. \tag{22.18}$$

An expression of the form $\langle \mathbf{w}, f(\mathbf{v}) \rangle_W$ for $\mathbf{v} \in V$ and $\mathbf{w} \in W$ is sometimes called a *matrix element* of the linear map f. In this language, the result (22.18) states that the entries of a representing matrix relative to ortho-normal bases are given by the matrix elements with respect to the basis vectors. It is worth noting that linear maps are uniquely determined by their matrix elements.

Proposition 22.2 *If two linear maps $f, g : V \to W$ between inner product vector spaces V, W have the same matrix elements for all $\mathbf{v} \in V$ and $\mathbf{w} \in W$ then $f = g$.*

Proof   Having the same matrix elements means that $\langle \mathbf{w}, f(\mathbf{v}) \rangle_W = \langle \mathbf{w}, g(\mathbf{v}) \rangle_W$ for all $\mathbf{v} \in V$ and $\mathbf{w} \in W$. From linearity this implies $\langle \mathbf{w}, (f - g)(\mathbf{v}) \rangle_W = 0$ and if we choose $\mathbf{w} = (f - g)(\mathbf{v})$ it follows that $|(f - g)(\mathbf{v})| = 0$. From positivity of the norm this implies $f(\mathbf{v}) = g(\mathbf{v})$ for all $\mathbf{v} \in V$ which is the required statement. $\square$

Problem 22.7 (*Matrix representing a linear map relative to an ortho-normal basis*)

For a fixed unit length vector $\mathbf{n} \in \mathbb{R}^3$, consider the linear maps $f, g : \mathbb{R}^3 \to \mathbb{R}^3$ defined by $f(\mathbf{v}) = (\mathbf{n} \cdot \mathbf{v})\mathbf{n}$ and $g = \mathrm{id}_{\mathbb{R}^3} - f$. Verify that both maps are projectors and compute the matrices Q, P which represent f and g relative to the standard unit vector basis.

Solution: The maps f satisfies the projector property $f \circ f = f$ since

$$f \circ f(\mathbf{v}) = f((\mathbf{n} \cdot \mathbf{v})\mathbf{n}) = \underbrace{(\mathbf{n} \cdot \mathbf{n})}_{=1}(\mathbf{n} \cdot \mathbf{v})\mathbf{n} = f(\mathbf{v}) \,.$$

and g is the complementary projector to f (see Section 20.3).

The three standard unit vectors form an ortho-normal basis of $\mathbb{R}^3$ so we can use Eq. (22.18) to compute the entries of Q and P. This gives $Q_{ij} = \mathbf{e}_i \cdot f(\mathbf{e}_j) = (\mathbf{n} \cdot \mathbf{e}_i)(\mathbf{n} \cdot \mathbf{e}_j) = n_i n_j$ and $P_{ij} = \mathbf{e}_i \cdot g(\mathbf{e}_j) = \delta_{ij} - n_i n_j$ or, in matrix notation

$$Q = \begin{pmatrix} n_1^2 & n_1 n_2 & n_1 n_3 \\ n_1 n_2 & n_2^2 & n_2 n_3 \\ n_1 n_3 & n_2 n_3 & n_3^2 \end{pmatrix} \,, \quad P = \begin{pmatrix} 1 - n_1^2 & -n_1 n_2 & -n_1 n_3 \\ -n_1 n_2 & 1 - n_2^2 & -n_2 n_3 \\ -n_1 n_3 & -n_2 n_3 & 1 - n_3^2 \end{pmatrix} \,. \tag{22.19}$$

Geometrically, f represents the projection of vectors into the direction of $\mathbf{n}$ and g is the projection onto the plane orthogonal to $\mathbf{n}$.

22.3.5   Orthogonal spaces

The notion of orthogonality for two vectors can be extended to vector subspaces. For a vector subspace $W \subset V$ of an inner product vector space V we can define the orthogonal complement $W^\perp$ by

$$W^\perp = \{\mathbf{v} \in V \,|\, \langle \mathbf{w}, \mathbf{v} \rangle = 0 \text{ for all } \mathbf{w} \in W\} \,. \tag{22.20}$$

In other words, $W^\perp$ consists of all vectors which are orthogonal to all vector in W. For example, if $W \subset \mathbb{R}^3$ is a plane through the origin then $W^\perp$ is the line through the origin perpendicular to this plane.

Proposition 22.3 *Let V be an inner product vector space and $W \subset V$ a vector subspace. Then we have the following statements.*

(i) $W^\perp$ is a vector subspace of V.
(ii) $V = W \oplus W^\perp$.

Proof   (i) First, $W^\perp$ is not empty since $\mathbf{0} \in W^\perp$. For $\mathbf{v}_1, \mathbf{v}_2 \in W^\perp$ we have $\langle \mathbf{w}, \mathbf{v}_1 \rangle = \langle \mathbf{w}, \mathbf{v}_2 \rangle = 0$ for all $\mathbf{w} \in W$. It follows that $\langle \mathbf{w}, \alpha_1 \mathbf{v}_1 + \alpha_2 \mathbf{v}_2 \rangle = \alpha_1 \langle \mathbf{w}, \mathbf{v}_1 \rangle + \alpha_2 \langle \mathbf{w}, \mathbf{v}_2 \rangle = 0$, so $\alpha_1 \mathbf{v}_1 + \alpha_2 \mathbf{v}_2 \in W^\perp$. This means $W^\perp$ is closed under vector addition and scalar multiplication and is, hence, a vector subspace.

(ii) We begin by showing that $W \cap W^\perp = \{\mathbf{0}\}$. A vector $\mathbf{v} \in W \cap W^\perp$ must be orthogonal to itself so must satisfy $\langle \mathbf{v}, \mathbf{v} \rangle = 0$ but from Def. 22.2 this implies $\mathbf{v} = \mathbf{0}$. We conclude that the sum $W + W^{\perp}$ is direct.

To show that $V = W \oplus W^{\perp}$ all we need to do is show equality of dimensions. To this end, we choose an ortho-normal basis $(\boldsymbol{\epsilon}_1, \dots, \boldsymbol{\epsilon}_k)$ of W and define the linear map $p_W : V \to V$ by $p_W(\mathbf{v}) = \sum_{i=1}^{k} \langle \boldsymbol{\epsilon}_i, \mathbf{v} \rangle \boldsymbol{\epsilon}_i$. Clearly $\text{Im}(p_W) \subset W$. For $\mathbf{w} \in W$ it follows from Eq. (22.15) that $p_W(\mathbf{w}) = \mathbf{w}$ so that $\text{Im}(p_W) = W$. Moreover, $\text{Ker}(p_W) = W^{\perp}$ and the claim follows from the dimension formula (14.13) applied to the map p_W. □

From Prop. 8.1, this result means that every vector $\mathbf{v} \in V$ can be written as a unique sum

$$\mathbf{v} = \mathbf{w}^{\|} + \mathbf{w}^{\perp} \qquad (22.21)$$

of a vector $\mathbf{w}^{\|} \in W$ and a vector $\mathbf{w}^{\perp} \in W^{\perp}$ in the orthogonal complement of W. To make this more explicit we introduce an ortho-normal basis $(\boldsymbol{\epsilon}_1, \dots, \boldsymbol{\epsilon}_k)$ on W and consider the endomorphism

$$p_W(\mathbf{v}) = \sum_{i=1}^{k} \langle \boldsymbol{\epsilon}_i, \mathbf{v} \rangle \boldsymbol{\epsilon}_i \, , \qquad (22.22)$$

which we have already used in the previous proof. Obviously, for $\mathbf{w} \in W$ we have $p_W(\mathbf{w}) = \mathbf{w}$, while $p_W(\mathbf{u}) = 0$ for $\mathbf{u} \in W^{\perp}$. This means p_W is a projector with $\text{Im}(p_W) = W$ and $\text{Ker}(p_W) = W^{\perp}$. The complementary projector $p_{W^{\perp}} := \text{id}_V - p_W$ projects onto $\text{Ker}(p_W) = W^{\perp}$.

The orthogonal decomposition (22.21) of vectors $\mathbf{v} \in V$ into components $\mathbf{w}^{\|} \in W$ and $\mathbf{w}^{\perp} \in W^{\perp}$ can then we written explicitly as

$$\mathbf{v} = \mathbf{w}^{\|} + \mathbf{w}^{\perp} \quad \text{where} \quad \begin{cases} \mathbf{w}^{\|} = p_W(\mathbf{v}) = \sum_{i=1}^{k} \langle \boldsymbol{\epsilon}_i, \mathbf{v} \rangle \boldsymbol{\epsilon}_i \\ \mathbf{w}^{\perp} = p_{W^{\perp}}(\mathbf{v}) = \mathbf{v} - \mathbf{w}^{\|} \end{cases} . \qquad (22.23)$$

Problem 22.8 *(Orthogonal projections)*

Consider the subspace $W = \text{Span}(\mathbf{v}_1, \mathbf{v}_2)$ in $\mathbb{R}^4$ with the dot product, where $\mathbf{v}_1 = (1, 1, 1, 1)^T$ and $\mathbf{v}_2 = (2, 2, 1, -1)^T$. Find the projector p_W onto W and decompose $\mathbf{v} = (2, 2, -1, -1)^T$ as in Eq. (22.21).

Solution: We would like to use Eqs. (22.23) so our first step is to find an ortho-normal basis $(\boldsymbol{\epsilon}_1, \boldsymbol{\epsilon}_2)$ for W. This can, of course, be done by applying the Gram–Schmidt procedure to $\mathbf{v}_1, \mathbf{v}_2$ which gives

$$\boldsymbol{\epsilon}_1 = \frac{1}{2}(1, 1, 1, 1)^T \, , \quad \boldsymbol{\epsilon}_2 = \frac{1}{\sqrt{6}}(1, 1, 0, -2)^T \, .$$

Then we have $\mathbf{v} = \mathbf{w}^{\|} + \mathbf{w}^{\perp}$ with

$$\mathbf{w}^{\|} = (\boldsymbol{\epsilon}_1 \cdot \mathbf{v})\boldsymbol{\epsilon}_1 + (\boldsymbol{\epsilon}_2 \cdot \mathbf{v})\boldsymbol{\epsilon}_2 = \tfrac{1}{2}(3, 3, 1, -3)^T$$
$$\mathbf{w}^{\perp} = \mathbf{v} - \mathbf{w}^{\|} = \tfrac{1}{2}(1, 1, -3, 1)^T \, .$$

Exercises

(†=challenging)

22.1 Which of the following maps are scalar products on $\mathbb{R}^2$?
(a) $\langle \mathbf{v}, \mathbf{w} \rangle = v_1^2 + v_2^2 + w_1^2 + w_2^2$
(b) $\langle \mathbf{v}, \mathbf{w} \rangle = (v_1 - v_2)(w_1 - w_2)$
(c) $\langle \mathbf{v}, \mathbf{w} \rangle = (v_1 - v_2)(w_1 - w_2) + v_2 w_2$
Here, $\mathbf{v} = (v_1, v_2)^T \in \mathbb{R}^2$ and $\mathbf{w} = (w_1, w_2)^T \in \mathbb{R}^2$. Provide reasoning in each case.

22.2 (a) For the vector space $\mathcal{M}_{n,n}(\mathbb{C})$ over $\mathbb{C}$ show that

$$\langle A, B \rangle = \mathrm{tr}(A^\dagger B)$$

defines a Hermitian scalar product, where $A, B \in \mathcal{M}_{n,n}(\mathbb{C})$.
(b) Show that the elementary matrices $E_{(ij)}$, where $i, j = 1, \ldots, n$, form an ortho-normal basis of $\mathcal{M}_{n,n}(\mathbb{C})$ with respect to this scalar product.

22.3 (a) For the vector space $\mathcal{H}_n$ over $\mathbb{R}$, which consists of Hermitian $n \times n$ matrices, show that

$$\langle A, B \rangle = \mathrm{tr}(AB)$$

defines a scalar product, where $A, B \in \mathcal{H}_n$.
(b) Show that $\frac{1}{\sqrt{2}}(\mathbb{1}_2, \sigma_1, \sigma_2, \sigma_3)$, where σ_i are the Pauli matrices, is an ortho-normal basis of $\mathcal{H}_2$, relative to this scalar product.

22.4 *Gram–Schmidt procedure in $\mathbb{R}^3$*
Use the Gram–Schmidt procedure to find an ortho-normal basis of $\mathbb{R}^3$ (with the standard scalar product), starting with the basis

$$\mathbf{v}_1 = \begin{pmatrix} 1 \\ 1 \\ 0 \end{pmatrix}, \mathbf{v}_2 = \begin{pmatrix} 2 \\ 1 \\ 2 \end{pmatrix}, \mathbf{v}_3 = \begin{pmatrix} 0 \\ 2 \\ -1 \end{pmatrix}.$$

Check your result.

22.5 *Scalar product on a function space*
On the vector space V of at most quadratic polynomials over $\mathbb{R}$, define

$$\langle p, q \rangle = \int_{-\infty}^{\infty} dx \, e^{-x^2} p(x) q(x) \,,$$

where $p, q \in V$.
(a) Why does this define a scalar product?
(b) Consider the polynomials $p_0(x) = b_0$, $p_1(x) = 2b_1 x$ and $p_2(x) = b_2(4x^2 - 2)$, with $b_0, b_1, b_2 \in \mathbb{R}$. Show that these polynomials are orthogonal under the scalar product.
(c) Determine the constants b_a such that the polynomials p_a have unit length.
(Hint: Look up integrals of the form $\int_{-\infty}^{\infty} dx \, x^n e^{-x^2}$)
(d) Find the coordinates of $p(x) = 4x^2 - 2x + 3$ relative to the basis (p_0, p_1, p_2) in two ways, namely by matching monomial coefficients and by carrying out scalar products.

22.6 (a) Show that the vectors

$$\epsilon_1 = \tfrac{1}{\sqrt{2}}(1, i, 0)^T$$
$$\epsilon_2 = \tfrac{1}{\sqrt{6}}(1, -i, 2)^T$$
$$\epsilon_3 = \tfrac{1}{\sqrt{3}}(i, 1, -i)^T$$

form an orthonormal basis of $\mathbb{C}^3$, relative to the standard Hermitian scalar product.
(b) Find the coordinates of the vectors

$$\mathbf{v} = (2, 1 + i, -3)^T$$
$$\mathbf{w} = (1, 1, 1)^T$$

relative to the basis from part (a).

22.7 Let V be a vector space over $\mathbb{R}$ or $\mathbb{C}$ with basis $(\mathbf{v}_1, \ldots, \mathbf{v}_n)$. Show that there is a unique scalar product on V relative to which $(\mathbf{v}_1, \ldots, \mathbf{v}_n)$ is an ortho-normal basis.

22.8 For vector subspaces $U, W \subset V$ of an inner product vector space V show the following:
(a) $(W^\perp)^\perp = W$
(b) $(U + W)^\perp = U^\perp \cap W^\perp$
(c) $U^\perp + W^\perp = (U \cap W)^\perp$

22.9 *Parallelogram identity*[†]
Let V be a vector space over $\mathbb{R}$.
(a) Show that a norm $|\cdot|$ on V associated to a scalar product $\langle\cdot,\cdot\rangle$ on V satisfies the parallelogram identity

$$|\mathbf{v}+\mathbf{w}|^2+|\mathbf{v}-\mathbf{w}|^2=2(|\mathbf{v}|^2+|\mathbf{w}|^2)$$

for all $\mathbf{v},\mathbf{w}\in V$.
(b) For $V=\mathbb{R}^n$, show that

$$|\mathbf{v}|=\sum_{i=1}^{n}|v_i|$$

defines a norm.
(c) Show that the norm from part (b) is not associated to a scalar product. (Hint: Show that this norm violates the parallelogram identity.)

22.10 Let V be an n-dimensional inner product vector space, $\mathbf{v}_1,\ldots,\mathbf{v}_n\in V$ and P the matrix with entries $P_{ij}=\langle\mathbf{v}_i,\mathbf{v}_j\rangle$.
(a) Show that $(\mathbf{v}_1,\ldots,\mathbf{v}_n)$ is a basis of V iff P is invertible.
(b) If $(\mathbf{v}_1,\ldots,\mathbf{v}_n)$ is a basis of V show there is a unique basis $(\mathbf{u}_1,\ldots,\mathbf{u}_n)$ of V with $\langle\mathbf{u}_i,\mathbf{v}_j\rangle=\delta_{ij}$.
(c) Show that the coordinates of a vectors $\mathbf{v}\in V$ relative to the basis $(\mathbf{v}_1,\ldots,\mathbf{v}_n)$ are given by $\langle\mathbf{u}_i,\mathbf{v}\rangle$.
(d) Work out how the coordinate vectors of $\mathbf{v}\in V$ relative to the two bases from part (b) are related.

23
Adjoint and unitary maps

In the previous chapter, we have explored inner product vector spaces by studying the interplay between the scalar product and the vector space structure. But vector spaces come equipped with homomorphisms, so it is natural to ask about the relationship between a scalar product and linear maps. The short version of the story is that a scalar product facilitates defining specific linear maps and single out interesting subsets of linear maps. In this chapter, we will study two classes of such linear maps — the *adjoint* and *self-adjoint* linear maps and the *unitary* maps.

As we will see, taking the adjoint is the operation required if we want to move a linear map from one argument of a scalar product into the other without changing the value of the scalar product. A self-adjoint map can be moved between the two scalar product arguments without changing its value. The adjoint is an abstract versions of Hermitian conjugation (or transposition in the real case) for matrices. Correspondingly, self-adjoint linear operators correspond to Hermitian matrices (symmetric matrices in the real case).

Unitary maps are linear maps which leave a scalar product invariant. This means that unitary maps do not change the basic geometrical quantities which follow from a scalar product, including the length of vectors. Therefore, on $\mathbb{R}^n$, unitary maps should be interpreted as rotations and combinations of rotations and reflections. On $\mathbb{C}^n$ unitary maps are identified with unitary matrices which can be viewed as generalizations of rotations and reflections to the complex case.

23.1   Adjoint and self-adjoint maps

Summary 23.1 *The adjoint linear map $f^\dagger$ of an endomorphism $f \in \text{End}(V)$ is defined by the relation $\langle \mathbf{v}, f(\mathbf{w}) \rangle = \langle f^\dagger(\mathbf{v}), \mathbf{w} \rangle$ for all $\mathbf{v}, \mathbf{w} \in V$. For $V = \mathbb{R}^n$ ($V = \mathbb{C}^n$) with the standard scalar product, the adjoint operation corresponds to matrix transposition (Hermitian conjugation). A map is self-adjoint or Hermitian iff $f = f^\dagger$. For $\mathbb{R}^n$ ($\mathbb{C}^n$) with the standard scalar product self-adjoint maps are given by symmetric (Hermitian) matrices.*

23.1.1   Definition and basic properties of adjoint map

We start with two inner product vector spaces V and W, with scalar products $\langle \cdot, \cdot \rangle_V$ and $\langle \cdot, \cdot \rangle_W$, and a linear map $f : V \to W$.

Definition 23.1 *(Adjoint linear map) For a linear map $f : V \to W$ between two inner product vector spaces V and W, an adjoint linear map $f^\dagger : W \to V$ for f is a map which satisfies*

$$\langle \mathbf{w}, f(\mathbf{v}) \rangle_W = \langle f^\dagger(\mathbf{w}), \mathbf{v} \rangle_V \,, \tag{23.1}$$

for all $\mathbf{v} \in V$ and all $\mathbf{w} \in W$.

In other words, a linear map can be 'moved' into the other argument of the scalar product by taking its adjoint. The following properties of the adjoint map are relatively easy to show.

Proposition 23.1 *(Properties of adjoint) If the adjoint exists it is unique and it satisfies*

$$
\begin{array}{llll}
(i) & (f^\dagger)^\dagger = f & (ii) & (\alpha f + \beta g)^\dagger = \bar{\alpha} f^\dagger + \bar{\beta} g^\dagger \\
(iii) & (g \circ f)^\dagger = f^\dagger \circ g^\dagger & (iv) & (f^{-1})^\dagger = (f^\dagger)^{-1}
\end{array}
\tag{23.2}
$$

where $\alpha, \beta \in \mathbb{F}$ and provided the various maps exist.

Proof All proofs involve verifying the stated equality inside a scalar product and then using Prop. 22.2 to remove the scalar product.

(Uniqueness) Say $f_1, f_2 : W \to V$ are two adjoints for $f : V \to W$. From Eq. (23.1) they must satisfy $\langle f_1(\mathbf{w}), \mathbf{v} \rangle_V = \langle \mathbf{w}, f(\mathbf{v}) \rangle_W = \langle f_2(\mathbf{w}), \mathbf{v} \rangle_V$ for all $\mathbf{v} \in V$ and all $\mathbf{w} \in W$. It follows from Prop. 22.2 that $f_1 = f_2$.

(i) For all $\mathbf{v} \in V$ and all $\mathbf{w} \in W$ we have

$$\langle \mathbf{w}, f(\mathbf{v}) \rangle_W \overset{(23.1)}{=} \langle f^\dagger(\mathbf{w}), \mathbf{v} \rangle_V \overset{(S3)}{=} \overline{\langle \mathbf{v}, f^\dagger(\mathbf{w}) \rangle_V} \overset{(23.1)}{=} \overline{\langle (f^\dagger)^\dagger(\mathbf{v}), \mathbf{w} \rangle_W} \overset{(S3)}{=} \langle \mathbf{w}, (f^\dagger)^\dagger(\mathbf{v}) \rangle_W$$

and comparing the left and right-hand sides gives $f = (f^\dagger)^\dagger$ from Prop. 22.2.

(ii) A straightforward calculation gives

$$\langle (\alpha f + \beta g)^\dagger(\mathbf{w}), \mathbf{v} \rangle_V \overset{(23.1)}{=} \langle \mathbf{w}, (\alpha f + \beta g)(\mathbf{v}) \rangle_W \overset{(S2)}{=} \alpha \langle \mathbf{w}, f(\mathbf{v}) \rangle_W + \beta \langle \mathbf{w}, g(\mathbf{v})_W \rangle_W$$

$$\overset{(23.1)}{=} \alpha \langle f^\dagger(\mathbf{w}), \mathbf{v} \rangle_V + \beta \langle g^\dagger(\mathbf{w}), \mathbf{v} \rangle_V \overset{(22.2)}{=} \langle (\bar{\alpha} f^\dagger + \bar{\beta} g^\dagger)(\mathbf{w}), \mathbf{v} \rangle_V$$

and comparing the left- and right-hand sides implies the claim from Prop. 22.2.

(iii) With the two linear maps $f : V \to W$ and $g : W \to U$ we have

$$\langle (g \circ f)^\dagger(\mathbf{u}), \mathbf{v} \rangle_V \overset{(23.1)}{=} \langle \mathbf{u}, g(f(\mathbf{v})) \rangle_U \overset{(23.1)}{=} \langle g^\dagger(\mathbf{u}), f(\mathbf{v}) \rangle_W \overset{(23.1)}{=} \langle (f^\dagger \circ g^\dagger)(\mathbf{u}), \mathbf{v} \rangle_V$$

and hence, $(g \circ f)^\dagger = f^\dagger \circ g^\dagger$ from Prop. 22.2.

(iv) Since

$$\langle (f^\dagger \circ (f^{-1})^\dagger)(\mathbf{v}_1), \mathbf{v}_2 \rangle_V \overset{(23.1)}{=} \langle \mathbf{v}_1, (f^{-1} \circ f)\mathbf{v}_2 \rangle_V = \langle \mathrm{id}_V(\mathbf{v}_1), \mathbf{v}_2 \rangle_V$$

it follows that $f^\dagger \circ (f^{-1})^\dagger = \mathrm{id}_V$ and, analogously, that $(f^{-1})^\dagger \circ f^\dagger = \mathrm{id}_W$. Together, these equations imply that $(f^{-1})^\dagger = (f^\dagger)^{-1}$. $\qquad \square$

23.1.2 Adjoint map relative to a basis

To get a better understanding of the adjoint map it is useful to work this out relative to ortho-normal bases. Say that $(\epsilon_1, \ldots, \epsilon_n)$ and $(\tilde{\epsilon}_1, \ldots, \tilde{\epsilon}_m)$ are ortho-normal bases of V and W while the linear maps $f \in \mathrm{Hom}(V, W)$ and $f^\dagger \in \mathrm{Hom}(W, V)$ are represented by matrices A and B, relative to those bases. From Eq. (22.18), the entries of A and B are given by the matrix elements

$$A_{ij} = \langle \tilde{\epsilon}_i, f(\epsilon_j) \rangle_W, \qquad B_{ij} = \langle \epsilon_i, f^\dagger(\tilde{\epsilon}_j) \rangle_V.$$

A short calculation

$$B_{ij} = \langle \epsilon_i, f^\dagger(\tilde{\epsilon}_j) \rangle_V \overset{(S1)}{=} \overline{\langle f^\dagger(\tilde{\epsilon}_j), \epsilon_i \rangle_V} \overset{(23.1)}{=} \overline{\langle \tilde{\epsilon}_j, f(\epsilon_i) \rangle_W} = \bar{A}_{ji} \quad \Rightarrow \quad B = A^\dagger \quad (23.3)$$

shows that the matrices are, in fact, related by Hermitian conjugation (or by transposition in the real case). There are several lessons from this. Previously (see Section 13.3), we have introduced Hermitian conjugation merely as a 'mechanical' operation to be carried out for matrices. Now we understand its proper mathematical context — it corresponds to carrying out the adjoint for a linear map. For the case of real scalar products, complex conjugation can be dropped in the above equations and the matrices representing f and $f^\dagger$ are related by transposition. This means, for matrices with real entries, we have also found the mathematical interpretation of matrix transposition.

So far, it is actually not clear whether the adjoint map always exists. However, for finite-dimensional vector spaces this is easy to show by reversing the above argument and defining $f^\dagger$ as the linear map associated to $A^\dagger$.

Finally, the relationship between the adjoint and Hermitian conjugation (transposition) of matrices explains the similarity between the rules for the adjoint in Prop. 23.1 and the rules for Hermitian conjugation (transposition) in Prop. 13.2 (Prop. 13.1).

Problem 23.1 *(Adjoint and determinant)*

How are the determinants of a linear map $f \in \mathrm{End}(V)$ and its adjoint $f^\dagger$ related?

Solution: If f is represented by the matrix A, relative to an ortho-normal basis of V, then $f^\dagger$ is represented by $A^\dagger = \bar{A}^T$. Hence

$$\det(f^\dagger) \overset{\mathrm{Def.\ 18.2}}{=} \det(A^\dagger) \overset{\mathrm{Prop.\ 18.1}}{=} \overline{\det(A)} \overset{\mathrm{Def.\ 18.2}}{=} \overline{\det(f)}. \quad (23.4)$$

In conclusion, the determinants of f and $f^\dagger$ are related by complex conjugation.

23.1.3 Examples

Example 23.1 *(The adjoint for coordinate vector spaces)*

We start with coordinate vector spaces $V = \mathbb{R}^n$ and $W = \mathbb{R}^m$, each equipped with the dot product, and a linear map $A : \mathbb{R}^n \to \mathbb{R}^m$, given by an $m \times n$ matrix A. Working out the adjoint of the matrix A, relative to the dot product, is straightforward.

$$\langle \mathbf{w}, A\mathbf{v} \rangle_V = \mathbf{w} \cdot (A\mathbf{v}) = w_i (A\mathbf{v})_i = w_i A_{ij} v_j = (A^T)_{ji} w_i v_j = (A^T \mathbf{w}) \cdot \mathbf{v} = \langle A^T \mathbf{w}, \mathbf{v} \rangle_V$$

The conclusion is that the adjoint of a matrix A relative to the dot product is simply its transpose A^T.

For the complex case, we consider vector spaces $V = \mathbb{C}^n$ and $W = \mathbb{C}^m$, each with the standard scalar product (22.9) and a linear map $A : \mathbb{C}^n \to \mathbb{C}^m$. To calculate the adjoint of A we proceed as above, except that complex conjugation has to be included.

$$\langle \mathbf{w}, A\mathbf{v} \rangle_V = \bar{w}_i (A\mathbf{v})_i = \bar{w}_i A_{ij} v_j = (A^T)_{ji} \bar{w}_i v_j = \overline{(A^\dagger \mathbf{w})}_i v_i = \langle A^\dagger \mathbf{w}, \mathbf{v} \rangle_V$$

Hence, in the complex case the adjoint of a matrix A relative to the standard scalar product is its Hermitian conjugate $A^\dagger$. Of course neither of these results is surprising in view of the discussion in Section 23.1.2. $\qquad\qquad\square$

Example 23.2 *(Adjoint of a derivative map)*

For a more abstract example of an adjoint linear map, consider the vector space V of (infinitely many times) differentiable functions $\varphi : [a, b] \to \mathbb{C}$, satisfying $\varphi(a) = \varphi(b)$, with scalar product

$$\langle \varphi, \psi \rangle = \int_a^b dx \, \overline{\varphi(x)} \psi(x) \,.$$

The derivative operator $D = d/dx : V \to V$ defines a linear map on this space and we would like to find its adjoint. Performing an integration by parts leads to

$$\langle \varphi, D\psi \rangle - \int_a^b dx \, \overline{\varphi(x)} \frac{d\psi}{dx}(x) - \left[\overline{\varphi(x)} \psi(x) \right]_a^b - \int_a^b dx \, \overline{\frac{d\varphi}{dx}(x)} \psi(x)$$

$$= \int_a^b dx \, \overline{(-D\varphi)(x)} \psi(x) = \langle -D\varphi, \psi \rangle \,.$$

Note that the boundary term vanishes due to the boundary condition on our functions. We conclude that $D^\dagger = -D$. $\qquad\qquad\square$

23.1.4 Kernel and image of the adjoint map

The kernel and the image are two vector subspaces naturally associated to a linear map f. It is, therefore, natural to ask about the image and kernel of the adjoint $f^\dagger$ and how they relate to their counterparts for f. Fortunately, there is a simple and beautiful answer.

Theorem 23.1 *For a linear map $f \in \mathrm{Hom}(V, W)$ between finite-dimensional vector spaces V and W we have the following equations.*

$$(i) \; \mathrm{Ker}(f^\dagger) = \mathrm{Im}(f)^\perp \,, \quad (ii) \; \mathrm{Im}(f^\dagger) = \mathrm{Ker}(f)^\perp \,, \quad (iii) \; \mathrm{rk}(f^\dagger) = \mathrm{rk}(f) \quad (23.5)$$

Proof (i) $\mathbf{w} \in \mathrm{Ker}(f^\dagger) \Leftrightarrow f^\dagger(\mathbf{w}) = \mathbf{0} \Leftrightarrow 0 = \langle f^\dagger(\mathbf{w}), \mathbf{v} \rangle_V = \langle \mathbf{w}, f(\mathbf{v}) \rangle_W \; \forall \mathbf{v} \in V \Leftrightarrow \mathbf{w} \in \mathrm{Im}(f)^\perp$.

(iii) $\mathrm{rk}(f^\dagger) \overset{(14.13)}{=} \dim_{\mathbb{F}}(W) - \dim_{\mathbb{F}}(\mathrm{Ker}(f^\dagger)) \overset{(i)}{=} \dim_{\mathbb{F}}(W) - \dim_{\mathbb{F}}(\mathrm{Im}(f)^\perp) \overset{\text{Prop. 22.3}}{=} \mathrm{rk}(f)$

(ii) We start by showing the inclusion $\mathrm{Im}(f^\dagger) \subset \mathrm{Ker}(f)^\perp$.

$\mathbf{v} \in \text{Im}(f^\dagger) \;\Rightarrow\; \mathbf{v} = f^\dagger(\mathbf{w})$ for a $\mathbf{w} \in W \;\Rightarrow\;$ for all $\mathbf{u} \in \text{Ker}(f)$ we have $\langle \mathbf{v}, \mathbf{u} \rangle_V = \langle f^\dagger(\mathbf{w}), \mathbf{u} \rangle_V = \langle \mathbf{w}, f(\mathbf{u}) \rangle_W = 0 \;\Rightarrow\; \mathbf{v} \in \text{Ker}(f)^\perp$

But the two spaces have the same dimensions since $\dim_{\mathbb{F}}(\text{Im}(f^\dagger)) = \text{rk}(f^\dagger) \overset{(ii)}{=} \text{rk}(f) = \dim_{\mathbb{F}}(V) - \dim_{\mathbb{F}}(\text{Ker}(f)) = \dim_{\mathbb{F}}(\text{Ker}(f)^\perp)$, so they must be equal. $\qquad\square$

In passing, we have learned that a linear map and its adjoint have the same rank. In view of Example 23.1, this implies that the rank of a matrix and its Hermitian conjugate (transpose for real matrices) are the same. A related statement has been shown, by more elementary methods, in Theorem 16.1.

23.1.5   Self-adjoint maps

The adjoint operation singles out a particular class of endomorphisms, which are invariant under taking the adjoint.

Definition 23.2 *An endomorphism $f \in \text{End}(V)$ on an inner product vector space V is called self-adjoint or Hermitian if*

$$\langle \mathbf{v}, f(\mathbf{u}) \rangle = \langle f(\mathbf{v}), \mathbf{u} \rangle \tag{23.6}$$

for all $\mathbf{v}, \mathbf{u} \in V$ or, equivalently, if $f = f^\dagger$.

A Hermitian linear map can be moved from one argument of the scalar product to the other without changing the scalar product's value. We emphasize that being self-adjoint is a property which is defined, and only makes sense, in relation to a scalar product.

In analogy with what we have done for matrices we can also define *anti-Hermitian* endomorphisms by the condition $f^\dagger = -f$. As Prop. 23.1 shows, multiplication by a factor of $\pm i$ converts between the Hermitian and anti-Hermitian case, so

$$f = f^\dagger \qquad \Leftrightarrow \qquad (\pm if)^\dagger = -(\pm if) . \tag{23.7}$$

Further, just as for matrices (see Example 13.2), every endomorphisms $f \in \text{End}(V)$ can be written as a (unique) sum of a Hermitian endomorphism f_+ and an anti-Hermitian endomorphism f_- as

$$f = f_+ + f_- \quad \text{where} \quad f_\pm = \frac{1}{2}(f \pm f^\dagger) . \tag{23.8}$$

For two Hermitian endomorphisms $f, g \in \text{End}(V)$ we can ask if the composition $f \circ g$ is Hermitian. Since $(f \circ g)^\dagger = g^\dagger \circ f^\dagger = g \circ f$ this is the case iff $f \circ g = g \circ f$, so iff f and g commute. So for $f = f^\dagger$ and $g = g^\dagger$ we have

$$f \circ g = (f \circ g)^\dagger \qquad \Leftrightarrow \qquad [f, g] = 0 . \tag{23.9}$$

Example 23.3 *(Hermitian maps for coordinate vector spaces)*

From Example 23.1, it is clear that the (anti-) Hermitian linear maps A on $\mathbb{R}^n$ with the dot product are precisely the (anti-) symmetric matrices. In the complex case, the (anti-) Hermitian maps A on $\mathbb{C}^n$ with the standard scalar product are the (anti-) Hermitian matrices. $\qquad\square$

Example 23.4 *(Hermitian differential operators)*

From Example 23.2 we know that the differential map $D = d/dx$ (on the vector space V of infinitely times differentiable functions $\varphi : [a, b] \to \mathbb{C}$ with $\varphi(a) = \varphi(b)$) is anti-Hermitian, so $D^\dagger = -D$. From Eq. (23.7) this means that $\pm iD$ is Hermitian. What about the multiplication map $X : V \to V$ defined by $X\varphi(x) = x\varphi(x)$? Using the scalar product from Example 23.2 we have

$$\langle \varphi, X\psi \rangle = \int_a^b dx\, \overline{\varphi(x)}(x\psi(x)) = \int_a^b dx\, \overline{x\varphi(x)}\psi(x) = \langle X\varphi, \psi \rangle\ ,$$

so that X is Hermitian, $X^\dagger = X$. $\qquad\square$

Problem 23.2 *(Hermitian operators on function vector spaces)*

Consider the vector space V of infinitely times differentiable functions $\varphi : [a, b] \to \mathbb{C}$ with $\varphi(a) = \varphi(b)$), the linear derivative map $P = -iD = -id/dx$ and the multiplication map X, all as defined in Examples 23.2 and (23.4). Show that any power P^k and X^k is Hermitian. Work out the commutator $[X, P]$ and show that $X \circ P$ is not Hermitian.

Soluion: From Eq. (23.9) we know that the composition of two Hermitian maps is Hermitian iff they commute. Since both X and P are Hermitian and commute with themselves, the powers X^k and P^k are Hermitian. Since $[X, P] = X \circ P - P \circ X = -ixD + iD \circ x = i \neq 0$ it follows that $X \circ P$ is not Hermitian.

In quantum mechanics, physical operators are represented by Hermitian linear maps. In this context, the above maps X and P correspond to the linear maps for position and momentum, respectively.

23.2 Unitary maps

Summary 23.2 *A map $f \in \text{End}(V)$ is called unitary if it leaves the scalar product on V invariant. This is equivalent to the condition $f^\dagger \circ f = \text{id}_V$. The unitary maps form a sub-group of $\text{GL}(V)$, called the unitary group $\text{U}(V)$. Unitary maps with determinant one form a sub-group of $\text{U}(V)$ called the special unitary group $\text{SU}(V)$. For $V = \mathbb{R}^n$ this leads to the orthogonal group $\text{O}(n)$ and the special orthogonal group (or rotation group) $\text{SO}(n)$. In the complex case, $V = \mathbb{C}^n$, we have the unitary and special unitary groups $\text{U}(n)$ and $\text{SU}(n)$.*

23.2.1 Definition of unitary maps

Another important class of linear maps which relate to a scalar product in a particular way are *unitary maps*. They are the linear maps which leave a scalar product unchanged in the sense of the following definition.

Definition 23.3 *An endomorphism $f \in \mathrm{End}(V)$ on an inner product vector space V with scalar product $\langle \cdot, \cdot \rangle$ is called unitary iff*

$$\langle f(\mathbf{v}), f(\mathbf{w}) \rangle = \langle \mathbf{v}, \mathbf{w} \rangle \tag{23.10}$$

for all $\mathbf{v}, \mathbf{w} \in V$.

In particular, unitary maps f leave lengths of vectors unchanged, so $|f(\mathbf{v})| = |\mathbf{v}|$ for all $\mathbf{v} \in V$. In fact, this property is already sufficient for the map to be unitary since the scalar product is determined by its associated norm (see Problem 22.1). In the real case, we can use the scalar product to defines angles between vectors as in Eq. (9.11). These angles are left unchanged by a unitary map f, that is, $\sphericalangle(\mathbf{v}, \mathbf{w}) = \sphericalangle(f(\mathbf{v}), f(\mathbf{w}))$. In short, we can think of unitary maps intuitively as those linear maps which leave basic geometrical characteristics of vectors invariant.

23.2.2 Unitary groups

Unitary maps have a number of interesting properties which are listed in the following proposition.

Proposition 23.2 *(Properties of unitary maps) For linear maps $f, g \in \mathrm{End}(V)$ on a (finite-dimensional) inner product vector space V have the following properties.*

(i)   *The identity map id_V is unitary.*
(ii)   *If f, g are unitary then so is $f \circ g$.*
(iii) *Unitary maps f are invertible.*
(iv)   *f is unitary iff $f^\dagger \circ f = \mathrm{id}_V$.*
(v)   *If f is unitary then so is the adjoint $f^\dagger = f^{-1}$.*

Proof (i) This is clear since $\mathrm{id}_V(\mathbf{v}) = \mathbf{v}$ for all $\mathbf{v} \in V$.

(ii) If f, g are unitary we have $\langle f(\mathbf{v}), f(\mathbf{w}) \rangle = \langle \mathbf{v}, \mathbf{w} \rangle$ and $\langle g(\mathbf{v}), g(\mathbf{w}) \rangle = \langle \mathbf{v}, \mathbf{w} \rangle$ for all $\mathbf{v}, \mathbf{w} \in V$, which implies

$$\langle f \circ g(\mathbf{v}), f \circ g(\mathbf{w}) \rangle = \langle f(\mathbf{v}), f(\mathbf{w}) \rangle = \langle \mathbf{v}, \mathbf{w} \rangle \ .$$

Hence, $f \circ g$ is unitary.

(iii) If $\mathbf{v} \in \mathrm{Ker}(f)$, so that $f(\mathbf{v}) = \mathbf{0}$ it follows from unitarity of f that $0 = \langle f(\mathbf{v}), f(\mathbf{v}) \rangle = \langle \mathbf{v}, \mathbf{v} \rangle$, so $\mathbf{v} = \mathbf{0}$. Hence, $\mathrm{Ker}(f) = \{\mathbf{0}\}$ which implies that f is invertible (see Cor. 14.2).

(iv) Unitarity and the definition of the adjoint implies

$$\langle f^\dagger \circ f(\mathbf{v}), \mathbf{w} \rangle \overset{(23.1)}{=} \langle f(\mathbf{v}), f(\mathbf{w}) \rangle \overset{(23.10)}{=} \langle \mathbf{v}, \mathbf{w} \rangle = \langle \mathrm{id}_V(\mathbf{v}), \mathbf{w} \rangle$$

for all $\mathbf{v}, \mathbf{w} \in V$. From Prop. 22.2 this is equivalent to $f^\dagger \circ f = \mathrm{id}_V$.

(v) A unitary f is invertible from (iii) and, from (iv), its left-inverse in the general linear group $\mathrm{GL}(V)$ is $f^\dagger$. But in a group the left-inverse is also the right-inverse (see Prop. 3.1), so $\mathrm{id}_V = f \circ f^\dagger = (f^\dagger)^\dagger \circ f^\dagger$. From (iv), the last relation implies that $f^\dagger = f^{-1}$ is unitary. $\qquad\square$

One way to summarize the content of the above lemma is to say that the set of unitary maps on V forms a sub-group of the general linear group $GL(V)$. This group is called the *unitary group* of V and it is denoted by

$$U(V) := \{f \in GL(V) \mid f \text{ is unitary}\} . \tag{23.11}$$

Problem 23.3 *Determinant of unitary maps*

Show that unitary maps f have a unit modulus determinant, so $|\det f)| = 1$.

Solution: The unitarity condition $f^\dagger \circ f = \mathrm{id}_V$ implies that

$$1 = \det(\mathrm{id}_V) = \det(f \circ f^\dagger) \overset{\text{Thm. (18.2)}}{=} \det(f)\det(f^\dagger) \overset{(23.4)}{=} |\det(f)|^2$$

An important sub-group of the unitary group $U(V)$ is the *special unitary group*, denoted $SU(V)$, which consists of all unitary maps with determinant one, so

$$SU(V) := \{f \in U(V) \mid \det(f) = 1\} . \tag{23.12}$$

To see that this set is indeed a group we check the standard conditions for a sub-group (see Def. 3.2). First, since $\det(\mathrm{id}_V) = 1$ the identity is an element of $SU(V)$. For two special unitary maps $f, g \in SU(V)$ it follows that $\det(f \circ g) = \det(f)\det(g) = 1$, so that $f \circ g \in SU(V)$ and, finally, $\det(f^{-1}) = \det(f)^{-1} = 1$, so $f^{-1} \in SU(V)$.

23.2.3 Orthogonal matrices

To get a better intuition for unitary maps it is useful to discuss coordinate vector spaces in more detail and we begin with the real case, that is, $\mathbb{R}^n$ with the standard scalar product (the dot product). In this case, the adjoint of an $n \times n$ matrix A is its transpose, A^T, so from Def. 23.3 and Prop. 23.2, unitarity of A is equivalent to any of the following conditions.

$$(A\mathbf{v}) \cdot (A\mathbf{w}) = \mathbf{v} \cdot \mathbf{w} \ \forall \mathbf{v}, \mathbf{w} \in \mathbb{R}^n \quad \Leftrightarrow \quad A^T A = \mathbb{1}_n \quad \Leftrightarrow \quad A^{-1} = A^T$$
$$\Leftrightarrow \quad \mathbf{A}^i \cdot \mathbf{A}^j = \delta_{ij} \quad i, j = 1, \ldots, n . \tag{23.13}$$

Matrices A satisfying this condition are called *orthogonal matrices* and they can be characterized, equivalently, by either one of the four conditions above. The simplest way to check if a given matrix is orthogonal is usually to verify the second condition, $A^T A = \mathbb{1}_n$. The third condition tells us it is easy to compute the inverse of an orthogonal matrix — it is simply the transpose. And, finally, the condition in the second row says that the column vectors of an orthogonal matrix form an ortho-normal basis with respect to the dot product.

The group formed by the orthogonal matrices (the unitary group of $\mathbb{R}^n$ with the dot product) is also called the *orthogonal group* and it is denoted by

$$O(n) := \{A \in GL(\mathbb{R}^n) \,|\, A^T A = \mathbb{1}_n\} \,. \tag{23.14}$$

Taking the determinant of Eq. (23.13) leads to $\det(A)^2 = 1$, so $\det(A) \in \{\pm 1\}$. The orthogonal matrices with determinant $+1$ are also called *rotations* and they form the *special orthogonal group* (the special unitary group of $\mathbb{R}^n$ with the dot product) denoted by

$$SO(n) := \{R \in O(n) \,|\, \det(R) = 1\} \,. \tag{23.15}$$

Note that the term 'rotation' is indeed appropriate for those matrices. Since they leave the dot product invariant they do not change lengths of vectors and angles between vectors and the $\det(A) = +1$ conditions excludes orthogonal matrices which contain reflections. To understand this last statement better, it is useful to look at the orthogonal matrices with determinant -1.

Consider an orthogonal matrix A with $\det(A) = -1$ and the specific orthogonal matrix $F = \mathrm{diag}(1, \ldots, 1, -1)$ with $\det(F) = -1$, which corresponds to a reflection in the last coordinate direction. Then the matrix $R = AF$ is a rotation since $\det(R) = \det(A) \det(F) = (-1)^2 = 1$. This means every orthogonal matrix A can be written as a product

$$A = RF \tag{23.16}$$

of a rotation R and a reflection F.

Problem 23.4 *(Orthogonal matrices in $\mathbb{R}^2$)*

Show that the two-dimensional rotation group $SO(2)$ consists of matrices $R(\theta)$, parametrized by $\theta \in [0, 2\pi)$, which satisfy $R(\theta_1) R(\theta_2) = R(\theta_1 + \theta_2)$. What is the interpretation of θ? Also, find the orthogonal matrices in $O(2)$. Show that $SO(2)$ is Abelian and that $O(2)$ is non-Abelian.

Solution: To find the explicit form of two-dimensional rotation matrices we start with a general 2×2 matrix

$$R = \begin{pmatrix} a & b \\ c & d \end{pmatrix} \,,$$

where $a, b, c, d \in \mathbb{R}$ and impose the conditions $R^T R = \mathbb{1}_2$ and $\det(R) = 1$. This gives

$$R^T R = \begin{pmatrix} a^2 + c^2 & ab + cd \\ ab + cd & b^2 + d^2 \end{pmatrix} \overset{!}{=} \begin{pmatrix} 1 & 0 \\ 0 & 1 \end{pmatrix} \,, \qquad \det(R) = ad - bc \overset{!}{=} 1 \,,$$

and, hence, the equations $a^2 + c^2 = b^2 + d^2 = 1$, $ab + cd = 0$ and $ad - bc = 1$. It is easy to show that a solution to these equations can always be written as $a = d = \cos(\theta)$, $c = -b = \sin(\theta)$, for $\theta \in [0, 2\pi)$ so that two-dimensional rotation matrices can be written in the form

$$R(\theta) = \begin{pmatrix} \cos\theta & -\sin\theta \\ \sin\theta & \cos\theta \end{pmatrix} \,. \tag{23.17}$$

For the rotation of a vector $\mathbf{x} = (x, y)^T \in \mathbb{R}^2$ we get

$$\mathbf{x}' = R\mathbf{x} = \begin{pmatrix} x\cos\theta - y\sin\theta \\ x\sin\theta + y\cos\theta \end{pmatrix} . \tag{23.18}$$

It is straightforward to verify that $|\mathbf{x}'| = |\mathbf{x}|$, as must be the case, and that the cosine of the angle between $\mathbf{x}$ and $\mathbf{x}'$ is given by

$$\cos(\sphericalangle(\mathbf{x}',\mathbf{x})) = \frac{\mathbf{x}' \cdot \mathbf{x}}{|\mathbf{x}'||\mathbf{x}|} = \frac{(x\cos\theta - y\sin\theta)x + (x\sin\theta + y\cos\theta)y}{|\mathbf{x}|^2} = \cos\theta . \tag{23.19}$$

This result means we should interpret $R(\theta)$ as a rotation by an angle θ. From the addition theorems of sine and cosine it also follows that

$$R(\theta_1)R(\theta_2) = R(\theta_1 + \theta_2) , \tag{23.20}$$

that is, the rotation angle adds up under composition of rotations, as one would expect. Eq. (23.20) implies that two-dimensional rotations commute since the right-hand side remains the same when θ_1 and θ_2 are exchanged. With the reflection $F = \mathrm{diag}(1, -1)$ the orthogonal group O(2) consists of the rotations $R(\theta)$ and the products

$$R(\theta)F = \begin{pmatrix} \cos\theta & \sin\theta \\ \sin\theta & -\cos\theta \end{pmatrix} \tag{23.21}$$

of F with rotations, where $\theta \in [0, 2\pi)$. Since $R(\theta)F \neq FR(\theta)$ (for $\theta \neq 0$) it follows that O(2) is non-Abelian.

Problem 23.5 *(Rotations in $\mathbb{R}^3$)*

Build three-dimensional rotations by using the two-dimensional rotations from Exercise 23.4. Show that the group SO(3) is non-Abelian.

Solution: The idea is to construct a 3×3 block matrix with a 1 in one diagonal entry and a two-dimensional rotation $R(\theta)$ as the complementary 2×2 block. There are three ways to do this, corresponding to where the 1 is placed along the diagonal, namely

$$R_1(\theta_1) = \begin{pmatrix} 1 & 0 & 0 \\ 0 & \cos\theta_1 & -\sin\theta_1 \\ 0 & \sin\theta_1 & \cos\theta_1 \end{pmatrix}, \ R_2(\theta_2) = \begin{pmatrix} \cos\theta_2 & 0 & -\sin\theta_2 \\ 0 & 1 & 0 \\ \sin\theta_2 & 0 & \cos\theta_2 \end{pmatrix}, \ R_3(\theta_3) = \begin{pmatrix} \cos\theta_3 & -\sin\theta_3 & 0 \\ \sin\theta_3 & \cos\theta_3 & 0 \\ 0 & 0 & 1 \end{pmatrix} .$$

We need to check that these matrices satisfy $R^T R = \mathbb{1}_3$ and $\det(R) = 1$, but this follows immediately from the fact that their 2×2 blocks of the form $R(\theta)$ satisfy these conditions. The above matrices satisfy $R_i(\theta_i)\mathbf{e}_i = \mathbf{e}_i$ so $\mathbf{e}_i$ is left invariant (it is an eigenvector with eigenvalue 1). Hence, we should interpret $R_i(\theta_i)$ as a rotation around the axis $\mathbf{e}_i$ with angle θ_i.

More general three-dimensional rotations can be obtained by multiplying the above matrices. Writing $s_i = \sin(\theta_i)$ and $c_i = \cos(\theta_i)$ for convenience of notation, their product is given by

$$R_1(\theta_1)R_2(\theta_2)R_3(\theta_3) = \left(\begin{array}{c|c|c} c_2 c_3 & -c_2 s_3 & -s_2 \\ \hline c_1 s_3 - c_3 s_1 s_2 & c_1 c_3 + s_1 s_2 s_3 & -c_2 s_1 \\ \hline c_1 c_3 s_2 + s_1 s_3 & c_3 s_1 - c_1 s_2 s_3 & c_1 c_2 \end{array} \right)$$

It turns out every three-dimensional rotation matrix can be written in this form, so three-dimensional rotations depend on three real parameters. This can be shown by solving the equations $R^T R = \mathbb{1}_3$ and $\det(R) = 1$, as we have done for two-dimensional rotations, although the explicit calculation is tedious.

Finally, we note that, unlike their two-dimensional counterparts, three-dimensional rotations do not, in general, commute. For example, apart from special choices for the angles $R_1(\theta_1)R_2(\theta_2) \neq R_2(\theta_2)R_1(\theta_1)$. Hence, SO(3) is non-Abelian.

Application 23.1 *Rotating physical systems*

Suppose we have a stationary coordinate system with coordinates $\mathbf{x} \in \mathbb{R}^3$ and another coordinate system with coordinates $\mathbf{y} \in \mathbb{R}^3$, which is rotating relative to the first one. Such a set-up can be used to describe the mechanics of objects in rotating systems and has many applications, for example to the physics of tops or the laws of motion in rotating systems such as the earth. Mathematically, the relation between these two coordinate system can be described by the equation

$$\mathbf{x} = R(t)\mathbf{y} \,, \tag{23.22}$$

where $R(t)$ are time-dependent rotation matrices. This means the matrices $R(t)$ satisfy

$$R(t)^T R(t) = \mathbb{1}_3 \,, \tag{23.23}$$

(as well as $\det(R(t)) = 1$)) for all times t. In practice, we can write rotation matrices in terms of rotation angles, as we have done in Problem 23.5. The time-dependence of $R(t)$ then means that the rotation angles are functions of time. For example, a rotation around the z-axis with constant angular speed ω can be written as

$$R(t) = \begin{pmatrix} \cos(\omega t) & -\sin(\omega t) & 0 \\ \sin(\omega t) & \cos(\omega t) & 0 \\ 0 & 0 & 1 \end{pmatrix} \,. \tag{23.24}$$

In physics, a rotation is often described by the angular velocity $\boldsymbol{\omega}$, a vector whose direction indicates the axis of rotation and whose length gives the angular speed. It is very useful to understand the relation between $R(t)$ and $\boldsymbol{\omega}$. To do this, define the matrix

$$W = R^T \dot{R} \,, \tag{23.25}$$

where the dot denotes the time derivative and observe, by differentiating Eq. (23.23) with respect to time, that

$$\underbrace{R^T \dot{R}}_{=W} + \underbrace{\dot{R}^T R}_{=W^T} = 0 \,. \tag{23.26}$$

Hence, W is an anti-symmetric matrix and can be written in the form

$$W = \begin{pmatrix} 0 & -\omega_3 & \omega_2 \\ \omega_3 & 0 & -\omega_1 \\ -\omega_2 & \omega_1 & 0 \end{pmatrix} \quad \text{or} \quad W_{ij} = \epsilon_{ikj}\omega_k \,. \tag{23.27}$$

The three independent entries ω_i of this matrix define the angular velocity $\boldsymbol{\omega} = (\omega_1, \omega_2, \omega_3)^T$. To see that this makes sense consider the example (23.24) and work out the matrix W.

$$W = \omega \begin{pmatrix} \cos(\omega t) & \sin(\omega t) & 0 \\ -\sin(\omega t) & \cos(\omega t) & 0 \\ 0 & 0 & 1 \end{pmatrix} \begin{pmatrix} -\sin(\omega t) & -\cos(\omega t) & 0 \\ \cos(\omega t) & -\sin(\omega t) & 0 \\ 0 & 0 & 0 \end{pmatrix} = \begin{pmatrix} 0 & -\omega & 0 \\ \omega & 0 & 0 \\ 0 & 0 & 0 \end{pmatrix} \,. \tag{23.28}$$

Comparison with the general form (23.27) of W then shows that the angular velocity for this case is given by $\boldsymbol{\omega} = (0, 0, \omega)$, indicating a rotation with angular speed ω around the z-axis, as expected.

In Problem 15.3 we have seen that the multiplication of an anti-symmetric 3×3 matrix with a vector can be written as a cross-product, so that

$$W\mathbf{b} = \boldsymbol{\omega} \times \mathbf{b} \tag{23.29}$$

for any vector $\mathbf{b} = (b_1, b_2, b_3)^T$. This can also be directly verified using the matrix form of W together with the definition, Eq. (10.6), of the cross product or, more elegantly, by the index calculation $W_{ij}b_j = \epsilon_{ikj}\omega_k b_j = (\boldsymbol{\omega} \times \mathbf{b})_i$, using the index form, Eq. (10.8), of the cross product. This relation can be used to re-write expressions which involve W in terms of the angular velocity $\boldsymbol{\omega}$.

For a simple application of this formalism, consider an object moving with velocity $\dot{\mathbf{y}}$ relative to the rotating system. What is its velocity relative to the stationary coordinate system? Differentiating Eq. (23.22) gives

$$\dot{\mathbf{x}} = R\dot{\mathbf{y}} + \dot{R}\mathbf{y} = R(\dot{\mathbf{y}} + W\mathbf{y}) = R(\dot{\mathbf{y}} + \boldsymbol{\omega} \times \mathbf{y}) \ . \tag{23.30}$$

The velocity $\dot{\mathbf{x}}$ in the stationary system has, therefore, two contribution, namely the velocity $\dot{\mathbf{y}}$ relative to the rotating system and the velocity $\boldsymbol{\omega} \times \mathbf{y}$ due to the rotation itself. For more on the mechanics of rotating systems (see, for example, Goldstein 2013; Landau and Lifshitz 1982).

23.2.4   Unitary matrices

We now carry out the analogous discussion in the complex case, so we would like to analyse the unitary maps on $\mathbb{C}^n$ with the standard scalar product (22.9). In this case, the adjoint of an $n \times n$ matrix A is its Hermitian conjugate $A^\dagger$, so that unitarity of A is equivalent to any of the following conditions.

$$(A\mathbf{v})^\dagger A\mathbf{w}) = \mathbf{v}^\dagger \mathbf{w} \ \forall \mathbf{v}, \mathbf{w} \in \mathbb{R}^n \quad \Leftrightarrow \quad A^T A = \mathbb{1}_n \quad \Leftrightarrow \quad A^{-1} = A^T$$
$$\Leftrightarrow \quad (\mathbf{A}^i)^\dagger \mathbf{A}^j = \delta_{ij} \quad i, j = 1, \ldots, n \ . \tag{23.31}$$

Matrices satisfying these conditions are called *unitary matrices*. As for orthogonal matrices, checking whether a given matrix is unitary is usually easiest accomplished using the second condition, $A^\dagger A = \mathbb{1}_n$. The third condition states that the inverse of a unitary matrix is simply its Hermitian conjugate and the condition in the second row says that the column vectors of a unitary matrix form an ortho-normal basis under the standard Hermitian scalar product on $\mathbb{C}^n$.

The group formed by the unitary matrices is called the *unitary group* and it is denoted by

$$\mathrm{U}(n) = \{A \in \mathrm{GL}(\mathbb{C})^n \,|\, A^\dagger A = \mathbb{1}_n\} \ . \tag{23.32}$$

It is clear from our general discussion (see Problem 23.3) that unitary matrices satisfy $|\det(A)| = 1$. Unitary matrices with determinant $+1$ are called *special unitary matrices*. They form the *special unitary group* denoted by

$$\mathrm{SU}(n) = \{U \in \mathrm{U}(n) \,|\, \det(U) = 1\} \,. \tag{23.33}$$

The relation between unitary and special unitary matrices is easy to understand. For a unitary matrix $A \in \mathrm{U}(n)$ we can always find a complex number ζ with $|\zeta| = 1$ such that $\zeta^n = \det(A)$. Then, the matrix $U = \zeta^{-1}A$ is special unitary since $\det(U) = \det(\zeta^{-1}A) = \zeta^{-n}\det(A) = 1$. This means every unitary matrix A can be written as a product

$$A = \zeta U \tag{23.34}$$

of a special unitary matrix U and a complex number ζ with $|\zeta| = 1$. We should think of unitary and special unitary matrices as the complex generalization of orthogonal matrices and rotations, respectively. In fact, unitary (special unitary) matrices with real entries are orthogonal matrices (rotations) since the unitarity condition $A^\dagger A = \mathbb{1}_n$ turns into the orthogonality condition $A^T A = \mathbb{1}_n$ if A is real. This means the (special) orthogonal groups are sub groups

$$\mathrm{O}(n) \subset \mathrm{U}(n)\,, \qquad \mathrm{SO}(n) \subset \mathrm{SU}(n)\,, \tag{23.35}$$

of the (special) unitary groups.

Problem 23.6 *(Special unitary matrices in two dimensions)*

Find the two-dimensional special unitary matrices and, hence, determine the group $\mathrm{SU}(2)$. Show that this group is non-Abelian.

Solution: We start with an arbitrary complex 2×2 matrix

$$U = \begin{pmatrix} \alpha & \beta \\ \gamma & \delta \end{pmatrix}\,,$$

where $\alpha, \beta, \gamma, \delta \in \mathbb{C}$ and impose the conditions $U^\dagger U = \mathbb{1}_2$ and $\det(U) = 1$. After a short calculation we find the group $\mathrm{SU}(2)$ is given by

$$\mathrm{SU}(2) = \left\{ \begin{pmatrix} \alpha & \beta \\ -\bar{\beta} & \bar{\alpha} \end{pmatrix} \,|\, \alpha, \beta \in \mathbb{C},\ |\alpha|^2 + |\beta|^2 = 1 \right\}\,. \tag{23.36}$$

This shows that two-dimensional special unitary matrices depend on two complex parameters α, β subject to the (real) constraint $|\alpha|^2 + |\beta|^2 = 1$ and, hence, on three real parameters. Inserting the special choice $\alpha = \cos\theta$, $\beta = -\sin\theta$ into (23.36) we recover the two-dimensional rotation matrices (23.4), so that $SO(2) \subset SU(2)$, as expected from our general discussion.

It is easy to choose matrices of the form (23.36) which do not commute, so $\mathrm{SU}(2)$ is non-Abelian.

The general study of orthogonal and unitary groups is part of the theory of *Lie groups*, a more advanced mathematical discipline which is beyond the scope of this introductory text (see, for example, Cornwell 1997; Fulton and Harris 2013).

Application 23.2 *Newton's equation in a rotating system*

Newton's law for the motion $\mathbf{x} = \mathbf{x}(t)$ of a mass point with mass m under the influence of a force $\mathbf{F}$ reads

$$m\ddot{\mathbf{x}} = \mathbf{F} , \qquad (23.37)$$

where the dot denotes the derivative with respect to time t. We would like to work out the form this law takes if we transform it to rotating coordinates $\mathbf{y}$, related to the original, non-rotating coordinates $\mathbf{x}$ by

$$\mathbf{x} = R(t)\mathbf{y} . \qquad (23.38)$$

Here $R(t)$ is a (generally time-dependent) rotation, that is, a 3×3 matrix satisfying

$$R(t)^T R(t) = \mathbb{1}_3 \qquad (23.39)$$

for all times t. For example, such a version of Newton's law is relevant to describing mechanics on earth.

To re-write Eq. (23.37) in terms of $\mathbf{y}$ we first multiply both sides with $R^T = R^{-1}$ so that

$$mR^T\ddot{\mathbf{x}} = \mathbf{F}_R , \qquad (23.40)$$

with $\mathbf{F}_R := R^T\mathbf{F}$ the force in the rotating coordinate system. If the rotation matrix is time-independent it can be pulled through the time derivatives on the LHS of Eq. (23.40) and we get $m\ddot{\mathbf{y}} = \mathbf{F}_R$. This simply says that Newton's law keeps the same form in any rotated (but not rotating!) coordinate system.

If R is time-dependent so that the system with coordinates $\mathbf{y}$ is indeed rotating relative to the coordinate system $\mathbf{x}$ we have to be more careful. Taking two time derivatives of Eq. (23.38) gives

$$\dot{\mathbf{x}} = R\dot{\mathbf{y}} + \dot{R}\mathbf{y} , \qquad \ddot{\mathbf{x}} = R\ddot{\mathbf{y}} + 2\dot{R}\dot{\mathbf{y}} + \ddot{R}\mathbf{y} . \qquad (23.41)$$

Using the second of these equations to replace $\ddot{\mathbf{x}}$ in Eq. (23.40) leads to

$$m\ddot{\mathbf{y}} = \mathbf{F}_R - 2mR^T\dot{R}\dot{\mathbf{y}} - mR^T\ddot{R}\mathbf{y} . \qquad (23.42)$$

Compared to Newton's equation in the standard form (23.37) we have acquired the two additional terms on the RHS which we should work out further. From Eq. (23.25), recall the definition $W = R^T\dot{R}$ and further note that $\dot{W} = R^T\ddot{R} + \dot{R}^T\dot{R} = R^T\ddot{R} + (\dot{R}^T R)(R^T\dot{R}) = R^T\ddot{R} - W^2$, so that

$$R^T\ddot{R} = \dot{W} + W^2 . \qquad (23.43)$$

With these results we can re-write Newton's equation (23.42) as

$$m\ddot{\mathbf{y}} = \mathbf{F}_R - 2mW\dot{\mathbf{y}} - mW^2\mathbf{y} - m\dot{W}\mathbf{y} . \qquad (23.44)$$

Also, recall that the matrix W is anti-symmetric, encodes the angular velocity $\boldsymbol{\omega}$, as in Eq. (23.27) and its action on vectors can be re-written as a cross product with the angular velocity $\boldsymbol{\omega}$ (see Eq. (23.29)). Then, Newton's equation (23.44) in a rotating system can be written in its final form

$$m\ddot{\mathbf{y}} = \mathbf{F}_R \underbrace{-2m\boldsymbol{\omega} \times \dot{\mathbf{y}}}_{\text{Coriolis force}} \underbrace{-m\boldsymbol{\omega} \times (\boldsymbol{\omega} \times \mathbf{y})}_{\text{centrifugal force}} \underbrace{-2m\dot{\boldsymbol{\omega}} \times \mathbf{y}}_{\text{Euler force}} . \qquad (23.45)$$

The three terms on the RHS represent the additional forces a mass point experiences in a rotating system. The centrifugal force is well-known. The Coriolis force is proportional to the velocity, $\dot{\mathbf{y}}$, and, hence, vanishes for mass points which rest in the rotating frame. It is, for example, responsible for the rotation of a Foucault pendulum. Finally, the Euler force is proportional to the angular acceleration, $\dot{\boldsymbol{\omega}}$. For the earth's rotation, $\boldsymbol{\omega}$ is approximately constant so the Euler force is quite small in this case. For more details on mechanics in rotating systems (see, for example, Goldstein 2013; Landau and Lifshitz 1982).

Exercises

(†=challenging)

23.1 Determine whether the matrices

$$A = \begin{pmatrix} \frac{1}{\sqrt{2}} & -\frac{1}{\sqrt{2}} \\ \frac{1}{\sqrt{2}} & \frac{1}{\sqrt{2}} \end{pmatrix},$$

$$B = \begin{pmatrix} \frac{2}{\sqrt{5}} & \frac{1}{\sqrt{5}} \\ \frac{1}{\sqrt{5}} & -\frac{2}{\sqrt{5}} \end{pmatrix}, \quad C = \begin{pmatrix} 1 & 2 \\ 0 & 1 \end{pmatrix}$$

are orthogonal or special orthogonal.

23.2 Determine whether the matrices

$$A = \begin{pmatrix} \frac{1}{\sqrt{2}} & 0 & \frac{1}{\sqrt{2}} \\ 0 & 1 & 0 \\ \frac{1}{\sqrt{2}} & 0 & -\frac{1}{\sqrt{2}} \end{pmatrix},$$

$$B = \begin{pmatrix} -\frac{1}{3} & -\frac{2}{3} & \frac{2}{3} \\ -\frac{2}{3} & \frac{2}{3} & \frac{1}{3} \\ -\frac{2}{3} & -\frac{1}{3} & -\frac{2}{3} \end{pmatrix}$$

are orthogonal or special orthogonal.

23.3 Verify that the matrix

$$U = \begin{pmatrix} \alpha & \beta \\ -\bar{\beta} & \bar{\alpha} \end{pmatrix}$$

where $\alpha, \beta \in \mathbb{C}$ and $|\alpha|^2 + |\beta|^2 = 1$ is special unitary.

23.4 Consider the linear map $f : \mathbb{R}^3 \to \mathbb{R}^3$ defined in Exercise 20.5.
(a) Find an explicit expression for $|f(\mathbf{v})|^2$, where $\mathbf{v} \in \mathbb{R}^3$.
(b) For which values of $\alpha, \beta \in \mathbb{R}$ is f a unitary (special unitary) linear map?
(c) What is the geometric interpretation of f in case it is unitary (special unitary)?

23.5 A vector space V over $\mathbb{C}$ with Hermitian scalar product $\langle \cdot, \cdot \rangle$ has a basis $(\boldsymbol{\epsilon}_1, \ldots, \boldsymbol{\epsilon}_n)$ and the vectors $\boldsymbol{\epsilon}'_j$ are defined by $\boldsymbol{\epsilon}'_j = \sum_i U_{ij} \boldsymbol{\epsilon}_i$, where $U_{ij} \in \mathbb{C}$. Show that
(a) $U_{ij} = \langle \boldsymbol{\epsilon}_i, \boldsymbol{\epsilon}'_j \rangle$.
(b) the matrix U with entries U_{ij} is unitary iff $(\boldsymbol{\epsilon}'_1, \ldots, \boldsymbol{\epsilon}'_n)$ is an orthonormal basis of V.

23.6 Let $f \in \text{End}(V)$ be a Hermitian map on an inner product vector space V. Show that
(a) $\text{Ker}(f) \perp \text{Im}(f)$
(b) $V = \text{Ker}(f) \oplus \text{Im}(f)$

23.7 *(Hermitian projectors)*
Show that a projector $p \in \text{End}(V)$ on an inner product vector space V is Hermitian iff $\text{Ker}(p) = \text{Im}(p)^{\perp}$.

23.8 For a linear map $f : V \to W$ between two inner product vector space V and W show that
(a) $\text{Ker}(f^{\dagger} \circ f) = \text{Ker}(f)$ and $\text{Ker}(f \circ f^{\dagger}) = \text{Im}(f)^{\perp}$.
(b) $\text{rk}(f^{\dagger} \circ f) = \text{rk}(f \circ f^{\dagger}) = \text{rk}(f)$.

23.9 *Unitary maps and orthogonal matrices*†
For a map $f \in \text{End}(V)$ on a vector space V over $\mathbb{R}$, denote by A_f the matrix which describes f relative to an ortho-normal basis $(\boldsymbol{\epsilon}_1, \ldots, \boldsymbol{\epsilon}_n)$ of V.
(a) Show that f is (special) unitary iff A_f is a (special) orthogonal matrix.
(b) Show that the map $\imath : U(V) \to O(n)$ defined by $f \mapsto A_f$ is a group isomorphism and that its restriction to $SU(V)$ gives a group isomorphism

$SU(V) \to SO(n)$.

23.10 *Unitary maps and matrices*[†]
Repeat the discussion from Exercise 23.9 for a vector space V over $\mathbb{C}$.

23.11 *Small rotations in three dimensions*
Using the notation from Problem 23.5, consider a rotation $R(\boldsymbol{\theta}) = R_1(\theta_1)R_2(-\theta_2)R_3(\theta_3) \in SO(3)$.
(a) By approximating $\sin(x) = x + \cdots$ and $\cos(x) = 1 + \cdots$, show that $R(\boldsymbol{\theta}) = \mathbb{1}_3 + \sum_i \theta_i T_i + \cdots$, where $\boldsymbol{\theta} = (\theta_1, \theta_2, \theta_3)^T$, the dots stand for terms quadratic and higher in θ_i and (T_1, T_2, T_3) is the basis for the space $\mathcal{A}_3$ of anti-symmetric 3×3 matrices from Exercise 13.9.
(b) For $\mathbf{x} \in \mathbb{R}^3$ define $\delta\mathbf{x} = R\mathbf{x} - \mathbf{x}$. Show that $\delta\mathbf{x} = \boldsymbol{\theta} \times \mathbf{x} + \cdots$, where the dots stand for terms quadratic and higher in θ_i.
(c) Work out $\delta\mathbf{x}$ for $\boldsymbol{\theta} = (0, 0, \theta)^T$ and $\mathbf{x} = (x, y, 0)^T$ and interpret this result geometrically.

23.12 *Hermitian differential operators*
On the inner product vector space V from Example 23.2 define the linear maps $D = d/dx$, $P = -iD$ and X, with $X(\varphi)(x) := x\varphi(x)$.
(a) Which of the maps X^2, P^2, XP, X^2P and XPX is Hermitian?
(b) For $XP + c$ find the values of $c \in \mathbb{C}$ so that the map is Hermitian.

23.13 *Rotating systems in two dimensions*[†]

Derive the analogue of the results from Applications 23.1 and 23.2 for two-dimensional rotations

$$R(\theta(t)) = \begin{pmatrix} \cos(\theta(t)) & \sin(\theta(t)) \\ -\sin(\theta(t)) & \cos(\theta(t)) \end{pmatrix}$$

where $\theta(t)$ is a time-dependent angle of rotation. Proceed as follows.
(a) Show that $W := R^T \dot{R} = \dot{\theta}\epsilon$ where ϵ is the matrix in Eq. (10.2).
(b) If $\mathbf{x}(t), \mathbf{y}(t) \in \mathbb{R}^2$ and $\mathbf{x} = R\mathbf{y}$ show that $\dot{\mathbf{x}} = R(\dot{\mathbf{y}} + \dot{\theta}\mathbf{y}^\times)$.
(c) Transform Newton's equation $m\ddot{\mathbf{x}} = \mathbf{F}$ in $\mathbb{R}^2$ to the rotating coordinates $\mathbf{y} = R^T\mathbf{x}$.

23.14 *Endomorphisms commuting with groups*[†]
Let V be a vector space over $\mathbb{C}$ and G a subgroup of $GL(V)$. The group G is called *irreducible* if no non-trivial subspace $U \subset V$ exists which is left invariant under all endomorphisms in G.
(a) Suppose G is irreducible and $f \in \text{End}(V)$ satisfies $[f, g] = 0$ for all $g \in G$. Show that $f = \lambda \text{id}_V$ for $\lambda \in \mathbb{C}$. (Hint: Consider an eigenspace of f.)
(b) Which of the groups $SO(2)$ and $SU(2)$ is irreducible?
(c) A matrix $A \in \text{End}(\mathbb{C}^2)$ is unchanged under unitary basis transformations. What is the most general form of A?

24
Diagonalization — again

With considerably more structure on our vector spaces, it is worth revisiting eigenvalues, eigenvectors and diagonalization. We have seen in Chapter 20 that not all endomorphisms can be diagonalized. Theorem 20.1 provides criteria for when this is possible but checking these usually amounts to calculating all the eigenvalues and eigenvectors.

The main problem we will tackle in this chapter is whether the presence of a scalar product simplifies matters and leads to more straightforward criteria for when an endomorphism can be diagonalized. In fact, we will show that two classes of endomorphisms — self-adjoint and *normal* endomorphisms — can always be diagonalized. Section 24.4 discusses functions of matrices, a topic somewhat outside the main narrative of linear algebra but one with many applications. As we explain, diagonalization of matrices is a powerful computational tool in this context.

The final section is looking at normal forms for homomorphisms $V \to W$, rather than at endomorphisms. This leads to the *singular value decomposition* of linear maps, a technique widely used in applications.

24.1 Hermitian maps

Summary 24.1 *Hermitian maps have real eigenvalues and eigenvectors for different eigenvalues are always orthogonal. They can always be diagonalized relative to an ortho-normal basis.*

24.1.1 Eigenvectors and eigenvalues of Hermitian maps

The following theorem shows that eigenvalues and eigenvectors of self-adjoint endomorphisms have special properties.

Theorem 24.1 *For an inner product vector space V and a self-adjoint endomorphism $f \in \mathrm{End}(V)$, we have the following statements:*

(i)   All eigenvalues of f are real.
(ii) Eigenvectors for different eigenvalues are orthogonal.

Proof (i) For the case of real scalar products the statement is trivial. For Hermitian scalar products, we start with an eigenvalue λ of f with eigenvector $\mathbf{v}$, so that $f(\mathbf{v}) = \lambda\mathbf{v}$. The self-adjoint property of f then implies

$$\lambda\langle\mathbf{v},\mathbf{v}\rangle \stackrel{(S1)}{=} \langle\mathbf{v},\lambda\mathbf{v}\rangle = \langle\mathbf{v},f(\mathbf{v})\rangle \stackrel{(23.6)}{=} \langle f(\mathbf{v}),\mathbf{v}\rangle = \langle\lambda\mathbf{v},\mathbf{v}\rangle \stackrel{(22.2)}{=} \bar{\lambda}\langle\mathbf{v},\mathbf{v}\rangle$$

Eigenvectors are non-zero, so $\langle\mathbf{v},\mathbf{v}\rangle \neq 0$, and it follows that $\lambda \in \mathbb{R}$.

(ii) If $f(\mathbf{v}_i) = \lambda_i\mathbf{v}_i$ for $i = 1,2$, where, from part (i), $\lambda_i \in \mathbb{R}$, it follows that

$$\lambda_2\langle\mathbf{v}_1,\mathbf{v}_2\rangle \stackrel{(S1)}{=} \langle\mathbf{v}_1,\lambda_2\mathbf{v}_2\rangle = \langle\mathbf{v}_1,f(\mathbf{v}_2)\rangle \stackrel{(23.6)}{=} \langle f(\mathbf{v}_1),\mathbf{v}_2\rangle = \langle\lambda_1\mathbf{v}_2,\mathbf{v}_2\rangle \stackrel{(22.2)}{=} \lambda_1\langle\mathbf{v}_1,\mathbf{v}_2\rangle \,.$$

For $\lambda_1 \neq \lambda_2$ this implies that $\langle\mathbf{v}_1,\mathbf{v}_2\rangle = 0$. □

24.1.2 Diagonalizing Hermitian maps

Orthogonality of eigenvectors for Hermitian maps is an important feature which facilitates the proof of the following statement.

Theorem 24.2 *A self-adjoint map $f \in \mathrm{End}(V)$ on a (finite-dimensional) inner product vector space V can be diagonalized and has an ortho-normal basis of eigenvectors.*

Proof We proof this by induction in $n = \dim_{\mathbb{F}}(V)$. For $n = 1$ the assertion is trivial. The induction assumption is that the statement holds for all dimensions $k < n$ and we need to show that it is true for dimension n.

The characteristic polynomial χ_f has at last one zero, λ, over the complex numbers but from Theorem 24.1 λ must, in fact, be real. Hence, even if V is a vector space over $\mathbb{R}$, the Hermitian map f has an eigenvalue λ and an associated non-trivial eigenspace $W = \mathrm{Eig}_f(\lambda)$. For any vectors $\mathbf{w} \in W$ and $\mathbf{v} \in W^\perp$ we have

$$\langle\mathbf{w},f(\mathbf{v})\rangle = \langle f(\mathbf{w}),\mathbf{v}\rangle = \langle\lambda\mathbf{w},\mathbf{v}\rangle = \lambda\langle\mathbf{w},\mathbf{v}\rangle = 0 \,.$$

This means that $f(\mathbf{v})$ is perpendicular to $\mathbf{w}$ so that, whenever $\mathbf{v} \in W^\perp$, then also $f(\mathbf{v}) \in W^\perp$. As a result, $W^\perp$ is invariant under f and we can consider the restriction $g = f|_{W^\perp}$ of f to $W^\perp$. Since $\dim(W^\perp) < n$, there is an ortho-normal basis $\boldsymbol{\epsilon}_1,\ldots,\boldsymbol{\epsilon}_k$ of $W^\perp$ consisting of eigenvectors of g (which are also eigenvectors of f) by the induction assumption. Add to this ortho-normal basis of $W^\perp$ an ortho-normal basis of W (which, by definition of W, consists of eigenvectors of f with eigenvalue λ). Since $V = W \oplus W^\perp$ (see Prop. 22.3) this list of vectors forms an ortho-normal basis of V which consists of eigenvectors of f. □

In summary, a self-adjoint endomorphism $f \in \mathrm{End}(V)$ on a finite dimensional inner product space V can be diagonalized, it has real eigenvalues and an ortho-normal basis of eigenvectors. In practice, the ortho-normal basis of eigenvectors can be found by first computing all the eigenspaces $\mathrm{Eig}_f(\lambda_i)$ and then constructing an ortho-normal basis on each eigenspace, using the Gram–Schmidt procedure. If the eigenvalue is non-degenerate, so that the eigenspace is one-dimensional, this simply amounts to normalizing the eigenvector. Then combine the bases for all eigenspaces into a basis of V. Theorem 24.1 guarantees that this basis is ortho-normal.

For a Hermitian matrix $A \in \mathrm{End}(\mathbb{F}^n)$ with an ortho-normal basis of eigenvectors $(\boldsymbol{\epsilon}_1,\ldots,\boldsymbol{\epsilon}_n)$ and corresponding eigenvectors $(\hat{\lambda}_1,\ldots,\hat{\lambda}_n)$, the diagonalizing basis transformation $P = (\boldsymbol{\epsilon}_1,\ldots,\boldsymbol{\epsilon}_n)$ is, in fact, a unitary matrix (orthogonal matrix in the real case), as follows from Eq. (23.31) (Eq. (23.13) in the real case). This means that $P^{-1} = P^\dagger$ and that diagonalization of A can be accomplished by

$$P^\dagger A P = \hat{A} = \text{diag}(\hat{\lambda}_1, \ldots, \hat{\lambda}_n) \,. \tag{24.1}$$

24.1.3 Examples

It is worth setting up an algorithm for how to diagonalize a self-adjoint endomorphism.

Algorithm *(Diagonalizing self-adjoint endomorphisms)*
To diagonalize a self-adjoint endomorphism $f \in \text{End}(V)$ carry out the following steps.

(1) *Determine the (pairwise distinct) eigenvalues $\lambda_1, \ldots, \lambda_k$ of f and their multiplicities m_i, by finding the zeros of the characteristic polynomial χ_f.*

(2) *For each eigenvalue λ_i, find the corresponding eigenspace $\text{Eig}_f(\lambda_i)$ by solving the linear system $(f - \lambda_i \,\text{id}_V)\mathbf{v} = \mathbf{0}$. The degeneracy $d_i = \dim_\mathbb{F}(\text{Eig}_f(\lambda_i))$ equals the multiplicity m_i.*

(3) *For each eigenspace $\text{Eig}_f(\lambda_i)$, choose an arbitrary basis and apply the Gram–Schmidt procedure to convert this into an ortho-normal basis of $\text{Eig}_f(\lambda_i)$. (If the eigenvalue is non-degenerate, so if $d_i = 1$, all this requires is normalizing the eigenvector for λ_i.)*

(4) *Combine the bases for all eigenspaces found in (3) into a single basis $(\epsilon_1, \ldots, \epsilon_n)$. This is an ortho-normal basis of V which diagonalizes f. The diagonal matrix describing f is $\hat{A} = \text{diag}(\hat{\lambda}_1, \ldots, \hat{\lambda}_n)$, where $\hat{\lambda}_i$ are the eigenvalues of ϵ_i, that is, the values λ_i repeated with multiplicities d_i.*

(5) *If f is a matrix $A \in \text{End}(\mathbb{F}^n)$, then form the unitary (or, for $\mathbb{F} = \mathbb{R}$, orthogonal) matrix $P = (\epsilon_1, \ldots, \epsilon_n)$ and diagonalize with the basis transformation $P^\dagger A P = \hat{A}$.*

Problem 24.1 *(Diagonalizing symmetric matrices)*

Diagonalize the symmetric matrix $A \in \text{End}(\mathbb{R}^3)$ given by

$$A = \frac{1}{4}\begin{pmatrix} 2 & 3\sqrt{2} & 3\sqrt{2} \\ 3\sqrt{2} & -1 & 3 \\ 3\sqrt{2} & 3 & -1 \end{pmatrix} \,.$$

Solution: The characteristic polynomial $\chi_A(\lambda) = \det(A - \lambda\mathbb{1}_3) = -\lambda^3 + 3\lambda + 2 = (2 - \lambda)(1 + \lambda)^2$ shows that there are two eigenvalues, $\lambda_1 = 2$ and $\lambda_2 = -1$, with multiplicities (=degeneracies) $m_1 = 1$ and $m_2 = 2$. For the eigenvector $\mathbf{v} = (x, y, z)^T$ for $\lambda_1 = 2$ we find

$$(A - 2\mathbb{1}_3)\mathbf{v} = \frac{3}{4}\begin{pmatrix} -2 & \sqrt{2} & \sqrt{2} \\ \sqrt{2} & -3 & 1 \\ \sqrt{2} & 1 & -3 \end{pmatrix}\begin{pmatrix} x \\ y \\ z \end{pmatrix} = \frac{3}{4}\begin{pmatrix} -2x + \sqrt{2}y + \sqrt{2}z \\ \sqrt{2}x - 3y + z \\ \sqrt{2}x + y - 3z \end{pmatrix} \overset{!}{=} \mathbf{0} \quad \Rightarrow \quad y = z = \frac{x}{\sqrt{2}} \,.$$

So, this eigenvalue is indeed non-degenerate. We should normalize the eigenvector using the dot product (since A is symmetric and, hence, self-adjoint relative to the dot product) which leads to $\epsilon_1 = (\sqrt{2}, 1, 1)^T/2$. For $\lambda_2 = -1$ a similar calculation gives

$$(A + \mathbb{1}_3)\mathbf{v} = \frac{3}{4}\begin{pmatrix} 2 & \sqrt{2} & \sqrt{2} \\ \sqrt{2} & 1 & 1 \\ \sqrt{2} & 1 & 1 \end{pmatrix}\begin{pmatrix} x \\ y \\ z \end{pmatrix} = \frac{3}{4}\begin{pmatrix} 2x + \sqrt{2}y + \sqrt{2}z \\ \sqrt{2}x + y + z \\ \sqrt{2}x + y + z \end{pmatrix} \overset{!}{=} \mathbf{0} \quad \Rightarrow \quad z = -\sqrt{2}x - y \,.$$

Since there is only one condition on x, y, z there are two linearly independent eigenvectors, so this eigenvalue has degeneracy 2, as expected. Obvious choices for the two eigenvectors are obtained by setting $(x, y) = (1, 0)$ and $(x, y) = (0, 1)$ which gives $\mathbf{v}_2 = (1, 0, -\sqrt{2})^T$ and $\mathbf{v}_3 = (0, 1, -1)^T$. Both of these vectors are orthogonal to $\boldsymbol{\epsilon}_1$ above, as they must be, but they are not orthogonal to one another. However, they do form a basis of the two-dimensional eigenspace $\mathrm{Eig}_A(-1) = \mathrm{Span}(\mathbf{v}_2, \mathbf{v}_3)$ to which we can apply the Gram–Schmidt procedure. The first step is to normalize $\mathbf{v}_2$ which gives

$$\boldsymbol{\epsilon}_2 = \frac{\mathbf{v}_2}{|\mathbf{v}_2|} = \frac{1}{\sqrt{3}} \begin{pmatrix} 1 \\ 0 \\ -\sqrt{2} \end{pmatrix}.$$

Next, we need to subtract from $\mathbf{v}_3$ its projection onto $\boldsymbol{\epsilon}_2$ and normalize, resulting in

$$\mathbf{v}_3' = \mathbf{v}_3 - (\boldsymbol{\epsilon}_2 \cdot \mathbf{v}_3)\boldsymbol{\epsilon}_2 = \frac{1}{3} \begin{pmatrix} -\sqrt{2} \\ 3 \\ -1 \end{pmatrix}, \qquad \boldsymbol{\epsilon}_3 = \frac{\mathbf{v}_3'}{|\mathbf{v}_3'|} = \frac{1}{2\sqrt{3}} \begin{pmatrix} -\sqrt{2} \\ 3 \\ -1 \end{pmatrix}.$$

The vectors $(\boldsymbol{\epsilon}_1, \boldsymbol{\epsilon}_2, \boldsymbol{\epsilon}_3)$ form an ortho-normal basis of eigenvectors, so the matrix

$$P = (\boldsymbol{\epsilon}_1, \boldsymbol{\epsilon}_2, \boldsymbol{\epsilon}_3) = \begin{pmatrix} \frac{1}{\sqrt{2}} & \frac{1}{\sqrt{3}} & -\frac{1}{\sqrt{6}} \\ \frac{1}{2} & 0 & \frac{\sqrt{3}}{2} \\ \frac{1}{2} & -\sqrt{\frac{2}{3}} & -\frac{1}{2\sqrt{3}} \end{pmatrix}$$

is orthogonal, as can be checked explicitly by verifying that $P^T P = \mathbb{1}_3$. Diagonalization is accomplished by $P^T A P = \mathrm{diag}(2, -1, -1)$.

Problem 24.2 *(Diagonalizing Hermitian matrices)*

Diagonalize the Hermitian matrix $A \in \mathrm{End}(\mathbb{C}^2)$ given by

$$A = \begin{pmatrix} 2 & 1 + 2i \\ 1 - 2i & -2 \end{pmatrix}.$$

Solution: The characteristic polynomial $\chi_A(\lambda) = \det(A - \lambda \mathbb{1}_2) = (\lambda + 3)(\lambda - 3)$ indicates we have two non-degenerate eigenvalues $\lambda_\pm = \pm 3$. Note, they are real as must be the case from Theorem 24.1. For the eigenvalues $\mathbf{v} = (x, y)^T$ we have

$$(A - \lambda_\mp \mathbb{1}_2)\mathbf{v} = (A \pm 3\,\mathbb{1}_2)\mathbf{v} = \begin{pmatrix} 2 \pm 3 & 1 + 2i \\ 1 - 2i & -2 \pm 3 \end{pmatrix}\begin{pmatrix} x \\ y \end{pmatrix} = 0 \quad \Rightarrow \quad \begin{cases} \mathbf{v}_- = (1 + 2i, -5)^T \\ \mathbf{v}_+ = (1 + 2i, 1)^T \end{cases}.$$

Note that $\mathbf{v}_-$ and $\mathbf{v}_+$ are automatically orthogonal, as they correspond to different eigenvalues. All we need to do is normalize them, relative to the standard scalar product on $\mathbb{C}^2$. Since $|\mathbf{v}_1|^2 = \mathbf{v}_1^\dagger \mathbf{v}_1 = 30$ and $|\mathbf{v}_2|^2 = \mathbf{v}_2^\dagger \mathbf{v}_2 = 6$ this gives

$$\boldsymbol{\epsilon}_- = \frac{1}{\sqrt{30}}\begin{pmatrix} 1 + 2i \\ -5 \end{pmatrix}, \quad \boldsymbol{\epsilon}_+ = \frac{1}{\sqrt{6}}\begin{pmatrix} 1 + 2i \\ 1 \end{pmatrix}, \quad U = (\boldsymbol{\epsilon}_-, \boldsymbol{\epsilon}_+) = \frac{1}{\sqrt{30}}\begin{pmatrix} 1 + 2i & \sqrt{5}(1 + 2i) \\ -5 & \sqrt{5} \end{pmatrix}.$$

It is easily verified that U is unitary by checking that $U^\dagger U = \mathbb{1}_2$. Diagonalization is accomplished by $U^\dagger A U = \mathrm{diag}(-3, 3)$.

Problem 24.3 *A self-adjoint differential operator*

Let V be the vector space of infinitely many times differentiable functions $\varphi : [-\pi, \pi] \to \mathbb{R}$ which satisfy $\varphi(-\pi) = \varphi(\pi)$, with the scalar product

$$\langle \varphi, \psi \rangle = \int_{-\pi}^{\pi} dx \, \varphi(x)\psi(x) .$$

Show that the linear map $D^2 = d^2/dx^2 : V \to V$ is self-adjoint. Find its eigenvalues and eigenvectors and convince yourself that eigenvectors for different eigenvalues are indeed orthogonal.

Solution: The proof that D^2 is self-adjoint is based on integration by parts, where the boundary terms vanish due to the 'periodicity condition' $\varphi(-\pi) = \varphi(\pi)$ on functions in V. Simplifying notation by dropping the argument x, we have

$$\langle \varphi, D^2\psi \rangle = \int_{-\pi}^{\pi} \varphi \, D^2\psi = -\int_{-\pi}^{\pi} (D\varphi)(D\psi) = \int_{-\pi}^{\pi} (D^2\varphi)\psi = \langle D^2\varphi, \psi \rangle ,$$

so D^2 is indeed self-adjoint. To find the eigenvalues and eigenvectors we solve the equation $D^2\varphi = \lambda\varphi$, subject to $\varphi(-\pi) = \varphi(\pi)$. The eigenvalues are $\lambda = 0$ and $\lambda = -k^2$, where $k = 1, 2, \ldots$, and defining the functions

$$c_0(x) = \frac{1}{\sqrt{2\pi}} , \qquad s_k(x) = \frac{1}{\sqrt{\pi}} \sin(kx) , \qquad c_k(x) = \frac{1}{\sqrt{\pi}} \cos(kx) ,$$

the corresponding eigenspaces can be written as

$$\mathrm{Eig}_{D^2}(0) = \mathrm{Span}(c_0) , \qquad \mathrm{Eig}_{D^2}(-k^2) = \mathrm{Span}(s_k, c_k) .$$

Hence, each eigenvalue $\lambda = -k^2$ for $k = 1, 2, \ldots$ has degeneracy two and the eigenvalue $\lambda = 0$ is non-degenerate. By using standard integrals for sine and cosine it is straightforward to check orthogonality, $\langle s_k, s_l \rangle = \langle c_k, c_l \rangle = \delta_{kl}$, of eigenvectors for different eigenvalues, as predicted by Theorem 24.1. Orthogonality of the functions s_k and c_k does not follow from Theorem 24.1. They are nevertheless orthogonal, $\langle s_k, c_k \rangle = 0$, as can be seen by direct integration. Altogether, this means that the functions c_k, for $k = 0, 1, \ldots$ and s_k for $k = 1, 2, \ldots$ are ortho-normal.

The functions s_k and c_k are at the heart of the *Fourier series* expansion. The idea is to expand a periodic function φ as

$$\varphi(x) = \sum_{k=0}^{\infty} \alpha_k c_k(x) + \sum_{k=1}^{\infty} \beta_k s_k(x) ,$$

in analogy with how we expand a vector in a finite-dimensional vector space in terms of an ortho-normal basis. Of course, the infinite summations in this expression raises issues of convergence. They are the subject of an analysis course and will not be tackled here. All we would like to point out is that a naive application of Eq. (22.15) for the coordinates relative to an ortho-normal basis leads to formulae for the *Fourier coefficients* α_k and β_k, namely

$$\alpha_k = \langle c_k, \varphi \rangle = \int_{-\pi}^{\pi} dx \, c_k(x)\varphi(x) , \qquad \beta_k = \langle s_k, \varphi \rangle = \int_{-\pi}^{\pi} dx \, s_k(x)\varphi(x) .$$

In the theory of Fourier expansion, these are key equations which are used to compute the Fourier coefficients of a function φ.

24.2 Normal maps*

Summary 24.2 *Normal maps are, by definition, maps which commute with their adjoint. The class of normal maps is larger than the one of Hermitian maps and it includes anti-Hermitian and unitary maps. Over the complex numbers, a map can be diagonalized relative to an ortho-normal basis if and only if it is normal.*

24.2.1 Definition of normal maps

Self-adjoint endomorphisms are not the most general class of endomorphisms which can be diagonalized relative to an ortho-normal basis. It turns out, the slightly weaker condition of *normality* is sufficient (and indeed necessary) for the existence of an ortho-normal basis of eigenvectors, provided the characteristic polynomial factorises. We begin by defining normal linear maps.

Definition 24.1 *An endomorphism $f \in \mathrm{End}(V)$ on an inner product vector space V is called normal if $[f, f^\dagger] = 0$.*

In other words, normal maps are those that commute with their adjoint. Clearly, self-adjoint maps are normal since $f = f^\dagger$ implies $[f, f^\dagger] = [f, f] = 0$. But there are other classes of normal endomorphisms. Anti self-adjoint maps, so maps f with $f = -f^\dagger$, are normal as well since $[f, f^\dagger] = -[f, f] = 0$. More importantly, unitary maps are normal since the unitarity conditions $f \circ f^\dagger = f^\dagger \circ f = \mathrm{id}_V$ leads to $[f, f^\dagger] = 0$. So it is clear that normal maps cover significantly more ground than self-adjoint ones.

Self-adjoint maps have real eigenvalues. This is not necessarily the case for normal maps but they satisfy a weaker property, namely that the eigenvalues of f and $f^\dagger$ are related by complex conjugation.

Lemma 24.1 *Let V be an inner product vector space and $f \in \mathrm{End}(V)$ normal. If λ is an eigenvalue of f with eigenvector $\mathbf{v}$ then $\bar{\lambda}$ is an eigenvalue of $f^\dagger$ for the same eigenvector $\mathbf{v}$.*

Proof First, we show that the map $g = f - \lambda\,\mathrm{id}_V$ is also normal. This follows from the straightforward calculation

$$[g, g^\dagger] = [f - \lambda\,\mathrm{id}_V, f^\dagger - \bar{\lambda}\,\mathrm{id}_V] = [f, f^\dagger] - \lambda[\mathrm{id}_V, f^\dagger] - \bar{\lambda}[f, \mathrm{id}_V] + |\lambda|^2[\mathrm{id}_V, \mathrm{id}_V] = 0\,,$$

where some of the properties of the adjoint map from Prop. 23.1 have been used. Now consider an eigenvalue λ of f with eigenvector $\mathbf{v}$, so that $f(\mathbf{v}) = \lambda\mathbf{v}$ or, equivalently, $g(\mathbf{v}) = 0$. Then we have

$$0 = \langle g(\mathbf{v}), g(\mathbf{v})\rangle = \langle \mathbf{v}, g^\dagger \circ g(\mathbf{v})\rangle = \langle \mathbf{v}, g \circ g^\dagger(\mathbf{v})\rangle = \langle g^\dagger(\mathbf{v}), g^\dagger(\mathbf{v})\rangle\,,$$

and it follows that $g^\dagger(\mathbf{v}) = \mathbf{0}$. Since $g^\dagger = f - \bar{\lambda}\,\mathrm{id}_V$ this, in turn, means that $f^\dagger(\mathbf{v}) = \bar{\lambda}\mathbf{v}$. Hence, $\bar{\lambda}$ is indeed an eigenvalue of $f^\dagger$ with eigenvector $\mathbf{v}$. $\qquad\square$

24.2.2 Diagonalization of normal maps

Now we are ready to prove the key result for normal maps.

Theorem 24.3 *(Diagonalization of normal maps) Let V be an inner product vector space and $f \in \text{End}(V)$ an endomorphism with a fully factorizing characteristic polynomial. Then f is normal if and only if it has an ortho-normal basis of eigenvectors.*

Proof '$\Leftarrow$': This is the 'easy' direction. Start with an ortho-normal basis $(\epsilon_1, \ldots, \epsilon_n)$ of eigenvector of f, so that $f(\epsilon_i) = \lambda_i \epsilon_i$. Then, Lemma (24.1) implies that $f^\dagger(\epsilon_i) = \bar\lambda_i \epsilon_i$ and it follows that

$$[f, f^\dagger](\epsilon_i) = (f \circ f^\dagger - f^\dagger \circ f)(\epsilon_i) = (|\lambda_i|^2 - |\lambda_i|^2)\epsilon_i = \mathbf{0} \ .$$

Since $[f, f^\dagger]$ vanishes on a basis it vanishes and, hence, f is normal.

'$\Rightarrow$': Conversely, assume that f is normal. We will show that f has an ortho-normal basis of eigenvectors by induction in $n = \dim(V)$. For $n = 1$ the statement is trivial. Assume that it is valid for all dimensions $k < n$. Since χ_f fully factorizes, f has at least one eigenvalue λ so that the associated eigenspace $W = \text{Eig}_f(\lambda)$ is non-trivial. For $\mathbf{w} \in W$ and $\mathbf{v} \in W^\perp$ we have

$$\langle f(\mathbf{v}), \mathbf{w} \rangle = \langle \mathbf{v}, f^\dagger(\mathbf{w}) \rangle \stackrel{\text{Lm. (24.1)}}{=} \langle \mathbf{v}, \bar\lambda \mathbf{w} \rangle = \bar\lambda \langle \mathbf{v}, \mathbf{w} \rangle = 0$$
$$\langle f^\dagger(\mathbf{v}), \mathbf{w} \rangle = \langle \mathbf{v}, f(\mathbf{w}) \rangle = \langle \mathbf{v}, \lambda \mathbf{w} \rangle = \lambda \langle \mathbf{v}, \mathbf{w} \rangle = 0 \ ,$$

so it follows that the orthogonal complement $W^\perp$ is invariant under f and $f^\dagger$. The restriction $f|_{W^\perp}$ is normal as well and, from the induction assumption has an ortho-normal basis of eigenvectors. Combining this basis with a basis of W gives a ortho-normal basis of eigenvectors for $V = W \oplus W^\perp$. $\square$

Diagonalizing normal maps by finding first their eigenvalues and then an ortho-normal basis of eigenvectors proceeds exactly as for Hermitian maps and we will refrain from further explicit examples. It is, however, useful to discuss the diagonalization of unitary maps in more detail.

24.2.3 Diagonalizing unitary maps

Proposition 24.1 *The eigenvalues of a unitary map have complex modulus one.*

Proof For an eigenvalue λ with eigenvector $\mathbf{v}$ of f the eigenvalue equations $f(\mathbf{v}) = \lambda \mathbf{v}$ together with unitarity condition (23.10) gives

$$|\lambda|^2 \langle \mathbf{v}, \mathbf{v} \rangle = \langle \lambda \mathbf{v}, \lambda \mathbf{v} \rangle = \langle f(\mathbf{v}), f(\mathbf{v}) \rangle = \langle \mathbf{v}, \mathbf{v} \rangle \ .$$

Since $\mathbf{v} \neq \mathbf{0}$ it follows that $|\lambda| = 1$. $\square$

Hence, in a complex inner product vector space a unitary map $f \in \text{U}(V)$ can be described, relative to an ortho-normal basis $(\epsilon_1, \ldots, \epsilon_n)$ of V, by a matrix $\hat{A} = \text{diag}(z_1, \ldots, z_n)$, where $z_i \in \mathbb{C}$ have complex length one, $|z_i| = 1$. In particular, for a unitary matrix $A \in \mathrm{U}(n)$ we have a unitary basis transformation $U = (\epsilon_1, \ldots, \epsilon_n) \in \mathrm{U}(n)$ which diagonalizes A, so that

$$\hat{A} = \mathrm{diag}(z_1, \ldots, z_n) = U^\dagger A U \quad \text{where} \quad |z_i| = 1 . \tag{24.2}$$

The eigenvalues of a unitary map $f \in \mathrm{U}(V)$ over a real inner product space can only assume the values ± 1. However, such a map can only be diagonalized if the characteristic polynomial fully factorizes and this is not necessarily the case over $\mathbb{R}$. If we assume full factorization, there is an ortho-normal basis $(\epsilon_1, \ldots, \epsilon_n)$ relative to which f is described by the matrix

$$\hat{A} = \mathrm{diag}(1, \ldots 1, -1, \ldots, -1) = R^T A R , \quad \text{where} \quad R = (\epsilon_1, \ldots, \epsilon_n) \in \mathrm{O}(n) . \tag{24.3}$$

Problem 24.4 *(Orthogonal matrices in two dimensions)*

Show that two-dimensional rotations $R \neq \pm \mathbb{1}_2$ cannot be diagonalized over $\mathbb{R}$. Diagonalize them over $\mathbb{C}$. Also show that two-dimensional orthogonal matrices A with $\det(A) = -1$ can be diagonalized over $\mathbb{R}$.

Solution: We know from Eq. (23.17) that two-dimensional rotations can be written as

$$R(\theta) = \begin{pmatrix} \cos\theta & -\sin\theta \\ \sin\theta & \cos\theta \end{pmatrix} , \tag{24.4}$$

where $\theta \in [0, 2\pi)$. For the characteristic polynomial we have $\chi_{R(\theta)}(\lambda) = (\lambda - \exp(i\theta))(\lambda - \exp(-i\theta))$, so the eigenvalues are $\lambda_\pm = \exp(\pm i\theta)$. These eigenvalues are not real unless $\theta = 0, \pi$ which corresponds to $R \neq \pm \mathbb{1}_2$, so apart from those special cases R cannot be diagonalized over the real numbers. However, it can be diagonalized over $\mathbb{C}$. The eigenvectors, normalized relative to the standard scalar product on $\mathbb{C}$, are $\mathbf{v}_\pm = (\pm i, 1)^T / \sqrt{2}$ and with the unitary basis transformation $U = (\mathbf{v}_-, \mathbf{v}_+) \in \mathrm{U}(2)$ we have

$$U^\dagger R(\theta) U = \mathrm{diag}(\exp(-i\theta), \exp(i\theta)) .$$

On the other hand, from Eq. (23.21), two-dimensional orthogonal matrices with negative determinant can be written as

$$A(\theta) = \begin{pmatrix} \cos\theta & \sin\theta \\ \sin\theta & -\cos\theta \end{pmatrix} .$$

The subtle change of sign, compared to rotations, is important since the characteristic polynomial $\chi_{A(\theta)}(\lambda) = (\lambda - 1)(\lambda + 1)$ now fully factorizes over $\mathbb{R}$. The eigenvalues are $\lambda_\pm = \pm 1$ with the corresponding eigenvectors $\mathbf{v}_+ = (\cos(\theta/2), \sin(\theta/2))^T$ and $\mathbf{v}_- = (-\sin(\theta/2), \cos(\theta/2))^T$. The diagonalizing matrix $R(\theta/2) = (\mathbf{v}_+, \mathbf{v}_-)$ is a rotation with angle $\theta/2$ and we have

$$R(\theta/2)^T A(\theta) R(\theta/2) = \mathrm{diag}(1, -1) .$$

24.2.4   Orthogonal matrices

The previous problem illustrates that not all orthogonal matrices can be diagonalized over $\mathbb{R}$. What is the simplest 'normal form' we can achieve in this case?

Theorem 24.4   *Let $A \in \mathrm{O}(n)$ be an orthogonal matrix. Then there exists a basis transformation $P \in \mathrm{O}(n)$ such that*

$$P^T A P = \mathrm{diag}(1, \ldots, 1, -1, \ldots, -1, R(\theta_1), \ldots, R(\theta_k)) , \tag{24.5}$$

where $R(\theta_i)$ are two-dimensional rotations as in Eq. (24.4).

Proof   We start by showing that there is a vector subspace $U \subset \mathbb{R}^n$ with $\dim_\mathbb{R}(U) \in \{1, 2\}$ which is invariant under A. If we consider A as a map on $\mathbb{C}^n$ then it is unitary and, over $\mathbb{C}$, there exists an eigenvalue λ with eigenvector $\mathbf{v}$, so $A\mathbf{v} = \lambda\mathbf{v}$. Complex conjugating this equation and taking into account that A is real we have $A\bar{\mathbf{v}} = \bar{\lambda}\bar{\mathbf{v}}$, so that $\bar{\mathbf{v}}$ is also an eigenvector of A with eigenvalue $\bar{\lambda}$. Define the two real vectors

$$\mathbf{v}_R = \frac{1}{2}(\mathbf{v} + \bar{\mathbf{v}}) , \qquad \mathbf{v}_I = \frac{1}{2i}(\mathbf{v} - \bar{\mathbf{v}}) ,$$

and define $U = \mathrm{Span}(\mathbf{v}_R, \mathbf{v}_I)$. It is clear that U is at least one-dimensional and at most two-dimensional. Invariance of U under A is easily checked and follows from the fact that $\mathbf{v}$ and $\bar{\mathbf{v}}$ are eigenvectors of A.

Let $W = U^\perp$ be the orthogonal complement. For $\mathbf{w} \in W$ and $\mathbf{u} \in U$ we have

$$(A\mathbf{w}) \cdot \mathbf{u} = (A^{-1}A\mathbf{w}) \cdot (A^{-1}\mathbf{u}) = \mathbf{w} \cdot \underbrace{(A^{-1}\mathbf{u})}_{\in U} = 0 .$$

Hence, W is invariant under A and we have a orthogonal decomposition $\mathbb{R}^n = U \oplus W$ into A-invariant subspaces. The restriction $A|_W$ is also orthogonal, so we can continue this process to find an orthogonal decomposition $\mathbb{R}^n = U_1 \oplus U_2 \oplus \cdots \oplus U_m$ into one- or two-dimensional A-invariant subspaces U_i. The restriction of A to each of those subspaces is orthogonal. If $\dim_\mathbb{R}(U_i) = 1$ then $A|_{U_i} \in \{\pm 1\}$ and if $\dim_\mathbb{R}(U_i) = 2$ then $A|_{U_i}$ must be a two-dimensional orthogonal matrix. From Exercise 24.4 this means either $A|_{U_i}$ is a two-dimensional rotation so can be written as $A|_{U_i} = R(\theta)$, or it is orthogonal with negative determinant and it can be diagonalized to $\mathrm{diag}(1, -1)$. $\square$

24.2.5   Three-dimensional rotations — again

Three-dimensional rotations are important for many applications and this justifies having a closer look. From Theorem 24.4 we know that a three-dimensional rotation R can be basis-transformed to the normal form in Eq. (24.5). Since $\det(R) = 1$ this means there must be a basis transformation $P = (\boldsymbol{\epsilon}_1, \boldsymbol{\epsilon}_2, \boldsymbol{\epsilon}_3) \in \mathrm{O}(3)$ such that

$$\hat{R} = P^T R P = \begin{pmatrix} 1 & \mathbf{0}^T \\ \mathbf{0} & R(\theta) \end{pmatrix} = \begin{pmatrix} 1 & 0 & 0 \\ 0 & \cos(\theta) & -\sin(\theta) \\ 0 & \sin(\theta) & \cos(\theta) \end{pmatrix} . \tag{24.6}$$

We conclude that a three-dimensional rotation always has at least one eigenvector $\mathbf{n} = \boldsymbol{\epsilon}_1$ with eigenvalue one,

$$Rn = n \; . \tag{24.7}$$

This eigenvector is called the *axis of rotation* and the angle θ which appears in Eq. (24.6) is called the *angle of rotation*. Basis-independence of the trace means that $\mathrm{tr}(R) = \mathrm{tr}(\tilde{R}) = 1 + 2\cos(\theta)$ and this leads to the interesting and useful formula

$$\cos(\theta) = \frac{1}{2}\left(\mathrm{tr}(R) - 1\right) \tag{24.8}$$

which allows for an easy computation of the angle of rotation, even if the rotation matrix is not in the simple form (24.6). The axis of rotation $\mathbf{n}$, on the other hand, can be found as the eigenvector for eigenvalue one, that is, by solving Eq. (24.7).

It can be useful to have a more explicit form of rotations in terms of the axis of rotation and the rotation angle available. To this end, we can write the matrix $\hat{R}$ from Eq. (24.6) as

$$\hat{R}_{ij} = (1 - \cos(\theta))\hat{n}_i\hat{n}_j + \cos(\theta)\delta_{ij} + \sin(\theta)\epsilon_{ikj}\hat{n}_k$$

where $\hat{n} = \mathbf{e}_1$ is the axis of rotation relative to the basis $(\epsilon_1, \epsilon_2, \epsilon_3)$. The reason for this somewhat contrived-looking form is that it is easily transformed back to the original basis by carrying out $R = P\hat{R}P^T$. In this case, all that happens is that $\hat{n}$ is replaced by $\mathbf{n} = P\hat{n}$, so that

$$R_{ij} = (1 - \cos(\theta))n_i n_j + \cos(\theta)\delta_{ij} + \sin(\theta)\epsilon_{ikj}n_k \; . \tag{24.9}$$

This is the desired form of a three-dimensional rotation in terms of the axis of rotation $\mathbf{n}$ with $|\mathbf{n}| = 1$ and the rotation angle $\theta \in [0, 2\pi)$. $\qquad\square$

Problem 24.5 *(Three-dimensional rotations)*

Check that the matrix $R \in \mathcal{M}_{3,3}(\mathbb{R})$ given by

$$R = \frac{1}{2}\begin{pmatrix} \sqrt{2} & -1 & -1 \\ 0 & \sqrt{2} & -\sqrt{2} \\ \sqrt{2} & 1 & 1 \end{pmatrix}$$

is a rotation matrix and find its axis of rotation and the (cosine of the) angle of rotation.

Solution: It is easy to verify that $R^T R = \mathbb{1}_3$ and $\det(R) = 1$ so this is indeed a rotation. By solving Eq. (24.7) for this matrix (and normalizing the eigenvector) we find for the axis of rotation

$$\mathbf{n} = \frac{1}{\sqrt{5 - 2\sqrt{2}}}(1, -1, \sqrt{2} - 1)^T \; .$$

Also, we have $\mathrm{tr}(R) = \sqrt{2} + 1/2$, so from Eq. (24.8) the angle of rotation satisfies

$$\cos(\theta) = \frac{1}{4}(2\sqrt{2} - 1) \; .$$

24.3 Singular value decomposition*

Summary 24.3 *A homomorphism $f \in \mathrm{Hom}(V, W)$ can always be diagonalized, relative to basis choices on V and W, to a matrix with $r = \mathrm{rk}(f)$ entries 1 along the diagonal and all other entries zero. If V and W are inner product vector spaces, f can also be diagonalized relative to ortho-normal bases on V and W and this leads to the singular value decomposition.*

So far we have been concerned with normal forms and diagonalization of endomorphisms, so, in practical terms, with square matrices. What about homomorphisms $V \to W$ between two different vector spaces, not necessarily with the same dimension? Finding a normal form for such homomorphisms is, in fact, a much easier problem than for endomorphisms. This is because we have two choices of bases, one on V and one on W, which we can adjust in order to find a simple representing matrix.

24.3.1 General bases

If we allow arbitrary basis choices on V and W every homomorphism $V \to W$ can be brought to a rather simple form.

Theorem 24.5 *For a homomorphisms $f \in \mathrm{Hom}(V, W)$ there exist bases $(\mathbf{v}_1, \dots, \mathbf{v}_n)$ of V and $(\mathbf{w}_1, \dots, \mathbf{w}_m)$ of W relative to which f is represented by an $m \times n$ matrix of the form*

$$\hat{A} = \left(\begin{array}{c|c} \mathbb{1}_r & 0 \\ \hline 0 & 0 \end{array} \right) \begin{array}{c} r \\ \\ m-r \end{array} \qquad , \tag{24.10}$$

where $r = \mathrm{rk}(f)$, with block sizes indicated on top and to the right.

Proof With $r = \mathrm{rk}(f)$ we can choose a basis $(\mathbf{v}_{r+1}, \dots, \mathbf{v}_n)$ of $\mathrm{Ker}(f)$ and complete this to a basis $(\mathbf{v}_1, \dots, \mathbf{v}_r, \mathbf{v}_{r+1}, \dots, \mathbf{v}_n)$ of V. The images of these basis vector are

$$f(\mathbf{v}_i) = \begin{cases} \mathbf{w}_i & \text{for } i \leq r \\ \mathbf{0} & \text{for } i > r \end{cases} , \tag{24.11}$$

where $(\mathbf{w}_1, \dots, \mathbf{w}_r)$ forms a basis of $\mathrm{Im}(f)$. This basis can be completed to a basis $(\mathbf{w}_1, \dots, \mathbf{w}_r, \dots, \mathbf{w}_m)$ of W. Comparing Eqs. (24.11) and (15.5) shows that, relative to the bases $(\mathbf{v}_1, \dots, \mathbf{v}_n)$ and $(\mathbf{w}_1, \dots, \mathbf{w}_m)$, f is described by the matrix (24.10). $\square$

The normal form (24.10) is completely characterized by the rank, $r = \mathrm{rk}(f)$, of the homomorphism, so can be immediately written down once the rank is known. The relevant bases may also be required and an algorithm for their computation can be easily extracted from the proof of the previous theorem.

Algorithm *(Compute the normal form of a homomorphism)*

To find the normal form (24.10) and associated bases for a homomorphism $f \in \text{Hom}(V, W)$, with $n = \dim_{\mathbb{F}}(V)$, $m = \dim_{\mathbb{F}}(W)$, and $r = \text{rk}(f)$, perform the following steps:

(1) Find $\text{Ker}(f)$ by solving the homogeneous linear system $f(\mathbf{v}) = \mathbf{0}$.
(2) Choose a basis $(\mathbf{v}_{r+1}, \ldots, \mathbf{v}_n)$ of $\text{Ker}(f)$ and complete to a basis $(\mathbf{v}_1, \ldots, \mathbf{v}_n)$ of V.
(3) Compute the images $\mathbf{w}_i = f(\mathbf{v}_i)$ for $i \leq r$ and complete to a basis $(\mathbf{w}_1, \ldots, \mathbf{w}_m)$ of W. Relative to this basis and the basis for V from (2) f is represented by the matrix (24.10).
(4) If $f = A \in \text{Hom}(\mathbb{F}^n, \mathbb{F}^m)$ is a matrix, then a basis transform with $Q = (\mathbf{w}_1, \ldots, \mathbf{w}_m)$ and $P = (\mathbf{v}_1, \ldots, \mathbf{v}_n)$ brings A into the normal form, that is, $\hat{A} = Q^{-1}AP$.

Problem 24.6 *(Normal form of a homomorphism)*

For the linear map $A \in \text{Hom}(\mathbb{R}^3, \mathbb{R}^2)$ given by

$$A = \begin{pmatrix} 1 & 1 & -2 \\ 0 & 1 & 1 \end{pmatrix}$$

find the normal form (24.10) and the associated bases.

Solution: We clearly have $r = \text{rk}(A) = 2$ (the first two rows are linearly independent) so we can immediately write down the normal form as

$$\hat{A} = \begin{pmatrix} 1 & 0 & 0 \\ 0 & 1 & 0 \end{pmatrix}.$$

In order to find the bases (and basis transformation) associated to this normal form, we follow the above algorithm.
(1) Solving $A\mathbf{v} = \mathbf{0}$ shows that $\text{Ker}(A) = \text{Span}(\mathbf{v}_3)$, where $\mathbf{v}_3 = (3, -1, 1)^T$.
(2) Setting $\mathbf{v}_1 = \mathbf{e}_1$ and $\mathbf{v}_2 = \mathbf{e}_2$ the vectors $(\mathbf{v}_1, \mathbf{v}_2, \mathbf{v}_3)$ form a basis of $\mathbb{R}^3$.
(3) With $\mathbf{w}_1 = A\mathbf{v}_1 = (1, 0)^T$ and $\mathbf{w}_2 = A\mathbf{v}_2 = (1, 1)^T$ we have basis $(\mathbf{w}_1, \mathbf{w}_2)$ of $\mathbb{R}^2$.
(4) The explicit basis transformation is accomplished with the matrices

$$P = (\mathbf{v}_1, \mathbf{v}_2, \mathbf{v}_3) = \begin{pmatrix} 1 & 0 & 3 \\ 0 & 1 & -1 \\ 0 & 0 & 1 \end{pmatrix}, \qquad Q = (\mathbf{w}_1, \mathbf{w}_2) = \begin{pmatrix} 1 & 1 \\ 0 & 1 \end{pmatrix}$$

and it is easy to verify that $\hat{A} = Q^{-1}AP$.

24.3.2   Ortho-normal bases

Allowing generic choices of bases on the domain and co-domain involves considerable freedom and consequently leads to a rather simple normal form (24.10) for homomorphisms $f \in \text{Hom}(V, W)$. What if we constrain the choice of bases to ortho-normal ones? Of course for this to make sense, we have to assume that V and W are inner product vector spaces with inner products $\langle \cdot, \cdot \rangle_V$ and $\langle \cdot, \cdot \rangle_W$. The first step is to define the analogue of eigenvalues. But an equation of the form $f(\mathbf{v}) = \sigma\mathbf{w}$, where $\mathbf{v} \in V$ and $\mathbf{w} \in W$, is not sufficient to characterize the number σ since it involves two vectors which can be arbitrarily re-scaled. Rather, we have to impose two equations and a natural choice is

$$f(\mathbf{v}) = \sigma \mathbf{w} \,, \qquad f^\dagger(\mathbf{w}) = \sigma \mathbf{v} \,, \qquad (24.12)$$

for $\mathbf{v} \in V$ and $\mathbf{w} \in W$ both non-zero. These two equations imply that $f^\dagger \circ f(\mathbf{v}) = \sigma^2 \mathbf{v}$ and, hence,

$$|\sigma|^2 \langle \mathbf{w}, \mathbf{w} \rangle_W = \langle f(\mathbf{v}), f(\mathbf{v}) \rangle_W = \langle \mathbf{v}, f^\dagger \circ f(\mathbf{v}) \rangle_V = \langle \mathbf{v}, \sigma^2 \mathbf{v} \rangle_V = \sigma^2 \langle \mathbf{v}, \mathbf{v} \rangle_V \,.$$

There are two immediate conclusions from this. First, $\sigma \in \mathbb{R}$ and we can always achieve $\sigma \geq 0$ by performing a re-scaling $\mathbf{v} \mapsto -\mathbf{v}$, if required. Secondly, for $\sigma > 0$ we learn that $|\mathbf{v}|_V = |\mathbf{w}|_W$. All this motivates the following definition.

Definition 24.2 *(Singular values and vectors) For a homomorphisms $f \in \mathrm{Hom}(V, W)$ between inner product vector spaces V and W, the number $\sigma \in \mathbb{R}^{\geq 0}$ is called a singular value of f if there are vectors $\mathbf{v} \in V$ and $\mathbf{w} \in W$ with $|\mathbf{v}|_V = |\mathbf{w}|_W \neq 0$ such that the Eqs. (24.12) are satisfied. In this case the vectors $(\mathbf{v}, \mathbf{w})$ are called singular vectors for σ.*

Eigenvalues are computed by finding the zeros of the characteristic polynomial. How can the singular values be determined?

Proposition 24.2 *For a homomorphisms $f \in \mathrm{Hom}(V, W)$ between inner product vector spaces V and W the following statements are equivalent.*

(i)   *$\sigma > 0$ is a singular values of f with singular vectors $(\mathbf{v}, \mathbf{w})$.*
(ii) *$\sigma^2 > 0$ is an eigenvalue of $f^\dagger \circ f$ with eigenvector $\mathbf{v}$ and $\mathbf{w} = \sigma^{-1} f(\mathbf{v})$.*
(iii) *$\sigma^2 > 0$ is an eigenvalue of $f \circ f^\dagger$ with eigenvector $\mathbf{w}$ and $\mathbf{v} = \sigma^{-1} f(\mathbf{w})$.*

Proof The proofs are straightforward and based on Eqs. (24.12) which imply $f^\dagger \circ f(\mathbf{v}) = \sigma^2 \mathbf{v}$ and $f \circ f^\dagger(\mathbf{w}) = \sigma^2 \mathbf{w}$. $\qquad\square$

Hence, we can find the singular values and vectors by working out the eigenvalues and eigenvectors of $f^\dagger \circ f$ or $f \circ f^\dagger$. Since both of these endomorphisms are Hermitian we know that they have an ortho-normal basis of eigenvectors. This observation is the basis for constructing the singular value normal form.

Theorem 24.6 *(Singular value normal form) Let $f \in \mathrm{Hom}(V, W)$ be a homomorphism over inner product vector spaces V and W with dimensions $n = \dim_\mathbb{F}(V)$ and $m = \dim_\mathbb{F}(W)$. Then, there are ortho-normal bases $(\boldsymbol{\epsilon}_1, \dots, \boldsymbol{\epsilon}_n)$ of V and $(\tilde{\boldsymbol{\epsilon}}_1, \dots, \tilde{\boldsymbol{\epsilon}}_m)$ of W relative to which f is described by the matrix*

$$\hat{A} = \left(\begin{array}{c|c} D & 0 \\ \hline 0 & 0 \end{array} \right) \begin{array}{l} r \\ m-r \end{array} \qquad , \qquad D = \mathrm{diag}(\sigma_1, \dots, \sigma_r) \,, \qquad (24.13)$$

where $r = \mathrm{rk}(f)$, $\sigma_1 \geq \sigma_2 \geq \cdots \geq \sigma_r > 0$ are the non-zero singular values of f and σ_i^2 the non-zero eigenvalues of $f^\dagger f$, repeated in line with degeneracies.

Proof First of all the eigenvalues λ of $f^\dagger \circ f$ must be real since $f^\dagger \circ f$ is Hermitian. They are also non-negative since the eigenvalue equation $f^\dagger \circ f(\mathbf{v}) = \lambda \mathbf{v}$ implies that

$$0 \leq \langle f(\mathbf{v}), f(\mathbf{v}) \rangle = \langle \mathbf{v}, f^\dagger \circ f(\mathbf{v}) \rangle = \lambda \langle \mathbf{v}, \mathbf{v} \rangle \ .$$

We conclude that $f^\dagger \circ f$ has an ortho-normal basis $(\boldsymbol{\epsilon}_1, \dots, \boldsymbol{\epsilon}_n)$ of eigenvectors with non-negative eigenvalues $(\sigma_1^2, \dots, \sigma_n^2)$. We can assume that all $\sigma_i \geq 0$ and adopt the ordering $\sigma_1 \geq \cdots \geq \sigma_r > 0$ and $\sigma_i = 0$ for $i > r$, where $r = \mathrm{rk}(f)$. This means, the last $n-r$ eigenvectors $\boldsymbol{\epsilon}_i$ for $i > r$ have zero eigenvalues and they span $\mathrm{Ker}(f^\dagger \circ f) = \mathrm{Ker}(f)$ (see Exercise 23.8). For the other eigenvectors, $\boldsymbol{\epsilon}_i$ for $i \leq r$, we can define the re-scaled images $\tilde{\boldsymbol{\epsilon}}_i = \sigma_i^{-1} f(\boldsymbol{\epsilon}_i)$. Since

$$\sigma_i \sigma_j \langle \tilde{\boldsymbol{\epsilon}}_i, \tilde{\boldsymbol{\epsilon}}_j \rangle_W = \langle f(\boldsymbol{\epsilon}_i), f(\boldsymbol{\epsilon}_j) \rangle_W = \langle \boldsymbol{\epsilon}_i, f^\dagger \circ f(\boldsymbol{\epsilon}_j) \rangle_V = \sigma_j^2 \langle \boldsymbol{\epsilon}_i, \boldsymbol{\epsilon}_j \rangle_V = \sigma_j^2 \delta_{ij}$$

these images form, in fact, an ortho-normal system which we can completed to an ortho-normal basis $(\tilde{\boldsymbol{\epsilon}}_1, \dots, \tilde{\boldsymbol{\epsilon}}_r, \dots, \tilde{\boldsymbol{\epsilon}}_m)$ of W. In summary, f acts on the ortho-normal basis vectors as

$$f(\boldsymbol{\epsilon}_i) = \begin{cases} \sigma_i \tilde{\boldsymbol{\epsilon}}_i & \text{for } i \leq r \\ \mathbf{0} & \text{for } i > r \end{cases} ,$$

which means the matrix which describes f relative to the bases $(\boldsymbol{\epsilon}_1, \dots, \boldsymbol{\epsilon}_n)$ and $(\tilde{\boldsymbol{\epsilon}}_1, \dots, \tilde{\boldsymbol{\epsilon}}_m)$ is indeed $\hat{A}$ in Eq. (24.13). $\qquad\square$

The above proof is constructive and can be directly translated into an algorithm for computing the singular value form and its associated bases.

Algorithm *(Singular value form) In order to compute the singular value form and associated bases of a homomorphism $f \in \mathrm{Hom}(V, W)$ between two inner product vector spaces V and W with dimensions $n = \dim_{\mathbb{F}}(V)$ and $m = \dim_{\mathbb{F}}(W)$ proceed as follows.*

(1) Find an ortho-normal basis of eigenvectors $(\boldsymbol{\epsilon}_1, \dots, \boldsymbol{\epsilon}_n)$ with eigenvalues $(\lambda_1, \dots, \lambda_n)$ of $f^\dagger \circ f \in \mathrm{End}(V)$, ordered such that $\lambda_1 \geq \cdots \geq \lambda_r > 0$ and $\lambda_i = 0$ for $i > r$, where $r = \mathrm{rk}(f)$. The singular values are $\sigma_i = \sqrt{\lambda_i}$.

(2) Find the re-scaled images $\tilde{\boldsymbol{\epsilon}}_i := \sigma_i^{-1} f(\boldsymbol{\epsilon}_i)$, for $i = 1, \dots, r$, which are automatically ortho-normal. Complete them to an ortho-normal basis $(\tilde{\boldsymbol{\epsilon}}_1, \dots, \tilde{\boldsymbol{\epsilon}}_m)$ of W.

(3) Relative to the bases $(\boldsymbol{\epsilon}_1, \dots, \boldsymbol{\epsilon}_n)$ of V and $(\tilde{\boldsymbol{\epsilon}}_1, \dots, \tilde{\boldsymbol{\epsilon}}_m)$ of W the map f is represented by the matrix $\hat{A}$ in singular value form given in Eq. (24.13).

(4) If $f = A \in \mathrm{Hom}(\mathbb{F}^n, \mathbb{F}^m)$ is a matrix, define the unitary matrices $U = (\boldsymbol{\epsilon}_1, \dots, \boldsymbol{\epsilon}_n)$ and $\tilde{U} = (\tilde{\boldsymbol{\epsilon}}_1, \dots, \tilde{\boldsymbol{\epsilon}}_m)$ in order to write down the singular value decomposition $A = \tilde{U} \hat{A} U^\dagger$ of A.

Problem 24.7 *(Singular value form)*

Find the singular value form and decomposition of $A \in \mathrm{Hom}(\mathbb{R}^3, \mathbb{R}^2)$ defined by

$$A = \begin{pmatrix} 1 & 1 & -2 \\ 1 & 2 & 1 \end{pmatrix} ,$$

with the scalar product on $\mathbb{R}^3$ and $\mathbb{R}^2$ given by the dot product.

Solution: (1) We have

$$M = A^T A = \begin{pmatrix} 2 & 3 & -1 \\ 3 & 5 & 0 \\ -1 & 0 & 5 \end{pmatrix} \quad \Rightarrow \quad \chi_M(\lambda) = -(\lambda - 7)(\lambda - 5)\lambda$$

so the eigenvalues of M are $\lambda_1 = 7$, $\lambda_2 = 5$ and $\lambda_3 = 0$, with associated eigenvectors $\epsilon_1 = (2, 3, -1)^T/\sqrt{14}$, $\epsilon_2 = (0, 1, 3)/\sqrt{10}$ and $\epsilon_3 = (5, -3, 1)/\sqrt{35}$. The non-zero singular values are $\sigma_1 = \sqrt{\lambda_1} = \sqrt{7}$ and $\sigma_2 = \sqrt{\lambda_2} = \sqrt{5}$.

(2) The images $\tilde{\epsilon}_i = \sigma_i^{-1} A \epsilon_i$ for $i = 1, 2$ are given by $\tilde{\epsilon}_1 = (1, 1)^T/\sqrt{2}$ and $\tilde{\epsilon}_2 = (-1, 1)^T/\sqrt{2}$.

(3) Relative to the bases $(\epsilon_1, \epsilon_2, \epsilon_3)$ of $\mathbb{R}^2$ and $(\tilde{\epsilon}_1, \tilde{\epsilon}_2)$ of $\mathbb{R}^2$ the map A is described by the singular value form

$$\hat{A} = \begin{pmatrix} \sqrt{7} & 0 & 0 \\ 0 & \sqrt{5} & 0 \end{pmatrix}.$$

(4) With the unitary matrices

$$U = (\epsilon_1, \epsilon_2, \epsilon_3) = \begin{pmatrix} \sqrt{\frac{2}{7}} & 0 & \sqrt{\frac{5}{7}} \\ \frac{3}{\sqrt{14}} & \frac{1}{\sqrt{10}} & -\frac{3}{\sqrt{35}} \\ -\frac{1}{\sqrt{14}} & \frac{3}{\sqrt{10}} & \frac{1}{\sqrt{35}} \end{pmatrix}, \qquad \tilde{U} = (\tilde{\epsilon}_1, \tilde{\epsilon}_2) = \frac{1}{\sqrt{2}} \begin{pmatrix} 1 & -1 \\ 1 & 1 \end{pmatrix},$$

the singular value decomposition is $A = \tilde{U}\hat{A}U^T$ and this can be easily verified explicitly.

Application 24.1 *(Data compression with singular values)*

For data that can be represented as a matrix the singular value decomposition can be used for data compression. Consider, for example, a (black-and-white) picture with size $m \times n$. The underlying data can be stored in an $m \times n$ matrix A whose entries $A_{ij} \in [0, 1]$ represent the grayscale of each pixel.

We can think of this matrix as a map $A : \mathbb{R}^n \to \mathbb{R}^m$ and find its singular value decomposition. With $U = (\epsilon_1, \ldots, \epsilon_n)$ and $\tilde{U} = (\tilde{\epsilon}_1, \ldots, \tilde{\epsilon}_m)$ being the relevant ortho-normal bases of $\mathbb{R}^n$ and $\mathbb{R}^m$ we can write

$$A = \tilde{U}\hat{A}U^\dagger \tag{24.14}$$

where $\hat{A}$ has the structure indicated in Eq. (24.13). As usual, we assume that the non-zero singular values σ_i, where $i = 1, \ldots, r = \text{rk}(A)$, are ordered by size, so that $\sigma_1 \geq \sigma_2 \geq \cdots \geq \sigma_r > 0$. Suppose, instead of all r singular values we only consider the $s < r$ largest ones, together with their singular vectors. This leads to the reduced matrices

$$\underbrace{U_s = (\epsilon_1, \ldots, \epsilon_s)}_{n \times s}, \quad \underbrace{\hat{A}_s = \text{diag}(\sigma_1, \ldots, \sigma_s)}_{s \times s}, \quad \underbrace{\tilde{U}_s = (\tilde{\epsilon}_1, \ldots, \tilde{\epsilon}_s)}_{m \times s}. \tag{24.15}$$

which we can use to define a reduced version, A_s, of the matrix A by

$$A_s = \tilde{U}_s \hat{A}_s U_s^\dagger. \tag{24.16}$$

Note that the matrix A_s, for any choice of s, has the same size, $m \times n$, as the original matrix A. Since we can expect the largest singular values to dominate in the singular value decomposition (24.14) for A the matrix A_s in Eq. (24.16) can be viewed as an approximation of A. The quality of this approximation depends on the nature of the data and, of course, on the number, s, of singular values considered.

The matrices in Eq. (24.15) which determine A_s contain a total of

$$ns + s + ms = (m + n + 1)s \qquad (24.17)$$

real values and this can be considerably smaller than the mn real values in the original matrix A. This is why a compression of the data can be achieved.

As an example, consider the following picture of Emmie the Cat, that has size $(m, n) = (200, 250)$.

Converting this leads to a 200×250 matrix A with a total of $50{,}000$ entries $A_{ij} \in [0, 1]$. This matrix has the maximal possible rank, $\mathrm{rk}(A) = 200$, as one would expect. (Recall that matrices with reduced rank are non-generic and there is no reason why a matrix derived from a complicated picture should have this property.) This means the matrix A has 200 non-zero singular values. If we calculate the reduced matrices A_{10}, A_{20}, and A_{50}, based on considering the largest 10, 20, and 50 singular values, respectively, from Eq. (24.16) and convert the resulting matrices A_s back into pictures we find the following.

The picture on the left, based on 10 singular values, is a relatively poor version of the original but the outlines are still visible. The middle picture, based on 20 singular values is already a good approximation and the picture on the right, based on 50 values, is almost as good as the original. From Eq. (24.17) with $m = 200$, $n = 250$, and $s = 10, 20, 50$, the number of real values underlying these three compressions are 4510, 9020, and 22,550, respectively, so in either case a relevant reduction has been achieved, compared to the 50,000 entries in A.

Application 24.2 *(Quark masses and singular values)*

Understanding properties of elementary particles requires the formalism of quantum field theory which is a challenging and thorny subject, well beyond the scope of this book.

However, without too much effort and assuming the reader is willing to accept a few basic facts, we can get to an interesting connection between masses of elementary particles and linear algebra.

For simplicity we focus on the masses of the quarks, which are the main building blocks of matter. There is a total of six types of quarks which are organised into two groups of three families each, the u-type quarks $(u^i) = (u, c, t)$ ('up", 'charm' and 'top') and the d-type quarks $(d^i) = (d, s, b)$ ('down', 'strange' and 'bottom'), where $i = 1, 2, 3$ labels the three families. In addition, each quark comes in a left- and right-handed version which we label by subscripts L and R, respectively. Only the members u and d of the first family form the matter which surrounds us, that is, the protons and neutrons.

The masses of these quarks are described by a certain part of the *standard model of particle physics*. Schematically, this is given by a *Lagrangian density*, an expression bi-linear in the quarks and written as

$$\mathcal{L}_{\text{mass}} = \sum_{i,j=1}^{3} M_{ij}^u \bar{u}_L^i u_R^j + \sum_{i,j=1}^{3} M_{ij}^d \bar{d}_L^i d_R^j . \tag{24.18}$$

Here M^u and M^d are 3×3 matrices, also called *mass matrices*, with generally complex entries. There is currently no accepted theory which tells us what these matrices look like but we do know that they determine the masses of the quarks as well as the related property of *quark mixing*. The way this works is as follows.

Suppose we transform the quarks to another basis (denoted by a hat) via

$$u_L^i = U_{u,j}^i \hat{u}_R^j , \quad u_R^j = V_{u,l}^j \hat{u}_R^l , \quad d_L^i = U_{d,k}^i \hat{d}_L^k , \quad d_R^j = V_{d,l}^j \hat{d}_R^l , \tag{24.19}$$

where $U_u, V_u, U_d, V_d \in \mathrm{U}(3)$ are unitary matrices. (These matrices have to be unitary in order to keep some other parts of the theory unchanged under these transformations. This works because the relevant parts are, structurally, of the same form as the standard Hermitian scalar product on $\mathbb{C}^3$.) Re-writing the mass Lagrangian density in terms of the new basis by inserting Eqs. (24.19) into Eq. (24.18) gives

$$\mathcal{L}_{\text{mass}} = \sum_{i,j=1}^{3} \hat{M}_{ij}^u \bar{\hat{u}}_L^i \hat{u}_R^j + \sum_{i,j=1}^{3} \hat{M}_{ij}^d \bar{\hat{d}}_L^i \hat{d}_R^j \quad \text{where} \quad \hat{M}_u = U_u^\dagger M_u V_u, \ \hat{M}_d = U_d^\dagger M_d V_d .$$

This means the basis change transforms the mass matrices as

$$M_u = U_u \hat{M}_u V_u^\dagger , \qquad M_d = U_d \hat{M}_d V_d^\dagger . \tag{24.20}$$

Note these equations have precisely the structure of a singular value decomposition of M_u and M_d. If we choose the unitary matrices U_u, V_u, U_d and V_d so that the singular value decomposition is realized then the new mass matrices, $\hat{M}_u = \mathrm{diag}(m_u, m_c, m_t)$ and $\hat{M}_d = \mathrm{diag}(m_d, m_s, m_b)$ are diagonal. The singular values which appear along the diagonal are interpreted as the quark masses and the matrices M_u and M_d have to be chosen so that their singular values coincide (within errors) with the measured masses of the quarks.

Of course one has to be careful in tracking which other parts of the standard model of particle physics is changed by the transformation (24.19). It turns out only the weak force is affected and since only the left-handed quarks experience this force, the change depends on U_u and U_d. More precisely, it depends on the relative transformation given by the famous *Cabbibo–Kobayashi–Maskawa matrix*

$$V_{\mathrm{CKM}} = U_u^\dagger U_d \,,$$

a unitary 3×3 matric which described the 'misalignment' between the left-handed u and d quarks. The entries of this matrix have been experimentally measured and reproducing these values from the theory places further constraints on the mass matrices M_u and M_d. For a gentle introduction to particle physics see, for example, Halzen and Martin 2008.

24.4 Functions of matrices*

Summary 24.4 *Functions $g : \mathbb{F} \to \mathbb{F}$ with a power series expansion can be extended to functions $g : \mathcal{M}_{n,n}(\mathbb{F}) \to \mathcal{M}_{n,n}(\mathbb{F})$ between $n \times n$ matrices by replacing the numerical argument in the power series by a matrix. Such functions of matrices can be evaluated by diagonalizing the matrix argument. The matrix exponential is of particular importance as it relates certain matrix vector spaces to matrix groups. In particular, anti-symmetric matrices exponentiate to rotations while the anti-Hermitian matrices exponentiate to unitary matrices.*

24.4.1 Defining functions of matrices

Suppose we have a function $\mathbb{F} \ni x \mapsto g(x) \in \mathbb{F}$. Can we insert a square matrix, rather than a number, into this function, that is, can we make sense of the expression $g(A)$, where $A \in \mathcal{M}_{n,n}(\mathbb{F})$? Suppose g has a power series expansion

$$g(x) = a_0 + a_1 x + a_2 x^2 + a_3 x^3 + \cdots \,, \tag{24.21}$$

where $a_i \in \mathbb{F}$. In this case, we can attempt to define $g(A)$ by inserting the matrix A into the power series, so

$$g(A) = a_0 \mathbb{1}_n + a_1 A + a_2 A^2 + a_3 A^3 + \cdots \,. \tag{24.22}$$

Note that this expression makes perfect sense, since powers, sums and scalar multiples of a square matrix are well-defined. The value $g(A) \subset \mathcal{M}_{n,n}(\mathbb{F})$ is a matrix of the same size as its argument A. Of course, if we are dealing with an infinite series (rather than merely a polynomial) issues of convergence are involved, for the original series (24.21) as well as its matrix version (24.22). We will not attempt to address these, as they are the subject of an analysis course (see, for example, Lang 1997). Instead, we focus on how to evaluate such matrix functions.

How do matrix functions relate to basis matrix operations? For transposition, we clearly have

$$g(A^T) = g(A)^T \tag{24.23}$$

so transposition commutes with evaluating the function. If the function g is such that $g(\bar{z}) = \overline{g(z)}$ for all $z \in \mathbb{C}$ (that is, if all coefficients a_i in the series (24.21) are real) it follows from Eq. (24.23) that

$$g(\bar{z}) = \overline{g(z)} \quad \text{for all } z \in \mathbb{C} \quad \Rightarrow \quad g(\bar{A}) = \overline{g(A)} \,, \quad g(A^\dagger) = g(A)^\dagger \,. \tag{24.24}$$

It is important to realize that functional equations which g satisfies for numerical arguments might not continue to hold for matrices. For example, the functional equation $e^{x+y} = e^x e^y$ of the exponential function does not remain true in general when matrix arguments are inserted. To see why this happens, recall from analysis that such equations are often proven by using the series expansion of the relevant function and that the proof involves commuting the numerical arguments. Matrices do not commute in general so simply repeating the standard proof with matrices does not work. However, this can be done if the matrices are special and do commute, so in this case functional equations typically remain valid for matrix arguments.

24.4.2 Matrix functions and diagonalization

Computing the matrix function of a diagonal matrix $\hat{A} = \text{diag}(\lambda_1, \dots, \lambda_n)$ is easy since

$$g(\hat{A}) = g(\text{diag}(\lambda_1, \dots, \lambda_n)) = \text{diag}(g(\lambda_1), \dots, g(\lambda_n)) . \tag{24.25}$$

This suggests that diagonalizing matrices might be helpful in order to evaluate matrix functions. The key reason for why this works is that the two operations of basis transformations and evaluating matrix functions commute. This can be easily verified, first for matrix powers and then for power series.

$$(P^{-1}AP)^k = P^{-1}A \underbrace{PP^{-1}}_{=\mathbb{1}} AP \cdots P^{-1}AP = P^{-1}A^k P \qquad \Rightarrow$$

$$g(P^{-1}AP) = \sum_k a_k (P^{-1}AP)^k = \sum_k a_k P^{-1}A^k P = P^{-1}\left(\sum_k a_k A^k\right) P = P^{-1}g(A)P$$

In summary, we have

$$g(P^{-1}AP) = P^{-1}g(A)P . \tag{24.26}$$

Now suppose that the matrix $A \in \mathcal{M}_{n,n}(\mathbb{F})$ can be diagonalized to a matrix $\hat{A} = \text{diag}(\lambda_1, \dots, \lambda_n) = P^{-1}AP$ with eigenvalues λ_i. Then,

$$g(A) = g(P\hat{A}P^{-1}) \overset{(24.26)}{=} Pg(\hat{A})P^{-1} \overset{(24.25)}{=} P\,\text{diag}(g(\lambda_1), \dots, g(\lambda_n))P^{-1} \tag{24.27}$$

So, for matrices which can be diagonalized there is a simple recipe for how to compute their matrix functions. Form the diagonal matrix with entries given by the function evaluated on the eigenvalues, then transform this matrix back to the original basis.

Example 24.1 *(The matrix exponential)*
The matrix exponential function is defined as

$$\exp(A) = \mathbb{1} + A + \frac{1}{2}A^2 + \frac{1}{6}A^3 + \cdots = \sum_{k=0}^{\infty} \frac{1}{k!}A^k . \tag{24.28}$$

We remark that the matrix exponential series always converges, just as its standard counterpart (essentially thanks to the $1/k!$ factors). As we will see, the matrix exponential plays an important role for matrix groups such as orthogonal and unitary groups. As mentioned above, the functional equation of the exponential function is not generally valid for matrices. However, for two matrices $A, B \in \mathcal{M}_{n,n}(\mathbb{F})$ we have

$$[A, B] = 0 \quad \Rightarrow \quad e^{A+B} = e^A e^B \ . \tag{24.29}$$

What is the determinant of a matrix exponential? Suppose the matrix A can be diagonalized with $\hat{A} = \mathrm{diag}(\lambda_1, \dots, \lambda_n) = P^{-1}AP$. Eq. (24.27) then implies that

$$\exp(A) = P\mathrm{diag}(\exp(\lambda_1), \dots, \exp(\lambda_n))P^{-1} \ ,$$

and, using the basis invariance of the determinant, we find that

$$\det(\exp(A)) = \prod_i \exp(\lambda_i) = \exp\left(\sum_i \lambda_i\right) = \exp(\mathrm{tr}(A)) \ . \tag{24.30}$$

In particular, the determinant of $\exp(A)$ is non-zero (since the exponential function has no zeros) so $\exp(A)$ is always invertible.

For the matrix exponential $R = e^A$ of a (real) anti-symmetric matrix A we have

$$R^T R = \exp(A)^T \exp(A) \overset{(24.23)}{=} \exp(A^T) \exp(A) = \exp(-A) \exp(A) \overset{(24.29)}{=} \exp(0) = \mathbb{1} \ ,$$

where, in the second last step, we have used the fact that A and $-A$ commute. The conclusion is that the exponential of an anti-symmetric matrix is an orthogonal matrix. In fact, since the trace of an anti-symmetric matrix vanishes, Eq. (24.30) implies that $\det(R) = 1$, so R is, in fact, a rotation.

Something analogous happens for the exponential $U = \exp(A)$ of an anti-Hermitian matrix A. Since

$$U^\dagger U = \exp(A)^\dagger \exp(A) \overset{(24.24)}{=} \exp(A^\dagger) \exp(A) = \exp(-A) \exp(A) \overset{(24.29)}{=} \exp(0) = \mathbb{1} \ ,$$

such a matrix is unitary. It is not necessarily special unitary since the trace of an anti-Hermitian matrix can be non-zero. However, if we impose that A is anti-Hermitian and that $\mathrm{tr}(A) = 0$ then $U = \exp(A)$ is special unitary.

The above examples point to a relationship between certain vector spaces and groups of matrices, via the matrix exponential: the vector space $\mathcal{A}_n$ of anti-symmetric matrices exponentiates to rotations in $\mathrm{SO}(n)$, the vector space $\mathcal{H}_n$ of anti-Hermitian matrices to the unitary group $\mathrm{U}(n)$ and the vector space of anti-Hermitian, traceless matrices to the special unitary group $\mathrm{SU}(n)$. These are examples of a general relationship between *Lie algebras* and *Lie groups*. The theory of Lie groups plays an important role in many scientific applications but anything more than a discussion of examples is beyond the scope of the present text (see, for example, Fulton and Harris 2013; Cornwell 1997).
□

Problem 24.8 *(Matrix monomial)*

Work out the value of the function $g(x) = x^n$, where $n \in \mathbb{N}$, for the matrix

$$A = \begin{pmatrix} 1 & 2i \\ -2i & 1 \end{pmatrix}.$$

Solution: The characteristic polynomial is $\chi_A(\lambda) = (\lambda - 3)(\lambda + 1)$, so the eigenvalues are $\lambda_1 = 3$ and $\lambda_2 = -1$. The associated eigenvectors (normalized relative to the standard scalar product on $\mathbb{C}^2$) are $\epsilon_1 = (i, 1)^T / \sqrt{2}$ and $\epsilon_2 = (-i, 1)^T / \sqrt{2}$, so we have the unitary matrix $U = (\epsilon_1, \epsilon_2)$ with $U^\dagger A U = \mathrm{diag}(3, -1)$. Then, applying Eq. (24.27) gives

$$g(A) = U \, \mathrm{diag}(3^n, (-1)^n) U^\dagger = \frac{1}{2} \begin{pmatrix} (-1)^n + 3^n & -i((-1)^n - 3^n) \\ i((-1)^n - 3^n) & (-1)^n + 3^n \end{pmatrix}.$$

Problem 24.9 *(Matrix sine)*

Evaluate the function $g(x) = \sin(x)$ on the matrix

$$A = \frac{\pi}{2} \begin{pmatrix} 2 & 1 \\ 1 & 2 \end{pmatrix}.$$

Solution: The characteristic polynomial $\chi_A(\lambda) = (\lambda - 3\pi/2)(\lambda - \pi/2)$ shows that the eigenvalues are $\lambda_1 = 3\pi/2$ and $\lambda_2 = \pi/2$. The associated eigenvectors (normalized relative to the dot product) $\epsilon_1 = (1, 1)^T / \sqrt{2}$ and $\epsilon_2 = (1, -1)^T / \sqrt{2}$ give rise to the orthogonal matrix $P = (\epsilon_1, \epsilon_2)$ with $P^T A P = \frac{\pi}{2} \mathrm{diag}(3, 1)$. It follows from Eq. (24.27) that

$$\sin(A) = P \, \mathrm{diag}(\sin(3\pi/2), \sin(\pi/2)) P^T = P \, \mathrm{diag}(-1, 1) P^T = -\begin{pmatrix} 0 & 1 \\ 1 & 0 \end{pmatrix}.$$

24.4.3 Direct computation of matrix functions

Sometimes a matrix function can be computed directly, without diagonalizing. Typically, this is possible when the argument matrix has specific properties, such as matrix powers which are easily evaluated. For example, consider a nilpotent matrix A with order q, so that $A^q = 0$. In this case the series for $g(A)$ terminates and we have

$$g(A) = \sum_{k=0}^{q-1} a_k A^k.$$

Another example is a projection matrix P. Of course the projector condition $P^2 = P$ applied repeatedly implies that $P^k = P$ for all $k = 1, 2, \dots$. Hence,

$$g(P) = \sum_{k=0}^{\infty} a_k P^k = a_0 \mathbb{1} + \sum_{k=1}^{\infty} a_k P = g(0) \mathbb{1} + (g(1) - g(0)) P.$$

Some other interesting examples which can be dealt with in this way are related to matrix groups.

Example 24.2 *(Two-dimensional rotations from matrix exponentials)*

We have seen in Example 24.1 that the matrix exponential of anti-symmetric matrices leads to rotations. Let us explore this in more detail for the case of 2×2 matrices. A general 2×2 anti-symmetric matrix can be written in the form

$$A = \theta T , \qquad T = \begin{pmatrix} 0 & 1 \\ -1 & 0 \end{pmatrix} ,$$

where $\theta \in \mathbb{R}$. Its matrix exponential can be worked out by diagonalizing A but also more directly, starting from the observation that $A^2 = -\theta^2 \mathbb{1}_2$. As a result, the even and odd powers of A are given by

$$A^{2n} = (-1)^n \theta^{2n} \mathbb{1}_2 , \qquad A^{2n+1} = (-1)^n \theta^{2n+1} T .$$

With these results it is straightforward to work out the matrix exponential explicitly.

$$\exp(A) = \sum_{n=0}^{\infty} \frac{A^n}{n!} = \sum_{n=0}^{\infty} \frac{A^{2n}}{(2n)!} + \sum_{n=0}^{\infty} \frac{A^{2n+1}}{(2n+1)!} = \sum_{n=0}^{\infty} \frac{(-1)^n \theta^{2n}}{(2n)!} \mathbb{1}_2 + \sum_{n=0}^{\infty} \frac{(-1)^n \theta^{2n+1}}{(2n+1)!} T$$

$$= \cos(\theta) \mathbb{1}_2 + \sin(\theta) T = \begin{pmatrix} \cos\theta & \sin\theta \\ -\sin\theta & \cos\theta \end{pmatrix} .$$

We have used definitions of the sine and cosine functions in terms of a series in the second last step. As is evident, the exponential of anti-symmetric 2×2 matrices does indeed produce rotations. The interesting additional information we obtain from this explicit calculation is that we do obtain all two-dimensional rotations in this way.
□

Example 24.3 *(Two-dimensional special unitary matrices from matrix exponential)*

We would like to verify explicitly that the matrix exponential of 2×2 anti-Hermitian, traceless matrices leads to special unitary matrices. To do so, we note that this vector space is three-dimensional with a basis $(i\sigma_1, i\sigma_2, i\sigma_3)$, where σ_i are the Pauli matrices introduced in Problem 13.12.

Working out the matrix exponential explicitly is made possible by the properties of the Pauli matrices which we have to develop first. From the explicit matrices in Problem 13.12 it is easy to verify that the Pauli matrices square to the unit matrix, $\sigma_i^2 = \mathbb{1}_2$, and two of them multiply to $\pm i$ times the third, for example $\sigma_1 \sigma_2 = i\sigma_3$. These relations can be summarized by writing

$$\sigma_i \sigma_j = \mathbb{1}_2 \delta_{ij} + i\epsilon_{ijk}\sigma_k . \tag{24.31}$$

This equation includes practically everything one needs to know about Pauli matrices. For example, subtracting from Eq. (24.31) the same equation with indices (i, j) exchanged gives the commutator

$$[\sigma_i, \sigma_j] = 2i\epsilon_{ijk}\sigma_k \tag{24.32}$$

of the Pauli matrices. For the purpose of computing matrix exponentials, we introduce the formal vector $\boldsymbol{\sigma} = (\sigma_1, \sigma_2, \sigma_3)^T$ and, for $\mathbf{a} \in \mathbb{R}^3$, we write linear combinations of

Pauli matrices as a formal dot product $\mathbf{a} \cdot \boldsymbol{\sigma} = a_i \sigma_i$. Multiplying Eq. (24.31) with $a_i a_j$ shows that $(\mathbf{a} \cdot \boldsymbol{\sigma})^2 = |\mathbf{a}|^2 \mathbb{1}_2$ and, hence, for $n \in \mathbb{N}$ we have

$$(\mathbf{a} \cdot \boldsymbol{\sigma})^{2n} = |\mathbf{a}|^{2n} \mathbb{1}_2 , \quad (\mathbf{a} \cdot \boldsymbol{\sigma})^{2n+1} = |\mathbf{a}|^{2n} \mathbf{a} \cdot \boldsymbol{\sigma} . \tag{24.33}$$

Thanks to these relations it is now easy to work out the matrix exponential $U = \exp(i\theta \mathbf{n} \cdot \boldsymbol{\sigma})$, where $\mathbf{n}$ is a real unit length vector and $\theta \in \mathbb{R}$. Using Eqs. (24.33) with $\mathbf{a} = \mathbf{n}$ we find

$$U = \exp(i\theta \, \mathbf{n} \cdot \boldsymbol{\sigma}) = \sum_{n=0}^{\infty} \frac{(i\theta)^n}{n!} (\mathbf{n} \cdot \boldsymbol{\sigma})^n = \cos(\theta)\mathbb{1}_2 + i\sin(\theta)\mathbf{n} \cdot \boldsymbol{\sigma}$$

$$= \begin{pmatrix} \cos\theta + in_3 \sin\theta & (n_2 + in_1)\sin\theta \\ -(n_2 - in_1)\sin\theta & \cos\theta - in_3 \sin\theta \end{pmatrix} = \begin{pmatrix} \alpha & \beta \\ -\bar{\beta} & \bar{\alpha} \end{pmatrix} ,$$

where $\alpha = \cos\theta + in_3 \sin\theta$ and $\beta = (n_2 + in_1)\sin\theta$. Comparison with Eq. (23.36) shows that this is, in fact, the general form of an SU(2) matrix. $\qquad\square$

Application 24.3 *(Differential equations and matrix exponential)*

The matrix exponential can be used to solve systems of first order ordinary differential equations. To see how this work, start with a single first order ordinary differential equation

$$\frac{dx}{dt} = ax$$

for a function $x \in \mathcal{C}^1(\mathbb{R}, \mathbb{F})$ and an arbitrary constant $a \in \mathbb{F}$, where we allow x to be real-valued, $\mathbb{F} = \mathbb{R}$, or complex-valued, $\mathbb{F} = \mathbb{C}$. The general solution to this equation can of course be written as an exponential

$$x(t) = e^{at} x_0 , \tag{24.34}$$

with an arbitrary 'initial value' $x_0 = x(0)$.

What about its multi-dimensional generalization

$$\frac{d\mathbf{x}}{dt} = A\mathbf{x} , \tag{24.35}$$

where $\mathbf{x}(t) = (x_1(t), \ldots, x_n(t))^T$ is a vector of n functions $x_i \in \mathcal{C}^1(\mathbb{R}, \mathbb{F})$ and $A \in \mathcal{M}_{n,n}(\mathbb{F})$ is a constant $n \times n$ matrix? The straightforward generalization of the solution (24.34) to the multi-dimensional case reads

$$\mathbf{x}(t) = e^{At} \mathbf{x}_0 , \tag{24.36}$$

where $\mathbf{x}_0 = \mathbf{x}(0) \in \mathbb{F}^n$ is an arbitrary vector of initial values. Note that, given our definition of the matrix exponential, Eq. (24.36) makes perfect sense. But does it really solve the differential equation (24.35)? We verify this by simply inserting Eq. (24.36) into the differential equation (24.35), using the definition of the matrix exponential.

$$\frac{d\mathbf{x}}{dt} = \frac{d}{dt} e^{At} \mathbf{x}_0 = \frac{d}{dt} \sum_{n=0}^{\infty} \frac{1}{n!} A^n t^n \mathbf{x}_0 = \sum_{n=1}^{\infty} \frac{1}{(n-1)!} A^n t^{n-1} \mathbf{x}_0 = A \sum_{n=0}^{\infty} \frac{1}{n!} A^n t^n \mathbf{x}_0 = A\mathbf{x}$$

In conclusion, Eq. (24.36) is indeed a solution of the differential equation (24.35) for arbitrary vectors $\mathbf{x}_0 \in \mathbb{F}^n$.

If the matrix A in Eq. (24.35) is anti-Hermitian, we know from Example 24.1 that $U(t) := e^{At}$ is unitary. In this case, the solution (24.36) can be represented by the action of a unitary map

$$\mathbf{x}(t) = U(t)\,\mathbf{x}_0 \ , \quad U(t) = e^{At} \in \mathrm{U}(n) \ , \tag{24.37}$$

on the initial value $\mathbf{x}_0$. This is also sometimes expressed by saying that the differential equation (24.35), for anti-Hermitian A, leads to a unitary evolution. A unitary evolution has interesting properties — for example, the length $|\mathbf{x}(t)|$ of the solution vector is the same for all t. In a physical context, the variable t often has the interpretation of time, so in this case we talk about unitary time evolution. An important physical equation with this property is Schrödinger's equation in quantum mechanics.

Exercises

(†=challenging)

24.1 On $\mathbb{R}^3$ with the dot product, find the eigenvalues and an ortho-normal basis of eigenvectors for the linear map given by

$$A = \begin{pmatrix} 1 & 2 & 1 \\ 2 & 1 & 1 \\ 1 & 1 & 2 \end{pmatrix} \ .$$

Construct an orthogonal matrix R such that $R^T M R$ is diagonal.

24.2 On $\mathbb{C}^2$ with the standard scalar product, find the eigenvalues and an ortho-normal basis of eigenvectors for the linear map given by

$$H = \begin{pmatrix} 10 & 3i \\ -3i & 2 \end{pmatrix} \ .$$

Construct a unitary matrix U such that $U^\dagger H U$ is diagonal.

24.3 Consider the vector space V of at most quadratic polynomials $p : [0, \infty) \to \mathbb{R}$ with scalar product

$$\langle p, q \rangle = \int_0^\infty dx\, e^{-x} p(x) q(x) \ .$$

(a) Show that the linear map $L : V \to V$ defined by

$$L = x \frac{d^2}{dx^2} + (1 - x) \frac{d}{dx}$$

is Hermitian.

(b) Find the eigenvalues and an ortho-normal basis of eigenvectors for L.

24.4 Show that the matrix

$$R = \frac{1}{3\sqrt{2}} \begin{pmatrix} 3 & 0 & 3 \\ -1 & -4 & 1 \\ 2\sqrt{2} & -\sqrt{2} & -2\sqrt{2} \end{pmatrix}$$

is a rotation. Compute the characteristic polynomial of R and verify that 1 is an eigenvalue. Compute the axis of rotation, $\mathbf{n}$, and $\cos(\theta)$, where θ is the angle of rotation.

24.5 A football is placed at the centre point of the pitch before the start and after the end of a game. Argue that there are (at least) two points on the football's surface which are in the same location at these two times.

24.6 Use the explicit formula (24.9) for three-dimensional rotations to confirm the results of Exercise 23.11.

24.7 (a) Find the singular value decomposition of the matrix

$$A = \begin{pmatrix} \frac{11}{\sqrt{6}} & \frac{16}{3} & -\frac{5}{3\sqrt{2}} \\ \frac{11}{2\sqrt{3}} & -\frac{5}{3\sqrt{2}} & \frac{37}{6} \\ \frac{9}{2} & \frac{11}{\sqrt{6}} & \frac{11}{2\sqrt{3}} \end{pmatrix} \ .$$

(b) Following the procedure described in Application 24.1, find an approximation A_2 for A by only keeping the two largest singular values.

24.8 * Show that the formula (24.30) holds even if the matrix A cannot be diagonalized. (Hint: Use the Jordan normal form.)

24.9 *Pauli matrices*
(a) Show that the space $\mathcal{H}_2^0$ of Hermitian traceless matrices in $\mathcal{M}_{2,2}(\mathbb{C})$ is a vector space over $\mathbb{R}$.
(b) Show that $(\sigma_1, \sigma_2, \sigma_3)$ is a basis of $\mathcal{H}_2^0$, where σ_i are the Pauli matrices.
(c) Show that $[\sigma_i, \sigma_j] = 2i\epsilon_{ijk}\sigma_k$ and $\{\sigma_i, \sigma_j\} = 2\mathbb{1}_2\delta_{ij}$, where $\{A, B\} := AB + BA$ is the anti-commutator.
(d) For $\mathbf{a}, \mathbf{b} \in \mathbb{R}^3$, show that $a_i\sigma_i, b_i\sigma_i \in \mathcal{H}_2^0$ commute iff $\mathbf{a}$ and $\mathbf{b}$ are parallel.

24.10 *Geometric series for matrices[†]*
Assume that the matrix $A \in \mathcal{M}_{n,n}(\mathbb{R})$ can be diagonalized with eigenvalues $\lambda_1, \ldots, \lambda_n$.
(a) Show that

$$\sum_{k=0}^{\nu} A^k = (\mathbb{1}_n - A^{\nu+1})(\mathbb{1}_n - A)^{-1} .$$

(b) If $|\lambda_i| < 1$ for $i = 1, \ldots, n$, show that $\sum_{k=0}^{\infty} A^k = (\mathbb{1}_n - A)^{-1}$.
(c) A sequence of vectors $\mathbf{x}_\nu \in \mathbb{R}^2$, where $\nu \in \mathbb{N}$, is defined recursively by $\mathbf{x}_{\nu+1} = \mathbf{x}_0 + A\mathbf{x}_\nu$, where $A \in \mathcal{M}_{2,2}(\mathbb{R})$ can be diagonalized with eigenvalues $|\lambda_i| < 1$. Find a formula for $\mathbf{x}_\nu$ in terms of $\mathbf{x}_0$.
(d) Within the setting of part (c), work out $\mathbf{x}_\nu$ for $\nu \to \infty$ with $\mathbf{x}_0 = \mathbf{e}_1$ and

$$A = \frac{1}{4}\begin{pmatrix} 1 & 2 \\ 2 & 1 \end{pmatrix} .$$

24.11 (a) Work out the matrix exponential $\Lambda(\xi) := \exp(\xi T)$, where $\xi \in \mathbb{R}$ and

$$T = \begin{pmatrix} 0 & 1 \\ 1 & 0 \end{pmatrix} .$$

(b) Show that $\Lambda(\xi_1)\Lambda(\xi_2) = \Lambda(\xi_1 + \xi_2)$.
(c) Show that $\mathcal{L} := \{\Lambda(\xi) \mid \xi \in \mathbb{R}\}$ is an Abelian group.
(d) Show that $\Lambda^T \eta \Lambda = \eta$, where $\eta = \text{diag}(-1, 1)$.

24.12 Find the matrix exponential $\exp(i\alpha P)$ for a projector $P \in \mathcal{M}_{n,n}(\mathbb{C})$, where $\alpha \in \mathbb{R}$. Show that $\exp(i\alpha P)$ can be diagonalized and find its eigenvalues.

25

Bi-linear and sesqui-linear forms*

25.1 Basics definitions*

Summary 25.1 *Symmetric bi-linear and Hermitian sesqui-linear forms, collectively called linear forms, are closely related to real and Hermitian scalar product but they lack the positivity requirement. Linear forms have an associated quadratic form, which is the analogue of the norm (squared) associated to a scalar product. Relative to a basis, a linear form can be described in terms of a symmetric or Hermitian matrix A which transform as $A \mapsto P^{\dagger}AP$ under a change of basis. A linear form is positive iff its describing matrices are positive definite and it is non-degenerate iff its describing matrices are non-singular.*

25.1.1 Definition of bi-linear and sesqui-linear forms

Bi-linear and sesqui-linear forms are generalizations of scalar products which are important for a range of applications. Perhaps the most prominent example of a bi-linear form (which is not a scalar product) is the Minkowski product which plays a central role in the theory of special relativity.

For the definition of a symmetric bilinear or Hermitian sesqui-linear form we can copy the definition of a real or Hermitian scalar product, Def. 22.2, except that the positivity condition (S3) is omitted.

Definition 25.1 *A symmetric bilinear form (Hermitian sesqui-linear form) on a vector space V over $\mathbb{F} = \mathbb{R}$ (over $\mathbb{F} = \mathbb{C}$) is a map $(\cdot, \cdot) : V \times V \to \mathbb{F}$ which satisfies the following rules for all $\mathbf{v}, \mathbf{u}, \mathbf{w} \in V$ and all $\alpha, \beta \in \mathbb{F}$,*

(B1)	$(\mathbf{v}, \mathbf{w}) = (\mathbf{w}, \mathbf{v})$, *for a symmetric bi-linear form, $\mathbb{F} = \mathbb{R}$*	*(symmetry)*
	$(\mathbf{v}, \mathbf{w}) = \overline{(\mathbf{w}, \mathbf{v})}$, *for a Hermitian sesqui-linear form, $\mathbb{F} = \mathbb{C}$*	*(hermiticity)*
(B2)	$(\mathbf{v}, \alpha\mathbf{u} + \beta\mathbf{w}) = \alpha(\mathbf{v}, \mathbf{u}) + \beta(\mathbf{v}, \mathbf{w})$	*(linearity)*

In the real case, the symmetry condition (B1) together with linearity in the second argument implies bi-linearity of $(\cdot, \cdot)$, just as in the case of a real scalar product (see Eq. (22.1)) — hence the name symmetric bi-linear form. In the complex case, that is for Hermitian sesqui-linear forms, linearity in the second argument translates, via hermiticity, into

$$(\alpha\mathbf{v} + \beta\mathbf{u}, \mathbf{w}) = \bar{\alpha}(\mathbf{v}, \mathbf{w}) + \bar{\beta}(\mathbf{u}, \mathbf{w}) . \tag{25.1}$$

in analogy with Hermitian scalar products (see Eq. (22.2)). In the following, to avoid awkward case distinctions, we will take *linear form* to mean symmetric bi-linear form for a vector space over $\mathbb{R}$ and Hermitian sesqui-linear form for a vector space over $\mathbb{C}$. Calculations will usually be carried out for the complex case, on the understanding that the corresponding real version is obtained by omitting all complex conjugations. Clearly, such a linear form $(\cdot, \cdot)$ on V is a scalar product iff $(\mathbf{v}, \mathbf{v}) > 0$ for all non-zero $\mathbf{v} \in V$. In analogy with scalar products, we can define two vectors $\mathbf{v}, \mathbf{w} \in V$ as *orthogonal* if $(\mathbf{v}, \mathbf{w}) = 0$.

Definition 25.2 *A linear form $(\cdot, \cdot)$ on V is called non-degenerate if $(\mathbf{v}, \mathbf{w}) = 0$ for all $\mathbf{v} \in V$ implies that $\mathbf{w} = \mathbf{0}$. Otherwise is is called degenerate.*

In other words, degeneracy amounts to the existence of non-zero vectors which are orthogonal to the entire vector space. We can collect such vectors in the vector subspace

$$V_0 := \{\mathbf{u} \in V \,|\, (\mathbf{v}, \mathbf{u}) = 0 \ \forall \mathbf{v} \in V\} \subset V \,, \tag{25.2}$$

so that the linear form is degenerate iff $\dim_{\mathbb{F}}(V_0) > 0$.

All scalar products are non-degenerate linear forms, as follows easily from the positive condition of scalar products:

$$\langle \mathbf{v}, \mathbf{w} \rangle = 0 \quad \forall \mathbf{v} \in V \quad \Longrightarrow \quad \langle \mathbf{w}, \mathbf{w} \rangle = 0 \quad \overset{\text{Def. 22.1}}{\Longrightarrow} \quad \mathbf{w} = \mathbf{0} \,.$$

However, not every non-degenerate linear form is a scalar product, as we will see.

25.1.2   The associated quadratic form

The *quadratic form* $q : V \to \mathbb{F}$ associated to a linear form $(\cdot, \cdot)$ is defined by

$$q(\mathbf{v}) = (\mathbf{v}, \mathbf{v}) \tag{25.3}$$

and is the analogue of the norm (squared) of a scalar product. Note that the value of the quadratic form is real, even for Hermitian sesqui-linear forms, as a consequence of condition (B1) in Def. 25.1. The quadratic form uniquely determines the linear form much as a scalar product is determined by its associated norm. To see this, the argument from Problem 22.1 can be repeated with the replacements $\langle \cdot, \cdot \rangle \to (\cdot, \cdot)$ and $|\cdot|^2 \to q(\cdot)$.

25.1.3   Linear forms relative to a basis

On a finite-dimensional vector spaces V it is quite straightforward to get a systematic handle on linear forms $(\cdot, \cdot)$ if we express them relative to a basis $(\mathbf{v}_1, \ldots, \mathbf{v}_n)$. Consider two vectors $\mathbf{v}, \mathbf{w} \in V$ and their coordinate vectors $\mathbf{x} = (x_1, \ldots, x_n)^T \in \mathbb{F}^n$ and $\mathbf{y} = (y_1, \ldots, y_n)^T \in \mathbb{F}^n$, so that $\mathbf{v} = \sum_i x_i \mathbf{v}_i$ and $\mathbf{w} = \sum_j y_j \mathbf{v}_j$. Then we can write the linear form as

$$(\mathbf{v}, \mathbf{w}) = \sum_{i,j} \bar{x}_i A_{ij} y_j = \mathbf{x}^\dagger A \mathbf{y} \quad \text{where} \quad A_{ij} = (\mathbf{v}_i, \mathbf{v}_j) \,. \tag{25.4}$$

The $n \times n$ matrix A is Hermitian (or symmetric in the real case) since

$$(A^\dagger)_{ij} = \bar{A}_{ji} \overset{(25.4)}{=} \overline{(\mathbf{v}_j, \mathbf{v}_i)} \overset{(B1)}{=} (\mathbf{v}_i, \mathbf{v}_j) \overset{(25.4)}{=} A_{ij} \, .$$

This shows that, given a basis, every linear form can be expressed in terms of a Hermitian matrix, as in Eq. (25.4). Conversely, for a Hermitian matrix A, a linear form is defined by Eq. (25.4). We refer to the matrix A in Eq. (25.4) as the matrix which describes the linear form relative to the basis $(\mathbf{v}_1, \dots, \mathbf{v}_n)$. How does this matrix change under a basis transformation?

Proposition 25.1 *(Basis change for a linear form) Let $(\cdot, \cdot)$ be a linear form on a (real or complex) vector space V with bases $(\mathbf{v}_1, \dots, \mathbf{v}_n)$ and $(\mathbf{v}'_1, \dots, \mathbf{v}'_n)$. Then the matrices A and A' which describe the linear form relative to these bases are related by*

$$A' = P^\dagger A P \, , \tag{25.5}$$

where the entries P_{ij} of P are obtained from $\mathbf{v}'_j = \sum_i P_{ij} \mathbf{v}_i$.

Proof

$$A'_{ij} = (\mathbf{v}'_i, \mathbf{v}'_j) = \left(\sum_k P_{ki} \mathbf{v}_k, \sum_l P_{lj} \mathbf{v}_l \right) = \sum_{k,l} \bar{P}_{ki} P_{lj} A_{kl} = (P^\dagger A P)_{ij}$$

$\square$

Note that this is conceptually quite similar to the basis change of a matrix describing a linear map, which is governed by the formula $A' = P^{-1} A P$. However, the appearance of $P^\dagger$, as opposed to P^{-1}, in Eq. (25.5) is a crucial difference. The two transformation laws are identical if $P^{-1} = P^\dagger$, that is if the basis transformation P is a unitary matrix, but otherwise they differ and care has to be taken to use the correct formula.

25.1.4   Positive definiteness

A linear form is a scalar product iff it is positive and this property can be translated into a property of the matrix describing a linear form.

Definition 25.3 *A Hermitian $n \times n$ matrix $A \in \mathcal{M}_{n,n}(\mathbb{F})$ is called positive definite iff $\mathbf{v}^\dagger A \mathbf{v} > 0$ for all non-zero $\mathbf{v} \in \mathbb{F}^n$. It is called negative definite iff $-A$ is positive definite and indefinite otherwise.*

By comparing with Eq. (25.4) it is clear that a linear form is positive definite and, hence that it is a scalar product, iff it is described by positive definite matrices. Whether a Hermitian matrix is positive (negative) definite can be decided simply by looking at its eigenvalues.

Proposition 25.2 *Positive and negative definiteness are properties independent under the basis transformation (25.5). A Hermitian matrix A is positive (negative) definite iff all its eigenvalues are positive (negative).*

Proof It follows from Eq. (25.5) that

$$\mathbf{v}^\dagger A' \mathbf{v} = \mathbf{v}^\dagger P^\dagger A P \mathbf{v} = (P\mathbf{v})^\dagger A (P\mathbf{v}) .$$

Since P is invertible this implies A is positive (negative) definite iff A' is.

From the previous statement, we can decide positive (negative) definiteness in any basis. But we know that a Hermitian matrix A can be diagonalized by a unitary basis transformation P so that $\hat{A} = \mathrm{diag}(\lambda_1, \ldots, \lambda_n) = P^\dagger A P$ with eigenvalues λ_i. The matrix A is positive (negative) definite iff $\hat{A}$ is and, since

$$\mathbf{v}^\dagger \hat{A} \mathbf{v} = \sum_{i=1}^{n} \lambda_i |v_i|^2$$

this is the case iff all eigenvalues are positive (negative). □

Problem 25.1 *(Polar decomposition)*

Show that every invertible matrix $A \in \mathcal{M}_{n,n}(\mathbb{C})$ has a unique decomposition $A = PU$, where P is Hermitian positive definite and U is unitary. Specialize the statement to $n = 1$ and discuss.

Solution: To get an idea where to start is is useful to assume for a moment we have a decomposition $A = PU$ as stated and explore the implications. Since U is unitary, so $UU^\dagger = \mathbb{1}_n$, we have $AA^\dagger = PUU^\dagger P = P^2$. This indicates we should define P as the square root of $AA^\dagger$ and, from Exercise 25.4, we know this is indeed possible provided $AA^\dagger$ is positive definite. This is not hard to show.

Certainly $\mathbf{x}^\dagger AA^\dagger \mathbf{x} = (A^\dagger \mathbf{x})^\dagger (A^\dagger \mathbf{x}) \geq 0$ from the positivity of the standard Hermitian scalar product on $\mathbb{C}^n$. Moreover, if $\mathbf{x}^\dagger AA^\dagger \mathbf{x} = 0$ then $A^\dagger \mathbf{x} = \mathbf{0}$ but since $A^\dagger$ is invertible this implies $\mathbf{x} = \mathbf{0}$. So we can conclude that $\mathbf{x}^\dagger AA^\dagger \mathbf{x} > 0$ whenever $\mathbf{x} \neq \mathbf{0}$ and, hence, $AA^\dagger$ is indeed positive definite.

It is now clear, there is only one possibility to define P and U, namely

$$P = (AA^\dagger)^{1/2} , \qquad U = P^{-1}A . \tag{25.6}$$

From Exercise 25.4, the matrix P is Hermitian positive definite (and, hence, invertible) and the above definitions ensure that $A = PU$. What remains to be shown is that U, as defined in Eq. (25.6), is indeed unitary and this follows from

$$U^\dagger U = A^\dagger (PP^\dagger)^{-1} A = A^\dagger (AA^\dagger)^{-1} A = \mathbb{1}_4 .$$

Writing an invertible matrix as a product of a positive definite and a unitary matrix is also referred to as *polar decomposition*.

For $n = 1$ the polar decomposition says that every (non-zero) complex number $z \in \mathbb{C}$ can be written as $z = r\zeta$, where $r \in \mathbb{R}^{>0}$ and $\zeta \in \mathrm{U}(1)$, as we already know from Problem 4.2. The polar decomposition for matrices can be seen as a generalization of its counterpart for complex numbers.

25.1.5 Degeneracy

Non-degeneracy of a linear form can also be translated into a property of the describing matrix.

Proposition 25.3 *A linear form is non-degenerate iff it is described by non-singular matrices.*

Proof A basis transformation (25.5) does not affect whether the matrix is invertible or singular (as follows, for example, by taking the determinant of Eq. (25.5)) so all matrices describing a linear form are either invertible or singular. It is, therefore, sufficient to look at one such matrix. Start with any matrix A describing the linear form. We can carry out a basis transformation with a unitary matrix U so that $U^\dagger A U = \hat{A} = \mathrm{diag}(\lambda_1, \ldots, \lambda_n)$, with real eigenvalues λ_i, is the matrix describing the linear form relative to a basis $(\mathbf{v}_1, \ldots \mathbf{v}_n)$. In this new basis, the linear form can be written as

$$(\mathbf{v}, \mathbf{w}) = \sum_{i=1}^{n} \lambda_i x_i y_i \,, \tag{25.7}$$

where $\mathbf{x} = (x_1, \ldots, x_n)^T$ and $\mathbf{y} = (y_1, \ldots, y_n)^T$ are the coordinates of $\mathbf{v}$ and $\mathbf{w}$, respectively. If the matrix $\hat{A}$ is singular it has at least one zero diagonal entry λ_i. In this case, Eq. (25.7) implies that $(\mathbf{v}, \mathbf{v}_i) = \lambda_i x_i = 0$ for all $\mathbf{v} \in V$, and, hence, the linear form is degenerate. On the other hand, if $\hat{A}$ is non-singular then all $\lambda_i \neq 0$. If $(\mathbf{v}, \mathbf{w}) = 0$ for all $\mathbf{v} \in V$ then, in particular $(\mathbf{v}_i, \mathbf{w}) = \lambda_i y_i = 0$, so that $y_i = 0$ for all $i = 1, \ldots, n$. Hence, $\mathbf{w} = \mathbf{0}$ and the linear form is non-degenerate. $\qquad\square$

25.2 Classification of linear forms*

Summary 25.2 *Every linear form on an n-dimensional vector space V can be described by a diagonal matrix with n_+ entries ± 1 and $n - n_+ - n_-$ entries 0 along the diagonal. The numbers (n_+, n_-) are called the signature of the linear form and the theorem of Sylvester shows they are indeed characteristic quantities of the linear form. A linear form is positive iff $n_+ = n$ and it is non-degenerate iff $n_+ + n_- = n$. The endomorphisms which leave a linear form invariant form a group which can be identified with the generalized orthogonal groups $\mathrm{O}(n_+, n_-)$ in the real case and with the generalized unitary groups $\mathrm{U}(n_+, n_-)$ in the complex case.*

25.2.1 A normal form for the describing matrix

We can attempt to classify linear forms by finding suitable bases, or, equivalently, coordinate transformations (25.5), such that their describing matrices take a simple form. Since the describing matrices are Hermitian we already know they have real eigenvalues and can always be diagonalized with a unitary basis transformation. This is, in fact, the key step underlying the proof of the following theorem.

Theorem 25.1 *For a linear form $(\cdot, \cdot)$ on V there exists a basis $(\mathbf{v}_1, \ldots, \mathbf{v}_n)$ so its describing matrix is $\eta = \mathrm{diag}(1, \ldots, 1, -1, \ldots, -1, 0, \ldots 0)$, with $n_\pm$ entries ± 1 along the diagonal and $n - n_+ - n_-$ zeros.*

Proof Let A be the Hermitian matrix which describes $(\cdot, \cdot)$ relative to an arbitrary basis of V. As a first step, we can diagonalize this matrix A by a unitary basis transformation U so that $U^\dagger A U = \hat{A} = \mathrm{diag}(\lambda_1, \dots, \lambda_k, 0, \dots, 0)$, where the possible zero eigenvalue has been separated off and all other eigenvalues λ_i are real, non-zero, and ordered such that $\lambda_i > 0$ for $i = 1, \dots, n_+$ and $\lambda_i < 0$ for $i = n_+ + 1, \dots, n_+ + n_-$. Then, define the matrix $P = UB$, where $B = \mathrm{diag}(|\lambda_1|^{-1/2}, \dots, |\lambda_k|^{-1/2}, 1, \dots, 1)$ and it follows that

$$P^\dagger A P = (UB)^\dagger A(UB) = BU^\dagger A U B = B\hat{A}B = \mathrm{diag}(\underbrace{1, \dots 1}_{n_+}, \underbrace{-1, \dots, -1}_{n_-}, 0, \dots, 0) = \eta \,.$$

$\square$

The proof shows that the integers $n_\pm$ in the above theorem count the number of positive and negative eigenvalues of a matrix which describes the linear form. In, summary, for any linear form $(\cdot, \cdot)$ on an n-dimensional vector space V we have a basis $(\mathbf{v}_1, \dots, \mathbf{v}_n)$ of V and a matrix

$$\eta = \mathrm{diag}(\underbrace{1, \dots, 1}_{n_+}, \underbrace{-1, \dots, -1}_{n_-}, \underbrace{0, \dots, 0}_{n - n_+ - n_-}) \,, \tag{25.8}$$

such that the linear form with arguments $\mathbf{v} = \sum_i x_i \mathbf{v}_i$ and $\mathbf{w} = \sum_j y_j \mathbf{v}_j$ can be written as

$$(\mathbf{v}, \mathbf{w}) = \mathbf{x}^\dagger \eta \mathbf{y} = \sum_{i,j=1}^{n} x_i \eta_{ij} y_j \,. \tag{25.9}$$

Evidently the vector space can be decomposed into vector subspaces as

$$V = V_+ \oplus V_- \oplus V_0 \quad \text{where} \quad \begin{cases} V_+ = \mathrm{Span}(\mathbf{v}_1, \dots, \mathbf{v}_{n_+}) \\ V_- = \mathrm{Span}(\mathbf{v}_{n_+ + 1}, \dots, \mathbf{v}_{n_+ + n_-}) \,, \\ V_0 = \mathrm{Span}(\mathbf{v}_{n_+ + n_- + 1}, \dots, \mathbf{v}_n) \end{cases} \tag{25.10}$$

where any non-zero vector $\mathbf{v} \in V_+$ ($\mathbf{v} \in V_-$) has a positive (negative) quadratic form, $q(\mathbf{u}) > 0$ ($q(\mathbf{u}) < 0$) and V_0 is, in fact, the subspace defined in Eq. (25.2). As we will see below, (n_+, n_-), with $n_\pm = \dim_{\mathbb{F}}(V_\pm)$, are characteristic numbers associated to the linear form which are referred to as the *signature* of the linear form.

25.2.2 Theorem of Sylvester

In the previous subsection we have derived a simple normal form which is characterized by the integers $n_\pm$ which count the number of positive and negative eigenvalues of a describing matrix A. However, as stands, these numbers could still depend on which describing A we choose as a starting point in the proof of Theorem 25.1. Of course, we do know that eigenvalues are preserved under a transformation $A \mapsto P^{-1}AP$ but the relevant transformation is $A \mapsto P^\dagger A P$ (see Eq. (25.5)). Unless P is unitary it is not clear if this transformation preserves the number of positive and negative eigenvalues. The point of the following theorem is to show that this is indeed the case and that $n_\pm$ are characteristic numbers of the linear form.

The decomposition in Eq. (25.10) suggests we should be looking at the sets

$$S_{\pm} = \{\mathbf{v} \in V \mid \pm q(\mathbf{v}) > 0\} \cup \{\mathbf{0}\}\,, \quad S_0 = \{\mathbf{v} \in V \mid q(\mathbf{v}) = 0\} \tag{25.11}$$

on which the quadratic form is positive, negative, or vanishing. Unfortunately, these sets are not vector subspaces since their definition involves a quadratic form (rather than a linear map) and inequalities. Clearly, the vector subspaces in the decomposition (25.10) are subsets, $V_{\pm} \subset S_{\pm}$ and $V_0 \subset S_0$, but they are by no means unique, as the following example shows.

Example 25.1 *(Minkowski product in $\mathbb{R}^n$)*

The Minkowski product on $\mathbb{R}^n$ is a symmetric bi-linear form defined as

$$(\mathbf{v}, \mathbf{w}) = \mathbf{v}^T \eta \mathbf{w} = \sum_{\mu,\nu=0}^{n-1} \eta_{\mu\nu} v^{\mu} w^{\nu} = -v^0 w^0 + \sum_{i=1}^{n-1} v^i w^i\,, \tag{25.12}$$

where $\mathbf{v}, \mathbf{w} \in \mathbb{R}^n$ and $\eta = \mathrm{diag}(-1, 1, \ldots, 1)$ is called the *Minkowski metric*. The Minkowski product on $\mathbb{R}^4$ plays a central role in the theory of special relativity. In this context, vectors $\mathbf{v} \in \mathbb{R}^4$ are also referred to as *four-vectors* and their zeroth components are interpreted as time and their other components as spatial coordinates. The Minkowski product is a symmetric bi-linear form and the describing matrix η has $n_+ = n - 1$ positive and $n_- = 1$ negative eigenvalues. Since, $(\mathbf{e}_1, \mathbf{e}_1) = -1$ it is clear that the Minkowski product is not positive and, hence, not a scalar product. However, η is a non-singular matrix so it follows from Prop. 25.3 that the Minkowski product is non-degenerate.

Splitting a vector as $\mathbf{v} = (v^0, \hat{\mathbf{v}})$ into time and spatial components, the sets in Eq. (25.11) are explicitly given by

$$\begin{array}{lll} S_+ = \{\mathbf{v} \in V \mid |\hat{\mathbf{v}}| < |v^0|\} \cup \{\mathbf{0}\} & \text{(\textbf{v} is time-like)} & \\ S_- = \{\mathbf{v} \in V \mid |\hat{\mathbf{v}}| > |v^0|\} \cup \{\mathbf{0}\} & \text{(\textbf{v} is space-like)} & . \\ S_0 = \{\mathbf{v} \in V \mid |\hat{\mathbf{v}}| = |v^0|\} & \text{(\textbf{v} is light-like or null)} & \end{array} \tag{25.13}$$

For $\mathbb{R}^2$ these sets are shown in Fig. 25.1. It is clear from the figure that the one-dimensional vector subspaces $V_+ = \mathrm{Span}(\mathbf{e}_1)$ and $V_- = \mathrm{Span}(\mathbf{e}_2)$ are contained in S_+ and S_-, respectively, but also that there are other one-dimensional subspaces with this property. In fact, any one-dimensional subspace $\mathrm{Span}(\mathbf{v})$ spanned by a time-like (space-like) vector $\mathbf{v}$ is contained in S_+ (S_-). The point is that the maximal vector subspaces contained in S_+ (S_-) all have the same dimension n_+ (n_-). This is, in fact, the statement of Sylvester's theorem which will be proven below.

In the context of special relativity, the set S_+ is referred to as the *light cone*, with the part of S_+ for $v^0 > 0$ called the future light cone and the $v^0 < 0$ part of S_+ the past light cone. $\qquad\square$

While the above example shows the vector subspaces $W_{\pm} \subset S_{\pm}$ of maximal dimension are not unique, the following theorem shows that their dimensions are.

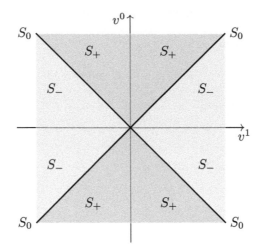

Fig. 25.1 The sets $S_\pm$ and S_0 in Eq. (25.13) for the Minkowski product in $\mathbb{R}^2$.

Theorem 25.2 *(Sylvester) Let* $(\cdot, \cdot)$ *be a linear form on V and* $V = W_+ \oplus W_- \oplus V_0 = \tilde{W}_+ \oplus \tilde{W}_- \oplus V_0$, *with* $W_\pm, \tilde{W}_\pm \subset S_\pm$ *and* $V_0 \subset S_0$ *as defined in Eq. (25.2). Then* $\dim_\mathbb{F}(W_\pm) = \dim_\mathbb{F}(\tilde{W}_\pm)$.

Proof   We assume that $\dim_\mathbb{F}(\tilde{W}_+) > \dim_\mathbb{F}(W_+)$ and show that this leads to a contradiction. In fact, from $\dim_\mathbb{F}(\tilde{W}_+) > \dim_\mathbb{F}(W_+) = \dim_\mathbb{F}(V) - \dim_\mathbb{F}(W_- \oplus V_0)$ it follows that $\dim_\mathbb{F}(\tilde{W}_+) + \dim_\mathbb{F}(W_- \oplus V_0) > \dim_\mathbb{F}(V)$ so that, by comparison with the dimension formula (8.2), the intersection $\tilde{W}_+ \cap (W_- \oplus V_0)$ must be non-trivial. Choose a non-zero $\mathbf{w} = \mathbf{w}_- + \mathbf{v}_0 \in \tilde{W}_+ \cap (W_- \oplus V_0)$, where $\mathbf{w}_- \in W_-$ and $\mathbf{v}_0 \in V_0$. Since $\mathbf{w} \in \tilde{W}_+$ it follows that $q(\mathbf{w}) > 0$ but, on the other hand, this is contradicted by

$$q(\mathbf{w}) = (\mathbf{w}_- + \mathbf{v}_0, \mathbf{w}_- + \mathbf{v}_0) = q(\mathbf{w}_-) \le 0 \,.$$

Hence, we conclude that $\dim_\mathbb{F}(\tilde{W}_+) \le \dim_\mathbb{F}(W_+)$. This argument can be repeated with the role of W_+ and $\tilde{W}_+$ reversed which leads to $\dim_\mathbb{F}(\tilde{W}_+) \ge \dim_\mathbb{F}(W_+)$ and, hence, $\dim_\mathbb{F}(\tilde{W}_+) = \dim_\mathbb{F}(W_+)$. In the same way, it can be shown that $\dim_\mathbb{F}(\tilde{W}_-) = \dim_\mathbb{F}(W_-)$ □

Hence, we can use any decomposition $V = W_+ \oplus W_- \oplus V_0$ with the properties stated in the theorem to define the dimensions $n_\pm := \dim_\mathbb{F}(W_\pm)$. The pair (n_+, n_-) is a characteristic of the linear form and its associated quadratic form and it is called the *signature*. Pulling together our various statements, we see that every linear form with signature (n_+, n_-) can be described by a matrix (25.8) and every matrix describing this linear form must have n_+ positive and n_- negative eigenvalues. The signature also provides straightforward criteria for positivity and non-degeneracy of a linear form.

Corollary 25.1 *For a linear form with signature* (n_+, n_-) *on an n-dimensional vector space we have the following statements.*

(i)   *The linear form is positive (it is a scalar product) iff $n_+ = n$.*

(ii)   *The linear form is non-degenerate iff $n_+ + n_- = n$.*

Proof (i) The linear form is positive iff any of its describing matrices A is positive definite and, from Prop. (25.2), this is the case iff all eigenvalues of A are positive. This in turn is equivalent to $n_+ = n$.

(ii) From Prop. (25.3) the linear form is non-degenerate iff a describing matrix A is non-singular. This is equivalent to all eigenvalues of A being non-zero and, hence, to $n_+ + n_- = n$. $\qquad\square$

25.2.3 Groups associated to linear forms

We have seen in Section 23.2 that the automorphisms which leave a scalar product invariant form a group, called the unitary group. For real and complex coordinate vector spaces this leads to the orthogonal and unitary matrix groups $O(n)$ and $U(n)$. This construction can be generalized to linear forms, as we will now explain.

Consider an n-dimensional vector space V with a non-degenerate linear form $(\cdot, \cdot)$ with signature (n_+, n_-). We say that an automorphism $f \in GL(V)$ leaves the linear form invariant if

$$(f(\mathbf{v}), f(\mathbf{w})) = (\mathbf{v}, \mathbf{w}) \quad \text{for all} \quad \mathbf{v}, \mathbf{w} \in V . \tag{25.14}$$

Clearly, these maps form a subgroup of $GL(V)$ which we can call $U(V, n_+, n_-)$. To get a more concrete description of this group we write $(\cdot, \cdot)$ in the standard form (25.9), relative to a suitable basis $(\mathbf{v}_1, \ldots, \mathbf{v}_n)$, so that the describing matrix $\eta_{ij} = (\mathbf{v}_i, \mathbf{v}_j)$ of the linear form is given by Eq. (25.8). We also introduce the matrix A which describes f relative to this basis, so that $f(\mathbf{v}_j) = \sum_i A_{ij}\mathbf{v}_i$. A map f leaves the linear form invariant iff

$$\eta_{ij} = (\mathbf{v}_i, \mathbf{v}_j) = (f(\mathbf{v}_i), f(\mathbf{v}_j)) = \left(\sum_k A_{ki}\mathbf{v}_k, \sum_l A_{lj}\mathbf{v}_l \right) = \sum_{k,l} \bar{A}_{ki}A_{jl}\eta_{kl} = (A^\dagger \eta A)_{ij} .$$

Hence, the group in question can be identified with the matrices A satisfying $A^\dagger \eta A = \eta$. Note that this is an obvious generalization of unitary groups which corresponds to the case $(n_+, n_-) = (n, 0)$ so that $\eta = \mathbb{1}_n$. In the real case (when the Hermitian conjugate becomes a transpose), these groups are denoted by

$$O(n_+, n_-) = \{A \in GL(\mathbb{R}^n) \,|\, A^T \eta A = \eta\} , \tag{25.15}$$

and are called *generalized orthogonal groups*. Matrices $A \in O(n_+, n_-)$ satisfy $\det(A) \in \{\pm 1\}$ from the multiplication theorem for determinants and we can define the subgroups

$$SO(n_+, n_-) = \{A \in O(n_+, n_-) \,|\, \det(A) = 1\} . \tag{25.16}$$

Proceeding analogously for the complex case, we can define the *generalized unitary groups*

$$U(n_+, n_-) = \{A \in GL(\mathbb{C}^n) \,|\, A^\dagger \eta A = \eta\} , \tag{25.17}$$

whose elements have unit modulus determinant, $|\det(U)| = 1$. Its subgroups of determinant one elements are denoted by

$$SU(n_+, n_-) = \{A \in U(n_+, n_-) \,|\, \det(A) = 1\} . \tag{25.18}$$

Problem 25.2 *(Lorentz group in two dimensions)*

Consider the Minkowski product on $\mathbb{R}^2$, defined by $(\mathbf{v}, \mathbf{w}) = \mathbf{v}^T \eta \mathbf{w}$, where $\eta = \mathrm{diag}(-1, 1)$. Show that the matrices $\tilde{\Lambda} \in \mathrm{SO}(1, 1)$ with $\tilde{\Lambda}_{11} \geq 0$ can be written in the form

$$\tilde{\Lambda}(\xi) = \begin{pmatrix} \cosh \xi & -\sinh \xi \\ -\sinh \xi & \cosh \xi \end{pmatrix}, \tag{25.19}$$

where $\xi \in \mathbb{R}$. In special relativity, the quantity ξ is also called *rapidity*. Show that multiplication of these matrices corresponds to addition of rapidities, that is, $\tilde{\Lambda}(\xi_1)\tilde{\Lambda}(\xi_2) = \tilde{\Lambda}(\xi_1 + \xi_2)$. Write $\tilde{\Lambda}(\xi)$ in terms of the quantity $\beta = \tanh \xi \in [-1, 1]$. How does addition of rapidity translate to the quantity β?

Solution: The story is analogous to the one for two-dimensional rotation, as developed in Exercise 23.4, but with trigonometric functions replaced by hyperbolic functions. We begin with a general 2×2 matrix

$$\tilde{\Lambda} = \begin{pmatrix} a & b \\ c & d \end{pmatrix},$$

where $a, b, c, d \in \mathbb{R}$, and insert this into the defining conditions, $\tilde{\Lambda}^T \eta \tilde{\Lambda} = \eta$ and $\det(\tilde{\Lambda}) = 1$, for $\mathrm{SO}(1, 1)$. This leads to

$$a^2 - c^2 = 1, \quad d^2 - b^2 = 1, \quad ab - cd = 0, \quad ad - cb = 1.$$

Since we assume that $\tilde{\Lambda}_{11} = a \geq 0$, the first of these equations implies that $a \geq 1$, so there must exist a $\xi \in \mathbb{R}$ such that $a = \cosh \xi$. After a suitable sign choice of ξ, we can then set $c = -\sinh \xi$. Solving the third equation for b, $b = cd/a$, and inserting into the fourth gives $d = a$ and using this in the third equation implies $b = c$. Hence, $\tilde{\Lambda}(\xi)$ has the stated form (25.19).

The relation $\tilde{\Lambda}(\xi_1)\tilde{\Lambda}(\xi_2) = \tilde{\Lambda}(\xi_1 + \xi_2)$ follows by carrying out the matrix multiplication on the left-hand side and then using the addition theorems for hyperbolic functions. With $\cosh \xi = \frac{1}{\sqrt{1-\beta^2}} =: \gamma$ and $\sinh \xi = \beta\gamma$ the matrices can be written in the form

$$\tilde{\Lambda}(\beta) = \begin{pmatrix} \gamma & -\beta\gamma \\ -\beta\gamma & \gamma \end{pmatrix}. \tag{25.20}$$

This is the familiar form for Lorentz transformations involving time and one spatial coordinate between two inertial systems with a relative velocity β (in units of the speed of light).

For two such transformations performed one after the other rapidities add up. The corresponding velocities $\beta_1 = \tanh \xi_1$, $\beta_2 = \tanh \xi_2$ and $\beta = \tanh(\xi_1 + \xi_2)$ are then related by

$$\beta = \frac{\beta_1 + \beta_2}{1 + \beta_1\beta_2}, \tag{25.21}$$

as follows immediately from the addition theorem $\tanh(\xi_1 + \xi_2) = \frac{\tanh \xi_1 + \tanh \xi_2}{1 + \tanh \xi_1 \tanh \xi_2}$. Eq. (25.21) is the standard formula for the addition of velocities in special relativity.

Application 25.1 *Lorentz group in four dimensions*

In Exercise 25.2 we have seen that the Lorentz group in two dimensions can rather easily be worked out explicitly, and that its structure is analogous to the group SO(2) of two-dimensional rotations. The physically relevant group is, of course, the Lorentz group in four dimensions. To define this group we introduce on $\mathbb{R}^4$ the Minkowski product

$$\langle \mathbf{v}, \mathbf{w} \rangle = \mathbf{v}^T \eta \mathbf{w} \,, \tag{25.22}$$

where $\mathbf{v}, \mathbf{w} \in \mathbb{R}^4$ and $\eta = \mathrm{diag}(-1, 1, 1, 1)$ is the *Minkowski metric*. The Minkowski product is a symmetric bi-linear form with signature $(n_+, n_-) = (3, 1)$. From Cor. 25.1 this means the Minkowski product is non-positive (it is not a scalar product) and non-degenerate. The group $\mathrm{O}(3, 1)$ which leaves the Minkowski product unchanged is called the *Lorentz group* and, from Eq. (25.15), it is explicitly given by

$$\mathcal{L} := \mathrm{O}(3,1) = \{\Lambda \in \mathrm{GL}(\mathbb{R}^4) \,|\, \Lambda^T \eta \Lambda = \eta\} = \{\Lambda \in \mathrm{GL}(\mathbb{R}^4) \,|\, \Lambda^\mu{}_\rho \Lambda^\nu{}_\sigma \eta_{\mu\nu} = \eta_{\rho\sigma}\} \,. \tag{25.23}$$

In Special Relativity the linear transformations

$$\mathbf{x} \mapsto \mathbf{x}' = \Lambda \mathbf{x} \tag{25.24}$$

generated by $\Lambda \in \mathcal{L}$ are interpreted as a transformations from one inertial system with space-time coordinates $\mathbf{x} = (t, x, y, z)^T$ to another one with space-time coordinates $\mathbf{x}' = (t', x', y', z')^T$.

The Lorentz group has an interesting global structure which can be seen as follows. First, taking the determinant of the defining equation, $\Lambda^T \eta \Lambda = \eta$ and using standard determinant properties, gives $\det(\Lambda)^2 = 1$, that is,

$$\det(\Lambda) = 1 \quad \text{or} \quad \det(\Lambda) = -1 \,. \tag{25.25}$$

Further, the $\rho = \sigma = 0$ component of the last Eq. (25.23) reads

$$-(\Lambda^0{}_0)^2 + \sum_{i=1}^3 (\Lambda^i{}_0)^2 = -1 \quad \Rightarrow \quad \Lambda^0{}_0 \geq 1 \quad \text{or} \quad \Lambda^0{}_0 \leq -1 \,. \tag{25.26}$$

Combining the sign choices in Eqs. 25.23 and (25.26) means the Lorentz group splits into four part, as summarized in the table below.

$\det(\Lambda)$	Λ_{00}	subset of $\mathcal{L}$	contains	given by
1	≥ 1	$\mathcal{L}_+^\uparrow$	$\mathbb{1}_4$	-
1	≤ -1	$\mathcal{L}_+^\downarrow$	$-\mathbb{1}_4$	$\mathcal{L}_+^\downarrow = -\mathcal{L}_+^\uparrow$
-1	≥ 1	$\mathcal{L}_-^\uparrow$	$P := \mathrm{diag}(1, -1, -1, -1)$	$\mathcal{L}_-^\uparrow = P\mathcal{L}_+^\uparrow$
-1	≤ -1	$\mathcal{L}_-^\downarrow$	$T := \mathrm{diag}(-1, 1, 1, 1)$	$\mathcal{L}_-^\downarrow = T\mathcal{L}_+^\uparrow$

The matrix P inverts the spatial coordinates under the transformation (25.24) and is called *parity* while T inverts time and is called *time inversion*. The matrices $(\mathbb{1}_4, -\mathbb{1}_4, P, T)$ are indeed Lorentz transformation (as they each satisfy $\Lambda^T \eta \Lambda = \eta$) and they realize the four possible choices of signs for $\det(\Lambda)$ and Λ_{00}. This shows that the four parts of the Lorentz group are indeed non-empty sets. What is more, the part $\mathcal{L}_+^\uparrow$, is a sub-group of the Lorentz group called the *proper, ortho-chronous Lorentz group*. It generates the other three parts via multiplication with $(-\mathbb{1}_4, P, T)$, as indicated in the last column of the table. If we think of a continuous path of Lorentz transformations in $\mathcal{L}$ then neither the sign of $\det(\Lambda)$ nor the sign of Λ_{00} can change from $+1$ to -1 or vice versa. This means that the four parts of the Lorentz group are disconnected and, schematically, the structure of the Lorentz group can be depicted as follows.

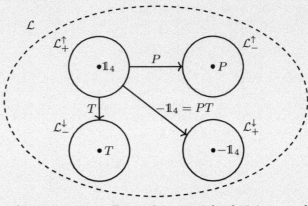

The Lorentz transformations normally used in special relativity are the proper, ortho-chronous Lorentz transformations $\mathcal{L}_+^\uparrow$. However, the other parts of the Lorentz group are relevant as well and it is an important question whether they constitute symmetries of nature in the same way that proper, ortho-chronous Lorentz transformations do. More to the point, the question is whether nature respects parity P and time-reversal T. One of the important and surprising discoveries of 20th century physics is that the weak interactions violate parity (Halzen and Martin, 2008).

As we have seen the entire Lorentz group $\mathcal{L}$ can be generated from $\mathcal{L}_+^\uparrow$ via multiplication with simple matrices. But what do proper ortho-chronous Lorentz transformation in $\mathcal{L}_+^\uparrow$ look like? To answer this question we basically have to solve Eq. (25.23) which is possible but somewhat difficult to do in full generality. However, some special Lorentz transformations are more easily obtained. First, we note that matrices of the type

$$\Lambda = \begin{pmatrix} 1 & 0 \\ 0 & R \end{pmatrix} \tag{25.27}$$

where R is a three-dimensional rotation matrix are elements of $\mathcal{L}_+^\uparrow$. Indeed, such matrices satisfy the defining relation for Lorentz transformations in Eq. (25.23) by virtue of $R^T R = \mathbb{1}_3$ and we also have $\det(\Lambda) = \det(R) = 1$ and $\Lambda^0{}_0 = 1$. In other words, regular three-dimensional rotations in the spatial directions are proper, ortho-chronous Lorentz transformations.

Another easy subset is constructed from the two-dimensional Lorentz transformation $\tilde{\Lambda}$ in Eq. (25.20), by writing down the block matrices

$$\Lambda = \begin{pmatrix} \tilde{\Lambda} & 0 \\ 0 & \mathbb{1}_2 \end{pmatrix} . \tag{25.28}$$

For these matrices, the transformations (25.24) become explicitly

$$t' = \gamma(t - \beta x) , \quad x' = \gamma(x - \beta t) , \quad y' = y , \quad z' = z . \tag{25.29}$$

These equations describe a transformation between two inertial systems which move relative to one another with velocity β in the x-direction. This is also called a boost in the x-direction with velocity β. A general boost depends on the velocity $\beta \in \mathbb{R}^3$ and is, hence, described by three parameters. (Its general form can, for example, be found in Jackson 1962). Every Lorentz transformation in $\mathcal{L}_{+}^{\uparrow}$ can be written in terms of a rotation and a general boost and can, hence, be parametrized by six quantities, the three angles describing a rotation and the velocity β.

25.3 Quadratic hyper-surfaces*

Summary 25.3 *A quadratic form q in $\mathbb{R}^n$ defines a quadratic hyper-surface via the equation $q(\mathbf{x}) = \text{const}$. The nature of this hyper-surface can be determined by diagonalizing the quadratic form and depends on the signature (n_+, n_-) of q. The quadratic curves in $\mathbb{R}^2$ are ellipses, hyperbolas and various degenerations of these. The quadratic surfaces in $\mathbb{R}^3$ consist of ellipsoids, hyperboloids, cylinders with elliptical or hyperbolic cross sections as well as various degenerations.*

25.3.1 Definition of quadratic hyper-surfaces

Linear forms and their associated quadratic forms lead to an important geometrical application, as they can be used to define *quadratic hyper-surfaces*. To see how this works, we start with a linear form $(\cdot, \cdot)$ with signature (n_+, n_-) on $\mathbb{R}^n$ and its associated quadratic form q. Following our earlier discussion (see Eq. (25.4)), we can write the linear and quadratic form, relative to the standard unit vector basis of $\mathbb{R}^n$, as

$$(\mathbf{x}, \mathbf{y}) = \sum_{i,j} x_i A_{ij} y_j = \mathbf{x}^T A \mathbf{y} \, , \qquad q(\mathbf{x}) = \sum_{i,j} x_i A_{ij} x_j = \mathbf{x}^T A \mathbf{x} \, , \qquad (25.30)$$

where A is a symmetric $n \times n$ matrix with entries $A_{ij} = (\mathbf{e}_i, \mathbf{e}_j)$.

A quadratic hyper-surface $S \subset \mathbb{R}^n$ is defined as the set of all vectors on which the quadratic form q has a fixed value k, so

$$S := \{\mathbf{x} \in \mathbb{R}^n \,|\, q(\mathbf{x}) = k\} = \{\mathbf{x} \in \mathbb{R}^n \,|\, \mathbf{x}^T A \mathbf{x} = k\} \, . \qquad (25.31)$$

This imposes one condition on the n components of the vector $\mathbf{x}$ so, intuitively, defines on object S of dimension $n-1$, that is, one less than the space it is embedded into – this is what the terminology 'hyper-surface' alludes to. We limit our discussion to values $k > 0$. A non-negative k can always be achieved by changing the matrix $A \mapsto -A$, if necessary, and we exclude $k = 0$ from our discussion since this leads to certain degenerate cases (see Exercise 25.7).

Problem 25.3 *(Quadratic forms and matrices)*

In $\mathbb{R}^3$ with coordinates $\mathbf{x} = (x_1, x_2, x_3)^T$ consider the quadratic form

$$q(\mathbf{x}) = 2x_1^2 + 4x_1 x_2 + x_2^2 - 6x_1 x_3 + 5x_3^2 \, .$$

Find the symmetric 3×3 matrix A such that $q(\mathbf{x}) = \mathbf{x}^T A \mathbf{x}$.

Solution: The matrix A can be easily read off from the coefficients in the quadratic form. The coefficients in front of the square terms x_i^2 become the diagonal entries A_{ii} while the off-diagonal entries A_{ij} and A_{ji} for $i < j$ are each given by half the coefficient of $x_i x_j$. Applying this to the above quadratic form gives

$$A = \begin{pmatrix} 2 & 2 & -3 \\ 2 & 1 & 0 \\ -3 & 0 & 5 \end{pmatrix}.$$

25.3.2 Diagonalization of quadratic hyper-surfaces

To understand the structure of quadratic surfaces better it is useful to diagonalize the quadratic form. As a symmetric matrix, A has an ortho-normal basis $R^T = (\epsilon_1, \ldots, \epsilon_n)$ of eigenvectors (relative to the dot product) and, as usual, we can diagonalize to $\hat{A} = \mathrm{diag}(\lambda_1, \ldots, \lambda_n) = RAR^T$, where $\lambda_i \in \mathbb{R}$ are the eigenvalues, for convenience ordered such that $\lambda_i > 0$ for $i = 1, \ldots, n_+$, $\lambda_i < 0$ for $i = n_+ + 1, \ldots, n_+ + n_-$ and $\lambda_i = 0$ for $i > n_+ + n_-$. Then, the defining equation of our quadratic hyper-surface from Eq. (25.31) can be re-written as

$$\mathbf{x}^T A \mathbf{x} = (R\mathbf{x})^T \hat{A} (R\mathbf{x}) = \mathbf{y}^T \hat{A} \mathbf{y} = \sum_i \lambda_i y_i^2 \overset{!}{=} k \,,$$

where $\mathbf{y} = R\mathbf{x}$ are the coordinates relative to the basis $(\epsilon_1, \ldots, \epsilon_n)$. Using slightly more suggestive notation, this means the defining equation for S in terms of the coordinates $\mathbf{y}$ and split into the parts with positive and negative eigenvalues has the form

$$\sum_{i=1}^{n_+} \frac{y_i^2}{R_i^2} - \sum_{i=n_++1}^{n_++n_-} \frac{y_i^2}{R_i^2} = 1 \quad \text{where} \quad R_i = \sqrt{\frac{k}{|\lambda_i|}} \,. \tag{25.32}$$

Evidently, the nature of the hyper-surface S depends on the signature (n_+, n_-) of the linear form in a crucial way.

If the signature is $(n_+, n_-) = (n, 0)$ then all eigenvalues are positive and Eq. (25.32) defines an *ellipsoid* whose half-axes have length R_i and point into the direction of the eigenvectors ϵ_i. If, in addition, all eigenvectors happen to be equal, $\lambda := \lambda_1 = \cdots = \lambda_n > 0$, then this ellipsoid degenerates into a sphere with radius $R = \sqrt{k/\lambda}$. If the signature is $(n_+, 0)$ with $n_+ < n$, then we have $n - n_+$ zero eigenvalues. In this case, the hyper-surface is an ellipsoid in the directions y_i with $i \leq n_+$ but, since Eq. (25.32) is independent of the coordinates y_i with $i > n_+$, it is cylindrical in those coordinates. On the other hand, for a signature (n_+, n_-) with $n_\pm > 0$ the quadratic hyper-surface is a *hyperboloid* in directions y_i with $1 \leq i \leq n_+ + n_-$ and it is cylindrical in the directions y_i with $i > n_+ + n_-$. For purely negative signatures, $(0, n_-)$, Eq. (25.32) does not have a solution since we are assuming that $k > 0$. If $k = 0$ is allowed various special cases arise but we will not discuss these in detail. To develop a better intuition it is useful to discuss the lowest-dimensional cases $n = 2$ and $n = 3$ in more detail.

Table 25.1 Quadratic curves in $\mathbb{R}^2$

(n_+, n_-)	Eigenvalues	Conditions on A	Curve	Equation
$(2,0)$	$\lambda_1 > 0, \lambda_2 > 0$	$\mathrm{tr}(A) > 0, \det(A) > 0$	ellipse	$\frac{y_1^2}{R_1^2} + \frac{y_2^2}{R_2^2} = 1$
	$\lambda_1 = \lambda_2 > 0$	$\mathrm{tr}(A) > 0,$ $4\det(A) = \mathrm{tr}(A)^2$	circle	$y_1^2 + y_2^2 = R_1^2$
$(1,1)$	$\lambda_1 > 0, \lambda_2 < 0$	$\det(A) < 0$	hyperbola	$\frac{y_1^2}{R_1^2} - \frac{y_2^2}{R_2^2} = 1$
$(1,0)$	$\lambda_1 > 0, \lambda_2 = 0$	$\mathrm{tr}(A) > 0, \det(A) = 0$	two lines	$y_1 = \pm R_1$

25.3.3 Quadratic curves in $\mathbb{R}^2$

A quadradic form q in $\mathbb{R}^2$ defines a hyper-surface of dimension one, that is, a quadratic curve. Explicitly, this can be written as

$$q(\mathbf{x}) = ax_1^2 + 2bx_1x_1 + cx_2^2 = \mathbf{x}^T A \mathbf{x} \overset{!}{=} k \quad \text{where} \quad A = \begin{pmatrix} a & b \\ b & c \end{pmatrix},$$

and $a, b, c \in \mathbb{R}$. In two dimensions, a good way to find the eigenvalues — and the signature — is to use the trace and determinant, that is, the relations

$$\lambda_1 + \lambda_2 = \mathrm{tr}(A) = a + c, \qquad \lambda_1 \lambda_2 = \det(A) = ac - b^2.$$

For example, the quadratic form q has signature $(2, 0)$ iff both eigenvalues of A are positive and this is the case iff $\mathrm{tr}(A) > 0$ and $\det(A) > 0$. For this signature, the quadratic curve is an ellipse described by the equation

$$\frac{y_1^2}{R_1^2} + \frac{y_2^2}{R_2^2} = 1 \quad \text{where} \quad R_i = \sqrt{\frac{k}{\lambda_i}}.$$

Recall that $\mathbf{y} = (y_1, y_2)^T$ are coordinates relative to the ortho-normal basis (ϵ_1, ϵ_2) of eigenvectors. This means that the half-axes of the ellipse point into the directions of the eigenvectors ϵ_i.

Going through the possible signatures systematically, leads to the four non-trivial cases listed in Table 25.1. The four types of curves in Table 25.1 are shown in Fig. 25.2.

Problem 25.4 *(Quadratic curves in $\mathbb{R}^2$)*

In $\mathbb{R}^2$ with coordinates $\mathbf{x} = (x_1, x_2)^T$ consider the quadratic form $q(\mathbf{x}) = 3x_1^2 + 2x_1x_2 + 2x_2^2$. Show that the equation $q(\mathbf{x}) = 1$ defines an ellipse and compute the directions and lengths of its half-axes.

Solution: The quadratic form can be written as $q(\mathbf{x}) = \mathbf{x}^T A \mathbf{x}$, with the matrix A given by

$$A = \begin{pmatrix} 3 & 1 \\ 1 & 2 \end{pmatrix}.$$

Since $\mathrm{tr}(A) = 5 > 0$ and $\det(A) = 5 > 0$ the case in the first row of Table 25.1 is realized. This means the signature is $(2, 0)$ and we are indeed dealing with an ellipse.

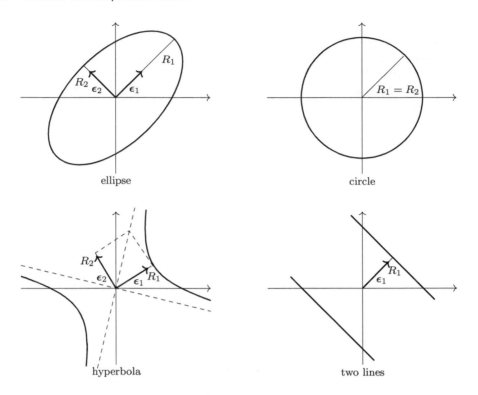

Fig. 25.2 The four types of quadratic curves in $\mathbb{R}^2$ from Table 25.1.

From the characteristic polynomial $\chi_A(\lambda) = \lambda^2 - 5\lambda + 5$ of A we find the eigenvalues $\lambda_\pm = (5 \pm \sqrt{5})/2$ and the (unnormalized) eigenvectors $\mathbf{v}_\pm = (1 \pm \sqrt{5}, 2)^T$. These are the directions of the ellipse's half-axes and their lengths are given by $R_\pm = 1/\sqrt{\lambda_\pm} = \sqrt{2/(5 \pm \sqrt{5})}$.

25.3.4   Quadratic surfaces in $\mathbb{R}^3$

A quadratic hyper-surface in $\mathbb{R}^3$ has dimension two, so it is a surface. Perhaps the simplest case is for signature $(3,0)$ which leads to the equation

$$\frac{y_1^2}{R_1^2} + \frac{y_2^2}{R_2^2} + \frac{y_3^2}{R_3^2} = 1 \quad \text{where} \quad R_i = \sqrt{\frac{k}{\lambda_i}},$$

written in coordinates $\mathbf{y} = (y_1, y_2, y_3)^T$ relative the ortho-normal basis $(\epsilon_1, \epsilon_2, \epsilon_3)$ of eigenvectors of A. This represent an ellipsoid with half-axes in the directions ϵ_i and lengths R_i. The other cases are listed in Table 25.2.

Problem 25.5 (*Quadratic surfaces in $\mathbb{R}^3$*)

For the quadratic form $q(\mathbf{x}) = x_1^2 + x_2^2 + x_3^2 - 2ax_1x_3$, determine the nature of the quadratic surface defined by $q(\mathbf{x}) = 1$, for all values of the parameter $a \in \mathbb{R}$.

Table 25.2 Quadratic surfaces in $\mathbb{R}^3$.

(n_+, n_-)	Eigenvalues	Surface	Equation
$(3,0)$	$\lambda_1 > 0, \lambda_2 > 0, \lambda_3 > 0$	ellipsoid	$\frac{y_1^2}{R_1^2} + \frac{y_2^2}{R_2^2} + \frac{y_3^2}{R_3^2} = 1$
	$\lambda_1 = \lambda_2 = \lambda_3 > 0$	sphere	$y_1^2 + y_2^2 + y_3^2 = R_1^2$
$(2,1)$	$\lambda_1 > 0, \lambda_2 > 0, \lambda_3 < 0$	hyperboloid, one sheet	$\frac{y_1^2}{R_1^2} + \frac{y_2^2}{R_2^2} - \frac{y_3^2}{R_3^2} = 1$
$(1,2)$	$\lambda_1 > 0, \lambda_2 < 0, \lambda_3 < 0$	hyperboloid, two sheets	$\frac{y_1^2}{R_1^2} - \frac{y_2^2}{R_2^2} - \frac{y_3^2}{R_3^2} = 1$
$(2,0)$	$\lambda_1 > 0, \lambda_2 > 0, \lambda_3 = 0$	elliptic cylinder	$\frac{y_1^2}{R_1^2} + \frac{y_2^2}{R_2^2} = 1$
$(1,1)$	$\lambda_1 > 0, \lambda_2 < 0, \lambda_3 = 0$	hyperbolic cylinder	$\frac{y_1^2}{R_1^2} - \frac{y_2^2}{R_2^2} = 1$
$(1,0)$	$\lambda_1 > 0, \lambda_2 = 0, \lambda_3 = 0$	two planes	$y_1^2 = R_1^2$

Solution: The above quadratic form can be written as $q(\mathbf{x}) = \mathbf{x}^T A \mathbf{x}$ with

$$A = \begin{pmatrix} 1 & 0 & -a \\ 0 & 1 & 0 \\ -a & 0 & 1 \end{pmatrix}.$$

The characteristic polynomial $\chi_A(\lambda) = -\lambda^3 + 3\lambda^2 + (a^2 - 3)\lambda + 1 - a^2$ leads to the eigenvalues $\lambda_1 = 1$, $\lambda_2 = 1 - a$ and $\lambda_3 = 1 + a$. For $-1 < a < 1$ all three eigenvalues are positive, so the signature of q is $(3, 0)$ and the surface is an ellipsoid. In particular, for $a = 0$ all three eigenvalues are equal and the ellipsoid degenerates to a sphere. For $|a| = 1$ two eigenvalues are positive and one vanishes so we have signature $(2, 0)$. Hence, the surface is a cylinder with elliptic cross section. Finally, for $|a| > 1$ two eigenvalues are positive and one is negative, so we have signature $(2, 1)$ and the surface is a one-sheeted hyperboloid.

Exercises

(†=challenging)

25.1 (a) For 2×2 Hermitian matrices A, formulate criteria for positive definiteness/negative definiteness/indefiniteness in terms of $\mathrm{tr}(A)$ and $\det(A)$.
(b) Are the following matrices positive definite, negative definite, or indefinite?

$$A_1 = \begin{pmatrix} 1 & -1 \\ -1 & 2 \end{pmatrix}, A_2 = \begin{pmatrix} 2 & 3 \\ 3 & -2 \end{pmatrix}$$
$$A_3 = \begin{pmatrix} -4 & 3 \\ 3 & -5 \end{pmatrix}, A_4 = \begin{pmatrix} 7 & 1 \\ 1 & -1 \end{pmatrix}$$

25.2 *Criterion for positive definiteness*

Let A be a Hermitian $n \times n$ matrix with characteristic polynomial $\chi_A(\lambda) = \sum_{k=0}^{n} c_k \lambda^k$. Show that A is positive definite iff $(-1)^k c_k > 0$ for $k = 0, \ldots, n - 1$. (Hint: Use the result from Exercise 4.17.)

25.3 *Another criterion for positive definiteness*†
For a symmetric matrix $A \in \mathcal{M}_{n,n}(\mathbb{R})$ denote by $A_{(k)}$ the $k \times k$ sub-matrices obtained by omitting the last $n-k$ rows and columns from A. Show that A is positive definite iff all minors $\det(A_{(k)})$ for $k = 1, \ldots, n$ are positive.

25.4 For a Hermitian positive definite matrix A show the following:
(a) A is invertible.
(b) There exists a sensible definition of the power A^x, where $x \in \mathbb{R}$, for which A^x is Hermitian positive definite as well.
(c) $A^x A^y = A^{x+y}$ for $x, y \in \mathbb{R}$.

25.5 *The group* $U(1,1)$
(a) Show that every $A \in U(1,1)$ can be written as $A = \zeta U$, where $U \in U(1,1)$ and $\zeta \in \mathbb{C}$ with $|\zeta| = 1$.
(b) Show that every $U \in U(1,1)$ can be written as

$$U = \begin{pmatrix} \alpha & \beta \\ \bar{\alpha} & \bar{\beta} \end{pmatrix},$$

where $\alpha, \beta \in \mathbb{C}$ and $|\alpha|^2 - |\beta|^2 = 1$.

25.6 A curve in $\mathbb{R}^2$ is defined by all vectors $\mathbf{x} = (x,y)^T$ which solve the equation

$x^2 + 3y^2 - 2xy = 1$.
(a) Show that this equation can be written in the form $\mathbf{x}^T A \mathbf{x} = 1$ and determine the matrix A.
(b) By diagonalizing the matrix A, show that this curve is an ellipse and determine the length of its two half-axes.

25.7 A curve in $\mathbb{R}^2$ is defined by all vectors $\mathbf{x}$ which satisfy $\mathbf{x}^T A \mathbf{x} = 0$ for a symmetric matrix $A \in \mathcal{M}_{2,2}(\mathbb{R})$. Discuss which types of curves arise depending on the signature of the associated linear form.

25.8 A linear form on $\mathbb{R}^3$ is defined by $(\mathbf{x}, \mathbf{y}) = \mathbf{x}^T A \mathbf{y}$, where A is the matrix from Exercise 24.1. What is the signature of this linear form and what type of surface is defined by $q(\mathbf{x}) = 1$, where q is the associated quadratic form? Find the length of half axes where appropriate.

Part VIII

Dual and tensor vector spaces*

In this final part we cover the more advanced topics of dual and tensor vector spaces. For the beginner, these topics often seem difficult and forbiddingly abstract. However, the idea of having universal constructions which allow us to create new vector spaces from given ones is a powerful one which has many applications in more advanced areas of mathematics and in science. We have already seen some examples of such constructions, namely the direct sum, $V \oplus W$ of two vector spaces (see Section 8.1.5) and the homomorphisms $\text{Hom}(V, W)$ of linear maps $V \to W$ (see Section 12.1.3).

We begin with a particularly important class of vector space homomorphisms, namely the *dual vector space* $V^* = \text{Hom}(V, \mathbb{F})$ whose elements consist of linear maps $V \to \mathbb{F}$, also called *linear functionals*. Vector spaces V and their duals V^* appear in many scientific applications but often implicitly or in disguise. For example, the presence of vectors with upper (=covariant) and lower (=contravariant) indices indicates that a vector space and its dual are in play. A prominent example is the theory of special relativity which uses upper and lower index objects. Vectors and dual vectors are also omnipresent in quantum mechanics where they are referred to as *ket vectors*, $|\cdot\rangle$, and *bra vectors*, $\langle\cdot|$, respectively. One reward for getting through the formal set-up for dual vector spaces will be a mathematical grounding for these various applications.

The dual vector space also underlies the definition of tensors. The tensor space $V \otimes W$ of two vector spaces V, W consists of all bi-linear forms $V^* \times W^* \to \mathbb{F}$. Relative to a choice of bases, such tensors can be represented by the two-index objects τ^{ia}. Tensoring can be repeated multiple times so we can, for example, consider the tensor space $V^{\otimes p} \otimes (V^*)^{\otimes q}$ which is built from p factors of V and q factors of V^*. Its elements are also referred to as (p, q) tensors and, relative to a choice of bases, they can be represented by objects $\tau^{i_1 \cdots i_p}{}_{j_1 \cdots j_q}$ with p upper and q lower indices. Of particular importance is the space $\Lambda^q V^*$ of completely anti-symmetric $(0, q)$ tensors, also called *alternating q forms*. The direct sum of the spaces $\Lambda^q V^*$ forms the *outer algebra* ΛV^* of V^*. Alternating q forms are important for some more advanced mathematical constructions, such as differential forms (see, for example, Lang 1997).

As we will see, many of the object in linear algebra can be described in terms of tensors. For example, we have an isomorphism $\text{Hom}(V, W) \cong W \otimes V^*$ which means that linear maps $V \to W$ can be described by tensors in $W \otimes V^*$. A symmetric bi-linear form on V can be identified with a tensor in $V^* \otimes V^*$. The determinant on $\mathbb{F}^n$ is, in fact, a tensor in $\Lambda^n \mathbb{F}^n$. We will also see that the various products, the dot product, the cross product and the triple product, introduced in a somewhat ad-hoc manner in Part III, have their natural home in the outer algebra $\Lambda \mathbb{R}^3$. By appealing to the index form of tensors we can also, finally, understand why objects such as the Levi-Civita or the Kronecker symbol are referred to as tensors. In short, the gain in understanding the structural aspects of linear algebra is well worth the effort of getting to grips with tensors.

In scientific applications, tensors are usually represented in index form. Conversely, the presence of such indexed objects usually indicates a mathematical formulation which involves tensors. A prominent example is the field-strength tensor $F_{\mu\nu}$ in covariant electromagnetism (see, for example, Jackson 1962) which can be viewed as a $(0, 2)$ tensor in $\Lambda^2 \mathbb{R}^4$. In the applications on machine learning (see Applications 11.1 and 13.3) we have explained the principle of neural networks in terms of vectors. More generally, neural networks can be seen as devices which process tensors.

We hope that all this provides the reader with sufficient motivation to carry on and tackle this final part.

26

The dual vector space*

As we have seen in Section 12.1.3, for two vector spaces V and W over $\mathbb{F}$, the space of homomorphisms $\mathrm{Hom}(V, W)$, which consists of all linear maps $V \to W$, is a vector space. A special case arises when we choose $W = \mathbb{F}$ (as a one-dimensional coordinate vector space over itself). The linear maps $V \to \mathbb{F}$ are called *linear functionals* and they form a vector space $V^* = \mathrm{Hom}(V, \mathbb{F})$, called the *dual vector space*, which is naturally associated to V.

It turns out that a vector space V and its dual V^* have the same dimension and are, hence, isomorphic. In general, this isomorphism depends on a choice of bases. However, a canonical isomorphism between V and V^* can be defined if V carries a non-degenerate linear form $(\cdot, \cdot)$. We will also see that transposition, introduced earlier as a somewhat ad-hoc operation on matrices, has a natural home in the context of dual vector spaces. In practice, relative to basis choices, vectors in V can be represented by objects v^i with an upper index, and dual vectors in V^* by objects v_i with a lower index. In this language, the canonical isomorphism between V and V^* is realized by lowering and raising indices, for example $v_i = g_{ij} v^j$, where g_{ij} is the matrix which describes the linear form.

26.1 Definition of dual vector space*

Summary 26.1 *For a vector space V over $\mathbb{F}$, linear functionals are linear maps $V \to \mathbb{F}$. The space of all linear functionals on V is called the dual vector space, $V^* = \mathrm{Hom}(V, \mathbb{F})$. For finite-dimensional vector spaces, V and V^* have the same dimension and for a basis of V there exists a unique dual basis of V^*. The double dual V^{**} is canonically isomorphic to V.*

26.1.1 Linear functionals

The formal definition of linear functionals and the dual vector space is as follows:

Definition 26.1 *For a vector space V over $\mathbb{F}$, a linear map $V \to \mathbb{F}$ is called a linear functional on V. The vector space $V^* := \mathrm{Hom}(V, \mathbb{F})$ of all linear functionals on V is called the dual vector space.*

Note that, in the above definition, the field $\mathbb{F}$ in $\mathrm{Hom}(V, \mathbb{F})$ is viewed as a one-dimensional vector space over itself. We emphasize that linear functionals $\varphi \in V^*$ are specific linear maps and, hence, satisfy the usual linearity property

$$\varphi(\alpha_1 \mathbf{v}_1 + \alpha_2 \mathbf{v}_2) = \alpha_1 \varphi(\mathbf{v}_1) + \alpha_2 \varphi(\mathbf{v}_2) \,, \tag{26.1}$$

for all vectors $\mathbf{v}_1, \mathbf{v}_2 \in V$ and all scalars $\alpha_1, \alpha_2 \in \mathbb{F}$. A wide variety of objects can be viewed as linear functionals, as the following examples show.

Example 26.1 *(Linear functionals on coordinate vector spaces)*

For the vector space $\mathbb{F}^n$ (seen as a vector space of columns) the dual vector space $(\mathbb{F}^n)^* = \mathrm{Hom}(\mathbb{F}^n, \mathbb{F}) = \mathcal{M}_{1,n}(\mathbb{F})$ consists of $1 \times n$ matrices, that is, of row vectors, with entries in $\mathbb{F}$ (see Eq. (13.12)). For such a row vector $\varphi = (\varphi_1, \ldots, \varphi_n)$ the action on a vector $\mathbf{v} \in V$ is given by matrix multiplication, so

$$\varphi(\mathbf{v}) = \varphi \mathbf{v} = \sum_{i=1}^{n} \varphi_i v_i \,. \tag{26.2}$$

Note that the matrix product of an $1 \times n$ matrix (a row vector) with an $n \times 1$ matrix (a column vector) is indeed a 1×1 matrix, so a number, as required. The expression on the right-hand side of Eq. (26.2) is, effectively, a dot product which we have now written as a matrix product.

To summarize, the elements of $\mathbb{F}^n$, as per our convention, can be viewed as column vectors while the elements of the dual, $(\mathbb{F}^n)^*$, can be viewed as row vectors, each with n entries in $\mathbb{F}$. In particular, $\mathbb{F}^n$ and $(\mathbb{F}^n)^*$ have the same dimension. $\qquad\square$

Example 26.2 *(Integrals as linear functionals)*

A prominent example of a linear functional on function vector spaces is the integral. For example, consider $\mathcal{C}([a, b])$, the space of continuous (real-valued) functions on the interval $[a, b] \subset \mathbb{R}$. The integral over $[a, b]$ defines a linear function $I : \mathcal{C}([a, b]) \to \mathbb{R}$ given by

$$I(g) = \int_a^b dx\, g(x) \,. \tag{26.3}$$

Linearity, $I(\alpha g + \beta h) = \alpha I(g) + \beta I(h)$, follows directly from linearity of the integral. $\square$

Example 26.3 *(Dirac delta)*

Another interesting functional on $\mathcal{C}([a, b])$ is the *Dirac delta functional* $\delta_c : \mathcal{C}([a, b]) \to \mathbb{R}$, defined by

$$\delta_c(g) := g(c) \,, \tag{26.4}$$

where $c \in [a, b]$ is fixed. The Dirac delta is indeed linear since

$$\delta_c(\alpha g + \beta h) = (\alpha g + \beta h)(c) = \alpha g(c) + \beta h(c) = \alpha \delta_c(g) + \beta \delta_c(h) \,.$$

As a mathematical object, δ_c is an example of a *distribution*. The mathematics of distributions is well beyond the scope of these lectures and cannot be pursued further (see, for example, Constantinescu 1980). However, the Dirac delta functional deserves a mention as it is widely used in applications. It is often (incorrectly) viewed as a 'function' $\delta(x - c)$ which vanishes everywhere except at $x = c$. In this language, Eq. (26.4) is written as

$$\int_a^b dx\, g(x)\delta(x - c) = g(c) \ . \tag{26.5}$$

An integral over a function which vanishes everywhere except at one point must be zero so Eq. (26.5) cannot be literally correct. It should be viewed as a symbolic equation which uses the integral to capture the linear nature of the Dirac delta functional and is written in place of the correct expression (26.4). □

26.1.2 Dual basis

In Example 26.1 we have seen that $\mathbb{F}^n$ and its dual, $(\mathbb{F}^n)^*$ have the same dimension. In fact, Eq. 12.7 shows that $\dim_{\mathbb{F}}(V) = \dim_{\mathbb{F}}(V^*)$ is generally true. The following theorem provides another proof of this fact which relies on constructing the dual basis.

Theorem 26.1 *Let V be a vector space with basis $(\epsilon_1, ..., \epsilon_n)$. Then there exists a unique basis $(\epsilon_*^1, ..., \epsilon_*^n)$ of V^*, called the dual basis, with*

$$\epsilon_*^i(\epsilon_j) = \delta_j^i \ , \tag{26.6}$$

for all $i, j - 1, \ldots, n$. In particular, $\dim_{\mathbb{F}}(V) = \dim_{\mathbb{F}}(V^*)$.

Proof Consider the coordinate map $\psi : \mathbb{F}^n \to V$ associated to the basis $(\epsilon_1, ..., \epsilon_n)$. We claim that the correct dual basis is defined by

$$\epsilon_*^i(\mathbf{v}) := v_i = \mathbf{e}_i \cdot \psi^{-1}(\mathbf{v}) \ , \tag{26.7}$$

where $\mathbf{v} = \sum_{i=1}^n v_i \epsilon_i$. First we check the duality property, Eq. (26.6).

$$\epsilon_*^i(\epsilon_j) = \mathbf{e_i} \cdot \psi^{-1}(\epsilon_j) \overset{(15.4)}{=} \mathbf{e_i} \cdot \mathbf{e}^j = \delta_i^j \ . \tag{26.8}$$

To verify that $(\epsilon_*^1, \ldots, \epsilon_*^n)$ forms a basis of V^* we first check linear independence. Applying the equation $\sum_i \beta_i \epsilon_*^i = 0$ to ϵ_j and using Eq. (26.8) immediately shows that all $\beta_j = 0$ and linear independence follows. To show that the ϵ_*^i span the dual space, we start with an arbitrary functional φ and define $\lambda_i = \varphi(\epsilon_i)$. It follows that

$$\varphi(\mathbf{v}) = \varphi\left(\sum_{i=1}^n v_i \epsilon_i\right) = \sum_{i=1}^n v_i \varphi(\epsilon_i) = \sum_{i=1}^n \lambda_i v_i = \sum_{i=1}^n \lambda_i \epsilon_*^i(\mathbf{v}) \ .$$

Dropping the argument $\mathbf{v}$ from either side shows that the arbitrary functional φ can indeed be written a a linear combination of the ϵ_*^i. Hence, $(\epsilon_*^1, \ldots, \epsilon_*^n)$ is a basis of V^*.

Finally, we need to proof uniqueness. Consider another basis $(\tilde{\epsilon}_*^1, \ldots, \tilde{\epsilon}_*^n)$ of V^* which satisfies $\tilde{\epsilon}_*^i(\epsilon_j) = \delta_j^i$. Equating this with Eq. (26.6) leads to $\tilde{\epsilon}_*^i(\epsilon_j) = \epsilon_*^i(\epsilon_j)$ for all $i, j = 1, \ldots, n$. Hence, $\tilde{\epsilon}_*^i$ and ϵ_*^i are equal since they coincide on a basis. □

A useful application of the dual basis is the calculation of the coordinates of a vector. For $\mathbf{v} \in V$ we have

$$\mathbf{v} = \sum_{i=1}^{n} v_i \epsilon_i \quad \Leftrightarrow \quad v_i = \epsilon_*^i(\mathbf{v}) \, , \tag{26.9}$$

as follows directly from the duality property (26.6).

Example 26.4 *(Dual basis for coordinate vector spaces)*

We start with $\mathbb{F}^n$ and its basis $(\mathbf{e}_1, \ldots, \mathbf{e}_n)$ of standard unit vectors. Its dual basis of $(\mathbb{F}^n)^*$ is given by $(\mathbf{e}_1^T, \ldots, \mathbf{e}_n^T)$, that is the standard unit vectors written as row vectors. Indeed, from Eq. (26.2) we have

$$\mathbf{e}_i^T(\mathbf{e}_j) = \mathbf{e}_i^T \mathbf{e}_j = \delta_{ij} \, ,$$

so that the defining property (26.6) of the dual basis is satisfied.

Next, consider an arbitrary basis $(\epsilon_1, \epsilon_2, \epsilon_3)$ of $\mathbb{R}^3$ and define the determinant $d := \det(\epsilon_1, \epsilon_2, \epsilon_3)$ which is non-zero from Theorem 10.1. We claim that the three vectors

$$\epsilon_*^1 := \frac{1}{d}(\epsilon_2 \times \epsilon_3)^T \, , \quad \epsilon_*^2 := \frac{1}{d}(\epsilon_3 \times \epsilon_1)^T \, , \quad \epsilon_*^3 := \frac{1}{d}(\epsilon_1 \times \epsilon_2)^T \tag{26.10}$$

form the dual basis. To see this just verify the defining property (26.6), for example

$$\epsilon_*^1(\epsilon_1) = \frac{1}{d}(\epsilon_2 \times \epsilon_3)^T \epsilon_1 = \frac{1}{d}(\epsilon_2 \times \epsilon_3) \cdot \epsilon_1 = 1 \, , \quad \epsilon_*^1(\epsilon_2) = \frac{1}{d}(\epsilon_2 \times \epsilon_3) \cdot \epsilon_2 = 0 \, ,$$

and similarly for the other combinations. Here, we have used standard properties of the triple product/the determinant, in particular the fact that a determinant with two same arguments vanishes (see Section 10.2). The dual basis (26.10) is, in fact, the transpose of the reciprocal basis introduced in Exercise 10.9. □

Problem 26.1 *(Computing the dual basis in $\mathbb{R}^3$)*

For the $\mathbb{R}^3$ basis $(\epsilon_1, \epsilon_2, \epsilon_3)$, where $\epsilon_1 = (1, 2, -1)^T$, $\epsilon_2 = (0, -1, 2)^T$ and $\epsilon_3 = (1, 0, 1)$, find the dual basis of $(\mathbb{R}^3)^*$. Use this result to compute the coordinates of the vector $\mathbf{v} = (2, 1, 3)^T \in \mathbb{R}^3$ relative to the basis $(\epsilon_1, \epsilon_2, \epsilon_3)$.

Solution: Inserting the vectors ϵ_i into Eq. (26.10), with $d = \det(\epsilon_1, \epsilon_2, \epsilon_3) = 2$, gives

$$\epsilon_*^1 = \frac{1}{2}(-1, 2, 1) \, , \quad \epsilon_*^2 = (-1, 1, 1) \, , \quad \epsilon_*^3 = \frac{1}{2}(3, -2, -1) \, .$$

As a check, it is useful to verify that $\epsilon_*^i(\epsilon_j) = \delta_j^i$.

To compute the coordinates v_i of the vector $\mathbf{v} = \sum_{i=1}^{3} v_i \epsilon_i$ we use Eq. (26.9) which gives

$$v_1 = \frac{1}{2}(-1, 2, 1) \begin{pmatrix} 2 \\ 1 \\ 3 \end{pmatrix} = \frac{3}{2} \, , \quad v_2 = (-1, 1, 1) \begin{pmatrix} 2 \\ 1 \\ 3 \end{pmatrix} = 2 \, , \quad v_3 = \frac{1}{2}(3, -2, -1) \begin{pmatrix} 2 \\ 1 \\ 3 \end{pmatrix} = \frac{1}{2} \, .$$

26.1.3 Index notation

Vectors and dual vectors are ubiquitous in scientific applications but they are rarely used in the abstract way in which they have been developed so far. To connect with applications, we have gradually slipped a notational refinement into our discussion, as the reader may have noticed. Basis vectors (ϵ_i) of V have been labelled by a lower index, while we have used an upper index for the basis vectors (ϵ_*^i) of the dual vector space V^*. Relative to these bases, elements of either vector space are written as

$$\mathbf{v} = v^i \epsilon_i \in V; , \qquad \varphi = \varphi_i \epsilon_*^i \in V^* . \tag{26.11}$$

The action of a linear functional on a vector then reads

$$\varphi(\mathbf{v}) = \sum_{i,j} \varphi_i v^j \underbrace{\epsilon_*^i(\epsilon_j)}_{=\delta_j^i} = \varphi_i v^i . \tag{26.12}$$

In writing these expressions, we have adopted a *refined Einstein summation convention* whereby a same upper (covariant) and lower (contravariant) index in a term is being summed over. In physics, such a summation of an upper and a lowed index is often referred to as a *contraction* of indices.

In applications, bases choices are usually implied (but rarely made explicit) and vectors and dual vectors are referred to by their coordinate vectors. In this case, the position of the index is used to distinguish the two types. Upper index objects v^i represent coordinate vectors relative to a basis (ϵ_i) of V and are also called *covariant vectors*. Lower index objects φ_i are the coordinates of dual vectors relative to the dual basis (ϵ_*^i) and are also called *contravariant vectors*. In this language, the action of a linear functional on a vector is simply written as on the right-hand side of Eq. (26.12), that is, as a contraction $\varphi_i v^i$. Whenever such upper and lowed index objects appear, implicit reference is being made to a vector space and its dual.

26.1.4 The double dual

What happens if we dualize the dual vector space, so if we form the *double dual* V^{**} of a vector space V? The duals of column vectors are row vectors obtained by transposition. Taking another transpose leads back to column vectors which suggests that the double dual is identical to the original vector space. The general statement is only slightly more complicated.

Theorem 26.2 *For a finite-dimensional vector space V, the linear map $\jmath : V \to V^{**}$ defined by $\jmath(\mathbf{v})(\varphi) := \varphi(\mathbf{v})$ is an isomorphism.*

Proof From Theorem 26.1 we know that $\dim(V) = \dim(V^*) = \dim(V^{**})$. Then, Corollary 14.2 tells us verifying $\mathrm{Ker}(\jmath) = \{\mathbf{0}\}$ is sufficient to show that $\jmath$ is an isomorphism. To do this we introduce a basis $(\epsilon_1, \ldots, \epsilon_n)$ on V with dual basis $(\epsilon_*^1, \ldots, \epsilon_*^n)$ and start with a vector $\mathbf{v} = v^i \epsilon_i \in \mathrm{Ker}(\jmath)$ in the kernel. It follows that $0 = \jmath(\mathbf{v})(\epsilon_*^i) = \epsilon_*^i(\mathbf{v}) = v_i$, so all coordinates v_i vanish and, hence, $\mathbf{v} = \mathbf{0}$. $\qquad\square$

Note that the definition of the above map $\jmath$ does not depend on a choice of basis. For this reason $\jmath$ is also referred to as a *canonical isomorphism* between V or V^{**}. In practice, we should think of V and V^{**} as the same space by identifying vectors $\mathbf{v} \in V$ with their images $\jmath(\mathbf{v}) \in V^{**}$. If we simplify our notation accordingly and write $\jmath(\mathbf{v})$ as $\mathbf{v}$, then the defining relation, $\jmath(\mathbf{v})(\varphi) = \varphi(\mathbf{v})$, for $\jmath$ turns into

$$\mathbf{v}(\varphi) = \varphi(\mathbf{v}) . \tag{26.13}$$

This means that the relation between V and V^* is 'symmetric'. Elements $\varphi \in V^*$ can act on vector $\mathbf{v} \in V$ but the converse is also possible and leads to the same scalar. This fact will become relevant for the discussion of tensors.

26.2 The dual map*

Summary 26.2 *For a vector subspace $U \subset V$ we define an orthogonal space $U^\perp \subset V^*$ which consists of all linear functionals which vanish on U. The dimensions of U and $U^\perp$ add up to the dimension of V. For a linear map $f : V \to W$ we can define a dual linear map $f^* : W^* \to V^*$. Relative to a choice of bases for V and W and their dual bases, the matrices which describe f and f^* are related by transposition. The linear map and its dual have the same rank, and their images and kernels are related.*

The two main structural features of vector spaces are vector subspaces and linear maps and it is natural to ask how they relate to dual vector spaces. We begin with vector subspaces.

26.2.1 The orthogonal space

For a vector subspace $W \subset V$ the *orthogonal space* $W^\perp$ associated to W is defined by

$$W^\perp = \{\varphi \in V^* \,|\, \varphi(\mathbf{w}) = 0 \ \forall \mathbf{w} \in W\} \subset V^* . \tag{26.14}$$

Linearity of the functionals φ means that $W^\perp$ is indeed a vector subspace. Earlier, we have introduced the concept of orthogonality based on a scalar product and have used this to define the orthogonal complement $W^\perp \subset V$ for a vector subspace $W \subset V$ (see Eq. (22.20)). Note that, despite the similar notation, the present notion is quite different. It does not rely on a scalar product and the orthogonal space $W^\perp$ in Eq. (26.14) is a subspace of V^*, rather than V.

Example 26.5 *(Orthogonal space for column vectors)*

For column vectors the interpretation of orthogonal spaces is quite straightforward. For example, consider a one-dimensional vector subspace $W = \mathrm{Span}(\mathbf{w}) \subset \mathbb{R}^2$. Linear functionals in $(\mathbb{R}^2)^*$ can be viewed as two dimensional row vectors, $\varphi = (\varphi_1, \varphi_2)$. The elements of $W^\perp$ are those for which $\varphi(\mathbf{w}) = 0$ and, hence, $W^\perp = \mathrm{Span}((\mathbf{w}^\times)^T)$, with the orthogonal $\mathbf{w}^\times$ of a two-dimensional vector defined in Eq. (10.1). So the orthogonal space to a line through $\mathbf{0}$ in $\mathbb{R}^2$ is indeed the orthogonal line through $\mathbf{0}$, but it is contained in the dual space $(\mathbb{R}^2)^*$. This is illustrated in Fig. 26.1.

For a one-dimensional sub vector space $W = \mathrm{Span}(\mathbf{w})$ in $\mathbb{R}^3$ there is an orthogonal

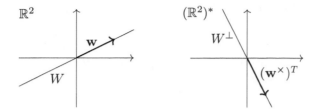

Fig. 26.1 The orthogonal space $W^\perp \subset (\mathbb{R}^2)^*$ to a sub space $W = \mathrm{Span}(\mathbf{w}) \subset \mathbb{R}^2$.

plane $U = \{\mathbf{v} \in \mathbb{R}^3 \mid \mathbf{v} \cdot \mathbf{w} = 0\}$, written down in Cartesian form. The orthogonal space is simple this same plane but written in term of row vectors, so $W^\perp = \{\mathbf{w}^T \mid \mathbf{w} \in U\} \subset (\mathbb{R}^3)^*$. $\qquad\square$

The above examples suggests that the dimensions of W and $W^\perp$ add up to the dimension of V. This is indeed the case.

Theorem 26.3 *For a finite-dimensional vector space V over $\mathbb{F}$ and a sub vector space $W \subset V$ we have*

$$\dim_{\mathbb{F}}(W) + \dim_{\mathbb{F}}(W^\perp) = \dim_{\mathbb{F}}(V) . \tag{26.15}$$

Proof We set $n = \dim_{\mathbb{F}}(V)$ and $m = \dim_{\mathbb{F}}(W)$ and choose a basis $(\epsilon_1, \ldots, \epsilon_m)$ for W which we complete to a basis $(\epsilon_1, \ldots, \epsilon_m, \epsilon_{m+1}, \ldots, \epsilon_n)$ for V. For a general functional $\varphi = \varphi_j \epsilon_*^j \in V^*$, written in terms of the dual basis $(\epsilon_*^1, \ldots, \epsilon_*^n)$, it follows

$$\varphi \in W^\perp \quad \Leftrightarrow \quad \varphi_i = \varphi(\epsilon_i) = 0 \text{ for } i = 1, \ldots, m .$$

As a result, $W^\perp = \mathrm{Span}(\epsilon_*^{m+1}, \ldots, \epsilon_*^n)$ and, being a subset of the dual basis, the vectors $\epsilon_*^{m+1}, \ldots, \epsilon_*^n$ are linearly independent and, hence, form a basis of $W^\perp$. This means $\dim_{\mathbb{F}}(W^\perp) = n - m$ which is what we wanted to show. $\qquad\square$

Another obvious guess from the above examples is that taking the orthogonal of an orthogonal space reverts to the original vector subspace.

Theorem 26.4 *For a finite-dimensional vector space V and a sub vector space $W \subset V$ we have $(W^\perp)^\perp = W$.*

Proof From the dimension formula (26.15) we conclude

$$\left. \begin{array}{l} \dim_{\mathbb{F}}(W) + \dim_{\mathbb{F}}(W^\perp) = \dim_{\mathbb{F}}(V) \\ \dim_{\mathbb{F}}((W^\perp)^\perp) + \dim_{\mathbb{F}}(W^\perp) = \dim_{\mathbb{F}}(V) \end{array} \right\} \quad \Rightarrow \quad \dim_{\mathbb{F}}(W) = \dim_{\mathbb{F}}((W^\perp)^\perp) .$$

Since the two spaces have the same dimension equality follows from Lemma 7.2 if we can show that $W \subset (W^\perp)^\perp$. A vector $\mathbf{w} \in W$ satisfies $\varphi(\mathbf{w}) = 0$ for all $\varphi \in W^\perp$. From Eq. (26.13) this means that $\mathbf{w}(\varphi) = 0$ for all $\varphi \in W^\perp$ so that $\mathbf{w} \in (W^\perp)^\perp$. $\square$

26.2.2   The dual map

Our next step is to discuss the relationship between dual vector spaces and linear maps. We start with a linear map $f : V \to W$, with V and W vector spaces over $\mathbb{F}$. Given this set-up, there is an obvious way to define a map $f^* : W^* \to V^*$, called the *dual map* of f, by the simple relation

$$f^*(\psi) := \psi \circ f \,, \tag{26.16}$$

where $\psi \in W^*$. Note that this makes sense: both sides of this definition are linear functionals in V^*, acting on vectors in V. Despite its simplicity, the definition (26.16) might seem painfully abstract. As often in linear algebra, we can get to a much more concrete picture if we introduce bases and work out the matrices associated to the linear maps. Our notation for the various basis sets is summarized in Table 26.1. Relative to these bases the linear map f and its dual map f^* are described by matrices

Table 26.1 Basis choices for V, W and their dual vector spaces.

Vector space	Dimension	Basis	Dual space	Dual basis
V	n	$(\mathbf{v}_1, \ldots, \mathbf{v}_n)$	V^*	$(\mathbf{v}_*^1, \ldots, \mathbf{v}_*^n)$
W	m	$(\mathbf{w}_1, \ldots, \mathbf{w}_m)$	W^*	$(\mathbf{w}_*^1, \ldots, \mathbf{w}_*^m)$

which we denote by A and B, respectively. From Theorem 15.1, these matrices satisfy

$$f(\mathbf{v}_j) = \sum_{i=1}^{m} A_{ij} \mathbf{w}_i \,, \qquad f^*(\mathbf{w}_*^j) = \sum_{i=1}^{n} B_{ij} \mathbf{v}_*^i \,. \tag{26.17}$$

To understand the relationship between A and B we simply evaluate the definition (26.16) with $\psi = \mathbf{w}_*^j$ and apply it to a basis vector $\mathbf{v}_k$.

$$f^*(\mathbf{w}_*^j)(\mathbf{v}_k) \overset{(26.16)}{=} w_*^j(f(\mathbf{v}_k)) \overset{(26.17)}{=} \mathbf{w}_*^j \left(\sum_{i=1}^{m} A_{ik} \mathbf{w}_i \right) = \sum_{i=1}^{m} A_{ik} \mathbf{w}_*^j(\mathbf{w}_i) \overset{(26.6)}{=} A_{jk}$$

$$\overset{(26.6)}{=} \sum_{i=1}^{n} A_{ji} \mathbf{v}_*^i(\mathbf{v}_k) = \sum_{i=1}^{n} (A^T)_{ij} \mathbf{v}_*^i(\mathbf{v}_k) \tag{26.18}$$

Eq. (26.18) holds on the basis $(\mathbf{v}_1, \ldots, \mathbf{v}_n)$ so we can drop the argument $\mathbf{v}_k$ on either side and conclude, by comparison with Eq. (26.17), that $B = A^T$. We summarize this result in the following theorem.

Theorem 26.5 *Let V and W be finite-dimensional vector space, $f : V \to W$ be a linear map and $f^* : W^* \to V^*$ its dual map, as defined in Eq. (26.16). If f is described by a matrix A relative to a choice of bases on V and W, then f^* is described by the transpose matrix A^T, relative to the dual choice of bases on W^* and V^*.*

Proof   This follows from the calculation (26.18). $\square$

In short, the dual of a map is simply the abstract version of the transpose of a matrix. This provides a practical interpretation of the abstract definition (26.16) but also a deeper understanding of matrix transposition, an operation which we have previously introduced in a somewhat ad-hoc manner[1].

26.2.3 Kernel and image of the dual map

Associated to each linear map are two vector subspaces, the kernel and image. The next theorem states an interesting relation between these spaces and their counterparts for the dual map.

Theorem 26.6 *For finite-dimensional vector spaces V and W, the kernel and image of the linear map $f : V \to W$ and its dual map $f^* : W^* \to V^*$ are related by*

$$(i) \ \mathrm{Ker}(f^*) = (\mathrm{Im}(f))^\perp , \qquad (ii) \ \mathrm{Im}(f^*) = (\mathrm{Ker}(f))^\perp . \qquad (26.19)$$

Proof (i) We use the definitions (26.16) and (26.14) of the dual map and the orthogonal space.

$$\psi \in \mathrm{Ker}(f^*) \overset{(26.16)}{\Leftrightarrow} f^*(\psi) = \psi \circ f = 0 \Leftrightarrow \psi(f(\mathbf{v})) = 0 \ \forall \mathbf{v} \in V \overset{(26.14)}{\Leftrightarrow} \psi \in (\mathrm{Im}(f))^\perp$$

(ii) This can be shown in a similar way and we leave it as an exercise. $\qquad \square$

In short, the orthogonal of the kernel and the image of f are the image and the kernel of the dual map f^*, respectively. Let us work this out for an example.

Problem 26.2 *(The dual map)*

Verify the relations (26.19) explicitly for the linear map $\mathbb{R}^3 \to \mathbb{R}^3$ specified by the matrix

$$A = \begin{pmatrix} -1 & 4 & 3 \\ 2 & -3 & -1 \\ 3 & 2 & 5 \end{pmatrix} .$$

Solution: The first two columns $\mathbf{A}^1$, $\mathbf{A}^2$, are linearly independent, while $\mathbf{A}^3 = \mathbf{A}^1 + \mathbf{A}^2$. This means that $\mathrm{Im}(A) = \mathrm{Span}(\mathbf{A}^1, \mathbf{A}^2)$, $\mathrm{rk}(A) = 2$ and, from the dimensional formula (14.13), $\dim_\mathbb{R}(\mathrm{Ker}(A)) = 3 - 2 = 1$. Since $A\mathbf{v} = \mathbf{0}$, where $\mathbf{v} = (1, 1, -1)^T$, it follows that $\mathrm{Ker}(A) = \mathrm{Span}(\mathbf{v})$.

The orthogonal spaces $\mathrm{Im}(A)^\perp$ and $\mathrm{Ker}(A)^\perp$ can be easily computed.

$$(\mathrm{Im}(A))^\perp = \mathrm{Span}(\psi) , \quad \psi = (\mathbf{A}^1 \times \mathbf{A}^2)^T = (13, 14, -5)$$
$$(\mathrm{Ker}(A))^\perp = \{\varphi \in \mathbb{R}^3 \,|\, \varphi\mathbf{v} = 0\} = \mathrm{Span}(\varphi_1, \varphi_2) , \quad \varphi_1 = (1, -1, 0) , \quad \varphi_2 = (1, 1, 2)$$

On the other hand, we can compute the kernel and image of the dual map, described by the transpose matrix

$$A^T = \begin{pmatrix} -1 & 2 & 3 \\ 4 & -3 & 2 \\ 3 & -1 & 5 \end{pmatrix} .$$

It is easy to see that $A^T \psi^T = \mathbf{0}$, so that $\mathrm{Ker}(A^T) = \mathrm{Span}(\psi) = (\mathrm{Im}(A))^\perp$. Further, since $\varphi_1^T = (5\mathbf{A}_2^T + \mathbf{A}_3^T)/13$ and $\varphi_2^T = (-\mathbf{A}_2^T + 5\mathbf{A}_3^t)/13$ the image of A^T is $\mathrm{Im}(A^T) = \mathrm{Span}(\varphi_1, \varphi_2) = (\mathrm{Ker}(A))^\perp$, all in accordance with Eqs (26.19).

---

[1]The scalar product has also provided us with a mathematical interpretation of transposition but only for vector spaces over $\mathbb{R}$. The dual map leads to an interpretation for vector spaces over an arbitrary field $\mathbb{F}$.

The following corollary is a simple but important conclusion from Theorem 26.6.

Corollary 26.1 *For finite-dimensional vector spaces V and W, a linear map $f : V \to W$ and its dual map $f^* : W^* \to V^*$ satisfy*

$$\mathrm{rk}(f^*) = \mathrm{rk}(f) \,. \tag{26.20}$$

Proof

$$\mathrm{rk}(f^*) \;=\; \dim_{\mathbb{F}}(\mathrm{Im}(f^*)) \overset{(26.19)}{=} \dim_{\mathbb{F}}(\mathrm{Ker}(f)^{\perp}) \overset{(26.15)}{=} \dim_{\mathbb{F}}(V) - \dim_{\mathbb{F}}(\mathrm{Ker}(f))$$
$$\overset{(14.13)}{=} \dim_{\mathbb{F}}(\mathrm{Im}(f)) \;=\; \mathrm{rk}(f) \,.$$

$\square$

A linear map and its dual map have the same rank! In particular, this implies that a matrix and its transpose have the same rank or, equivalently, that the row and column ranks of a matrix are always equal. This statement has already been proven, by more elementary methods, in Theorem 16.1.

26.3   Linear forms and dual space*

Summary 26.3 *A linear form $(\cdot, \cdot)$ on a vector space V allows us to define a map $V \to V^*$ between the vector space and its dual. This map is bijective iff the linear form is non-degenerate. In index notation, this map amounts to raising and lowering indices.*

26.3.1   The map between V and V^*

It is interesting to discuss how linear forms fit into the relation between vector spaces and their dual spaces. To this end, consider a vector space V with a linear form $(\cdot, \cdot)$, bi-linear symmetric in the real case and sesqui-linear Hermitian in the complex case. We can use this linear form to define a map $s : V \to V^*$ between the vector space and its dual by

$$s(\mathbf{v})(\mathbf{w}) = (\mathbf{v}, \mathbf{w}) \,, \tag{26.21}$$

where $\mathbf{v}, \mathbf{w} \in V$. Note that $s(\mathbf{v})$ is indeed a linear functional since the linear form $(\cdot, \cdot)$ is linear in its second argument. The map s inherits its properties from the properties of the linear form, with respect to its first argument. In the real case, $(\cdot, \cdot)$ is linear in its first argument, so that s is linear as well. However, in the complex case, the first argument of $(\cdot, \cdot)$ and s are semi-linear, that is,

$$s(\alpha \mathbf{v} + \beta \mathbf{w}) = \bar{\alpha}\, s(\mathbf{v}) + \bar{\beta}\, s(\mathbf{w}) \,, \tag{26.22}$$

for all $\mathbf{v}, \mathbf{w} \in V$ and all $\alpha, \beta \in \mathbb{C}$. The following proposition provides another useful link between the properties of s and the linear form.

Proposition 26.1 *The map $s : V \to V^*$, defined in Eq. (26.21), is bijective if and only if the linear form $(\cdot, \cdot)$ is non-degenerate.*

Proof The proof is simple in the real case when s is linear. Since $\dim(V) = \dim(V^*)$ we know that s is bijective iff its kernel is trivial. This is the same as saying that $s(\mathbf{v})(\mathbf{w}) = (\mathbf{v}, \mathbf{w}) = 0$ for all $\mathbf{w} \in V$ implies that $\mathbf{v} = 0$ which is indeed precisely the definition of non-degeneracy.

In the complex case, s is only semi-linear and we cannot immediately apply our various statements about linear maps. We leave the proof of this case as Exercise 26.4.

□

The proposition means that a non-degenerate linear form provides us with an identification of the vector space V and its dual V^*.

Application 26.1 *(Dirac notation)*

Dirac notation is a system of notation for inner product vector spaces which was invented for efficient calculation in quantum mechanics (see, for example, Dirac 2019; Sakurai and Napolitano 2017; Messiah 2014) but which can also be convenient in a purely mathematical context.

The setting is a vector space V with a Hermitian scalar product. Since a scalar product is a non-degenerate sesqui-linear form it defines a bijective map $s : V \to V^*$, as discussed in Section 26.3.1. From Eq. (26.21) we know that this map identifies the action of dual vector on vectors with the scalar product. *Dirac notation* makes this identification notationally manifest. By writing vectors $\mathbf{v} \in V$ as 'ket' vectors $|\mathbf{v}\rangle$ and dual vectors $s(\mathbf{w})$, obtained from vectors $\mathbf{w} \in V$ by the map s, as 'bra' vectors $\langle \mathbf{w}|$, the scalar product is written as a 'bra(c)ket':

$$\langle \mathbf{w} | \mathbf{v} \rangle := \langle \mathbf{w}, \mathbf{v} \rangle = s(\mathbf{w})(\mathbf{v}) \ . \tag{26.23}$$

In Dirac notation, matrix elements of Hermitian linear maps $h : V \to V$ are written as

$$\langle \mathbf{w} | h | \mathbf{v} \rangle := \langle \mathbf{w}, h(\mathbf{v}) \rangle = \langle h(\mathbf{w}), \mathbf{v} \rangle \ . \tag{26.24}$$

Note that having h sit symmetrically between the two arguments of the scalar product makes notational sense for a Hermitian map, since it does not matter which of the two arguments it acts on. Suppose we select on ortho-normal basis (ϵ_i) of V. These basis vectors are also written as $|i\rangle$ and their ortho-normality condition is

$$\langle i | j \rangle = \delta_{ij} \ . \tag{26.25}$$

The matrix elements of h relative to this basis are $h_{ij} = \langle i | h | j \rangle$ and the map itself can then be written as

$$h = \sum_{i,j} h_{ij} |i\rangle\langle j| \ . \tag{26.26}$$

This relation can be easily verified by computing the matrix elements of the right-hand side

$$\langle k | \sum_{i,j} h_{ij} |i\rangle\langle j|l\rangle = \sum_{i,j} h_{ij} \langle k|i\rangle\langle j|l\rangle \overset{(26.25)}{=} \sum_{i,j} h_{ij} \delta_{ki}\delta_{jl} = h_{kl} = \langle k|h|l\rangle$$

which are indeed identical to the matrix elements of h. In more mathematical terms, we can view Eq. (26.26) as an explicit realization of the isomorphism $\mathrm{End}(V) \cong V \otimes V^*$ defined in Eq. 27.22. The identity id_V has matrix elements δ_{ij} so Eq, (26.26) becomes

$$\mathrm{id}_V = \sum_i |i\rangle\langle i| \; . \tag{26.27}$$

This simple formula is an efficient tool for computation. For example, consider expanding a vector $|\psi\rangle \in V$ in terms of our ortho-normal basis.

$$|\psi\rangle = \mathrm{id}_V |\psi\rangle = \sum_i |i\rangle\langle i|\psi\rangle \tag{26.28}$$

This is of course the well-known result that the coordinates of $|\psi\rangle$ relative to the ortho-normal basis $(|i\rangle)$ are given by the scalar products $\langle i|\psi\rangle$ (see Eq. (22.15)) but here it follows simply by inserting the identity map in the form (26.27). Also note that the projector $P_i : V \to V$ which projects onto the one-dimensional space spanned by $|i\rangle$ can be written as

$$P_i = |i\rangle\langle i| \; . \tag{26.29}$$

Indeed, for any vector $|\psi\rangle \in V$ we have $P_i|\psi\rangle = |i\rangle\langle i|\psi\rangle$ which is the standard form (22.22) of an orthogonal projector onto a one-dimensional space.

Now suppose $|i\rangle$ is an ortho-normal basis of eigenvectors of h with eigenvalues λ_i. In this case, the basis vector are often labelled by the eigenvalues, so they are written as $|\lambda_i\rangle$ (Of course, this only makes sense if the eigenvalues are non-degenerate, or else an additional label is required.) The eigenvalue equation then takes the suggestive form

$$h|\lambda_i\rangle = \lambda_i|\lambda_i\rangle \; , \tag{26.30}$$

while Eq. (26.26) can be simplified to

$$h = \sum_i \lambda_i |\lambda_i\rangle\langle\lambda_i| = \sum_i \lambda_i P_{\lambda_i} \; , \tag{26.31}$$

where $P_{\lambda_i} = |\lambda_i\rangle\langle\lambda_i|$ is the projector onto the eigenspace spanned by $|\lambda_i\rangle$.

Application 26.2 *Quantum mechanics — a rough dictionary*

The close connection between mathematics and physics is particularly pronounced when a physical theory can be viewed as an instance or a special version of a mathematical discipline. The relationship between linear algebra and quantum mechanics is a case in point. Having learned linear algebra means having learned about many of the structural aspects of quantum mechanics. However, mathematicians' and physicists' language famously differs, so recognizing and using this relationship requires setting up a dictionary. A rough version of such a dictionary is given in the following table.

Mathematical structure	Interpretation in quantum mechanics
Inner product vector space V over $\mathbb{C}$	All states of a quantum system
A vector $|\psi\rangle \in V$	A specific quantum state of the system
A Hermitian map $h \in \mathrm{End}(V)$	A physical observable, such as space, position,...
Eigenvalues λ_i of h	Values that the observable can take
Ortho-normal eigenvectors $|\lambda_i\rangle$	States in which observable assume values λ_i
$p_i = |\langle\lambda_i|\psi\rangle|^2$, where $\langle\psi|\psi\rangle = 1$	Probability to measure λ_i in state $|\psi\rangle$
$\langle\psi|h|\psi\rangle$, where $\langle\psi|\psi\rangle = 1$	Expectation value for observable in state $|\psi\rangle$
$|\psi\rangle \mapsto P_i|\psi\rangle$, where $P_i = |\lambda_i\rangle\langle\lambda_i|$	Collapse of state after measurement

Let us discuss some of these correspondences in more detail. (For a proper introduction into quantum mechanics, see, for example, Dirac 2019; Sakurai and Napolitano 2017; Messiah 2014.) The mathematical area for quantum mechanics are vector spaces V over $\mathbb{C}$ with a Hermitian scalar product $\langle \cdot | \cdot \rangle$. The specific realization and dimension of the vector space V depends on the quantum system and its description. Some quantum systems are described by a finite-dimensional vector space, others require infinite-dimensional vector spaces, where methods of functional analysis, notably Hilbert spaces, become relevant (see, for example, Rynne and Youngson 2008). Naturally, within our context, we will focus on the finite-dimensional case, as illustrated by the spin system in Application 26.4.

Possible quantum states for a system are represented by vectors $|\psi\rangle \in V$. In order to extract measurements from such a quantum state we require observables, such as position, momentum, energy, angular momentum and so forth. In quantum mechanics, each of these observables is represented by a specific Hermitian map $h \in \mathrm{End}(V)$. A particularly important such map is the Hamilton operator $H \in \mathrm{End}(V)$ which corresponds to the energy of the system. The Hamilton operator also determines the time evolution $|\psi(t)\rangle$ of a quantum system via the time-dependent Schrödinger equation

$$H|\psi(t)\rangle = i\frac{d}{dt}|\psi(t)\rangle \ . \tag{26.32}$$

We recall from Section 24 that Hermitian maps h can be diagonalized and that they have an ortho-normal basis $(|\lambda_i\rangle)$ of eigenvectors with real eigenvalues λ_i. In quantum mechanics, the eigenvalues λ_i are the possible values which can be measured for the observable h — and it is, therefore, important that they are real — while the corresponding eigenvectors $|\lambda_i\rangle$ are the quantum states in which the observable assumes the values λ_i.

What is measured for the observable h in a more general quantum state $|\psi\rangle \in V$, which is not necessarily one of the eigenstates? In this case, the outcome is probabilistic in nature and the value λ_i is measured with probability $p_i = |\langle \lambda_i | \psi \rangle|^2$, provided $\langle \psi | \psi \rangle = 1$. Note that the p_i sum up to one,

$$\sum_i p_i = \sum_i |\langle \lambda_i | \psi \rangle|^2 = \sum_i \langle \psi | \lambda_i \rangle \langle \lambda_i | \psi \rangle \overset{\text{Eq. (26.27)}}{=} \langle \psi | \psi \rangle = 1$$

in line with their interpretation as probabilities. If λ_i is observed with probability p_i then the expectation value $\langle h \rangle_\psi$ for h in the state $|\psi\rangle$ is given by

$$\langle h \rangle_\psi = \sum_i p_i \lambda_i = \sum_i \lambda_i \langle \psi | \lambda_i \rangle \langle \lambda_i | \psi \rangle \overset{\text{Eq. (26.31)}}{=} \langle \psi | h | \psi \rangle \ ,$$

as claimed in the above table.

The process of measurement in quantum mechanics is described in terms of the mathematics of projectors. If the observable h is measured to be λ_i for a state $|\psi\rangle$ the state collapses from $|\psi\rangle$ to $P_i|\psi\rangle = |\lambda_i\rangle\langle\lambda_i|\psi\rangle$ immediately after the measurement. This collapse of the state is attributed to the interference of the measuring process with the system.

26.3.2   Index notation — again

It is instructive to work out the identification (26.21) of V and V^* relative to a basis $(\epsilon_1, \ldots, \epsilon_n)$ of V and its dual basis $(\epsilon_*^1, \ldots, \epsilon_*^n)$. Suppose the linear form is described by the matrix g with entries $g_{ij} = (\epsilon_i, \epsilon_j)$, so that

$$(\mathbf{v}, \mathbf{w}) = g_{ij}\bar{v}^i w^j \tag{26.33}$$

for $\mathbf{v} = \sum_i v^i \epsilon_i$ and $\mathbf{w} = \sum_j w^j \epsilon_j$ (see Eq. (25.4)). The matrix g is also sometimes referred to as a *metric*. For the action of the map s from Eq. (26.21) on the basis vectors we find

$$s(\epsilon_i)(\epsilon_k) = (\epsilon_i, \epsilon_k) = \bar{g}_{ki} = \bar{g}_{ji}\epsilon_*^j(\epsilon_k) \quad \Rightarrow \quad s(\epsilon_j) = \bar{g}_{ij}\epsilon_*^i$$

Hence, relative to a basis and its dual, the map s is described by the matrix $\bar{g}$. It is common to use the same symbol for the coordinates of a vector and the dual vector, related under s. If we adopt this convention and write $\mathbf{v} = v^j \epsilon_j$ and $s(\mathbf{v}) = v_i \epsilon_*^i$ then

$$v_i \epsilon_*^i = s(\mathbf{v}) = \bar{v}^j s(\epsilon_j) = \bar{g}_{ij}\bar{v}^j \epsilon_*^i$$

shows that the coordinates v_i of the dual vector are obtained from the coordinates v^j of the vector by

$$v_i = \bar{g}_{ij}\bar{v}^j . \tag{26.34}$$

If the linear form is non-degenerate then the matrix g is invertible and the entries of g^{-1} are often denoted by g^{ij}. Eq. (26.34) can then be inverted as

$$v^i = g^{ij}\bar{v}_j . \tag{26.35}$$

Physicists refer to the Eqs. (26.34) and (26.35) by saying that we can 'lower and raise indices' with the metric g_{ij} and its inverse g^{ij}. With this notation, the linear form (26.33) can be written as

$$(\mathbf{v}, \mathbf{w}) = g_{ij}\bar{v}^i w^j = \bar{v}^i \bar{w}_i = v_j w^j . \tag{26.36}$$

Application 26.3 *(Four vectors in special relativity)*

The structure we have discussed in Sections 26.3.1 and 26.3.2 is realized in special relativity, although this is not usually made explicit in expositions of the subject.

To see the connection, we introduce the Minkowski product on $\mathbb{R}^4$ by

$$(\mathbf{v}, \mathbf{w}) := \mathbf{v}^T \eta \mathbf{w} \tag{26.37}$$

just as we have done in Application 25.1. Here $\mathbf{v}, \mathbf{w} \in \mathbb{R}^4$ are four-vectors and $\eta = \mathrm{diag}(-1, 1, 1, 1)$. Recall that the Minkowski product has signature $(n_+, n_-) = (3, 1)$ and, from Prop. 25.1, it is non-degenerate.

We would like to represent vectors in $\mathbb{R}^4$ by coordinate vectors relative to the basis of standard unit vectors $(\mathbf{e}_\mu)$, where we use the index range $\mu, \nu, \ldots = 0, 1, 2, 3$, with $\mathbf{e}_0 = (1, 0, 0, 0)^T$ pointing in the time direction and $\mathbf{e}_i$, with $i = 1, 2, 3$ pointing in the three spatial directions, as is common in relativity. We also introduce the dual basis $(\mathbf{e}_*^\nu)$ for $(\mathbb{R}^4)^*$ so that vectors $\mathbf{v} \in \mathbb{R}^4$ and dual vectors $\mathbf{w}_* \in (\mathbb{R}^4)^*$ can be written as

$$\mathbf{v} = v^\mu \mathbf{e}_\mu , \qquad \mathbf{w}_* = w_\nu \mathbf{e}_*^\nu . \tag{26.38}$$

Hence, vectors are represented by coordinate vectors v^μ with upper indices and dual vectors by coordinate vectors w_ν with lower indices. By virtue of the duality relation, $\mathbf{e}^\mu_*(\mathbf{e}_\nu) = \delta^\mu_\nu$ the action of a dual vector an a vector can be written as

$$\mathbf{w}_*(\mathbf{v}) = w_\mu v^\mu \, . \tag{26.39}$$

The metric $g_{\mu\nu} = (\mathbf{e}_\mu, \mathbf{e}_\nu) = \eta_{\mu\nu}$ which describes the Minkowski product relative to our basis is, in fact, the Minkowski metric $\eta_{\mu\nu}$. Hence, the isomorphism between $\mathbb{R}^4$ and $(\mathbb{R}^4)^*$ induced by the Minkowski product is represented by lowering and raising indices with $\eta_{\mu\nu}$ and its inverse $\eta^{\mu\nu}$. This means for $\mathbf{v} = v^\mu \mathbf{e}_\mu \in V$ and $\mathbf{v}_* = v_\mu \mathbf{e}^\mu_* = s(\mathbf{v}) \in V^*$ related by the bijective map s in Eq. (26.21), we have

$$v_\mu = \eta_{\mu\nu} v^\nu \, , \qquad v^\mu = \eta^{\mu\nu} v_\nu \, . \tag{26.40}$$

With this notation, the Minkowski product can be expressed as

$$(\mathbf{v}, \mathbf{w}) = \eta_{\mu\nu} v^\mu w^\nu = v^\mu w_\mu = v_\nu w^\mu \, . \tag{26.41}$$

The Lorentz group $\mathcal{L} = O(3,1)$ is by definition the group which leaves the Minkowski product invariant (see Application 25.1). Eq. (26.41), therefore, indicates that Lorentz-invariant expressions are those with all indices contracted. (For an introduction to special relativity see, for example, Goldstein 2013.)

Application 26.4 *(A spin system)*

Many quantum mechanical systems are based on infinite dimensional vector spaces V, so their mathematics is somewhat beyond our scope. In this application we discuss a quantum system based on a single spin (such as the spin of an electron). The associated vector space V over $\mathbb{C}$ is two-dimensional and has an ortho-normal basis $(|\uparrow\rangle, |\downarrow\rangle)$ of two states which are interpreted as 'spin up' and 'spin down'. A general element $|\psi\rangle \in V$ has the form

$$|\psi\rangle = \sum_{s=\uparrow,\downarrow} \alpha_s |s\rangle = \alpha_\uparrow |\uparrow\rangle + \alpha_\downarrow |\downarrow\rangle \, ,$$

where $\alpha_\uparrow, \alpha_\downarrow \in \mathbb{C}$. If we normalize the state, $\langle\psi|\psi\rangle = |\alpha_\uparrow|^2 + |\alpha_\downarrow|^2 \overset{!}{=} 1$, the complex moduli $|\alpha_\uparrow|^2$ and $|\alpha_\downarrow|^2$ of the coordinates should be interpreted as the probabilities of 'spin up' and 'spin down' when measuring the state $|\psi\rangle$ (see Application 26.2).

Recall that in quantum mechanics physical quantities are represented by Hermitian linear maps $V \to V$. A particularly important such map is the *Hamilton operator* $H : V \to V$ which corresponds to the energy. Its matrix elements

$$H_{ss'} = \langle s|h|s'\rangle$$

must form a Hermitian 2×2 matrix $\mathcal{H}$. In Exercise 13.12 we have seen that the two-dimensional unit matrix and the three Pauli matrices form a basis of the vector space of 2×2 Hermitian matrices. This means that the matrix $\mathcal{H}$ can be written as

$$\mathcal{H} = a\mathbb{1}_2 + \mathbf{b} \cdot \boldsymbol{\sigma} \, , \tag{26.42}$$

where $\boldsymbol{\sigma} = (\sigma_1, \sigma_2, \sigma_3)$ is a formal vector which contains the Pauli matrices, $a \in \mathbb{R}$ and $\mathbf{b} \in \mathbb{R}^3$. In a physical context, the term $a\mathbb{1}_2$ represents on overall energy contribution which affects all spin states equally (for example a kinetic energy of the electron) while the term $\mathbf{b} \cdot \boldsymbol{\sigma}$ may describe the effect of a magnetic field proportional to $\mathbf{b}$.

What are the eigenvalues and eigenvectors of H? A quick calculation of the characteristic polynomial

$$\chi_{\mathcal{H}}(E) = \det \begin{pmatrix} a - E + b_3 & b_1 + ib_2 \\ b_1 - ib_2 & a - E - b_3 \end{pmatrix} = (a - E)^2 - |\mathbf{b}|^2 \stackrel{!}{=} 0$$

shows that the two energy eigenvalues are $E_\pm = a \pm |\mathbf{b}|$. The corresponding eigenstates $|E_\pm\rangle$ form an ortho-normal basis of V and satisfy the eigenvalue equation $H|E_\pm\rangle = E_\pm|E_\pm\rangle$. Let us consider two simple special cases.

(1) First assume that $b_1 = b_2 = 0$, so that $\mathcal{H} = \text{diag}(a + b_3, a - b_3)$ with energy eigenvalues $E_\pm = a \pm b_3$. The corresponding eigenvectors are the standard unit vector $\mathbf{e}_1$, $\mathbf{e}_2$, so the energy eigenstates

$$|E_+\rangle = |\uparrow\rangle \,, \quad |E_-\rangle = |\downarrow\rangle \,,$$

are the spin up and spin down states. Hence, for the energy eigenstate $|E_+\rangle$, the probability of measuring spin up is $|\langle\uparrow, E_+\rangle|^2 = 1$ while the probability for measuring spin down is $|\langle\downarrow|E_+\rangle|^2 = 0$. The situation is of course reversed for the energy eigenstate $|E_-\rangle$.

(2) As a second example consider $b_2 = b_3 = 0$ and $b_1 > 0$, so that $\mathcal{H}$, its eigenvalues and eigenvectors are given by

$$\mathcal{H} = \begin{pmatrix} a & b_1 \\ b_1 & a \end{pmatrix} \,, \quad E_\pm = a \pm b_1 \,, \quad \mathbf{v}_\pm = \frac{1}{\sqrt{2}}(1, \pm 1) \,.$$

Now the energy eigenstates are

$$|E_\pm\rangle = \frac{1}{\sqrt{2}}(|\uparrow\rangle \pm |\downarrow\rangle) \,, \tag{26.43}$$

so the probability of measuring a spin up state with energy E_+ is $|\langle\uparrow|E_+\rangle|^2 = 1/2$.

The evolution of a state $|\psi(t)\rangle$ with time t is governed by the time-dependent Schrödinger equation

$$H|\psi(t)\rangle = i\frac{d}{dt}|\psi(t)\rangle \,.$$

The simplest way to solve this equation is by writing the state $|\psi(t)\rangle$ as a linear combination $|\psi(t)\rangle = \alpha_+(t)|E_+\rangle + \alpha_-(t)|E_-\rangle$ of the energy eigenstates, with time-dependent coordinates $\alpha_\pm(t)$. Inserting this into the Schrödinger equation, using that $H|E_\pm\rangle = E_\pm|E_\pm\rangle$, gives the simple differential equations $\dot{\alpha}_\pm = -iE_\pm\alpha_\pm$. They are solved by $\alpha_\pm(t) = \beta_\pm \exp(-iE_\pm t)$, where $\beta_\pm \in \mathbb{C}$ are integration constants, so that the complete solution reads

$$|\psi(t)\rangle = \beta_+ e^{-iE_+ t}|E_+\rangle + \beta_- e^{-iE_- t}|E_-\rangle \,. \tag{26.44}$$

Consider the second case above where $b_2 = b_3 = 0$, $b_1 > 0$, $E_\pm = a \pm b_1$ and the energy eigenstates are given by Eq. (26.43). The constants $\beta_\pm$ allow us to specify a state at some initial time, say $t = 0$. Let us assume that the system is initially in a spin-up state, so $|\psi(0)\rangle = |\uparrow\rangle = (|E_+\rangle + |E_-\rangle)/\sqrt{2}$. This fixes the constants to $\beta_\pm = 1/\sqrt{2}$. Inserting into Eq. (26.44), the resulting time-dependent solution reads

$$|\psi(t)\rangle = \frac{e^{-iat}}{\sqrt{2}}\left(e^{-ib_1 t}|E_+\rangle + e^{ib_1 t}|E_-\rangle\right) \,.$$

Solutions such as these can be used to answer questions about the probability of measuring certain quantities as a function of time. For example, if we want to know the probability for the spin pointing downwards, we compute

$$\langle \downarrow |\psi(t)\rangle = \frac{e^{-iat}}{2} \left(\langle E_+| - \langle E_-| \right) \left(e^{-ib_1 t}|E_+\rangle + e^{ib_1 t}|E_-\rangle \right) = \sin(b_1 t) \, ,$$

where $|\downarrow\rangle = (|E_+\rangle - |E_-\rangle)/\sqrt{2}$ and the ortho-normality of the states $|E_\pm\rangle$ has been used. Hence, the probability for measuring a downward spin at time t is $|\langle \downarrow |\psi(t)\rangle|^2 = \sin^2(b_1 t)$. By a similar calculation, the probability for an upward spin is $|\langle \uparrow |\psi(t)\rangle|^2 = \cos^2(b_1 t)$

Exercises

(†=challenging)

26.1 Consider $\mathbb{R}^3$ with basis $(\mathbf{v}_1, \mathbf{v}_2, \mathbf{v}_3)$, where

$$\mathbf{v}_1 = (1, 0, 2)^T$$
$$\mathbf{v}_2 = (\ 3, 1, 0)^T$$
$$\mathbf{v}_3 = (1, -1, -2)^T \, .$$

(a) Find the dual basis of $(\mathbb{R}^3)^*$.
(b) Find the coordinates of $\mathbf{x} = (x, y, z)^T \in \mathbb{R}^3$ relative to the basis $(\mathbf{v}_1, \mathbf{v}_2, \mathbf{v}_3)$.

26.2 Consider the vector space V of at most quadratic polynomials over $\mathbb{R}$ on the interval $[-1, 1]$ and define the maps $\varphi_k : V \to \mathbb{R}$ by

$$\varphi_k(p) = \int_{-1}^{1} dx\, x^k p(x) \, ,$$

where $k = 0, 1, 2$.
(a) Show that the φ_k are linear functionals on V.
(b) Show that $(\varphi_0, \varphi_1, \varphi_2)$ is a basis of V^*.
(c) Find the basis of V^* dual to the monomial basis $(1, x, x^2)$.

26.3 Consider the vector space $V = \mathcal{M}_{n,n}(\mathbb{F})$ of $n \times n$ matrices.
(a) Show that the trace $\mathrm{tr} : V \to \mathbb{F}$ is a linear functional on V.
(b) Find the basis of V^* dual to the basis $(E_{(ij)})$ of standard unit matrices on V.

(c) Write the trace as a linear combination of the dual basis vectors from (b).

26.4 Prove Prop. 26.1 for a Hermitian sesqui-linear form.

26.5 Let V be a vector space over $\mathbb{R}$ with a non-degenerate symmetric bi-linear form $(\cdot, \cdot)$. Show that there is a natural non-degenerate symmetric bi-linear form $(\cdot, \cdot)_*$ on V^*. Relative to a basis of V and its dual basis, work out the matrices which describe these linear forms.

26.6 *Practice Dirac notation*
A complex inner product vector space V has two ortho-normal bases $(|i\rangle)$ and $(|a\rangle)$, where $i, a = 1, \ldots, n$. Use Dirac notation to do the following:
(a) Find the coordinates of $|\psi\rangle \in V$ relative to the basis $(|i\rangle)$.
(b) Show that the matrix U with entries $U_{ia} = \langle i|a\rangle$ is unitary.
(c) Find the relation between the coordinates of $|\psi\rangle \in V$ relative to the bases $(|i\rangle)$ and $(|a\rangle)$.
(d) Find the relation between the matrix elements $\langle i|f|j\rangle$ and $\langle a|f|b\rangle$ of a linear map $f \in \mathrm{End}(V)$.

26.7 *A three-state quantum system*†
The task is to generalize some of the discussion in Application 26.4 to a three-dimensional vector space V with ortho-normal basis $|s\rangle$, where $s = -1, 0, 1$. The Hamilton operator $H \in$

End(V) has matrix elements $\mathcal{H}_{ss'} = \langle s|H|s'\rangle$ where

$$\mathcal{H} = \begin{pmatrix} a & 0 & b \\ 0 & a & 0 \\ b & 0 & a \end{pmatrix},$$

and $a, b \in \mathbb{R}$.

(a) Find the eigenvalues and eigenvectors of H.

(b) What is the time evolution of $|\psi(t)\rangle \in V$?

(c) If $|\psi(0)\rangle = |s\rangle$ what is the probability of measuring $|s'\rangle$ at time t, where $s, s' = -1, 0, 1$?

27

Tensors*

In Section 8.1.5 we have seen that the Cartesian product $V \times W \cong V \oplus W$ of two vector spaces V, W over the same field $\mathbb{F}$ can be made into a vector space by defining vector addition and scalar multiplication component-wise, as in Eq. (8.8). However, there is a different and more ambitious way to build a vector space based on the Cartesian product $V \times W$ and this leads to the tensor product vector space $V \otimes W$. Rather than just considering all the pairs $(\mathbf{v}, \mathbf{w}) \in V \times W$ as we did for the direct sum, the tensor product $V \otimes W$ also includes linear combinations of such pairs. But this raises a problem. In the spirit of linearity, it seems desirable for $(\mathbf{v}, \alpha\mathbf{w})$ to equal $\alpha(\mathbf{v}, \mathbf{w})$ and for $(\mathbf{v}, \mathbf{w}_1 + \mathbf{w}_2)$ to equal $(\mathbf{v}, \mathbf{w}_1) + (\mathbf{v}, \mathbf{w}_2)$. However, the Cartesian product is not linear in its arguments. The tensor product $V \otimes W$ fixes this 'deficiency' by identifying $(\mathbf{v}, \alpha\mathbf{w})$ with $\alpha(\mathbf{v}, \mathbf{w})$ as well as $(\mathbf{v}, \mathbf{w}_1 + \mathbf{w}_2)$ with $(\mathbf{v}, \mathbf{w}_1) + (\mathbf{v}, \mathbf{w}_2)$ in $V \otimes W$ (with similar identifications in the first argument). In fact, the tensor product can be defined as the linear combinations of elements in $V \times W$ subject to these identifications, although we will follow a different path.

A much more practical approach is to start with bases $(\mathbf{v}_1, \ldots, \mathbf{v}_n)$ and $(\mathbf{w}_1, \ldots, \mathbf{w}_m)$ of V and W and define the tensor product $V \otimes W$ as the set of all formal linear combinations

$$\sum_{i,a} \tau^{ia} \mathbf{v}_i \otimes \mathbf{w}_a ,$$

where $\tau^{ia} \in \mathbb{F}$ are scalars and the pairs $(\mathbf{v}_i, \mathbf{w}_j) \in V \times W$ have been written as $\mathbf{v}_i \otimes \mathbf{w}_j$. This already points to the way tensors are frequently used in scientific applications. Relative to a choice of bases, they can be identified with multi-index objects, such as τ^{ia} above.

From a mathematical viewpoint it is, of course, not a good idea to introduce a new concept in a basis-dependent way, so we begin our discussion with a basis-independent definition of $V \otimes W$, as the space of bi-linear maps $V^* \times W^* \to \mathbb{F}$. This construction can be generalized to an arbitrary number of tensor products. In particular, this leads to (p, q) tensors which are elements of the tensor space $V^{\otimes p} \otimes (V^*)^{\otimes q}$ and can be represented by indexed objects with p upper and q lower indices. We also introduce anti-symmetric $(0, q)$ tensors, also referred to as *alternating q forms*, which form vector spaces denoted as $\Lambda^q V^*$. Their direct sum, $\Lambda V^* = \Lambda^0 V^* \oplus \cdots \oplus \Lambda^n V^*$, is called the *outer algebra* of V^*. As we will show, the dot, cross, and triple products on $\mathbb{R}^3$ have their natural home in the outer algebra $\Lambda\mathbb{R}^3$.

27.1 Tensor basics*

Summary 27.1 *The tensor space $V \otimes W$ of two vector spaces V and W over $\mathbb{F}$ consists of all bi-linear forms $V^* \times W^* \to \mathbb{F}$. Tensors can be built up from vectors by using the tensor product. For bases $(\mathbf{v}_i)$ and $(\mathbf{w}_a)$ of V and W the tensors $(\mathbf{v}_i \otimes \mathbf{w}_a)$ form a basis of $V \otimes W$, so that $\dim_{\mathbb{F}}(V \otimes W) = \dim_{\mathbb{F}}(V) \dim_{\mathbb{F}}(W)$. Relative to a choice of bases, tensors in $V \otimes W$ are represented by two-index objects. Under a basis change these two-index objects follow a characteristic transformation law, with a transformation matrix acting on each index.*

27.1.1 Definition of tensors

For now our discussion involves two (finite-dimensional) vector spaces V and W over the same field $\mathbb{F}$, and their dual vector spaces V^* and W^*. Before we start, it is useful to fix a suggestive notation for these vector spaces, their bases and typical elements as in Table 27.1. The bases are chosen to be dual to one another, so

Table 27.1 Bases, typical elements, and index notation for tensor products.

Vector space	V	V^*	W	W^*
Basis	$(\mathbf{v}_i)_{i=1,\ldots,n}$	$(\mathbf{v}_*^j)_{j=1,\ldots,n}$	$(\mathbf{w}_a)_{a=1,\ldots,m}$	$(\mathbf{w}_*^b)_{b=1,\ldots,m}$
Typical element	$\mathbf{v} = v^i \mathbf{v}_i$	$\mathbf{v}_* = v_j \mathbf{v}_*^j$	$\mathbf{w} = w^a \mathbf{w}_a$	$\mathbf{w}_* = w_b \mathbf{w}_*^b$
Index version	v^i	v_j	w^a	w_b

$$\mathbf{v}_*^j(\mathbf{v}_i) = \delta_i^j \,, \qquad \mathbf{w}_*^b(\mathbf{w}_a) = \delta_a^b \,, \tag{27.1}$$

and the action of dual vectors on vectors (and vice versa, see Eq. (26.13)) can be written in terms of index contractions as

$$\mathbf{v}_*(\mathbf{v}) = \mathbf{v}(\mathbf{v}_*) = v^i v_i \,, \qquad \mathbf{w}_*(\mathbf{w}) = \mathbf{w}(\mathbf{w}_*) = w^a w_a \,. \tag{27.2}$$

Throughout this chapter we will use the Einstein summation convention, whereby same upper and lower indices in the same term are summed over ('contracted'), in order to simplify notation. The abstract definition of the tensor space $V \otimes W$ is as follows:

Definition 27.1 *For two finite-dimensional vector spaces V and W over the same field $\mathbb{F}$ the tensor space $V \otimes W$ is the set of all bi-linear forms $V^* \times W^* \to \mathbb{F}$, so*

$$V \otimes W = \{\tau : V^* \times W^* \to \mathbb{F} \,|\, \tau \text{ is bi-linear}\} \,. \tag{27.3}$$

In other words, a tensor $\tau \in V \otimes W$ takes as its arguments two functionals, $\mathbf{v}_* \in V^*$ and $\mathbf{w}_* \in W^*$, such that $\tau(\mathbf{v}_*, \mathbf{w}_*) \in \mathbb{F}$ is a scalar and linear in each of its arguments. Addition and scalar multiplication of tensors are defined as usual by

$$(\tau_1 + \tau_2)(\mathbf{v}_*, \mathbf{w}_*) = \tau_1(\mathbf{v}_*, \mathbf{w}_*) + \tau_2(\mathbf{v}_*, \mathbf{w}_*) \,, \qquad (\alpha\tau)(\mathbf{v}_*, \mathbf{w}_*) = \alpha\tau(\mathbf{v}_*, \mathbf{w}_*) \,, \quad (27.4)$$

where $\tau, \tau_1, \tau_2 \in V \otimes W$ and $\alpha \in \mathbb{F}$. It is readily verified that $\tau_1 + \tau_2$ and $\alpha\tau$ are again bi-linear and are, hence, elements of $V \otimes W$ as well. We conclude that $V \otimes W$ is a vector space with addition and scalar multiplication given by Eq. (27.4).

27.1.2 The tensor product

Elements in $V \otimes W$ can be constructed by 'tensoring together' elements of V and W. To this end, we define the tensor product $\mathbf{v} \otimes \mathbf{w} \in V \otimes W$ for two vectors $\mathbf{v} \in V$ and $\mathbf{w} \in W$ by

$$\mathbf{v} \otimes \mathbf{w}(\mathbf{v}_*, \mathbf{w}_*) := \mathbf{v}(\mathbf{v}_*)\mathbf{w}(\mathbf{w}_*) \,. \tag{27.5}$$

Note that this definition makes sense since the elements of V and W can be viewed as linear functionals on V^* and W^* (see Eq. (26.13)). The right-hand side of Eq. (27.5) is linear in each of the arguments $\mathbf{v}_*$ and $\mathbf{w}_*$, so we conclude that $\mathbf{v} \otimes \mathbf{w}$ is indeed bi-linear and, hence, an element of $V \otimes W$. The tensor product itself is also linear in both arguments, as follows easily from the definition (27.5). This implies the following rules for calculating with the tensor product.

$$\begin{aligned}(\alpha_1\mathbf{v}_1 + \alpha_2\mathbf{v}_2) \otimes \mathbf{w} &= \alpha_1\mathbf{v}_1 \otimes \mathbf{w} + \alpha_2\mathbf{v}_2 \otimes \mathbf{w} \\ \mathbf{v} \otimes (\beta_1\mathbf{w}_1 + \beta_2\mathbf{w}_2) &= \beta_1\mathbf{v} \otimes \mathbf{w}_1 + \beta_2\mathbf{v} \otimes \mathbf{w}_2\end{aligned} \tag{27.6}$$

Tensors in $V \otimes W$ which can be written as a tensor product $\mathbf{v} \otimes \mathbf{w}$ are called *decomposable*. Not all tensor in $V \otimes W$ can be written in this way (see Exercise 27.1), but all tensors can be obtained from linear combinations of tensor products, as shown in the following theorem.

Theorem 27.1 *Let V and W be two vector spaces over the same field $\mathbb{F}$ with bases $(\mathbf{v}_i)_{i=1,\dots,n}$ and $(\mathbf{w}_a)_{a=1,\dots,m}$, respectively. Then the $n\,m$ vectors $(\mathbf{v}_i \otimes \mathbf{w}_a)$ form a basis of $V \otimes W$. In particular,*

$$\dim_{\mathbb{F}}(V \otimes W) = \dim_{\mathbb{F}}(V) \dim_{\mathbb{F}}(W) \,. \tag{27.7}$$

Proof To show linear independence we let the equation $\tau^{ia}\mathbf{v}_i \otimes \mathbf{w}_a = 0$ act on the dual basis vectors $(\mathbf{v}_*^j, \mathbf{w}_*^b)$ which, due to the duality relations (27.1), immediately leads to $\tau^{jb} = 0$. Since this holds for all $j = 1, \dots, n$ and $b = 1, \dots, m$ linear independence follows.

To show that our prospective basis spans $V \otimes W$ we start with an arbitrary tensor $\tau \in V \otimes W$ and introduce its 'components' $\tau^{ia} := \tau(\mathbf{v}_*^i, \mathbf{w}_*^a)$. Define the tensor $\tilde{\tau} = \tau^{ia}\mathbf{v}_i \otimes \mathbf{w}_a$. It follows that $\tau(\mathbf{v}_*^j, \mathbf{w}_*^b) = \tau^{jb} = \tilde{\tau}(\mathbf{v}_*^j, \mathbf{w}_*^b)$ for all $j = 1, \dots, n$ and all $a = 1, \dots, m$ and, hence, that $\tau = \tilde{\tau}$ is in the span of $(\mathbf{v}_i \otimes \mathbf{w}_a)$. $\qquad\square$

27.1.3 The universal property

We have started motivating the tensor product as an 'improved' version of the Cartesian product, consistent with linearity. Now that we have defined the tensor product, what precisely is its relation to the Cartesian product? First, we note that for two vector spaces V, W over $\mathbb{F}$ we have a map

$$\rho : V \times W \rightarrow V \otimes W , \qquad \rho((\mathbf{v}, \mathbf{w})) := \mathbf{v} \otimes \mathbf{w} , \qquad (27.8)$$

which is bi-linear due to Eq. (27.6). The *universal property* of the tensor product asserts that every bi-linear map $\phi : V \times W \rightarrow U$ from the Cartesian product $V \times W$ to another vector space U over $\mathbb{F}$ can be 'lifted' to a unique linear map $\psi : V \otimes W \rightarrow U$, in the sense that $\phi = \psi \circ \rho$. This can also be expressed by saying that there is a unique map ψ for which the diagram

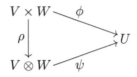

commutes. The term 'commute' in this context means both possible paths from $V \times W$ to U in the above diagram lead to the same result. We formulate and proof this in the following theorem.

Theorem 27.2 *(Universal property) For vector spaces V, W, U over $\mathbb{F}$ and a bi-linear map $\phi : V \times W \rightarrow U$, there exists a unique linear map $\psi : V \otimes W \rightarrow U$ such that $\phi = \psi \circ \rho$, where ρ is defined in Eq. (27.8).*

Proof We choose bases $(\mathbf{v}_1, \ldots, \mathbf{v}_n)$ and $(\mathbf{w}_1, \ldots, \mathbf{w}_m)$ on V and W. Then the given map $\phi : V \times W \rightarrow U$ is characterized by the images $\mathbf{u}_{ia} := \phi((\mathbf{v}_i, \mathbf{w}_a))$. From Theorem 12.1, there exists a linear map $\psi : V \otimes W \rightarrow U$ with $\psi(\mathbf{v}_i \otimes \mathbf{w}_a) = \mathbf{u}_{ia}$ and this map obviously satisfies $\phi = \psi \circ \rho$. To show uniqueness, we start with a map $\psi : V \otimes W \rightarrow U$ which satisfies the desired property $\phi = \psi \circ \rho$. It follows that

$$\mathbf{u}_{ia} = \phi((\mathbf{v}_i, \mathbf{w}_a)) = \psi(\rho((\mathbf{v}_i, \mathbf{w}_a))) = \psi(\mathbf{v}_i \otimes \mathbf{w}_a)$$

and since $(\mathbf{v}_i \otimes \mathbf{w}_a)$ is a basis of $V \otimes W$ (see Theorem 27.1) this determines ψ uniquely. $\square$

27.1.4 Indices

We can repeat the above discussion for the tensor product $V \otimes W$ with the replacements $V \rightarrow V^*$ or $W \rightarrow W^*$ (keeping in mind that $V \cong V^{**}$ and $W \cong W^{**}$) and in this way arrive at the tensors in Table 27.2. The last row of the table shows the two index

Table 27.2 Tensors from vector spaces V, W and their duals.

Tensor space	$V \otimes W$	$V \otimes W^*$	$V^* \otimes W$	$V^* \otimes W^*$
Basis	$(\mathbf{v}_i, \mathbf{w}_a)$	$(\mathbf{v}_i, \mathbf{w}^a_*)$	$(\mathbf{v}^i_*, \mathbf{w}_a)$	$(\mathbf{v}^i_*, \mathbf{w}^a_*)$
Typical element	$\tau^{ia} \mathbf{v}_i \otimes \mathbf{w}_a$	$\tau^i{}_a \mathbf{v}_i \otimes \mathbf{w}^a_*$	$\tau_i{}^a \mathbf{v}^i_* \otimes \mathbf{w}_a$	$\tau_{ia} \mathbf{v}^i_* \otimes \mathbf{w}^a_*$
Index version	τ^{ia}	$\tau^i{}_a$	$\tau_i{}^a$	τ_{ia}

objects which represent the different types of tensors, where a covariant (=upper) index refers to the vector spaces V and W and a contravariant (=lower) index refers to their duals V^* and W^*.

The action of a tensor on its arguments can be written in terms of index contractions. For example for a tensor $\tau \in V \otimes W$ we have, using linearity, the definition (27.5) of the tensor product and the duality relations (27.1), that

$$\tau(\mathbf{v}_*, \mathbf{w}_*) = \tau^{ia} v_j w_b \mathbf{v}_i(\mathbf{v}_*^j) \mathbf{w}_a(\mathbf{w}_*^b) = \tau^{ia} v_j w_b \delta_i^j \delta_a^b = \tau^{ia} v_i w_a . \qquad (27.9)$$

Similar results hold for the other types of tensors in Table 27.2.

27.1.5 Basis transformation of tensors

Suppose we have two choices of bases on V and W and their duals, as in Table 27.3. The matrices P and Q in the last row parametrize the change of basis. As usual, we

Table 27.3 Choices of bases for tensor transformations.

	V	V^*	W	W^*
First basis	$(\mathbf{v}_i)$	$(\mathbf{v}_*^i)$	$(\mathbf{w}_a)$	$(\mathbf{w}_*^a)$
Second basis	$(\tilde{\mathbf{v}}_i)$	$(\tilde{\mathbf{v}}_*^i)$	$(\tilde{\mathbf{w}}_a)$	$(\tilde{\mathbf{w}}_*^a)$
Relation	$\tilde{\mathbf{v}}_i = P_i{}^j \mathbf{v}_j$	$\tilde{\mathbf{v}}_*^i = P_{*j}^i \mathbf{v}_*^j$	$\tilde{\mathbf{w}}_a = Q_a{}^b \mathbf{w}_b$	$\tilde{\mathbf{w}}_*^a = Q_{*b}^a \mathbf{w}_*^b$

require that the bases for V and V^* (and for W and W^*) are dual, for either set of bases. For this to remain true as we pass from the first to the second basis, the matrices P and P_* (and Q and Q_*) need to be related. The short calculation

$$\delta_j^i = \tilde{\mathbf{v}}_*^i(\tilde{\mathbf{v}}_j) = P_{*k}^i P_j{}^l \mathbf{v}_*^k(\mathbf{v}_l) = P_{*k}^i P_j{}^l \delta_l^k = (P_* P^T)^i{}_j ,$$

shows that $P_* = (P^T)^{-1}$ and, similarly, $Q_* = (Q^T)^{-1}$. What we would like to know is the relation between the index versions of the tensors

$$\tilde{\tau}_{ia} = \tau(\tilde{\mathbf{v}}_i, \tilde{\mathbf{w}}_a) , \qquad \tau_{ia} = \tau(\mathbf{v}_i, \mathbf{w}_a)$$

which follows from

$$\tilde{\tau}_{ia} = \tau(\tilde{\mathbf{v}}_i, \tilde{\mathbf{w}}_a) = \tau(P_i{}^j \mathbf{v}_j, Q_a{}^b \mathbf{w}_b) = P_i{}^j Q_a{}^b \tau_{jb} .$$

Carrying out similar calculations for the other tensors in Table 27.2 leads to the transformation laws

$$\tilde{\tau}_{ia} = P_i{}^j Q_a{}^b \tau_{jb} , \quad \tilde{\tau}_i{}^a = P_i{}^j Q_{*b}^a \tau_j{}^b , \quad \tilde{\tau}^i{}_a = P_{*j}^i Q_a{}^b \tau^j{}_b , \quad \tilde{\tau}^{ia} = P_{*j}^i Q_{*b}^a \tau^{jb} . \qquad (27.10)$$

Note that these rules are quite systematic. A covariant index transforms with P_* (or Q_*) while a contravariant index transforms with P (or Q).

27.1.6   Induced maps on tensors

Suppose we have linear maps $f \in \mathrm{End}(V)$ and $g \in \mathrm{End}(W)$. These maps lead to an induced tensor map $f \otimes g \in \mathrm{End}(V \otimes W)$ which is defined by

$$(f \otimes g)(\mathbf{v} \otimes \mathbf{w}) := f(\mathbf{v}) \otimes g(\mathbf{w}) \,. \tag{27.11}$$

Note it is sufficient to make this definition on decomposable tensors $\mathbf{v} \otimes \mathbf{w}$ — linearity implies it extends uniquely to all tensors. The map tensor product is compatible with composition of maps,

$$(\tilde{f} \otimes \tilde{g}) \circ (f \otimes g) = (\tilde{f} \circ f) \otimes (\tilde{g} \circ g) \,, \tag{27.12}$$

for $f, \tilde{f} \in \mathrm{End}(V)$ and $g, \tilde{g} \in \mathrm{End}(W)$, as follows immediately from Eq. (27.11).

Suppose that the map f is represented by an $n \times n$ matrix A, relative to a basis $(\mathbf{v}_i)$ of V and that g is represented by an $m \times m$ matrix B, relative to a basis $(\mathbf{w}_a)$ of W. It is then natural to ask what the representing matrix for the tensor map $f \otimes g$ relative to the basis $(\mathbf{v}_i \otimes \mathbf{w}_a)$ of $V \otimes W$ is. First we note, from Theorem 27.1, that $\dim_{\mathbb{F}}(V \otimes W) = nm$, so the matrix we are looking for must be of size $(nm) \times (nm)$. As usual, a representing matrix is found by acting on the basis vectors and this leads to

$$\left. \begin{array}{l} f(\mathbf{v}_i) = A^j{}_i \mathbf{v}_j \\ f(\mathbf{w}_a) = B^b{}_a \mathbf{w}_b \end{array} \right\} \quad \Rightarrow \quad (f \otimes g)(\mathbf{v}_i \otimes \mathbf{w}_a) = f(\mathbf{v}_i) \otimes g(\mathbf{w}_a) = A^j{}_i B^b{}_a \mathbf{v}_i \otimes \mathbf{w}_b \,.$$

This means the desired representing matrix has as its entries all possible products $A^j{}_i B^b{}_a$ of entries of A and B. Note that this gives exactly the correct number, $(nm)^2$, of entries. How exactly these entries are arranged into a matrix depends on choosing an ordering for the tensor basis $(\mathbf{v}_i \otimes \mathbf{w}_a)$. Suppose we agree on the following ordering.

$$(\mathbf{v}_1 \otimes \mathbf{w}_1, \dots, \mathbf{v}_1 \otimes \mathbf{w}_m, \mathbf{v}_2 \otimes \mathbf{w}_1, \dots, \mathbf{v}_2 \otimes \mathbf{w}_m, \dots) \,.$$

Relative to this ordering the representing matrix for $f \otimes g$ is given by the *Kronecker product* $A \times B$ of matrices, defined by

$$A \times B := \begin{pmatrix} A_{11}B & A_{12}B & \cdots & A_{1n}B \\ A_{21}B & A_{22}B & \cdots & A_{2n}B \\ \vdots & \vdots & \vdots & \vdots \\ A_{n1}B & A_{n2}B & \cdots & A_{nn}B \end{pmatrix} \,. \tag{27.13}$$

In other words, the Kronecker product $A \times B$ is obtained by replacing every entry A_{ij} of A with the $m \times m$ matrix $A_{ij}B$. It is linear in each argument,

$$\begin{array}{l} (\alpha A + \tilde{\alpha}\tilde{A}) \times B = \alpha A \times B + \tilde{\alpha}\tilde{A} \times B \\ A \times (\beta B + \tilde{\beta}\tilde{B}) = \beta A \times B + \tilde{\beta}A \times \tilde{B} \end{array} \,, \tag{27.14}$$

a property which follows easily from its definition, Eq. (27.13). Eq. (27.12) also implies that the Kronecker product is compatible with matrix multiplication, that is,

$$(\tilde{A} \times \tilde{B})(A \times B) = (\tilde{A}A) \times (\tilde{B}B) \,. \tag{27.15}$$

Problem 27.1 *Tensor maps and Kronecker product*

The linear maps $f, g \in \text{End}(\mathbb{R}^2)$ are defined by $f(\mathbf{v}) = (\mathbf{n} \cdot \mathbf{v})\mathbf{n}$ and $g(\mathbf{v}) = \mathbf{v}^\times$, where $\mathbf{n} \in \mathbb{R}^2$. Find the matrices A and B which represent these maps relative to the standard unit vector basis $(\mathbf{e}_1, \mathbf{e}_2)$ of $\mathbb{R}^2$ and the matrix C which represents $f \otimes g$ relative to the basis $(\mathbf{e}_1 \otimes \mathbf{e}_1, \mathbf{e}_1 \otimes \mathbf{e}_2, \mathbf{e}_2 \otimes \mathbf{e}_1, \mathbf{e}_2 \otimes \mathbf{e}_2)$ of $\mathbb{R}^2 \otimes \mathbb{R}^2$. Check that $C = A \times B$.

Solution: Since $(f(\mathbf{e}_i))_j = (\mathbf{n} \cdot \mathbf{e}_i)n_j = n_i n_j$, $g(\mathbf{e}_1) = -\mathbf{e}_2$ and $g(\mathbf{e}_2) = \mathbf{e}_1$ we have

$$A = \begin{pmatrix} n_1^2 & n_1 n^2 \\ n_1 n_2 & n_2^2 \end{pmatrix}, \qquad B = \begin{pmatrix} 0 & 1 \\ -1 & 0 \end{pmatrix}.$$

From Eq. (27.11), the action of $f \otimes g$ on the tensor basis is

$$\left. \begin{aligned} (f \otimes g)(\mathbf{e}_1 \otimes \mathbf{e}_1) &= f(\mathbf{e}_1) \otimes g(\mathbf{e}_1) = -n_1 n_i \mathbf{e}_i \otimes \mathbf{e}_2 \\ (f \otimes g)(\mathbf{e}_1 \otimes \mathbf{e}_2) &= f(\mathbf{e}_1) \otimes g(\mathbf{e}_2) = n_1 n_i \mathbf{e}_i \otimes \mathbf{e}_1 \\ (f \otimes g)(\mathbf{e}_2 \otimes \mathbf{e}_1) &= f(\mathbf{e}_2) \otimes g(\mathbf{e}_1) = -n_2 n_i \mathbf{e}_i \otimes \mathbf{e}_2 \\ (f \otimes g)(\mathbf{e}_2 \otimes \mathbf{e}_2) &= f(\mathbf{e}_2) \otimes g(\mathbf{e}_2) = n_2 n_i \mathbf{e}_i \otimes \mathbf{e}_1 \end{aligned} \right\} \Rightarrow C = \begin{pmatrix} 0 & n_1^2 & 0 & n_1 n^2 \\ -n_1^2 & 0 & -n_1 n_2 & 0 \\ 0 & n_1 n_2 & 0 & n_2^2 \\ -n_1 n_2 & 0 & -n_2^2 & 0 \end{pmatrix}$$

Forming the Kronecker product by replacing the entries A_{ij} of A with $A_{ij}B$ does indeed lead to the matrix $C = A \times B$.

27.2 Further tensor properties*

Summary 27.2 *The symmetric and anti-symmetric tensors in $V \otimes V$ form vector subspaces S^2V and Λ^2V, respectively, and $V \otimes V = S^2V \oplus \Lambda^2V$. Symmetric and anti-symmetric tensors can be constructed from vectors by using the symmetrized tensor product and the wedge product. Linear maps $V \to W$ can be identified with tensors in $W \otimes V^*$.*

27.2.1 Symmetric and anti-symmetric tensors

For the tensor product $V \otimes V$ of V with itself we can consider symmetric and anti-symmetric tensors, which are defined as follows.

Definition 27.2 *A tensor $\tau \in V \otimes V$ is called symmetric if $\tau(\mathbf{v}_*, \tilde{\mathbf{v}}_*) = \tau(\tilde{\mathbf{v}}_*, \mathbf{v}_*)$ and anti-symmetric if $\tau(\mathbf{v}_*, \tilde{\mathbf{v}}_*) = -\tau(\tilde{\mathbf{v}}_*, \mathbf{v}_*)$, for all $\mathbf{v}_*, \tilde{\mathbf{v}}_* \in V^*$.*

From Eq. (27.4) it is clear that the sum and scalar multiple of symmetric (anti-symmetric) tensors are again symmetric (anti-symmetric). Therefore, symmetric and anti-symmetric tensor form vector subspaces of $V \otimes V$ which are also denoted by

$$\begin{aligned} S^2V &= \{\tau \in V \otimes V \mid \tau \text{ symmetric}\} \\ \Lambda^2V &= \{\tau \in V \otimes V \mid \tau \text{ anti-symmetric}\}. \end{aligned} \tag{27.16}$$

To construct symmetric and anti-symmetric tensors we can define the following refinements of the tensor product

$$\mathbf{v} \otimes_S \tilde{\mathbf{v}} := \mathbf{v} \otimes \tilde{\mathbf{v}} + \tilde{\mathbf{v}} \otimes \mathbf{v}, \quad \mathbf{v} \wedge \tilde{\mathbf{v}} := \mathbf{v} \otimes \tilde{\mathbf{v}} - \tilde{\mathbf{v}} \otimes \mathbf{v}, \tag{27.17}$$

where $\mathbf{v}, \tilde{\mathbf{v}} \in V$. Clearly, the symmetrized tensor product $\mathbf{v} \otimes_S \tilde{\mathbf{v}}$ produces a symmetric tensor, while the *wedge product* $\mathbf{v} \wedge \tilde{\mathbf{v}}$ produces an anti-symmetric tensor. Both products are linear in each argument, just as the tensor product itself, as follows immediately from the rules (27.6). In addition, the definitions in Eq. (27.17) imply that

$$\mathbf{v} \otimes_S \tilde{\mathbf{v}} = \tilde{\mathbf{v}} \otimes_S \mathbf{v} , \qquad \mathbf{v} \wedge \tilde{\mathbf{v}} = -\tilde{\mathbf{v}} \wedge \mathbf{v} . \tag{27.18}$$

Theorem 27.1 has the following analogue for symmetric and anti-symmetric tensors.

Theorem 27.3 *For a vector space V with basis $(\mathbf{v}_i)_{i=1,\ldots,n}$, the $n(n+1)/2$ vectors $(\mathbf{v}_i \otimes_S \mathbf{v}_j)_{i \leq j}$ form a basis of $S^2 V$, while the $n(n-1)/2$ vectors $(\mathbf{v}_i \wedge \mathbf{v}_j)_{i<j}$ form a basis of $\Lambda^2 V$. In particular,*

$$\dim_{\mathbb{F}}(S^2 V) = \frac{1}{2} n(n+1) , \qquad \dim_{\mathbb{F}}(\Lambda^2 V) = \frac{1}{2} n(n-1) , \tag{27.19}$$

and $V \otimes V = S^2 V \oplus \Lambda^2 V$.

Proof   We start with the anti-symmetric case. Letting the equation $\sum_{i<j} \tau^{ij} \mathbf{v}_i \wedge \mathbf{v}_j = 0$ act on $(\mathbf{v}_*^k, \mathbf{v}_*^l)$ for $k < l$ gives $\tau^{kl} = 0$ and this shows linear independence of $(\mathbf{v}_i \wedge \mathbf{v}_j)_{i<j}$.

Consider an anti-symmetric tensor τ, define $\tau^{ij} = \tau(\mathbf{v}_*^i, \mathbf{v}_*^j)$ and the anti-symmetric tensor $\tilde{\tau} = \sum_{i<j} \tau^{ij} \mathbf{v}_i \wedge \mathbf{v}_j$. It follows that $\tau(\mathbf{v}_*^i, \mathbf{v}_*^j) = \tau^{ij} = \tilde{\tau}(\mathbf{v}_*^i, \mathbf{v}_*^j)$ and, hence, $\tau = \tilde{\tau}$. This shows that the vectors $(\mathbf{v}_i \wedge \mathbf{v}_j)_{i<j}$ span $\Lambda^2 V$.

The proof of the symmetric case works analogously and the dimension formulae (27.19) follow immediately by counting the number of basis vectors.

If the tensor τ is symmetric and anti-symmetric at the same time we have $\tau(\mathbf{v}_*, \tilde{\mathbf{v}}_*) = \tau(\tilde{\mathbf{v}}_*, \mathbf{v}_*) = -\tau(\mathbf{v}_*, \tilde{\mathbf{v}}_*)$ for all $\mathbf{v}, \mathbf{v}_* \in V^*$ so that $\tau = 0$. This shows that $S^2 V \cap \Lambda^2 V = \{0\}$ and, hence, that the sum $S^2 V + \Lambda^2 V$ is direct (see Section 8.1.4). From Eq. (8.7) we find for the dimension of this direct sum

$$\dim_{\mathbb{F}}(S^2 V \oplus \Lambda^2 V) = \dim_{\mathbb{F}}(S^2 V) + \dim_{\mathbb{F}}(\Lambda^2 V) = \frac{n(n+1)}{2} + \frac{n(n-1)}{2}$$
$$= n^2 = \dim_{\mathbb{F}}(V \otimes V) .$$

This equality of dimensions together with Corollary 7.1 then shows that $S^2 V \oplus \Lambda^2 V = V \otimes V$. $\square$

From the previous theorem, symmetric and anti-symmetric tensors can be written as

$$\tau_+ = \frac{1}{2} \tau_+^{ij} \mathbf{v}_i \otimes_S \mathbf{v}_j , \qquad \tau_- = \frac{1}{2} \tau_-^{ij} \mathbf{v}_i \wedge \mathbf{v}_j , \tag{27.20}$$

where τ_+^{ij} is symmetric, so $\tau_+^{ij} = \tau_+^{ji}$ and τ_-^{ij} is anti-symmetric, $\tau_-^{ij} = -\tau_-^{ji}$. Furthermore, the decomposition $V \otimes V = S^2 V \oplus \Lambda^2 V$ tells us that every tensor $\tau = \tau^{ij} \mathbf{v}_i \otimes \mathbf{v}_j \in V \otimes V$ can be written as a unique sum, $\tau = \tau_+ + \tau_-$ of a symmetric and an anti-symmetric tensor. At the level of index notation, this translated into $\tau^{ij} = \tau_+^{ij} + \tau_-^{ij}$. This is in direct analogy to our earlier result that every square matrix can be written as a unique sum of a symmetric and an anti-symmetric matrix (see Example 13.1).

27.2.2   Linear maps as tensors

Many of the objects in linear algebra we have encountered can be phrased in the language of tensors and linear maps are a case in point. Consider the space $\mathrm{Hom}(V, W)$ of linear maps $V \to W$. This space has the same dimension, $\dim_{\mathbb{F}}(V) \dim_{\mathbb{F}}(W)$, as the tensor space $W \otimes V^*$, so we know these two spaces must be isomorphic. However, the relationship is even closer since there is a canonical isomorphism $\imath : \mathrm{Hom}(V, W) \to W \otimes V^*$ defined by

$$\imath(f)(\mathbf{w}_*, \mathbf{v}) := f(\mathbf{v})(\mathbf{w}_*) \,, \tag{27.21}$$

where $\mathbf{v} \in V$ and $\mathbf{w}_* \in W^*$. Note that the right-hand side of this definition makes sense: $f(\mathbf{v})$ is a vector in W which can act on a dual vector $\mathbf{w}_* \in W$.

Theorem 27.4 *The map defined in Eq. (27.21) is an isomorphism so we have a canonical isomorphism* $\mathrm{Hom}(V, W) \cong W \otimes V^*$.

Proof   Since $\mathrm{Hom}(V, W)$ and $W \otimes V^*$ have the same dimension all we have to do is show that $\mathrm{Ker}(\imath) = \{0\}$ (see Corollary 14.2). But a map $f \in \mathrm{Ker}(\imath)$ satisfies $\imath(f) = 0$ and, hence $\imath(f)(\mathbf{w}_*, \mathbf{v}) - f(\mathbf{v})(\mathbf{w}_*) = 0$ for all $\mathbf{v} \in V$ and $\mathbf{w}_* \in W^*$. This implies that f is the zero map, $f = 0$.   □

How does the map $\imath$ looks like more explicitly? Suppose $f \in \mathrm{Hom}(V, W)$ is described by a matrix A, relative to our choice of bases, so that $f(\mathbf{v}_i) = A^b{}_i \mathbf{w}_b$. Then we can find the components of the tensor $\imath(f)$ by letting it act on $(\mathbf{w}_*^a, \mathbf{v}_i)$ which gives

$$\imath(f)(\mathbf{w}_*^a, \mathbf{v}_i) \overset{(27.21)}{=} f(\mathbf{v}_i)(\mathbf{w}_*^a) = A^b{}_i \mathbf{w}_b(\mathbf{w}_*^a) = A^b{}_i \delta_b^a = A^a{}_i \,.$$

So, in summary, we have

$$f \text{ with } f(\mathbf{v}_i) = A^a{}_i \mathbf{w}_a \quad \overset{\imath}{\longrightarrow} \quad \imath(f) = A^a{}_i \mathbf{w}_a \otimes \mathbf{v}_*^i \,, \tag{27.22}$$

so, perhaps not surprisingly, both the map f and the associated tensor $\imath(f)$ are determined by the same matrix $A^a{}_i$. In particular, consider the identity map id_V on V (setting $W = V$ in the previous discussion). Its associated tensor is

$$\mathrm{id}_V \quad \overset{\imath}{\longrightarrow} \quad \imath(\mathrm{id}_V) = \delta_i^j \mathbf{v}_j \otimes \mathbf{v}_*^i = \mathbf{v}_i \otimes \mathbf{v}_*^i \,, \tag{27.23}$$

with index version given by the Kronecker delta δ_i^j. This is the sense in which we can think about the Kronecker delta as (the index version of) a tensor in $V \otimes V^*$.

Problem 27.2 *(Linear maps and decomposable tensors)*

For a non-zero linear map $f : V \to W$, show that the corresponding tensor $\imath(f) \in W \otimes V^*$ is decomposable iff $\mathrm{rk}(f) = 1$. In this case, write down a simple representing matrix for f.

Solution: We start by assuming that $\mathrm{rk}(f) = 1$, so that $\mathrm{Im}(f) = \mathrm{Span}(\mathbf{w})$ for a non-zero vector $\mathbf{w} \in W$. Then every image of f is a multiple of $\mathbf{w}$ and we can write $f(\mathbf{v}) = \varphi(\mathbf{v})\mathbf{w}$, where $\varphi \in V^*$. From Eq. (27.21) we have

$$\imath(f)(\mathbf{w}_*, \mathbf{v}) = f(\mathbf{v})(\mathbf{w}_*) = \mathbf{w}(\mathbf{w}_*)\varphi(\mathbf{v}) = \mathbf{w} \otimes \varphi(\mathbf{v}, \mathbf{w}_*) \,,$$

and this shows that $\imath(f) = \mathbf{w} \otimes \varphi$ is decomposable. Conversely, if $\imath(f)$ is decomposable it can be written as $\imath(f) = \mathbf{w} \otimes \varphi$ and we have

$$f(\mathbf{v})(\mathbf{w}_*) = \imath(f)(\mathbf{w}_*, \mathbf{v}) = \varphi(\mathbf{v})\mathbf{w}(\mathbf{w}_*) \,.$$

Dropping the argument $\mathbf{w}_*$ on either side gives $f(\mathbf{v}) = \varphi(\mathbf{v})\mathbf{w}$. This shows that $\mathrm{Im}(f) = \mathrm{Span}(\mathbf{w})$ and, hence, $\mathrm{rk}(f) = 1$.

Since maps $f \in \mathrm{Hom}(V, W)$ with $\mathrm{rk}(f) = 1$ are quite special we conclude that tensors in $W \otimes V^*$ are 'typically' not decomposable.

To find the representing matrix we introduce bases $(\mathbf{v}_1, \ldots, \mathbf{v}_n)$ on V and $(\mathbf{w}_1, \ldots, \mathbf{w}_m)$ on W, define $\alpha_i = \varphi(\mathbf{v}_i)$ and expand $\mathbf{w} = \beta^a \mathbf{w}_a$. For the representing matrix A of f we find

$$A^a{}_i \mathbf{w}_a = f(\mathbf{v}_i) = \varphi(\mathbf{v}_i)\mathbf{w} = \alpha_i \beta^a \mathbf{w}_a$$

so that $A^a{}_i = \alpha_i \beta^a$. In matrix-vector notation this can be written as

$$A = \boldsymbol{\alpha}\boldsymbol{\beta}^T \,,$$

where $\boldsymbol{\alpha} = (\alpha_1, \ldots, \alpha_n)^T$ and $\boldsymbol{\beta} = (\beta^1, \ldots, \beta^m)^T$.

27.2.3 Bi-linear forms and tensors

Consider a symmetric bi-linear form $(\cdot, \cdot)$ on V. Comparing Definitions 22.2 and 27.1 shows that this bi-linear form is the same as a symmetric tensor in $V^* \otimes V^*$. More precisely, if we introduce the matrix $g_{ij} = (\mathbf{v}_i, \mathbf{v}_j)$ which described the linear form (see Eq. (25.4)), we can write

$$(\cdot, \cdot) = \frac{1}{2}g_{ij}\mathbf{v}^i_* \otimes_S \mathbf{v}^j_* \,, \tag{27.24}$$

so the index version of this tensor is simply g_{ij}. We have seen in Section 26.3.2 that a non-degenerate linear form on V leads to a canonical injective map $V \to V^*$. Relative to a basis choice, this isomorphism is described by g_{ij} and its inverse g^{ij} via lowering and raising indices. We can now extend this operation to tensors. For example, we can convert a tensor τ^{ij} in $V \otimes V$ into a tensor $\tau^i{}_j$ in $V \otimes V^*$ via

$$\tau^i{}_j = g_{jk}\tau^{ik} \,, \qquad \tau^{ij} = g^{jk}\tau^i{}_k \,. \tag{27.25}$$

27.3 Multi-linearity*

Summary 27.3 *The tensor product can be generalized to an arbitrary number of vector spaces. Of particular importance are (p, q) tensors which are elements of $V^{\otimes p} \otimes (V^*)^q$. Relative to a basis they are described by objects with p covariant (upper) indices and q contravariant (lower) indices. Alternating forms are $(0, q)$ tensors and, due to anti-symmetry, they are non-trivial for $q = 0, 1, \ldots, \dim_{\mathbb{F}}(V)$ only. The alternating*

> *forms over a vector space V, together with the wedge product, form the outer algebra, ΛV^*, of V^*. The outer algebra of $\mathbb{R}^3$ is the natural framework for the cross, dot, and triple products.*

So far we have discussed tensor products of two vector spaces which lead to what is also referred to as rank two tensors. But the discussion is easily generalized to an arbitrary number, k, of vector spaces. This leads to *multi-linear algebra*, which deals with objects linear in k arguments. These are also called *rank k tensors* and in index notation, they are represented by objects with k indices.

27.3.1 Higher-rank tensors

Consider k vector spaces V_p, where $p = 1, \dots, k$, over the same field $\mathbb{F}$ and also introduce bases $(\mathbf{v}_{p,i})$ for V_p and dual bases $(\mathbf{v}_*^{p,i})$ for V_p^*. The tensor space of rank k tensors is defined as

$$V_1 \otimes \cdots \otimes V_k := \{\tau : V_1^* \times \cdots \times V_k^* \to \mathbb{F} \mid \tau \text{ is linear in each argument}\} . \quad (27.26)$$

Vector addition and scalar multiplication on this space is defined by the obvious generalisation of Eq. (27.4) to multiple arguments. As before, we can introduce the tensor product

$$\mathbf{v}_1 \otimes \cdots \otimes \mathbf{v}_k(\mathbf{v}_*^1, \dots, \mathbf{v}_*^k) := \prod_{p=1}^{k} \mathbf{v}_p(\mathbf{v}_*^p) , \quad (27.27)$$

where $\mathbf{v}_p \in V_p$ and $\mathbf{v}_*^p \in V_p^*$ for $p = 1, \dots, k$. Theorem 27.1 then generalizes to the following statement.

Theorem 27.5 *Let V_p, where $p = 1, \dots, k$, be vector spaces over the same field $\mathbb{F}$ with bases $(\mathbf{v}_{p,i})$. Then the tensors $(\mathbf{v}_{1,i_1} \otimes \cdots \otimes \mathbf{v}_{k,i_k})$ form a basis of $V_1 \otimes \cdots \otimes V_k$. In particular,*

$$\dim_{\mathbb{F}}(V_1 \otimes \cdots \otimes V_k) = \prod_{p=1}^{k} \dim_{\mathbb{F}}(V_p) . \quad (27.28)$$

Proof The proof is a straightforward generalization of the proof of Theorem 27.1.
□

In particular, this means the tensors $\tau \in V_1 \otimes \cdots \otimes V_k$ can be written as

$$\tau = \tau^{i_1 \cdots i_k} \mathbf{v}_{1,i_1} \otimes \cdots \otimes \mathbf{v}_{k,i_k} , \quad (27.29)$$

and correspond to objects $\tau^{i_1 \cdots i_k}$ with k indices.

27.3.2 (p, q) tensors

From the k vector spaces $V_1, \dots, V_k$ we can choose p spaces to be equal to V and $q = k - p$ spaces to be equal to V^*, for a given vector space V. In other words, we can consider the tensor spaces

$$\underbrace{V \otimes \cdots \otimes V}_{p} \otimes \underbrace{V^* \otimes \cdots \otimes V^*}_{q} =: V^{\otimes p} \otimes (V^*)^{\otimes q} . \tag{27.30}$$

The tensors τ in this space are also referred to as (p, q) tensors and they can be written as

$$\tau = \tau^{i_1 \cdots i_p}{}_{j_1 \cdots j_q} \mathbf{v}_{i_1} \otimes \cdots \otimes \mathbf{v}_{i_p} \otimes \mathbf{v}_*^{j_1} \otimes \cdots \otimes \mathbf{v}_*^{j_q} , \tag{27.31}$$

so they are represented, in index form, by objects $\tau^{i_1 \cdots i_p}{}_{j_1 \cdots j_q}$ with p upper and q lower indices. The (p, q) tensors for low values of p and q correspond to objects we have already encountered. For example, $(0, 0)$ tensors carry no indices and are, hence, scalars in $\mathbb{F}$. Further, $(1, 0)$ tensors carry one covariant index and are vectors in V while $(0, 1)$ tensors carry one contravariant index and are dual vectors in V^*. The discussion in Section 27.2.2 shows that $\operatorname{End}(V) \cong V \otimes V^*$, so $(1, 1)$ tensors can be identified with linear maps $V \to V$.

The transformation laws for tensors under the basis change in Table 27.3 are easily generalized to (p, q) tensors. From Eq. (27.18) every covariant index transforms with P and every covariant index with P_*. This leads to

$$\tilde{\tau}^{i_1 \cdots i_p}{}_{j_1 \cdots j_q} = P_*^{i_1}{}_{k_1} \cdots P_*^{i_p}{}_{k_p} P_{j_1}{}^{l_1} \cdots P_{j_q}{}^{l_q} \tau^{k_1 \cdots k_p}{}_{l_1 \cdots l_q} . \tag{27.32}$$

If we have a non-degenerate bi-linear form $(\cdot, \cdot)$, represented by a metric g_{ij}, available on V then we can lower and raise indices of (p, q) tensors, in analogy with Eq. (27.25), thereby converting to a different (p, q) type.

For $(p, 0)$ or $(0, q)$ tensors we can also consider totally symmetric and totally anti-symmetric tensors, which generalizes the discussion of anti-symmetric tensors from Section 27.2.1. A particularly important class of tensors are completely anti-symmetric $(0, q)$ tensors, also called *alternating q forms* which we discuss next.

27.3.3 Alternating q forms

An alternating q form is a $(0, q)$ tensor $\omega \in (V^*)^{\otimes q}$ which is completely anti-symmetric, that is

$$\omega(\ldots, \mathbf{v}, \ldots, \tilde{\mathbf{v}}, \ldots) = -\omega(\ldots, \tilde{\mathbf{v}}, \ldots, \mathbf{v}, \ldots) , \tag{27.33}$$

for all $\mathbf{v}, \tilde{\mathbf{v}} \in V$ and the dots stand for arguments which remain unchanged. The above anti-symmetry condition is preserved under addition and scalar multiplication of $(0, q)$ tensors, so the alternating q forms constitute a vector space, denoted by

$$\Lambda^q V^* := \{\omega \in (V^*)^{\otimes q} \mid \omega \text{ completely anti-symmetric}\} , \tag{27.34}$$

for $q = 1, 2, \ldots$. It is also customary and convenient to define $\Lambda^0 V^* = \mathbb{F}$. Note that the alternating one forms $\Lambda^1 V^* = V^*$ are simply the linear functionals on V so we can think of alternating q forms as generalizing the notion of linear functionals.

We can produce alternating q forms from linear functionals in V^* by defining a generalization of the wedge product in Eq. (27.17). For functionals $\mathbf{w}_*^1, \ldots, \mathbf{w}_*^q \in V^*$ and vectors $\mathbf{u}_1, \ldots, \mathbf{u}_q \in V$ we can define the wedge product $\mathbf{w}_*^1 \wedge \cdots \wedge \mathbf{w}_*^q$ by

$$\mathbf{w}_*^1 \wedge \cdots \wedge \mathbf{w}_*^q(\mathbf{u}_1, \ldots, \mathbf{u}_q) := \det \begin{pmatrix} \mathbf{w}_*^1(\mathbf{u}_1) & \cdots & \mathbf{w}_*^1(\mathbf{u}_q) \\ \vdots & \ddots & \vdots \\ \mathbf{w}_*^q(\mathbf{u}_1) & \cdots & \mathbf{w}_*^q(\mathbf{u}_q) \end{pmatrix}. \tag{27.35}$$

That this does indeed define an alternating q form follows immediately from the properties of the determinant. More specifically, linearity in each argument $\mathbf{u}_i$ is a direct consequence of linearity of the determinant in its arguments (the columns of the matrix). Complete anti-symmetry follows from the anti-symmetry of the determinant under exchange of its arguments.

The wedge product is linear in each argument and totally anti-symmetric, so the rules for calculation are

$$\begin{aligned} \cdots \wedge (\alpha \mathbf{v}_* + \beta \tilde{\mathbf{v}}_*) \wedge \cdots &= \alpha(\cdots \wedge \mathbf{v}_* \wedge \cdots) + \beta(\cdots \wedge \tilde{\mathbf{v}}_* \wedge \cdots) \\ \cdots \wedge \mathbf{v}_* \wedge \cdots \wedge \tilde{\mathbf{v}}_* \wedge \cdots &= -(\cdots \wedge \tilde{\mathbf{v}}_* \wedge \cdots \wedge \mathbf{v}_* \wedge \cdots) \\ \cdots \wedge \mathbf{v}_* \wedge \cdots \wedge \mathbf{v}_* \wedge \cdots &= 0 \end{aligned} \tag{27.36}$$

where $\mathbf{v}_*, \tilde{\mathbf{v}}_* \in V^*$, $\alpha, \beta \in \mathbb{F}$ and the dots denote arguments which remain unchanged. The first two of these rules follow since the determinant is also linear and anti-symmetric in its rows (remembering that the determinant does not change under transposition) and the third rule is a direct consequence of the second.

Just as for the case of anti-symmetric rank two tensors (Theorem 27.3) we can use the wedge product to construct a basis for the space of alternating q forms.

Theorem 27.6 *For a vector space V with basis $(\mathbf{v}_i)_{i=1,\ldots,n}$ and dual basis $(\mathbf{v}_*^i)_{i=1,\ldots,n}$, the tensors $(\mathbf{v}_*^{i_1} \wedge \cdots \wedge \mathbf{v}_*^{i_q})$, where $1 \le i_1 < i_2 < \cdots < i_q \le n$ form a basis of the alternating q forms $\Lambda^q V^*$. In particular,*

$$\dim_{\mathbb{F}}(\Lambda^q V^*) = \binom{n}{q}. \tag{27.37}$$

Proof The proof that the stated tensors form a basis is a direct generalization of the proof of Theorem 27.3. The number of tensors $(\mathbf{v}_*^{i_1} \wedge \cdots \wedge \mathbf{v}_*^{i_q})$ subject to the index restriction $1 \le i_1 < i_2 < \cdots < i_q \le n$ is given by the binomial in Eq. (27.37) (the number of ways to choose q different indices from n) and this proofs the dimension formula. $\qquad\square$

The theorem tells us that every alternating q form $\omega \in \Lambda^q V^*$ can be written as

$$\omega = \sum_{i_1 < \cdots < i_q} \omega_{i_1 \cdots i_q} \mathbf{v}_*^{i_1} \wedge \cdots \wedge \mathbf{v}_*^{i_q} = \frac{1}{q!} \sum_{i_1, \ldots, i_q} \omega_{i_1 \cdots i_q} \mathbf{v}_*^{i_1} \wedge \cdots \wedge \mathbf{v}_*^{i_q} \tag{27.38}$$

where we have converted the first, restricted sum into an unrestricted sum using the anti-symmetry of the wedge product. Hence, in index form, an alternating q form is represented by an object $\omega_{i_1 \cdots i_q}$ with q totally anti-symmetric indices. The above theorem also tells us that there are no non-trivial alternating q forms for $q > \dim_{\mathbb{F}}(V)$. In this case, every wedge product in Eq. (27.38) contains at least two same basis vectors, which means, from the last Eq. (27.36), that it vanishes. The highest non-trivial forms for an n-dimensional space V are the alternating n forms in $\Lambda^n V^*$, a one-dimensional space spanned by $\mathbf{v}_*^1 \wedge \cdots \wedge \mathbf{v}_*^n$.

The wedge product can be extended to alternating forms of arbitrary degree. To do so we introduce a p form $\nu \in \Lambda^p V^*$ as

$$\nu = \frac{1}{p!} \nu_{j_1 \cdots j_p} \mathbf{v}_*^{j_1} \wedge \cdots \wedge \mathbf{v}_*^{j_p}$$

and define the generalization $\wedge : \Lambda^q V^* \times \Lambda^p V^* \to \Lambda^{q+p} V^*$ of the wedge product by

$$\omega \wedge \nu := \frac{1}{q!\, p!} \omega_{i_1 \cdots i_q} \nu_{j_1 \cdots j_p} \mathbf{v}_*^{i_1} \wedge \cdots \wedge \mathbf{v}_*^{i_q} \wedge \mathbf{v}_*^{j_1} \wedge \cdots \wedge \mathbf{v}_*^{j_p} . \tag{27.39}$$

By using the anti-symmetry, Eq. (27.36), repeatedly is is easy to show that

$$\omega \wedge \nu = (-1)^{qp} \nu \wedge \omega \tag{27.40}$$

for a q form ω and a p form ν. For a vector space V with $\dim_{\mathbb{F}}(V) = n$, the direct sum

$$\Lambda V^* = \Lambda^0 V^* \oplus \Lambda^1 V^* \oplus \cdots \oplus \Lambda^n V^* \tag{27.41}$$

together with the wedge product is also called the *outer algebra* of V^*. Since the wedge product is linear in each argument, this is, in fact, an algebra in the sense of Def. 6.4. The dimension of the outer algebra is

$$\dim_{\mathbb{F}}(\Lambda V^*) \stackrel{(27.37)}{=} \sum_{q=0}^{n} \binom{n}{k} = 2^n . \tag{27.42}$$

27.3.4   The determinant as an alternating form

Consider the coordinate vector space $\mathbb{F}^n$ with standard unit vector basis $\mathbf{e}_i$ and dual basis $\mathbf{e}_*^i$. For n vectors $\mathbf{u}_i = u_i^j \mathbf{e}_j$, with $\mathbf{e}_*^j(\mathbf{u}_i) = u_i^j$, we have

$$\mathbf{e}_*^1 \wedge \cdots \wedge \mathbf{e}_*^n (\mathbf{u}_1, \ldots, \mathbf{u}_n) \stackrel{(27.35)}{=} \det \begin{pmatrix} \mathbf{e}_*^1(\mathbf{u}_1) & \cdots & \mathbf{e}_*^1(\mathbf{u}_n) \\ \vdots & \ddots & \vdots \\ \mathbf{e}_*^n(\mathbf{u}_1) & \cdots & \mathbf{e}_*^n(\mathbf{u}_n) \end{pmatrix} = \det \begin{pmatrix} u_1^1 & \cdots & u_n^1 \\ \vdots & \ddots & \vdots \\ u_1^n & \cdots & u_n^n \end{pmatrix} .$$

$$= \det(\mathbf{u}_1, \ldots, \mathbf{u}_n)$$

Hence we learn that the determinant is, in fact, an alternating n form

$$\det = \mathbf{e}_*^1 \wedge \cdots \wedge \mathbf{e}_*^n = \frac{1}{n!} \epsilon_{i_1 \cdots i_n} \mathbf{e}_*^{i_1} \wedge \cdots \wedge \mathbf{e}_*^{i_n} \tag{27.43}$$

whose index version is the Levi-Civita tensor $\epsilon_{i_1 \cdots i_n}$.

27.3.5 The outer algebra of $\mathbb{R}^3$

As an example, we would like to develop the outer algebra of the vector space $\mathbb{R}^3$ in detail. We introduce the standard unit vectors basis $(\mathbf{e}_i)$ as well as the following two and three forms.

$$\nu_1 = \mathbf{e}_2 \wedge \mathbf{e}_3 , \quad \nu_2 = \mathbf{e}_3 \wedge \mathbf{e}_1 , \quad \nu_3 = \mathbf{e}_1 \wedge \mathbf{e}_2 , \quad \Omega = \mathbf{e}_1 \wedge \mathbf{e}_2 \wedge \mathbf{e}_3$$

Some of the properties of the outer algebra

$$\Lambda \mathbb{R}^3 = \Lambda^0 \mathbb{R}^3 \oplus \Lambda^1 \mathbb{R}^3 \oplus \Lambda^2 \mathbb{R}^3 \oplus \Lambda^3 \mathbb{R}^3$$

are summarized in the table below.

Forms	Space	Basis	Dimension	Typical element
0 forms	$\Lambda^0 \mathbb{R}^3 = \mathbb{R}$	(1)	1	$a,\ a \in \mathbb{R}$
1 forms	$\Lambda^1 \mathbb{R}^3 = \mathbb{R}^3$	$(\mathbf{e}_1, \mathbf{e}_2, \mathbf{e}_3)$	3	$v^i \mathbf{e}_i,\ v^i \in \mathbb{R}$
2 forms	$\Lambda^2 \mathbb{R}^3$	(ν_1, ν_2, ν_3)	3	$u^i \nu_i,\ u^i \in \mathbb{R}$
3 forms	$\Lambda^3 \mathbb{R}^3$	(Ω)	1	$b \Omega,\ b \in \mathbb{R}$

Note that three forms are the highest non-trivial forms and that the dimensions of the four parts add up to $8 = 2^3$, in accordance with Eq. (27.42). The table makes it clear that zero and three forms can be interpreted as scalars while one and two forms can be identified with vectors in $\mathbb{R}^3$.

Let us work out the wedge product explicitly, starting with two one forms $\mathbf{v} = v^i \mathbf{e}_i$ and $\mathbf{w} = w^j \mathbf{e}_j$. A quick calculation, based on the rules in Eq. (27.36) gives

$$\mathbf{v} \wedge \mathbf{w} = (v^1 \mathbf{e}_1 + v^2 \mathbf{e}_2 + v^3 \mathbf{e}_3) \wedge (w^1 \mathbf{e}_1 + w^2 \mathbf{e}_2 + w^3 \mathbf{e}_3)$$
$$= (v^2 w^3 - v^3 w^2)\nu_1 + (v^3 w^1 - v^1 w^3)\nu_2 + (v^1 w^2 - v^2 w^1)\nu_3 = (\mathbf{v} \times \mathbf{w})^i \nu_i$$

This shows that the wedge product of one forms on $\mathbb{R}^3$ is, in fact, the cross product which we have introduced, in a somewhat ad-hoc fashion, in Chapter 10. The outer algebra is the proper mathematical context in which to discuss the cross product. Within this framework it is also evident why the cross product can only be defined in the three-dimensional case. The wedge product of two one forms gives two forms and these can only be re-interpreted as vectors if the dimension of $\Lambda^2 \mathbb{R}^n$ equals n. From the dimension formula (27.37) this is only the case for $n = 3$.

What about the wedge product of three one forms $\mathbf{v} = v^i \mathbf{e}_i$, $\mathbf{u} = u^j \mathbf{e}_j$ and $\mathbf{w} = w^k \mathbf{e}_k$?

$$\mathbf{v} \wedge \mathbf{u} \wedge \mathbf{w} = v^i u^j w^k \mathbf{e}_i \wedge \mathbf{e}_j \wedge \mathbf{e}_k = \epsilon_{ijk} v^i u^j w^k \mathbf{e}_1 \wedge \mathbf{e}_2 \wedge \mathbf{e}_3 = \det(\mathbf{v}, \mathbf{u}, \mathbf{w})\, \Omega$$

So this is the three-dimensional determinant or, equivalently, the triple product.

The final non-trivial case is the wedge product of a one form $\mathbf{v} = v^i \mathbf{e}_i$ and a two form $\mathbf{u} = u^j \nu_j$.

$$\mathbf{v} \wedge \mathbf{u} = (v^1 \mathbf{e}_1 + v^2 \mathbf{e}_2 + v^3 \mathbf{e}_3) \wedge (w^1 \nu_1 + w^2 \nu_2 + w^3 \nu_3) = (\mathbf{v} \cdot \mathbf{u})\, \Omega$$

So this corresponds to the dot product on $\mathbb{R}^3$.

In summary, we have seen that the products between vectors introduced in Part III, that is, the cross, triple and dot products, have a natural home in the outer algebra and are really all special versions of the wedge product.

Problem 27.3 *(The outer algebra of $\mathbb{R}^2$)*

Explicitly work out the outer algebra of $\mathbb{R}^2$ by introducing a suitable basis and compute the wedge product.

Solution: The outer algebra is $\Lambda\mathbb{R}^2 = \Lambda^0\mathbb{R}^2 \oplus \Lambda^1\mathbb{R}^2 \oplus \Lambda^2\mathbb{R}^2$. With the standard unit vector basis $(\mathbf{e}_i)_{i=1,2}$ and $\Omega = \mathbf{e}_1 \wedge \mathbf{e}_2$, we can write the three pieces of the outer algebra as

$$\Lambda^0\mathbb{R}^2 = \mathrm{Span}(1) , \quad \Lambda^1\mathbb{R}^2 = \mathrm{Span}(\mathbf{e}_1, \mathbf{e}_2) , \quad \Lambda^2\mathbb{R}^2 = \mathrm{Span}(\Omega) .$$

The total dimension is $\dim_\mathbb{R}(\Lambda\mathbb{R}^2) = 1+2+1 = 4$, in accordance with Eq. (27.42). The wedge product of two one forms $\mathbf{v} = v^i\mathbf{e}_i$ and $\mathbf{w} = w^j\mathbf{e}_j$ gives

$$\mathbf{v} \wedge \mathbf{w} = (v^1\mathbf{e}_1 + v^2\mathbf{e}_2) \wedge (w^1\mathbf{e}_1 + w^2\mathbf{e}_2) = (v^1w^2 - v^2w^1)\omega = \det(\mathbf{v}, \mathbf{w})\omega .$$

So the wedge product of two one forms in $\mathbb{R}^2$ gives the two-dimensional determinant.

Problem 27.4 *(The outer algebra of $\mathbb{R}^n$)*

Show that the wedge products between elements of $\mathbb{R}^n$ and $\Lambda^{n-1}\mathbb{R}^n$ leads to the dot product. Use this observation to define a generalization of the cross product which produces a vector orthogonal, with respect to the dot product, to given vectors $\mathbf{v}_1, \ldots, \mathbf{v}_{n-1} \in \mathbb{R}^n$.

Solution: The key is that $\mathbb{R}^n$ and $\Lambda^{n-1}\mathbb{R}^n$ have the same dimension, n, so they can be identified. To do this, we use the standard unit vector basis $(\mathbf{e}_1, \ldots, \mathbf{e}_n)$ on $\mathbb{R}^n$ and on $\Lambda^{n-1}\mathbb{R}^n$ we introduce the basis $(\nu_1, \ldots, \nu_n)$ with

$$\nu_k = (-1)^{k+1}\mathbf{e}_1 \wedge \cdots \wedge \widehat{\mathbf{e}_k} \wedge \cdots \wedge \mathbf{e}_n ,$$

where the hat indicates that the vector underneath should be omitted. We also introduce $\Omega = \mathbf{e}_1 \wedge \cdots \wedge \mathbf{e}_n$ which spans $\Lambda^n\mathbb{R}^n$. Then we can identify $n-1$ forms $w = w^k\nu_k$ with vectors $\mathbf{w} = w^k\mathbf{e}_k$ and we find that the wedge product of $\mathbf{v} = v^i\mathbf{e}_i$ and w,

$$\mathbf{v} \wedge w = v^iw^k\mathbf{e}_i \wedge \nu_k = (-1)^{k+1}v^iw^k\mathbf{e}_i \wedge \mathbf{e}_1 \wedge \cdots \wedge \widehat{\mathbf{e}_k} \wedge \cdots \wedge \mathbf{e}_n = \delta_{ik}v^iw^k\Omega = (\mathbf{v} \cdot \mathbf{w})\Omega ,$$

is indeed proportional to the dot product.

For the second part, define $w = w^i\nu_i := \mathbf{v}_1 \wedge \cdots \wedge \mathbf{v}_{n-1}$. But it is clear that $(\mathbf{v}_k \cdot \mathbf{w})\Omega = \mathbf{v}_k \wedge w = 0$ vanishes from from anti-symmetry of the wedge product (see Eq. (27.36)). Hence, $\mathbf{w} \cdot \mathbf{v}_k = 0$ for all $k = 1, \ldots, n-1$, as desired.

Application 27.1 *(Quantum bits and computation)*

As explained in Application 26.1, the set of possible states of a quantum system forms a vector space V over $\mathbb{C}$. Time-evolution of states $|\psi(t)\rangle \in V$ is governed by the time-dependent Schrödinger equation

$$H|\psi(t)\rangle = i\frac{d}{dt}|\psi(t)\rangle , \tag{27.44}$$

where the Hermitian linear map $H : V \to V$, called the Hamilton operator, is the observable which corresponds to energy. One way to solve the time-dependent Schrödinger equation is in terms of the energy eigenstates $|E_i\rangle$, as we have done in Eq. (26.44).

For another approach, start with any ortho-normal basis $(|i\rangle)$, where $i = 1, \ldots, n$, on V. Then introduce the matrix $\mathcal{H}$ with entries $\mathcal{H}_{ij} = \langle i|H|j\rangle$ which describes the Hamilton operator relative to this basis and write states $|\psi(t)\rangle$ as a linear combination of this basis.

$$|\psi(t)\rangle = \sum_i \alpha_i(t)|i\rangle , \quad \alpha_i(t) = \langle i|\psi(t)\rangle , \quad \boldsymbol{\alpha}(t) = (\alpha_1(t), \alpha_2(t), \ldots)^T . \tag{27.45}$$

By acting with a dual basis vector $\langle i|$ and inserting the identity $\mathrm{id}_V = \sum_j |j\rangle\langle j|$ (see Eq. (26.27)) on the left-hand side, the Schrödinger equation can be converted into the vector/matrix equation:

$$\sum_j \langle i|H|j\rangle|j\langle\psi(t)\rangle = i\frac{d}{dt}\langle i|\psi(t)\rangle \quad \Leftrightarrow \quad \mathcal{H}\boldsymbol{\alpha} = i\frac{d\boldsymbol{\alpha}}{dt} . \tag{27.46}$$

We have already solved differential equations of this type in Application 24.3. Setting $A = -i\mathcal{H}$ in Eq. (24.36), the solution can be written down in terms of the matrix exponential.

$$\boldsymbol{\alpha}(t) = U(t)\boldsymbol{\alpha}_0 , \quad U(t) = e^{-i\mathcal{H}t} . \tag{27.47}$$

Note that $A = -i\mathcal{H}$ is anti-Hermitian (since $\mathcal{H}$ is Hermitian) so that the above matrix $U(t)$ is indeed unitary, as suggested by the notation (see Application 24.3). The conclusion is that time-evolution in Quantum Mechanics is realized by the action of unitary linear maps which depend on the Hamilton operator and, hence, on the physical properties of the quantum system.

A classical computation proceeds by performing a sequence of logical operations on some information, typically given by a number of bits. For a quantum computation, the information consists of an initial quantum state $|\psi_0\rangle$ which is processed by successive physical manipulations, each of which is described by a unitary linear map U_a, as in Eq. (27.47). Schematically, a quantum computation can, therefore, be represented as

$$|\psi_0\rangle = \sum_i \alpha_{0,i}|i\rangle \quad \to \quad \boldsymbol{\alpha}_{\mathrm{final}} = U_1 U_2 \cdots U_d\, \boldsymbol{\alpha}_0 \quad \to \quad |\psi_{\mathrm{final}}\rangle = \sum_i \alpha_{\mathrm{final},i}|i\rangle . \tag{27.48}$$

Each step in the calculation corresponds to a multiplication with a unitary matrix $U_a \in \mathrm{U}(n)$, also called a quantum logic gate. Let us discuss in more detail how this works for the simple spin system from Application 26.4.

Recall in this case the vector space V is two-dimensional and its ortho-normal basis of up and down spin states is denoted by $(|\downarrow\rangle, |\uparrow\rangle)$. The information which can be stored in such a two-dimensional quantum state is also called a quantum bit or qubit for short. For the general Hamilton operator, Eq. (26.42), of this system the matrix exponential can be worked out explicitly.

$$U(t) = e^{-i\mathcal{H}t} \overset{(26.42)}{=} e^{-iat}e^{-i\mathbf{b}\cdot\boldsymbol{\sigma}t} = e^{-iat}\left(\cos(|\mathbf{b}|t)\mathbb{1}_2 - i\sin(|\mathbf{b}|t)\frac{\mathbf{b}\cdot\boldsymbol{\sigma}}{|\mathbf{b}|}\right) . \tag{27.49}$$

In the last step we have used the result from Exercise 24.3 with $\theta = -|\mathbf{b}|t$ and $\mathbf{n} = \mathbf{b}/|\mathbf{b}|$. We also know from Exercise 24.3 and Eq. (23.34) that all U(2) matrices can be obtained from Eq. (27.49), for suitable values of $a \in \mathbb{R}$, $\mathbf{b} \in \mathbb{R}^3$ and $t \in \mathbb{R}$.

Common choices for quantum logic gates for a single qubit are the identity, the not gate and the Hadamard gate, with the following unitary maps, actions on the basis states and diagrams.

$$\text{unitary matrix} \qquad \mathbb{1}_2 \qquad U_N := \begin{pmatrix} 0 & 1 \\ 1 & 0 \end{pmatrix} \qquad U_{Hd} := \tfrac{1}{\sqrt{2}} \begin{pmatrix} 1 & 1 \\ 1 & -1 \end{pmatrix}$$

$$\text{action} \qquad \begin{array}{l} \mathrm{id}_V |\downarrow\rangle = |\downarrow\rangle \\ \mathrm{id}_V |\uparrow\rangle = |\uparrow\rangle \end{array} \qquad \begin{array}{l} N|\downarrow\rangle = |\uparrow\rangle \\ N|\uparrow\rangle = |\downarrow\rangle \end{array} \qquad \begin{array}{l} \mathrm{Hd}|\downarrow\rangle = \tfrac{1}{\sqrt{2}}(|\downarrow\rangle + |\uparrow\rangle) \\ \mathrm{Hd}|\uparrow\rangle = \tfrac{1}{\sqrt{2}}(|\downarrow\rangle - |\uparrow\rangle) \end{array}$$

diagram ——————   ——⊕——   ——$\boxed{H}$——

A collection of p qbits is described by states in the tensor space $V^{\otimes p}$ of $(p, 0)$ tensors, which has dimension 2^p. Let us explicitly discuss the case of two qubits, so the space $V \otimes V$ of $(2, 0)$ tensors which, from Theorem 27.1, has a basis $(|\downarrow\downarrow\rangle, |\downarrow\uparrow\rangle, |\uparrow\downarrow\rangle, |\uparrow\uparrow\rangle)$. Here, the notation $|\uparrow\uparrow\rangle := |\uparrow\rangle \otimes |\uparrow\rangle$ is a convenient shorthand for the tensor product. Simple logic gates on $V \otimes V$ can be obtained by tensoring maps on V, following Section 27.1. For example, the map $N \otimes \mathrm{id}_V$ acts as a not on the first qubit and as an identity on the second one, so

$$N \otimes \mathrm{id}_V |\uparrow s\rangle = |\downarrow s\rangle , \quad N \otimes \mathrm{id}_V |\downarrow s\rangle = |\uparrow s\rangle ,$$

where $s = \uparrow, \downarrow$. Of course, there are unitary maps on $V \otimes V$ which cannot be obtained as a tensor maps in this way, for example the conditional not map $CN : V \otimes V \to V \otimes V$ defined by

$$U_{CN} = \begin{pmatrix} 1 & 0 & 0 & 0 \\ 0 & 1 & 0 & 0 \\ 0 & 0 & 0 & 1 \\ 0 & 0 & 1 & 0 \end{pmatrix}, \qquad \begin{array}{l} CN|\downarrow\downarrow\rangle = |\downarrow\downarrow\rangle \\ CN|\downarrow\uparrow\rangle = |\downarrow\uparrow\rangle \\ CN|\uparrow\uparrow\rangle = |\uparrow\downarrow\rangle \\ CN|\uparrow\downarrow\rangle = |\uparrow\uparrow\rangle \end{array} .$$

The CN map inverts the second spin provided the first one points upwards and it has no effect otherwise. Consider a quantum circuit $CN \circ (Hd \otimes \mathrm{id}_V)$ which can also be represented by the diagram

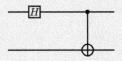

Here the vertical line indicates that the not gate on the second qubit is controlled by the first qubit. For example, the action of this circuit on an initial state $|\uparrow\uparrow\rangle$ is

$$CN \circ (Hd \otimes \mathrm{id}_V)|\uparrow\uparrow\rangle = \frac{1}{\sqrt{2}} CN(|\downarrow\uparrow\rangle - |\uparrow\uparrow\rangle) = \frac{1}{\sqrt{2}}(|\downarrow\uparrow\rangle - |\uparrow\downarrow\rangle)$$

Evidently, the circuit has converted a decomposable tensor product $|\uparrow\uparrow\rangle = |\uparrow\rangle \otimes |\uparrow\rangle$ into a linear combination of decomposable tensors. Such linear combinations are also referred to as *entangled states*. Useful application of quantum computing rely on more complicated circuits which are beyond our present scope. (An introduction to quantum computing can, for example, be found in Jones and Jaksch 2012.)

Exercises

(†=challenging, ††=difficult, wide-ranging)

27.1 Let V be a two-dimensional vector space over $\mathbb{R}$.
(a) Show that not all tensors in $V \otimes V$ can be written in the form $\mathbf{v} \otimes \mathbf{w}$, where $\mathbf{v}, \mathbf{w} \in V$.
(b) Show that the set of tensors in $V \otimes V$ which can be written as $\mathbf{v} \otimes \mathbf{w}$ can be viewed as a quadratic hypersurface in $\mathbb{R}^4$ with signature $(2, 2)$.

27.2 (a) For two inner product vector spaces V and W show that there is a natural scalar product that can be defined on $V \otimes W$.
(b) If $(\mathbf{v}_i)$ and $(\mathbf{w}_a)$ are ortho-normal bases of V and W show that $(\mathbf{v}_i \otimes \mathbf{w}_a)$ is an ortho-normal basis of $V \otimes W$.
(c) Relative to he bases in part (b), how are the matrix elements of $f \subset \text{End}(V)$ and $g \in \text{End}(W)$ related to the matrix elements of $f \otimes g$.

27.3 *Tensors and eigenvalues*
Consider linear maps $f \in \text{End}(V)$, $g \in \text{End}(W)$ and the tensor map $f \otimes g \in \text{End}(V \otimes W)$.
(a) How are basis transformations for matrices representing f and g related to basis transformations of matrices which represent $f \otimes g$?
(b) Show, if λ is an eigenvalue of f with eigenvector $\mathbf{v}$ and μ an eigenvalue of g with eigenvector $\mathbf{w}$, then $\lambda\mu$ is an eigenvalue of $f \otimes g$ with eigenvector $\mathbf{v} \otimes \mathbf{w}$.
(c) Show, if f and g can be diagonalized then so can $f \otimes g$.

27.4 A linear map $A \in \text{End}(\mathbb{R}^2)$ is given by

$$A = \begin{pmatrix} 0 & 1 \\ 1 & 0 \end{pmatrix} .$$

(a) Find its eigenvalues and eigenvectors and diagonalize A.
(b) Do the same for $A \otimes A \in \text{End}(\mathbb{R}^2 \otimes \mathbb{R}^2)$.
(c) Generalize the result from part (b) to an arbitrary number of tensor factors, so to $A \otimes \cdots \otimes A \in \text{End}(\mathbb{R}^2 \otimes \cdots \otimes \mathbb{R}^2)$.

27.5 *Pauli matrices and Kronecker product*[†]
Define the anti-commutator of two square matrices A, B by $\{A, B\} := AB + BA$.
(a) Show that the Pauli matrices σ_i satisfy $\{\sigma_i, \sigma_j\} = 2\mathbb{1}_2 \delta_{ij}$.
(b) Define the 4×4 matrices γ_m, where $m = 1, 2, 3, 4$, by

$$\begin{aligned} \gamma_1 &= \sigma_1 \times \sigma_3 & \gamma_2 &= \sigma_2 \times \sigma_3 \\ \gamma_3 &= \mathbb{1}_2 \times \sigma_1 & \gamma_4 &= \mathbb{1}_2 \times \sigma_2 . \end{aligned}$$

Show that $\{\gamma_m, \gamma_n\} = 2\mathbb{1}_4 \delta_{mn}$.
(c) For the matrix $\gamma_5 := \gamma_1 \gamma_2 \gamma_3 \gamma_4$ show that $\gamma_5^2 = \mathbb{1}_4$ and that $\text{tr}(\gamma_5) = 2$.
(d) Show that the maps $P_\pm \in \text{End}(\mathbb{C}^4)$ defined by $P_\perp - \frac{1}{2}(\mathbb{1}_4 \pm \gamma_5)$ are projectors onto two-dimensional subspaces.
(e) Based on the results so far, define 8×8 matrices Γ_m, where $m = 1, \ldots, 6$ with $\{\Gamma_m, \Gamma_n\} = 2\mathbb{1}_8 \delta_{mn}$.

27.6 *Symmetric tensors*[††]
(a) Consider symmetric tensors in $S^2 V^*$, for a vector space V over $\mathbb{R}$, represented, relative to a basis, by symmetric two-index objects τ_{ij}. We say two such objects, τ_{ij} and $\tilde{\tau}_{ij}$, are conjugate if there exists a basis transformation (27.10) which relates them. Find the conjugacy classes and a simple representative tensor in each class.
(b) For $\dim_\mathbb{R}(V) = 2$, repeat the discussion from part (a) for symmetric tensors in $S^3 V^*$, represented by completely symmetric three-index objects τ_{ijk}.

27.7 *Outer algebra of $\mathbb{R}^4$*[†]
Following Section 27.3.5, work out the outer algebra of $\mathbb{R}^4$.

27.8 *Quantum circuits*
(a) Following the set-up in Application 27.1, find the action of the quantum circuit $\text{CN} \circ (\text{Hd} \otimes \text{id}_V)$ on all basis vectors $(\downarrow\downarrow), |\downarrow\uparrow\rangle, |\uparrow\downarrow\rangle, |\uparrow\uparrow\rangle)$ of the two qubit system.

(b) Using the basic gates from Application 27.1, build a quantum circuit C which acts as

$$C|\downarrow\downarrow\rangle = \tfrac{1}{\sqrt{2}}(|\downarrow\uparrow\rangle + |\uparrow\uparrow\rangle)$$
$$C|\downarrow\uparrow\rangle = \tfrac{1}{\sqrt{2}}(|\downarrow\downarrow\rangle + |\uparrow\downarrow\rangle)$$
$$C|\uparrow\downarrow\rangle = \tfrac{1}{\sqrt{2}}(|\downarrow\downarrow\rangle - |\uparrow\downarrow\rangle)$$
$$C|\uparrow\uparrow\rangle = \tfrac{1}{\sqrt{2}}(|\downarrow\uparrow\rangle - |\uparrow\uparrow\rangle).$$

References

Armstrong, M.A. (2013). *Groups and Symmetries*. Springer, New York.

Bennett, M.K. (2011). *Affine and Projective Geometry*. Wiley, New York.

Blyth, T.S. and Robertson, E.F. (1975). *Basic Linear Algebra*. Springer, New York.

Constantinescu, F. (1980). *Distributions and Their Application in Physics*. Pergamon, Oxford.

Cornwell, J.F. (1997). *Group Theory in Physics*. Elsevier Science.

Curtis, C.W. (1996). *Linear Algebra. An Introductory Approach*. Springer, New York.

da Silva, I.N. (2017). *Artificial Neural Networks. A Practical Course*. Springer, Basel.

Dirac, P.A.M. (2019). *Principles of Quantum Mechanics*. Oxford University Press, Oxford.

Fischer, G. (2010). *Linear Algebra*. Vieweg+Teubner Verlag/Springer.

Fulton, W. and Harris, J. (2013). *Representation Theory: A First Course*. Springer, New York.

Goldstein, H. (2013). *Classical Mechanics*. Pearson, Reading, MA.

Goodfellow, I., Bengio, Y., and Courville, A. (2016). *Deep Learning*. MIT Press, Cambridge MA and London.

Halmos, P.R. (2017). *Finite-dimensional Vector Spaces*. Dover Publications, New York.

Halzen, F. and Martin, A.D. (2008). *Quarks and Leptons: An Introductory Course in Modern Particle Physics*. Wiley, New Delhi.

Jackson, J.D. (1962). *Classical Electromagnetism*. John Wiley & Sons, New York and London.

Janich, K. (1994). *Linear Algebra*. Springer, New York.

Jones, J.A. and Jaksch, D. (2012). *Quantum Information, Computation and Communication*. Cambridge University Press, Cambridge, UK.

Landau, L.D. and Lifshitz, E.M. (1982). *Mechanics*. Elsevier Science.

Lang, S. (1996). *Linear Algebra* (3rd edn). Springer, New York.

Lang, S. (1997). *Undergraduate Analysis*. Springer, New York.

Lang, S. (1998). *Basic Mathematics*. Springer, New York.

Lang, S. (2000). *Undergraduate Algebra* (3rd edn). Springer, New York.

Lang, S. (2013). *Complex Analysis*. Springer, New York.

Manin, Y.I. and Kostrikin, A..I. (1989). *Linear Algebra and Geometry*. Gordon and Breach, Reading.

Messiah, A. (2014). *Quantum Mechanics*. Dover Publications, New York.

Rynne, B.P. and Youngson, M.A. (2008). *Linear Functional Analysis* (2nd edn). Springer, London.

Sakurai, J.J. and Napolitano, J. (2017). *Modern Quantum Mechanics*. Cambridge University Press, Cambridge, UK.

Strang, G. (1988). *Linear Algebra and Its Applications.* Thomson Learning, Belmont, CA.

Wybourne, B.G. (1974). *Classical Groups for Physicists.* Wiley & Sons, London.

Index